Studies in Computational Intelligence

Volume 1038

Series Editor

Janusz Kacprzyk, Polish Academy of Sciences, Warsaw, Poland

The series "Studies in Computational Intelligence" (SCI) publishes new developments and advances in the various areas of computational intelligence—quickly and with a high quality. The intent is to cover the theory, applications, and design methods of computational intelligence, as embedded in the fields of engineering, computer science, physics and life sciences, as well as the methodologies behind them. The series contains monographs, lecture notes and edited volumes in computational intelligence spanning the areas of neural networks, connectionist systems, genetic algorithms, evolutionary computation, artificial intelligence, cellular automata, self-organizing systems, soft computing, fuzzy systems, and hybrid intelligent systems. Of particular value to both the contributors and the readership are the short publication timeframe and the world-wide distribution, which enable both wide and rapid dissemination of research output.

Indexed by SCOPUS, DBLP, WTI Frankfurt eG, zbMATH, SCImago.

All books published in the series are submitted for consideration in Web of Science.

More information about this series at https://link.springer.com/bookseries/7092

Essam Halim Houssein · Mohamed Abd Elaziz ·
Diego Oliva · Laith Abualigah
Editors

Integrating Meta-Heuristics and Machine Learning for Real-World Optimization Problems

Springer

Editors
Essam Halim Houssein
Faculty of Computers and Information,
Minia University
Minia, Egypt

Diego Oliva
Department of Computer Sciences
University of Guadalajara
Guadalajara, Jalisco, Mexico

Mohamed Abd Elaziz
Faculty of Computer Science &
Engineering
Galala University
Suze, Egypt

Department of Mathematics
Faculty of Science, Zagazig University
Zagazig, Egypt

Laith Abualigah
Faculty of Computer Sciences
and Informatics
Amman Arab University
Amman, Jordan

ISSN 1860-949X ISSN 1860-9503 (electronic)
Studies in Computational Intelligence
ISBN 978-3-030-99078-7 ISBN 978-3-030-99079-4 (eBook)
https://doi.org/10.1007/978-3-030-99079-4

This Springer imprint is published by the registered company Springer Nature Switzerland AG
The registered company address is: Gewerbestrasse 11, 6330 Cham, Switzerland

About This Book

In recent years, metaheuristics (MHs) have become essential tools for solving challenging optimization problems encountered in industry, engineering, biomedical, image processing, and the theoretical field. Several different metaheuristics exist, and new methods are under constant development. One of the most fundamental principles in our world is the search for an optimal state. Therefore, choose the right, and correct solution technique for an optimization problem can be crucially important in finding the right solutions for a given optimization problem (unconstrained and constrained optimization problems). There exist a diverse range of MHs for optimization. Optimization techniques have been used for many years in the formulation and solution of computational problems.

This book brings together outstanding research and recent developments in metaheuristics (MHs), machine learning (ML), and their applications in the industrial world. Among the subjects to be considered are theoretical developments in MHs; performance comparisons of MHs; suitable methods combining different types of approaches such as constraint programming and mathematical programming techniques; parallel and distributed MHs for multi-objective optimization; adaptation of discrete MHs to continuous optimization; dynamic optimization; software implementations; and real-life applications. Besides, machine learning (ML) is a data analytics technique to use computational methods. Therefore, recently, MHs have been combined with several ML techniques to deal with different global and engineering optimization problems, also real-world applications. Finding an optimal solution or even sub-optimal solutions is not an easy task. Chapters published in this book describe original works in different topics in science and engineering, such as metaheuristics, machine learning, soft computing, neural networks, multi-criteria decision-making, energy efficiency, sustainable development, etc.

Before digging deeper into the matter, we will attempt to classify these algorithms as an overview and discuss some basic use cases. In this book, a classification of metaheuristic algorithms and a rough taxonomy of global optimization methods were presented. Generally, optimization algorithms can be divided into two basic classes: deterministic and probabilistic algorithms. We will briefly introduce optimization algorithms such as particle swarm optimization, harmony search, firefly algorithm, and cuckoo search. It also presents a variety of solution techniques for optimization problems, emphasizing concepts rather than rigorous mathematical details and proofs.

Minia, Egypt
Prof. Dr. Essam Halim Houssein
essam.halim@mu.edu.eg

Zagazig, Egypt
Dr. Mohamed Abd Elaziz
abd_el_aziz_m@yahoo.com

Guadalajara, México
Dr. Diego Oliva
diego.oliva@cucei.udg.mx
doliva@ucm.es

Amman, Jordan
Dr. Laith Abualigah
aligah.2020@gmail.com

Contents

Combined Optimization Algorithms for Incorporating DG in Distribution Systems

Hussein Abdel-mawgoud, Salah Kamel, and Ahmad Eid

Abstract The electric grid faces many challenges to face the growth in customer demands overall the world. Therefore, incorporating DG especially natural sources such as photovoltaic (PV) and wind turbine (WT) in distribution network is increased to reduce these challenges. Installing PV and WT in Radial Distribution System (RDS) enhance the system reliability and system voltage, decrease the system losses and global warming, and increases the system capacity. The total system real loss is applied as objective function with system constraints. The sensitivity analysis is utilized to obtain the preferable locations for installing DG in RDS. In this study, single and multiple of WT and PV is integrated in RDS using moth flame optimization (MFO) algorithm, salp swarm algorithm (SSA) and chaotic moth flame optimization (CMFO) algorithm. Also, installing DG is applied in IEEE 33 and IEEE 69 bus RDS. From results, installing multiple DG achieves better results than installing one DG in RDS. Also, the presented algorithms proved their effectively to obtain the best sizes and locations of DG that lead to better reduction in system losses than other efficient algorithms.

Keywords Optimization · DG · MFO · CMFO · SSA · PLSF · RDS

H. Abdel-mawgoud
Electrical Engineering Department, Aswan Faculty of Engineering, Aswan University, Aswan, Egypt

S. Kamel (✉) · A. Eid
Electrical Engineering Department, Faculty of Engineering, Aswan University, Aswan 81542, Egypt
e-mail: skamel@aswu.edu.eg

A. Eid
e-mail: ahmadeid@aswu.edu.eg; a.mohamad@qu.edu.sa

A. Eid
Department of Electrical Engineering, College of Engineering, Qassim University, Unaizah 56452, Saudi Arabia

E. H. Houssein et al. (eds.), *Integrating Meta-Heuristics and Machine Learning for Real-World Optimization Problems*, Studies in Computational Intelligence 1038,
https://doi.org/10.1007/978-3-030-99079-4_1

1 Introduction

The electrical load demand is increasing rapidly and increasing the capacity of central power station is not encouraged to avoid the increasing in carbon dioxide emission, replacing the distribution feeders and installing a new substation. Therefore installing DGs in RDS is the best solution to meet the required load demands. There are several expressions for Distributed energy resources (DGs) are still currently used such as decentralized generation, dispersed generation and distributed Generation (DG), etc. [1]. DGs are defined as an electric sources installed in distribution system or close to the customer locations. Natural resources such as WT and PV depends on natural sources, so they are encouraged to be installed in RDS. WT operates and pause generating power when the wind speed is more than the cut in and cut off velocities. PV produces a clean electricity from the sun radiation. Incorporating WT and PV in RDS decrease the system loss, increase the system capacity and enhance the system voltage. There are different types of DG that can be incorporated in RDS as follows [1]:

- DGs produce both real and reactive powers, such as double fed wind turbine and synchronous generator.
- DGs produce real power alone, such as PV.
- DGs draw reactive power and inject active power such as induction wind turbine.
- DGs produce reactive power alone such as capacitor bank and synchronous compensator.

Recently, metaheuristic optimization techniques are utilized to solve many optimization problems with less search agent and maximum iteration values than other techniques [2]. Many algorithms are utilized for obtaining the optimal allocation of DGs to decrease the system losses. These algorithms are categorized into four types: (1) optimization techniques like mixed integer nonlinear programming (MINLP) [3], (2) analytical techniques like the dual index analytical approach [4], (3) heuristic and metaheuristic method such as backtracking search Algorithm (BSA) [5], Genetic algorithm (GA) [6], Ant Lion Optimizer (ALO) [7], Artificial Bee Colony algorithm (ABC) [8], Particle Swarm Optimization (PSO) [9], harmony search algorithm [10], flower pollination algorithm (FPA) [11] and Krill herd algorithm (KHA) [12], (4) hybrid method like harmony search and particle ant bee colony [13]. In this chapter, DG allocation in RDS is obtained by efficient metaheuristic algorithms such as MFO, CMFO, SSA and HMFOSCA and the best buses are determined using sensitivity analysis. The contribution of this study can be represented as follows:

- DG integration in RDS using MFO, CMFO and SSA algorithms.
- Installing three, two and one PV and WT in IEEE 69 and IEEE 33 bus RDS.
- The total system real loss is utilized as single function.
- The preferable locations for DG integration are obtained by the sensitivity analysis.
- Studding the effect on system performance by incorporating DG in RDS.

- Studding the convergence characteristics of the presented algorithms with other algorithms.

2 Objective Function

WT and PV are utilized as DG to be installed in RDS to decrease the power flows drawn from substation as shown in Fig. 1. The distribution power flows can be evaluated by forward-backward method [14].

The reactive and real load flows are evaluated by Eqs. (1) and (2).

$$P_{S1} = P_{S2} + P_{L,S2} + R\left(\frac{(P_{s2} + P_{L,S2})^2 + (Q_{s2} + Q_{L,S2})^2}{|V_{S2}|^2}\right) \tag{1}$$

$$Q_{S1} = Q_{S2} + Q_{L,S2} + X\left(\frac{(P_{S2} + P_{L,S2})^2 + (Q_{S2} + Q_{L,S2})^2}{|V_{S2}|^2}\right) \tag{2}$$

The voltage magnitude of buses is calculated using Eq. (3)

$$V_{S2}^2 = \left(V_{s1}^2 - 2(P_{S1}R + Q_{S1}X) + \left(R^2 + X^2\right)\frac{(P_{S1}^2 + Q_{s1}^2)}{V_{s1}^2}\right) \tag{3}$$

where, V_{S2} and V_{S1} are the voltage of buses $S2$ and $S1$, respectively. Q_{S1} and P_{S2} are the reactive and real power flow among bus $S1$ and bus $S2$, respectively. R and X are the resistance and reactance of the line between buses $S1$ and $S2$, respectively. The system load flows are changed by installing DG at bus r and can be calculated

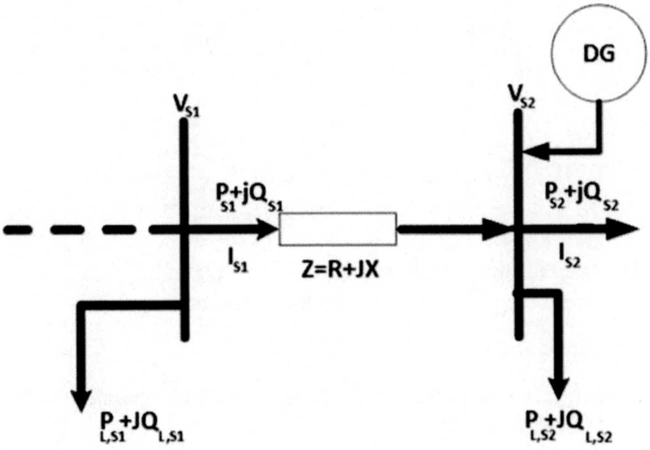

Fig. 1 Portion of RDS with integration of DG at bus (r)

using Eqs. (4) and (5):

$$P_{S1} = P_{S2} + P_{L,S2} + R\left(\frac{(P_{s2} + P_{L,S2})^2 + (Q_{s2} + Q_{L,S2})^2}{|V_{S2}|^2}\right) - P_{DER,r} \quad (4)$$

$$Q_{S1} = Q_{S2} + Q_{L,S2} + X\left(\frac{(P_{S2} + P_{L,S2})^2 + (Q_{S2} + Q_{L,S2})^2}{|V_{S2}|^2}\right) - Q_{DER,r} \quad (5)$$

Therefore, the reactive and real losses among bus $S1$ and bus $S2$ can be formulated from equations below

$$P_{loss(S1,S2)} = R_{S1,S2}\left(\frac{P_{S1}^2 + jQ_{S1}^2}{|V_{S1}|^2}\right) \quad (6)$$

$$Q_{loss(S1)} = x_{S1,S2}\left(\frac{P_{S1}^2 + jQ_{S1}^2}{|V_{S1}|^2}\right) \quad (7)$$

The total real losses is represented as objective function by Eq. (8)

$$F_{obj} = \sum_{c=1}^{Nbr}(P_{loss}(c)) \quad (8)$$

where Nbr is the whole system lines.

The operation and load flow constraints such as branches current, bus voltages and DG sizing as shown next.

A. Equality constraints

The energy consumption should be equivalent to the energy generation.

$$P_s + \sum_{w=1}^{NDG} P_{DG}(w) = \sum_{w=1}^{L} P_{L,w} + \sum_{w=1}^{Nbr} P_{loss}(w) \quad (9)$$

$$Q_s + \sum_{b=1}^{NDG} Q_{DG}(b) = \sum_{b=1}^{L} Q_{L,b} + \sum_{b=1}^{Nbr} Q_{loss}(b) \quad (10)$$

where, *NDG* represents the overall numbers of DGs. Q_s and P_s represent the reactive and real power from the grid (slack bus) in RDS, respectively.

B. Inequality constraints

Branches current, bus voltages and DG sizing should be operated below the high allowable limits as shown below:

(1) Voltage limits

The bus voltage has to be operated among minimum voltage (V_{min}) and maximum voltage (V_{max}) values as illustrated by Eq. (11).

$$V_{min} \leq V_i \leq V_{max} \tag{11}$$

(2) DG sizing limits

$$\sum_{w=1}^{NDG} P_{DG}(w) \leq \left(\sum_{w=1}^{L} P_{L,w} + \sum_{w=1}^{Nbr} P_{loss}(w) \right) \tag{12}$$

$$P_{DG,S2}(min) \leq P_{DG,S2} \leq P_{DG,S2}(max) \tag{13}$$

$$\sum_{w=1}^{NDG} Q_{DG}(w) \leq \left(\sum_{w=1}^{L} Q_L(w) + \sum_{w=1}^{Nbr} Q_{loss}(w) \right) \tag{14}$$

(3) Line capacity limits

The current of branch k should be operated below the high allowable limits ($I_{max,k}$) [11].

$$I_k \leq I_{max,k} \qquad k = 1, 2, 3, \ \ldots, \ N.b \tag{15}$$

3 Active Power Loss Sensitivity Factor (PLSF)

The sensitivity of bus ($i + 1$) to the injection power and its effect on real loss can be obtained by PLSF as illustrated by Eq. (16) [13]. From Fig. 2, the best half locations of IEEE 33-bus RDS using PLSF from worst to best are 2, 30, 25, 20, 23, 31, 29, 10, 13, 24, 9, 5, 4, 28, 3, 8 and 6. From Fig. 3, the best half locations of IEEE 69-bus RDS using PLSF from worst to best are 62, 20, 34, 68, 63, 41, 19, 21, 48, 5, 16, 65, 17, 9, 11, 49, 64, 8, 53, 15, 54, 14, 13, 12, 56, 55, 59, 10, 60, 61, 6, 7, 58 and 57. This chapter applies PLSF to decrease the simulation time and the search agents of the presented algorithms.

$$PLSF = \frac{\partial P_{loss(S1,S2)}}{\partial P_{S2}} = R\left(\frac{2P_{S2}}{|V_{S2}|^2}\right) \tag{16}$$

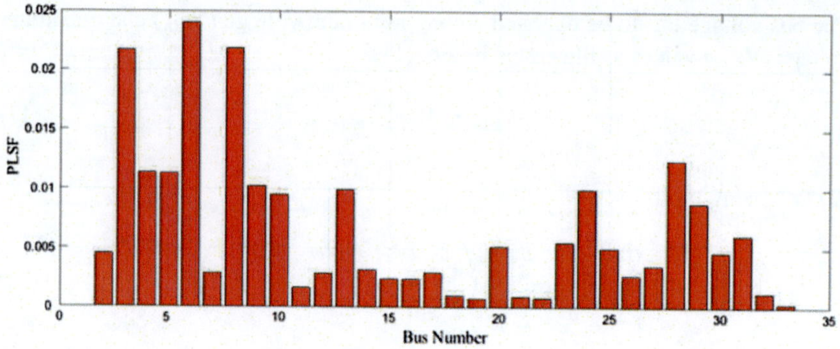

Fig. 2 Sensitivity analysis of IEEE 33-bus RDS

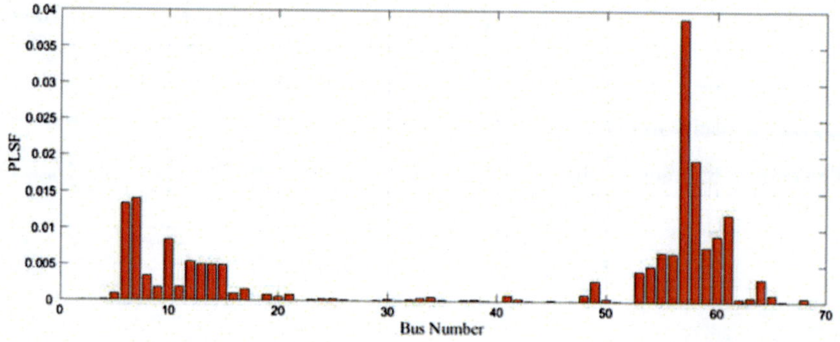

Fig. 3 Sensitivity analysis of IEEE 69-bus RDS

4 Voltage Stability Index (VSI)

In this chapter, VSI is used to indicate the security level of distribution network. VSI is utilized to determine the bus sensitivity to voltage collapse as illustrated by Eq. (17) [15]. When the bus has high value of VSI, the bus become more stable. The optimal allocation of DG in RDS increases the value of VSI for each bus in RDS. The summation of VSI for all buses in RDS is defined as summation of voltage stability index.

$$VSI_{(S2)} = |V_{S1}|^4 - 4(P_{S2}X - Q_{S2}R)^2 - 4(P_{S2}X + Q_{S2}R)|V_{S1}|^2 \tag{17}$$

where, $VSI_{(S2)}$ represents the VSI for bus $S2$ and V_{S1} represents the voltage at bus $S1$ while P_{S2} and Q_{S2} are the real and reactive power generation at bus $S2$. But, X and R are the reactance and the resistance among buses $S1$ and $S2$, respectively.

5 Voltage Deviation

The security index and power equality of power system are determined by the system voltage. Therefore, voltage deviation (VD) is utilized to measure the system voltage that affects the performance of power system. Summation of voltage deviation (SVD) is determined by the total value of VD as shown in Eq. (18).

$$SVD = \sum_{u=1}^{nb} \left(V_u - V_{ref}\right)^2 \tag{18}$$

where, V_{ref} is the value of reference voltage that is equal to 1 pu and V_u is the bus system voltage.

6 Metaheuristic Optimization Techniques

This chapter discusses different efficient methods based on modern metaheuristic optimization algorithms and sensitivity analysis to obtain the best sizes and locations of DGs in RDS under equality and inequality constraints.

6.1 Moth Flame Optimization (MFO) Algorithm

The natural behavior of moths is mathematically modeled to create MFO algorithm. The moths move in spiral path by taking a constant angle to the artificial light as illustrated in Fig. 4 [16]. Also, the moths moves in straight line by taking a constant angle to the moon light. MFO starts with random solutions (moths), then obtain the value and the number of flames and finally, the logarithmic spiral function is applied to obtain the new position of moth according to the flame position. The number of

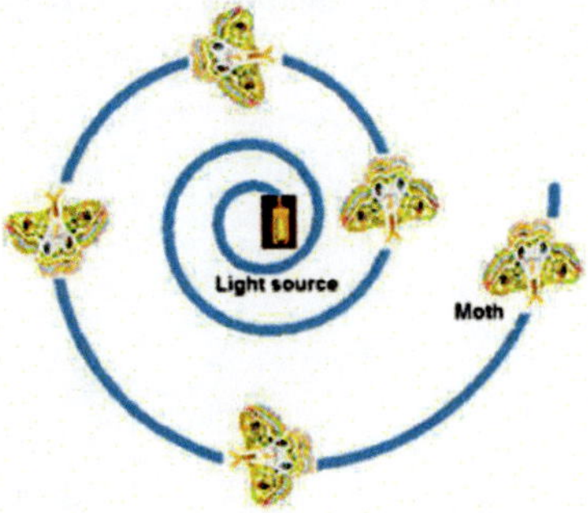

Fig. 4 The moths revolves around the light

flames are changed from maximum to minimum with increasing the iteration until reach to the maximum iteration through the simulation algorithm.

The summarized steps of MFO are illustrated as shown next.

Step 1: Generate initial search agents (moths) between the lower and upper values of dimensions as follow:

$$X(n,d) = rand \times (Up(n,d) - Lp(n,d)) + Lp(n,d) \tag{19}$$

where, n and d are the total moths and variables, respectively.

Step 2: The obtained moths are formulated as follows:

$$A = \begin{bmatrix} A_{1,1} & A_{1,2} & \cdots & A_{1d} \\ A_{2,1} & A_{22} & \cdots & A_{2,1} \\ \vdots & \vdots & \ddots & \vdots \\ A_{n,1} & A_{n,2} & \ldots & A_{n,d} \end{bmatrix} \tag{20}$$

Step 3: Obtain the fitness of all moths

$$OA = [OA_1 OA_2 OA_3 \ \ldots \ OA_N]^T \tag{21}$$

Step 4: Determining the flame matrix by sorting the moths according to their fitness.

$$F = \begin{bmatrix} F_{1,1} & F_{1,2} & \cdots & F_{1d} \\ F_{2,1} & F_{22} & \cdots & F_{2,1} \\ \vdots & \vdots & \ddots & \vdots \\ F_{n,1} & F_{n,2} & \ldots & F_{n,d} \end{bmatrix} \tag{22}$$

Step 5: Evaluate the fitness of all flames that are formulated as follow:

$$\mathrm{OF} = [\mathrm{OF_1 OF_2 OF_3} \ \ldots \ \mathrm{OF_N}]^\mathrm{T} \tag{23}$$

Step 6: Evaluate the position of moths by the logarithmic spiral function as follow:

$$D_i = |F_j - A_i| \tag{24}$$

$$A_{i\_new} = D_i e^{bt} \cos(2\pi t) + F_j \tag{25}$$

$$t = rand \times (a - 1) + 1 \tag{26}$$

$$a = \left(-1 - \left(\frac{T}{T_{max}}\right)\right) \tag{27}$$

$$\text{Flames} = \text{round}\left(M - T \times \frac{M-1}{T_{max}}\right) \tag{28}$$

where, M represents the whole flames.

Step 7: Go to steps 3 when the maximum iteration is not satisfied.

Step 8: Obtain the preferable fitness and the best flame (sizes and locations of DGs).

6.2 *Chaotic Moth-Flame Optimization (CMFO) Algorithm*

CMFO algorithm is created by adding Chaos function to MFO algorithm to enhance the characteristics of MFO algorithm [17, 18]. The logistic map is used as chaos function to control the rates of exploitation and exploration of MFO algorithm. Exploration phase is applied to determine the preferable area of search space, then the exploitation phase is utilized to determine the global superior results in the preferable area. The summarized steps of CMFO algorithm are illustrated as shown next.

Step 1: Generate initial search agents (moths) between the lower and upper values of dimensions as follow:

$$X(n,d) = rand \times (Up(n,d) - Lp(n,d)) + Lp(n,d) \tag{29}$$

where, n and d are the total moths and variables, respectively.

Step 2: The obtained moths position are formulated as follows:

$$A = \begin{bmatrix} A_{1,1} & A_{1,2} & \cdots & A_{1d} \\ A_{2,1} & A_{22} & \cdots & A_{2,1} \\ \vdots & \vdots & \ddots & \vdots \\ A_{n,1} & A_{n,2} & \ldots & A_{n,d} \end{bmatrix} \tag{30}$$

Step 3: Obtain the values of chaos function (CF)

Step 4: Obtain the fitness of all moths

$$OA = [OA_1 OA_2 OA_3 \ \ldots \ OA_N]^T \tag{31}$$

Step 5: Determining the flame matrix by sorting the moths according to their fitness.

$$F = \begin{bmatrix} F_{1,1} & F_{1,2} & \cdots & F_{1d} \\ F_{2,1} & F_{22} & \cdots & F_{2,1} \\ \vdots & \vdots & \ddots & \vdots \\ F_{n,1} & F_{n,2} & \ldots & F_{n,d} \end{bmatrix} \tag{32}$$

Step 6: Evaluate the fitness of all flames that are formulated as follow:

$$\mathrm{OF} = [\mathrm{OF}_1 \mathrm{OF}_2 \mathrm{OF}_3 \ \ldots \ \mathrm{OF}_N]^T \tag{33}$$

Step 7: Evaluate the moth position by Eq. (35)

$$D_i = |F_j - A_i| \tag{34}$$

$$A_{i\_new} = D_i e^{bt} \cos(2\pi t) + F_j \tag{35}$$

$$t = rand \times (a - 1) + 1 \tag{36}$$

$$a = \left(\left(\frac{IS - IS_{min}}{IS_{max} - IS_{min}} \right) - 2 \right) T \tag{37}$$

$$\text{Flames} = \text{round}\left(M - T \times \frac{M - 1}{T_{max}} \right) \tag{38}$$

where, M represents the whole flames.

Step 8: Go to steps 3 when the maximum iteration is not satisfied.

Step 9: Obtain the preferable fitness and the best flame (sizes and locations of DGs).

6.3 Salp Swarm Algorithm (SSA)

SSA is a population metaheuristic algorithm that is created from the natural behavior of salps in oceans. Salps are similar to jelly fish and move in salp chain in oceans as shown in Fig. 5 [19]. The first salp at the front of salp chain is called as leading salp and the rest of salps are called as followers. SSA begins with random positions (salps), then obtain the value of fitness for all salps to obtain the leading salp and the follower salps. Finally, update the follower salps according to each other to become in a chain toward to the leading salp and update the leading salp position according to the food source (best solution). The summarized steps of SSA algorithm are illustrated as shown next.

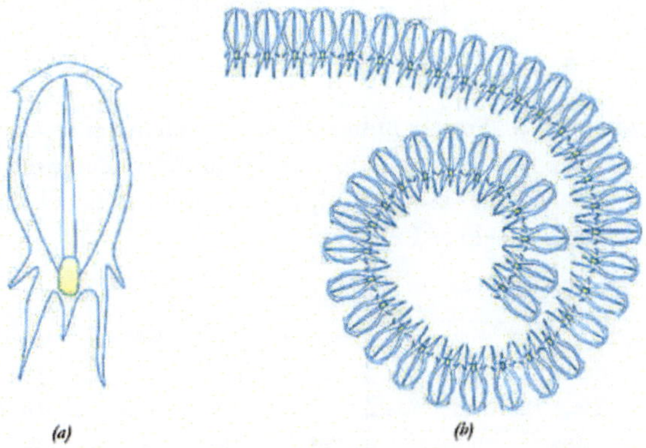

Fig. 5 **a** Single salp, **b** salp swarm

Step 1: Generate initial search agents (moths) between the lower and upper values of dimensions as follow:

$$S(n, d) = rand\ (UP(n, d) - LP(n, d)) + LP(n, d) \tag{39}$$

where, n and d are the total salps and variables, respectively.

Step 2: The obtained salps position are formulated as follows:

$$P = \begin{bmatrix} P_{1,1} & P_{1,2} & \cdots & P_{1,i} \\ P_{2,1} & P_{22} & \cdots & P_{2,\mathrm{i}} \\ \vdots & \vdots & \ddots & \vdots \\ P_{m,1} & P_{m,2} & \ldots & P_{m,i} \end{bmatrix} \tag{40}$$

Step 3: Obtain the fitness for all salps

$$OS = [OS_1 OS_2 OS_3\ \ldots\ OS_m] \tag{41}$$

Step 4: The salps are sorted according to their fitness as follows:

$$L = \begin{bmatrix} L_{1,1} & L_{1,2} & \cdots & L_{1d} \\ L_{2,1} & L_{22} & \cdots & L_{2,1} \\ \vdots & \vdots & \ddots & \vdots \\ L_{n,1} & L_{n,2} & \ldots & L_{n,d} \end{bmatrix} \tag{42}$$

Step 5: the leading salp position is obtained by Eq. (43).

$$P_{1,j\_new} = \begin{cases} F_j + c_1\big((Up_j - Lp_j)c_2 + Lp_j\big) & c_3 \geq 0 \\ F_j - c_1\big((Up_j - Lp_j)c_2 + Lp_j\big) & c_3 < 0 \end{cases} \tag{43}$$

$$c_1 = 2e^{-\left(\frac{4l}{L}\right)^2} \tag{44}$$

where, c_2 and c_3 are uniformly random number in the range of (0, 1),F_j is the food source position, L and l are the maximum and current iteration.

Step 6: The follower salps position are updated according to each other toward the leadind salp as follows:

$$P_{i,j\_new} = \frac{1}{2}\left(P_{i,j\_new} + P_{i-1,j\_new}\right) \tag{45}$$

Step 7: Go to steps 3 when the maximum iteration is not satisfied.

Step 8: Obtain the preferable fitness and the best flame (sizes and locations of DGs).

7 Simulation Results

7.1 IEEE 33-Bus RDS

This system has total load of 3715 kW and 2300 kVAr with 32 branches and 33 buses as illustrated in Fig. 6 [20]. The used parameters are given in Table 1. The power and

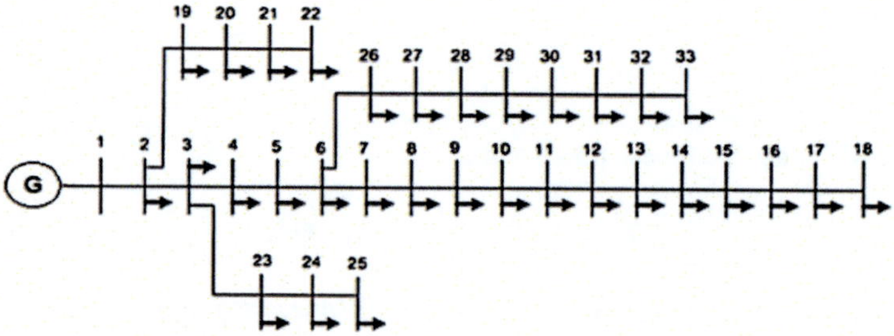

Fig. 6 IEEE 33-bus system diagram

Table 1 The utilized system limits and parameters

Item	Value
DG power factor limits	$1 \geq PF_{DG,i} \geq 0.7$
limits of DG sizing	$3MW \leq P_{DG,i} \geq 0.3\text{MW}$
Limits of voltage	$1.05\text{pu} \geq V_i \geq 0.9\text{pu}$
Maximum iteration	100
Number of population	50

voltage base values are 100 MVA and 12.66 kV, respectively. Without incorporating DG, the real loss is 210.997 kW with minimum voltage of 0.90377 pu.

7.1.1 The Optimal Allocation of DG Using CMFO

From Table 2, incorporating three, two and one PV in RDS decrease the total real loss to 72.786 kW, 87.166 kW and 111.027 kW respectively. Also, incorporating three, two and one WT in RDS decrease the total real loss to 11.742 kW, 28.505 kW and 67.868 kW, respectively as shown in Fig. 7. The obtained results proved that incorporating multiple DG gives superior results than incorporating one DG in RDS and integrating WT gives superior results than incorporating PV in RDS under system constraints. From Table 3, CMFO algorithm is applied to evaluate the optimal allocation of DGs in RDS and compared with other efficient algorithms. Fig. 8 illustrates that CMFO algorithm has an efficient convergence characteristic speed when compared to other efficient algorithms such as moth flame optimization (MFO) algorithm and grey wolf optimization (GWO) algorithm.

Table 2 Simulation results of total real loss using CMFO algorithm

Item	Without DG	1-WT	2-WT	3-WT	1-PV	2-PV	3-PV
Total real loss (kW)	210.997	67.868	28.505	11.742	111.027	87.166	72.786
Power loss reduction (%)	–	67.8	86.5	94.4	47.4	58.7	65.5

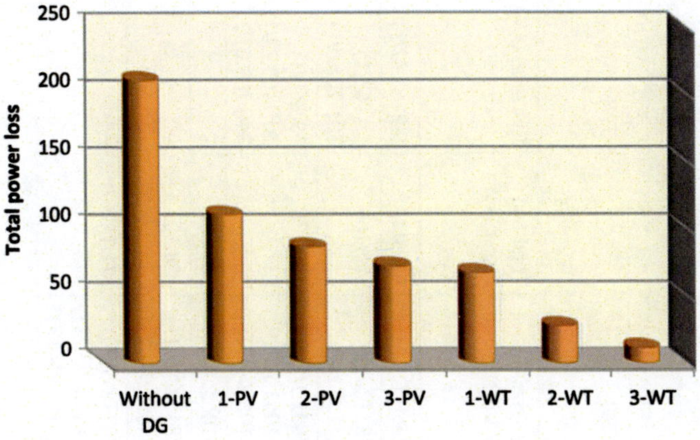

Fig. 7 Total real loss by incorporating single and multiple PV and WT using CMFO algorithm in RDS

Table 3 The comparison of simulation results using proposed method and other optimization algorithm

No. of different type of DG	CMFO		GWO [3]		EA [10]		Hybrid [13]	
	Bus (size (kVA/pf))	Total real losses (kW)	Bus (size (kVA/pf))	Total real loss (kW)	Bus (size (kVA/pf))	Total real loss (kW)	Bus (size (kVA/pf))	Active system power loss (kW)
1-PV	6(2590/1)	111.027	6(2761.82/1)	111.42	6(2530/1)	111.07	6(2530/1)	111.07
2-PV	13(851.5/1) 30(1157.6/1)	87.166	13(903/1) 30(1201.6/1)	87.43	13(844/1) 30(1149/1)	87.172	13(844/1) 30(1149/1)	87.172
3-PV	13(801.7/1) 30(1053.7/1) 6(3106.2/0.82)	72.786	13(850/1) 30(1100.77/1) 24(1103.87/1)	73.06	13(798/1) 30(1050) 24(1099)	72.787	13(798/1) 30(1050) 24(1099)	72.787
1-WT	6(3106.2/0.82)	67.868	–	–	6(3119/0.82)	67.87	6(3028/0.82)	67.937
2-WT	13(845.6/0.91) 30(1557.9/0.73)	28.505	–	–	13(938/0.90) 30(1573/0.73)	28.52	13(1039/0.91) 30(1508/0.72)	28.98
3-WT	13(877.8/0.90) 30(1443.9/0.71) 24(1183.95/0.9)	11.742	–	–	13(886/0.90) 30(1450/0.71) 24(1189/0.90)	11.8	13(873/0.9) 30(1439/0.71) 24(1186/0.89)	11.76

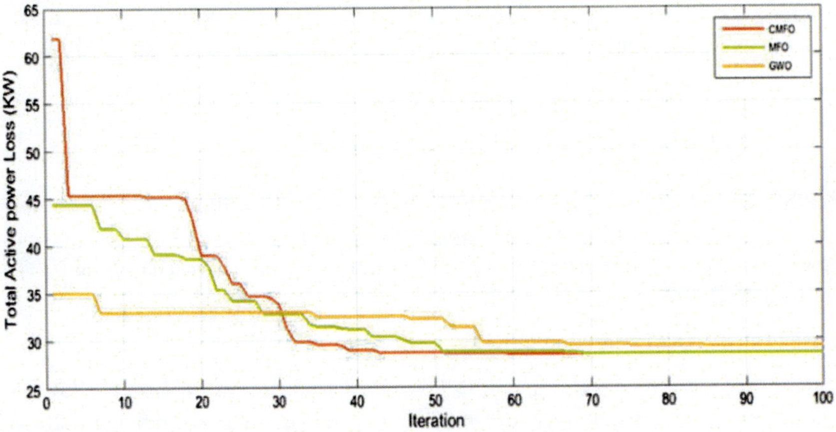

Fig. 8 Total real loss by incorporating two WT in RDS using CMFO, GWO and MFO

7.2 *IEEE 69-Bus RDS*

From Fig. 9, this system has total load of 2694.6 kVAR and 3801.49 kW with 69 buses and 68 branches [21]. The used parameters are given in Table 4. The power and

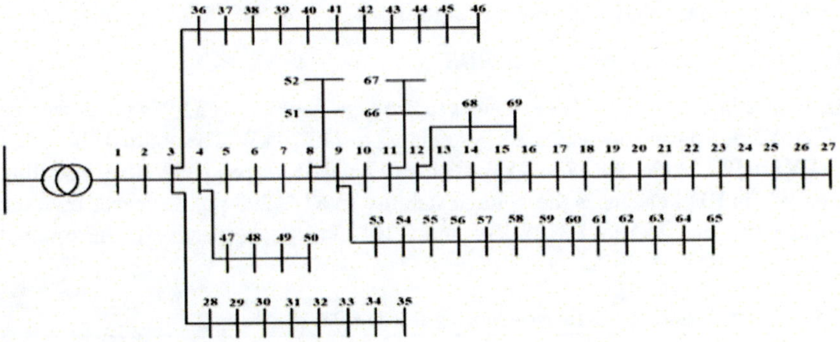

Fig. 9 The IEEE 69-bus system diagram

Table 4 The used parameters

Item	Value
DG power factor limits	$0.65 \leq PF_{DG,i} \leq 1$
DG sizing limits	$3\text{MW} \geq P_{DG,i} \geq 0\text{MW}$
Voltage limits	$1.05 \geq V_i \geq 0.95\text{pu}$
Maximum iteration	100
Number of population	50

voltage base values are 100 MVA and 12.66 kV, respectively. Without incorporating DG, the minimum voltage is 0.90919 pu at bus 65 with total loss of 224.999 kW.

7.2.1 The Optimal Allocation of DG Using MFO

The problem formulation is represented using multi-objective function. Active power loss sensitivity factor (PLSF) can be applied to determine the sensitivity of bus system and choose the best candidate buses for integrating of DG in RDS. The studied cases are presented next.

Case (1): One DG Integration in RDS

The optimal placement and sizing of WT and PV based DG units re listed in Table 5. Also, the minimum bus voltage, the obtained power loss, the maximum bus voltage, VD and VSI are summarized in Table 5. One PV is installed in RDS to decrease the system losses from 224.999 kW to 83.224 kW, enhance the voltage stability from 61.2181 to 64.61156 pu and decrease the VD from 1.8374 to 0.8755 pu. The optimal sizing of one PV is 1867.336 kW at bus 61. Installing WT in RDS enhances the voltage stability to 64.61156 pu and decreases both all the system loss and VD to 23.171 kW and 0.591456 pu, respectively. The system voltage is improved best results are obtained by incorporating WT than incorporating PV in RDS. Figure 10 illustrates that incorporating WT obtain preferable solutions than incorporating PV in RDS. Table 6 and Table 7 illustrate that the presented algorithm is an effective to determine better solutions than other algorithms.

Case (2): Two DGs Integration in RDS

Incorporating two PV in RDS enhances the voltage stability to 65.9954 pu, decreases VD to 0.5091 pu and decreases the system loss to 71.679 kW. The optimal allocation of PV are 522.74 kW and 1776.9 kW at buses 17 and 61, respectively. Also, installing two WT in RDS enhances the voltage stability to 67.66298 pu, decreases both the system loss and VD to 7.3122 kW and 0.1013545 pu, respectively. the optimal

Table 5 The obtained results for incorporating DGs in RDS

Item	Location (kVA/PF)	P_{loss} (kW)	Loss reduction	V_{max} (pu)	V_{min} (pu)	VD (pu)	VSI (pu)
Without DG	–	225.000	–	0.99997	0.90919	1.8374	61.2181
1-PV	61 (1867.336/1)	83.224	63.01171	0.99997	0.96826	0.8755	64.61156
1-PV	61(2235.92/0.81)	23.171	89.70182	0.99998	0.97242	0.591456	65.70429
2-PV	61 (1776.9/1) 17 (522.747/1)	71.679	68.14281	0.99997	0.97872	0.5091	65.9954
2-WT	61 (2173.3/0.81) 17 (661.56/0.83)	7.3122	96.7501	1.00	0.99426	0.1013545	67.66298

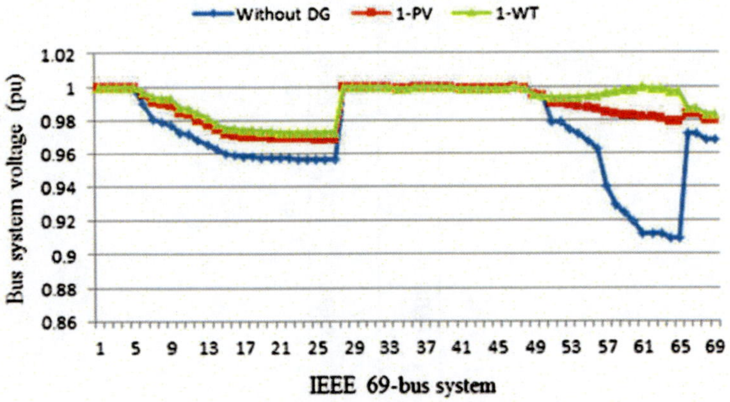

Fig. 10 Bus system voltages with inclusion single DG

allocation of two WT are 2173.3 kVA with 0.81 and 661.56 kVA with 0.83 power factor at buses 17 and 61, respectively. Installing two WT obtain preferable solutions than installing two PV in RDS as shown in Tables 8 and 9. The system voltage is enhanced by incorporating WT and PV in RDS as illustrated in Fig. 11. From Tables 8 and 9, the simulation results illustrated that the presented algorithm is an efficient to obtain preferable solutions than other techniques.

7.2.2 Incorporating DG in RDS Using SSA

Installing three, two and one PV in RDS decrease the system real loss to 69.427 kW, 71.675 kW and 83.222 kW, respectively. Also two and three WT are incorporated in RDS to decrease the system loss to 23.169 kW and 7.201 kW, respectively. The optimal allocation of three WT is the best case as it can minimize the total real loss to 4.269 kW. From Fig. 12, integration of two and one PV in RDS enhance the minimum voltage to 0.97893 pu at bus 65 and 0.96829 pu at bus 27, respectively. Also, the minimum voltage is enhanced to 0.97898 pu at bus 65 by installing three PV in RDS. Installing WT achieve better reduction in branches current than installing PV in RDS as shown in Fig. 13. Installing three and two WT in RDS enhance the system voltage to 0.98811 pu at bus 50 and 0.98808 pu at bus 50, respectively. Figure 14 illustrates the system real loss by incorporating DG in RDS. The obtained results using SSA algorithm are better than other efficient algorithms as shown in Tables 10 and 11. Figure 15 illustrates that SSA algorithm has an efficient convergence characteristic speed when compared to other efficient algorithms.

Table 6 The comparative study for incorporating one PV in RDS

	Item	Presented algorithm	GWO [22]	WOA [23]	Analytical [24]	GA [25]	MTLBO [26]	CSA [27]	SGA [27]
1-PV	P_{loss} (kW)	83.224	83.24	83.2279	92	83.4252	83.323	83.8	89.4
	Size (kVA/PF	1867.3357/1	1928.67/1	1872.82/1	1807.8/1	1794/1	819.691/1	2000/1	2300/1
	Location	61	61	61	61 ara>	61	61	61	61

Table 7 The comparative study for incorporating one WT in RDS

	Item	Presented algorithm	SGA [27]	CSA [27]	GA [28]	PSO [27]	WOA [23]
1-WT	P_{loss} (kW)	23.171	64.4	52.6	38.458	52.6	27.9649
	Size (kVA/PF)	2235.915/0.814	2600/NR	2300/NR	2155.6/NR	2300/NR	2217.39/0.9
	Location	61	61	61	61	61	61

Table 8 The comparative study for incorporating two PV in RDS

	Technique	Presented algorithm	GA [28]	CSA [27]	SGA [27]	MTLBO [26]	PSO [27]	GWO [22]
2-PV	P_{loss} (kW)	71.679	71.791	76.4	82.9	71.776	78.8	71.74
	Location	61 17	61 11	61 22	61 17	61 17	62 14	61 17
	Size (kVA/PF)	1776.9/1 522.742/1	1777/1 555/1	2100/1 600/1	2400/1 1000/1	1732/1 519.705/1	2100/1 700/1	1816.42/1 566.08/1

Table 9 The comparative study for incorporating two WT in RDS

	Item	Presented algorithm	PSO [27]	SGA [27]	CSA [27]
2-WT	P_{loss} (kW)	7.3122	42.4	44	39.9
	Location	61 17	62 18	62 18	61 18
	Size (kVA/PF)	2173.3/0.81 661.56/0.83	1900/NR 900/NR	2300/NR 600/NR	2000/NR 800/NR

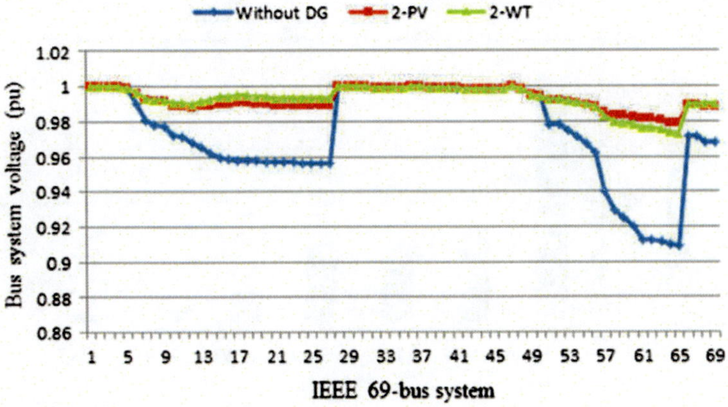

Fig. 11 Bus system voltages with inclusion two DGs

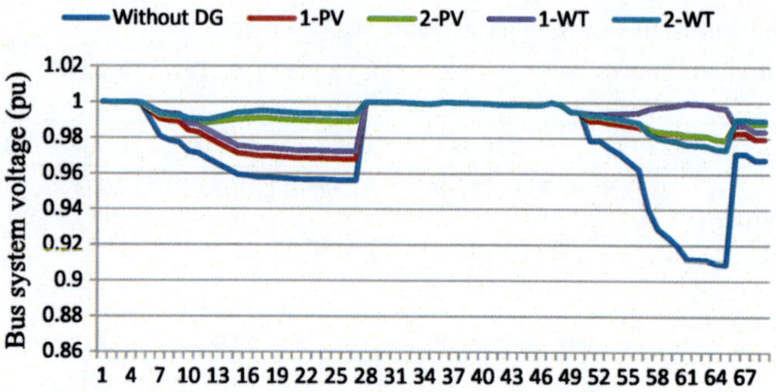

Fig. 12 Bus system voltages for different types of DG

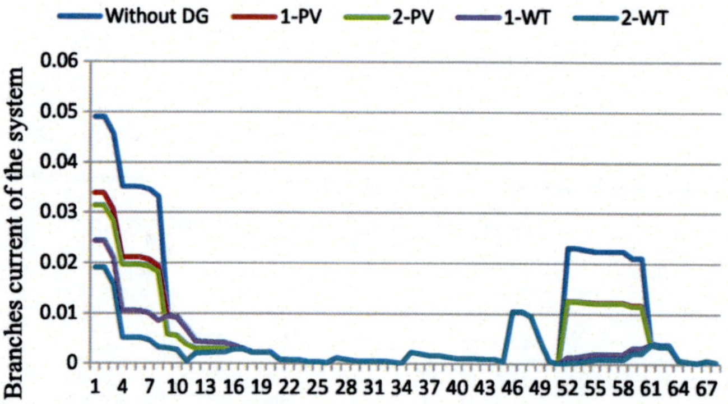

Fig. 13 Branches current for different types of DG

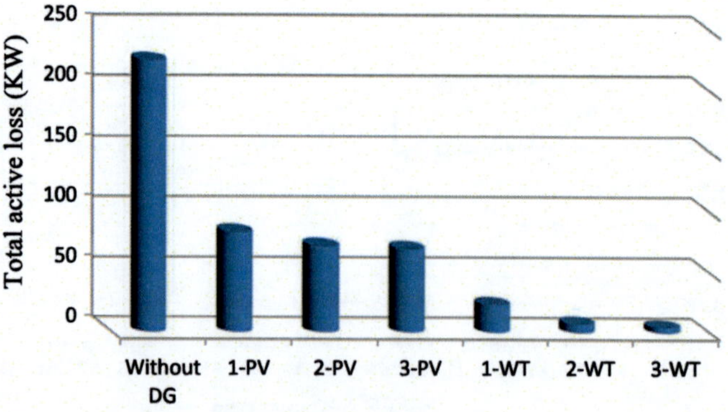

Fig. 14 The total real losses by installing DGs in RDS

Table 10 Simulation results obtained by different studied optimization techniques in case of incorporating DG (PV) in RDS

Method	1-PV		2-PV		3-PV	
	Power Loss (kW)	Bus no (kVA/PF)	Power Loss (kW)	Bus no (kVA/PF)	Power Loss (kW)	Bus no (kVA/PF)
Base Case	224.999	- -	224.999	- -	224.999	- -
SSA	83.222	61(1872.7/1)	71.6745	61(1781.5/1) 17(531.5/1)	69.4266	61(1719/1) 17(380.5/1) 11(526.7/1)
EA [29]	83.23	61(1878/1)	71.68	61(1795/1) 17(534/1)	69.62	61(1795/1) 18(380/1) 11(467/1)
Hybrid [30]	83.372	61(1810/1)	71.82	61(1720/1) 17(520/1)	69.52	61(1670/1) 17(380/1) 11(510/1)
KHA [12]	- -	- -	–	–	–	–

Table 11 Simulation results obtained by different optimization techniques in case of incorporating DG (WT) in RDS

Method	1-WT		2-WT		3-WT	
	Power Loss (kW)	Bus no (kVA/PF)	Power Loss (kW)	Bus no (kVA/PF)	Power Loss (kW)	Bus no (kVA/PF)
Base case	224.999	- -	224.999	- -	224.999	- -
SSA	23.1688	61(2243.9/0.82)	7.2013	61(2131.3/0.81) 17(630.7/0.83)	4.269	61(2057.1/0.81) 17(454.9/0.84) 11(608.9/0.82)
Hybrid [29]	23.92	61(2200/0.82)	7.21	61(2120/0.81) 17(630/0.82)	4.30	61(2060/0.83) 18(480/0.77) 66(530/0.82)
EA [30]	23.26	61(2290/0.82)	7.35	61(2189, 0.82) 17(643/0.83)	4.48	61(2113/0.82) 18(458/0.83) 11(668/0.82)
KHA [12]	23.22	61(2290/0.82)	–	–	5.91	61(1773/0.86) 22(357/0.86) 11(560/0.86)

8 Conclusion

This chapter presents an efficient algorithms to determine the preferable sizing and placement of PV and WT based DG in RDS. Sensitivity analysis is applied to obtain

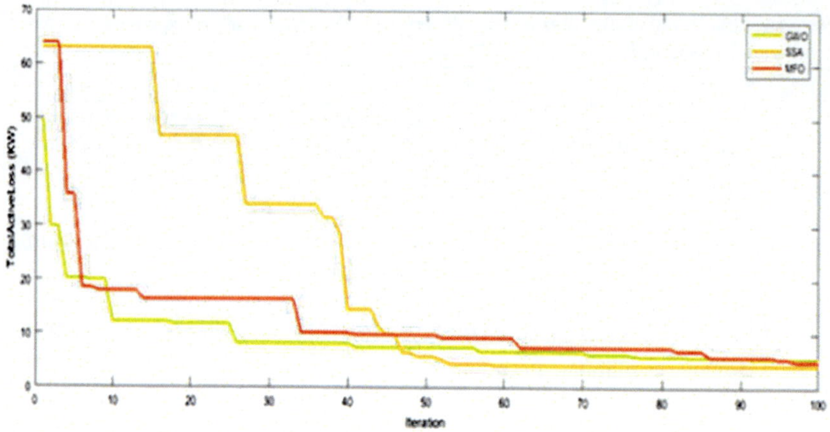

Fig. 15 Total real loss by incorporating three WT in RDS using SSA, GWO and MFO

the best locations for incorporating DG in RDS to decrease the search agents and the simulation time algorithm. The presented algorithms that are utilized for DG integration in RDS are MFO, CMFO and SSA algorithms. The results proved that integrating WT obtain preferable solutions than integrating PV and installing multiple DGs obtains preferable solutions than installing one DG in RDS. Also, installing DG in RDS enhances the system voltage, increases the system capacity, decreases the system loss and increases the system reliability. In the future, the presented algorithms can be utilized to solve difficult problems due to their efficient characteristics. Acknowledgements: The authors thank the support of the National Research and Development Agency of Chile (ANID), ANID/Fondap/15110019.

Acknowledgements The authors thank the support of the National Research and Development Agency of Chile (ANID), ANID/Fondap/15110019.

References

1. B. Singh, V. Mukherjee, P. Tiwari, A survey on impact assessment of DG and FACTS controllers in power systems. Renew. Sustain. Energy Rev. **42**, 846–882 (2015)
2. A. Kumar, P. V. Babu, V. Murty, Distributed generators allocation in radial distribution systems with load growth using loss sensitivity approach. J. Inst. Eng. (India) Ser. B **98**, 275–287 (2017)
3. S. Kaur, G. Kumbhar, J. Sharma, A MINLP technique for optimal placement of multiple DG units in distribution systems. Int. J. Electr. Power Energy Syst. **63**, 609–617 (2014)
4. D.Q. Hung, N. Mithulananthan, Loss reduction and loadability enhancement with DG: a dual-index analytical approach. Appl. Energy **115**, 233–241 (2014)
5. A. El-Fergany, Optimal allocation of multi-type distributed generators using backtracking search optimization algorithm. Int. J. Electr. Power Energy Syst. **64**, 1197–1205 (2015)
6. S. Biswas, S.K. Goswami, A. Chatterjee, Optimum distributed generation placement with voltage sag effect minimization. Energy Convers. Manag. **53**, 163–174 (2012)

7. A. Abdelaziz, E. Ali, S.A. Elazim, Flower pollination algorithm and loss sensitivity factors for optimal sizing and placement of capacitors in radial distribution systems. Int. J. Electr. Power Energy Syst. **78**, 207–214 (2016)
8. F.S. Abu-Mouti, M. El-Hawary, Optimal distributed generation allocation and sizing in distribution systems via artificial bee colony algorithm. IEEE Trans. Power Deliv. **26**, 2090–2101 (2011)
9. N. Kanwar, N. Gupta, K. Niazi, A. Swarnkar, R. Bansal, Simultaneous allocation of distributed energy resource using improved particle swarm optimization. Appl. Energy **185**, 1684–1693 (2017)
10. Z.W. Geem, J.H. Kim, G.V. Loganathan, A new heuristic optimization algorithm: harmony search. Simulation **76**, 60–68 (2001)
11. E.S. Oda, A.A. Abdelsalam, M.N. Abdel-Wahab, M.M. El-Saadawi, Distributed generations planning using flower pollination algorithm for enhancing distribution system voltage stability. Ain Shams Eng. J. **8**, 593–603 (2017)
12. S. Sultana, P.K. Roy, Krill herd algorithm for optimal location of distributed generator in radial distribution system. Appl. Soft Comput. **40**, 391–404 (2016)
13. K. Muthukumar, S. Jayalalitha, Optimal placement and sizing of distributed generators and shunt capacitors for power loss minimization in radial distribution networks using hybrid heuristic search optimization technique. Int. J. Electr. Power Energy Syst. **78**, 299–319 (2016)
14. U. Eminoglu, M.H. Hocaoglu, Distribution systems forward/backward sweep-based power flow algorithms: a review and comparison study. Electr. Power Compon. Syst. **37**, 91–110 (2008)
15. S. Sivanagaraju, N. Visali, V. Sankar, T. Ramana, Enhancing voltage stability of radial distribution systems by network reconfiguration. Electr. Power Compon. Syst. **33**, 539–550 (2005)
16. S. Mirjalili, Moth-flame optimization algorithm: a novel nature-inspired heuristic paradigm. Knowl. Based Syst. **89**, 228–249 (2015)
17. M. Wang, H. Chen, B. Yang, X. Zhao, L. Hu, Z. Cai et al., Toward an optimal kernel extreme learning machine using a chaotic moth-flame optimization strategy with applications in medical diagnoses. Neurocomputing **267**, 69–84 (2017)
18. E. Emary, H.M. Zawbaa, Impact of chaos functions on modern swarm optimizers. PLoS One **11**, e0158738 (2016)
19. S. Mirjalili, A.H. Gandomi, S.Z. Mirjalili, S. Saremi, H. Faris, S.M. Mirjalili, Salp swarm algorithm: a bio-inspired optimizer for engineering design problems. Adv. Eng. Softw. **114**, 163–191 (2017)
20. B. Venkatesh, R. Ranjan, Optimal radial distribution system reconfiguration using fuzzy adaptation of evolutionary programming. Int. J. Electr. Power Energy Syst. **25**, 775–780 (2003)
21. J. Savier, D. Das, Impact of network reconfiguration on loss allocation of radial distribution systems. IEEE Trans. Power Deliv. **22**, 2473–2480 (2007)
22. A. Sobieh, M. Mandour, E.M. Saied, M. Salama, Optimal number size and location of distributed generation units in radial distribution systems using Grey Wolf optimizer. Int. Electr. Eng. J **7**, 2367–2376 (2017)
23. P. Dinakara Prasasd Reddy, T. Reddy Dr, Optimal renewable resources placement in distribution. Electr. Power Energy Syst. **28**, 669–678, (2017)
24. T. Gözel, M.H. Hocaoglu, An analytical method for the sizing and siting of distributed generators in radial systems. Electr. Power Syst. Res. **79**, 912–918 (2009)
25. I. Pisica, C. Bulac, M. Eremia, Optimal distributed generation location and sizing using genetic algorithms, in *15th International Conference on Intelligent System Applications to Power Systems, 2009. ISAP'09* (2009), pp. 1–6
26. J.A.M. García, A.J.G. Mena, Optimal distributed generation location and size using a modified teaching–learning based optimization algorithm. Int. J. Electr. Power Energy Syst. **50**, 65–75 (2013)
27. W. Tan, M. Hassan, M. Majid, H.A. Rahman, Allocation and sizing of DG using cuckoo search algorithm, in *2012 IEEE International Conference on Power and Energy (PECon)* (2012), pp. 133–138

28. T. Shukla, S. Singh, V. Srinivasarao, K. Naik, Optimal sizing of distributed generation placed on radial distribution systems. Electr. Power Compon. Syst. **38**, 260–274 (2010)
29. S. Kansal, V. Kumar, B. Tyagi, Hybrid approach for optimal placement of multiple DGs of multiple types in distribution networks. Int. J. Electr. Power Energy Syst. **75**, 226–235 (2020)
30. K. Mahmoud, N. Yorino, A. Ahmed, Optimal distributed generation allocation in distribution systems for loss minimization. IEEE Trans. Power Syst. **31**, 960–969 (2018)

Intelligent Computational Models for Cancer Diagnosis: A Comprehensive Review

Essam Halim Houssein, Hager N. Hassan, Mustafa M. Al-Sayed, and Emad Nabil

Abstract Computational modeling can be defined as the use of computers for studying and simulating complex systems, such as the human brain, organisms, and Earth's global climate, using mathematics, physics, and computer science. Cancer is a complicated and heterogeneous illness that involves a large number of cells, reactions, and events that take place throughout time. Computational modelling, in combination with experimental studies and clinical testing, can help researchers better understand cancer and develop better treatment techniques. Computational modeling supplies tools for tackling the complexity of cancer and providing a detailed mechanistic insight that tells us which treatment is appropriate for a patient, whether a cancer medicine will stop a tumour from growing and whether a cancer drug will have an effect on healthy tissues in the human body. Many previous studies of Computational modeling are used for cancer diagnosis that is comparatively efficient and less weak. Gene expression microarray technology, which is generally used for cancer diagnosis, prediction, or prognosis, suffers from the curse of a dimensional problem. This problem can be solved using gene selection, which is implemented to the microarray for choosing the most useful features from the original features. The selecting of the optimal number of informative genes is considered an NP-hard problem. The meta-heuristic algorithms (MAs) which are robust at choosing the best solutions for complicated optimization problems for solving NP-hard problems, are also beneficial for solving big problems in sensible computational time. Meta-heuristic algorithms have become popular and powerful in computational modeling and many applications. The core of meta-heuristics Algorithms (MAs) is convenient for choosing the most informative and relative features as preprocessing step to Machine Learning (ML) step for achieving the highest classification accuracy for cancer diagnosis. This

E. H. Houssein (✉) · H. N. Hassan · M. M. Al-Sayed
Faculty of Computers and Information, Minia University, Minia, Egypt
e-mail: essam.halim@mu.edu.eg

E. Nabil
Faculty of Computers and Artificial Intelligence, Cairo University, Cairo, Egypt

Faculty of Computer and Information Systems, Islamic University of Madinah, Madinah, Saudi Arabia

E. H. Houssein et al. (eds.), *Integrating Meta-Heuristics and Machine Learning for Real-World Optimization Problems*, Studies in Computational Intelligence 1038,
https://doi.org/10.1007/978-3-030-99079-4_2

chapter introduces a comparative study among the most common MAs algorithms combined with ML computational modeling techniques for cancer diagnosis.

Keywords Support vector machines · Feature selection · Microarray cancer classification · Gene selection

1 Introduction

Computational models which be defined as mathematical models, can be simulated using computation for studying complicated systems. Computational models, in biology, can be used for studying an propagation of an contagious disease such as the flu. Computer simulations are used to adjust the mathematical model parameters for studying various possible outcomes. In biological systems, the use of computational models allows us for improving our understanding of the concealed molecular mechanisms, discovering the pathogenesis of complex diseases, and promoting an optimal treatment strategy, and discovering new drug [1].

Bioinformatics is a scientific field which applies computational models to the biological world [2]. Bioinformatics is an interdisciplinary field that develops ways and software tools for understanding biological data, specifically when the datasets are complex and large. As an interdisciplinary field of science, bioinformatics combines statistics, mathematics, computer science, biology, and information engineering for analyzing and interpreting the biological data [2].

As shown in Fig. 1, bioinformatic mechanisms derive their power from the combination of several important sciences, such as information engineering, biology, mathematics, statistics and computer science for analyzing and interpreting biological data.

Over the past two decades, the emergence of DNA microarray datasets has prompted a new line of research in both machine learning and bioinformatics [3]. In bioinformatics, there are some technologies used for genomic analyses such as DNA microarray, which is one of the most rapidly evolving innovative technologies in the field of genetic research [4]. The technology of DNA microarray is widely utilised to check for everything from cancer to scourge control [5]. It examines the gene expression profiles of hundreds of genes at once and distinguishes between cancer-related gene expression profiles and normal profiling.

DNA microarray datasets face the problem of a dimensional curse which caused by unrelated and redundant genes and the limited number of samples. So, the conventional classification of gene expression becomes more difficult for a given sample [6]. For overcoming the curse of dimensionality problem, gene selection to microarray gene expression is performed in order to select the most informative subset of features from the original feature set. Selecting the optimal number of informative genes can be deemed as NP-hard problem [7].

Several meta-heuristics algorithms [8] are applied in DNA microarray technology to select the most relative and informative genes, which helps for classifying cancer

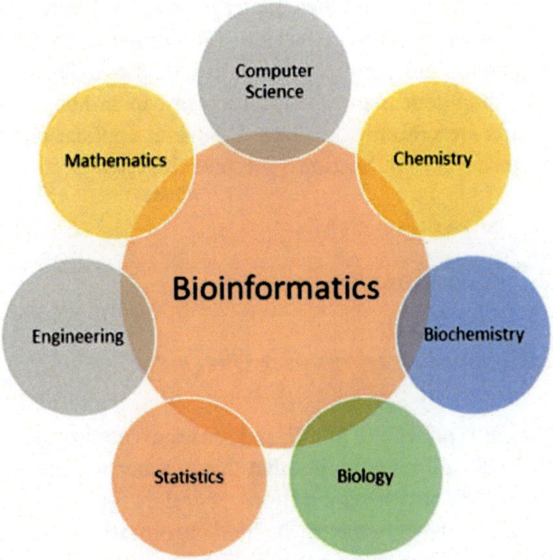

Fig. 1 Interdisciplinary bioinformatics: building bridges with other fields

types. Meta-heuristics algorithms can be combined with several Machine Learning (ML) techniques such as Artificial Neural Networks (ANN), Support Vector Machine (SVM) [9], k-Nearest Neighbors (kNN) and Deep Learning (DL) for detecting cancerous and normal humans and classifying among various types of cancers.

This chapter introduces a comparative study among the most common MAs combined with ML computational modeling techniques used for cancer diagnosis. The covered techniques include Barnacles Mating Optimizer (BMO) [10], Manta-Ray Foraging Optimization (MRFO) [11], Particle Swarm Optimization (PSO) [12], Genetic Algorithm (GA) [13], Tunicate Swarm Algorithm (TSA) [14], and Artificial Bee Colony (ABC) [15]. These techniques can be considered as meta-heuristic techniques used for selecting the most relative genes from microarray datasets. SVM, in this study, has been used as a classifier for measuring the performance of those techniques as gene selectors. In the domain of protein identification, cancer studies, and Microarray data, SVM is considered as the most widely used classification algorithm due to its outstanding results [16]. Also, As a preprocessing stage, the [17] approach of minimum Redundancy Maximum Relevance (mRMR) was applied. When optimization algorithms are implemented on complex and high-dimensional data like microarray data, they suffer from computational efficiency problems. Therefore, mRMR is used to reduce the high dimensional of microarray datasets and to select the most relative genes from cancer microarray datasets. According to the experimental comparative results, BMO and MRFO techniques have achieved 99.4% and 100% for the average accuracy of 100 iterations which are the highest averages comparing with the other comparative algorithms (i.e., PSO, GA, TSA, an ABC) which achieved (0.982537225, 0.974236125, 0.987986075, and 0.985617825) respectively.

The remainder of the chapter is presented as follows: Sect. 2 provides the review for cancer diagnosis overview, the computational models' background for cancer diagnosis, and the used microarray datasets. The overview of different techniques including gene expression microarray, meta-heuristics, and machine learning algorithms are presented in Sect. 3. Several applications for Cancer Diagnosis are presented in Sect. 4. Section 6 presents the chapter conclusion.

2 Preliminaries

2.1 Cancer Diagnosis Overview

Cancer is one of the deadliest diseases of our time, killing more than 9 million people in 2018, and it is thought that this number will rise in the future. While the war on cancer has cost several billions of dollars for curing cancer or improving the quality of life of cancer patients, its hidden formation mechanisms, progression, control or therapeutic cure have not been fully revealed. A multidisciplinary effort that brings together biologists, clinicians, and quantitative scientists is required. Computational simulations and mathematical modeling supply advanced tools in order to analyze experimental data [18].

Cancer represents a problem of complex systems involving interactions between cancer cells and their micro-tissue environments [19, 20]. Mathematical models can render as "virtual laboratories" which have fully controlled circumstances where clinicians and scientists can explore the arising clinical behaviors that outcome from hypotheses of basic cell and can straighten new treatment strategies [21].

In medicine, machine learning is also used for cancer diagnosis. Figure 2 presents the statistics for machine learning and cancer diagnosis studies from 2010 to 2020 based on the Scopus databases information. Figure 3 shows that medicine represents the highest machine learning-based area.

With the microarray techniques development, a large number of gene expression profiles have emerged on a large scale. These microarray datasets are applied to different application areas, like cancer classification, tumor detection, and prediction of several scarce diseases [22].

Using DNA microarray datasets for gathering information from cell and tissue samples as regards gene expression differences could be beneficial to identify particular types of tumor or for disease diagnosis. Although so minuscule samples (often fewer than 100 patients) exist for training and testing, the amount of features within the raw datasets ranges from 6000 to 60,000, as it collectively evaluates the expression of genes. The separation of patients with cancer versus healthy people based on their "profile" gene expression represents the exemplary classification task (binary approach). Also, there are other datasets whose goal is to distinguish between different types of cancers (multi-class approach), thus making the task more complex [3].

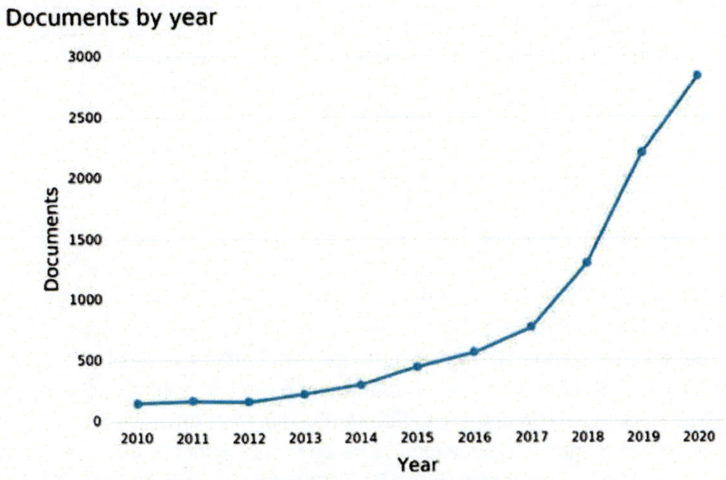

Fig. 2 Histogram of machine learning for diagnosis of cancer publication performed in the last decade [2010–2020] according to the Scopus database

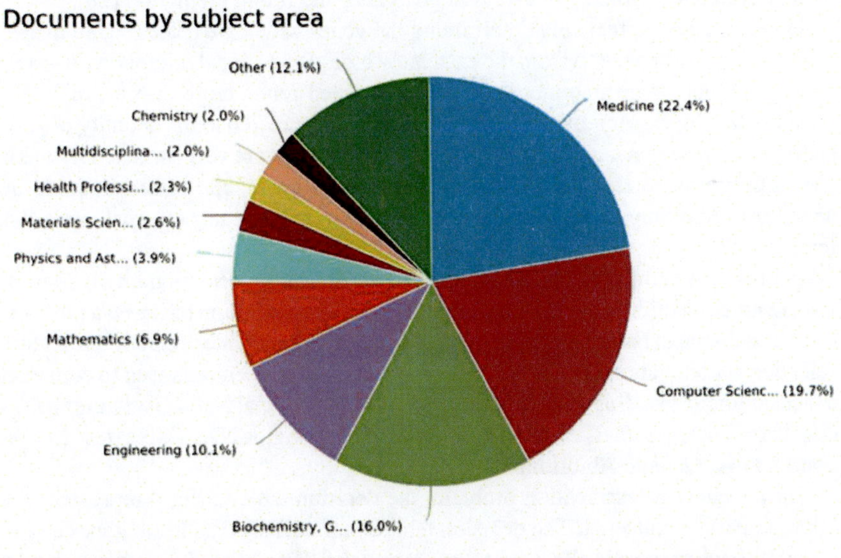

Fig. 3 Distribution of machine learning for research area according to the Scopus database

Samples	Condition attributes (Genes)				Decision attributes(Classes)
	Gene 1	Gene 2	...	Gene n	Class label
1	g(1, 1)	g(1, 2)	...	g(1, n)	Class (1)
2	g(2, 1)	g(2, 2)	...	g(2, n)	Class (2)
...	...	...	...	...	...
...	...	...	...	...	...
m	g(m, 1)	g(m, 2)	...	g(m, n)	Class (m)

Fig. 4 Decision table of microarray data

Various studies have demonstrated that most of the measured genes in a DNA microarray technology are irrelevant in the precise classification of several classes of the problem [23]. Given the use of memory and high computational time to classify high dimensional data and for avoiding the "curse of dimensionality" problem [24], appropriate feature (gene) selection procedure to improve the performance of classification plays a pivotal role in the DNA microarray analysis [25].

Feature (gene) selection is the procedure for determining and deleting unrelated features to the training data, as the learning algorithm can only focus on those sides of the training data which are useful in analyzing and future prediction [26].

In general, how effectively determining the genes which are more related to specific cancers' pathogenesis from the extremely high-dimensional microarray datasets which contains a huge amount of noisy and unrelated genes, represents the main difficulty in this topic. Moreover, the sample numbers compared to the quantity of gene expression levels that have been measured in experiments are very limited. Often this affects the prediction accuracy. The feature selection and regularization methods are necessary to investigate in this utmost of very few observations on very numerous features [27].

Because the number of unrelated genes used for cancer prediction might be massive, feature selection is particularly critical for cancer prediction based on a microarray. Even the most basic predictive models can achieve accurate prediction if feature selection is done rationally [28]. Different methods have been proposed to construct cancer predictors such as RSs (Rough Sets), ANNs (Artificial Neural Networks), DTs (Decision Trees), SVMs, Naive Bayes (NB), Clustering, k-NNs (k-Nearest Neighbours), GAs (Genetic Algorithms), et al. [29].

For the cancer classification problem, the decision table in Fig. 4 represents the microarray data collected. The decision table of microarray data has n genes and m samples. Each sample is allocated to one class label. The level of gene expression y in sample x is represented by g(x, y).

2.2 *Computational Modeling Overview*

Computational Modeling in Bioengineering and Bioinformatics promotes complementary specialities that hold great promise for the progression of research and development in complex biological and medical systems, and in the environment, public health, drug design, and so on.

Bioinformatics is an interdisciplinary field of growing interest in the fields of medicine, biology, and genetics. Bioinformatics can be also defined as a comprehensive concept that deals with all information technology applications in the molecular biology field [30].

Bioinformatics is also referred to as computational biology. However, computational biology deals mainly with biological systems modeling. The main bioinformatics components are developing algorithms and software tools and analyzing and interpreting biological data by using a set of software tools and specific algorithms [31].

In systems biology, the computational models are used for exploring the pathogenesis of complex diseases, improving the understanding of the latent molecular mechanisms, and promoting new drug discovery and treatment strategy optimization [1]. Various computational models have been built for elucidating the complex behaviors of cancers, like immune inert, drug resistance, and tumor progression, and for cancer diagnosis and prognosis, [1].

2.3 DNA Microarray Datasets

Traditional tumor diagnostic procedures which are based on the morphological manifestation of tumors are not always efficient because diagnostic errors often occur. Furthermore, an assortment of studies has exposed that cancer is a disease that involves dynamic changes to the genome. Therefore, the use of molecular markers for tumors may be an alternative way for diagnosing cancer. Rapid progress in DNA microarray technology which simultaneously enables the measurement of expression levels for tens of thousands of genes in one experiment, makes the discovery of cancerous molecular markers possible [32]. Even after the professional work in applying gene expression monitoring via DNA microarray technology for cancer classification by Golub et al. [23], numerous investigations about using gene expression microarray technology for building cancer diagnosis, prediction, or prognosis classifiers have been conducted.

Using microarray technology, researchers have attempted for simultaneously analyzing thousands of genes to obtain important information about particular cellular functions of gene(s), since an organism's physiology changes are mostly associated with gene expression patterns changes, this information can be used in cancer prognosis and diagnosis [25].

Figure 5 presents the common process for obtaining the gene expression data from a DNA microarray. Those gene expression profiles are used as inputs to analyze data on a large scale data, for example, in order to increase our comprehension of diseased and normal states [3].

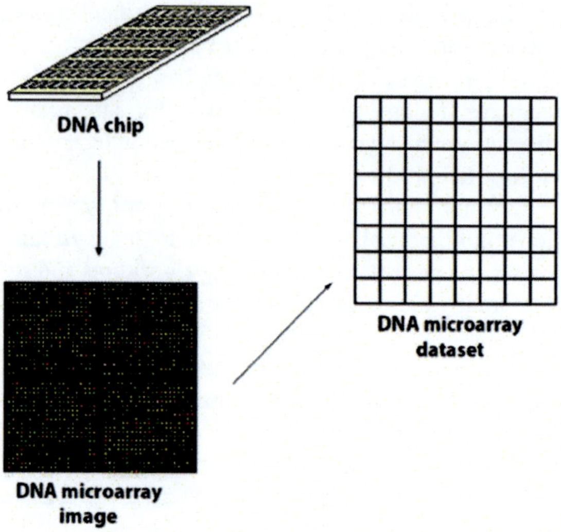

Fig. 5 Common procedure of obtaining the gene expression data using DNA microarray

The matrices are created from the DNA microarray pictures that arise from the microarray analysis. The samples are represented by the columns in the transformed matrices, while the genes are represented by the rows. The expression level of a single gene in a specific sample is represented by each cell value [23].

3 Common Techniques Used in Cancer Diagnosis

3.1 Machine Learning Techniques

Machine learning [33], which automatically enables one to learn and improve from previous experience without being clearly programmed, represents one of the AI applications. For making the best decisions, the process of learning starts with observations or data as direct, or instruction experience. Figure 6 presents the main applications of ML in medicine.

The supervised, unsupervised and semi-supervised represent the major types of ML. In supervised, the algorithms use the previously learned as new data by using labels for predicting future events. Semi-supervised is done between unsupervised and supervised learning as both unlabeled and labeled data are used for training. The system's capacity to derive a function to characterise the hidden structure of unlabeled data is referred to as unsupervised. The different machine learning techniques are illustrated in Fig. 7.

One of the popular classification techniques is SVM [34] which can be exemplified as linear and nonlinear. The linear model can be used to solve regression and

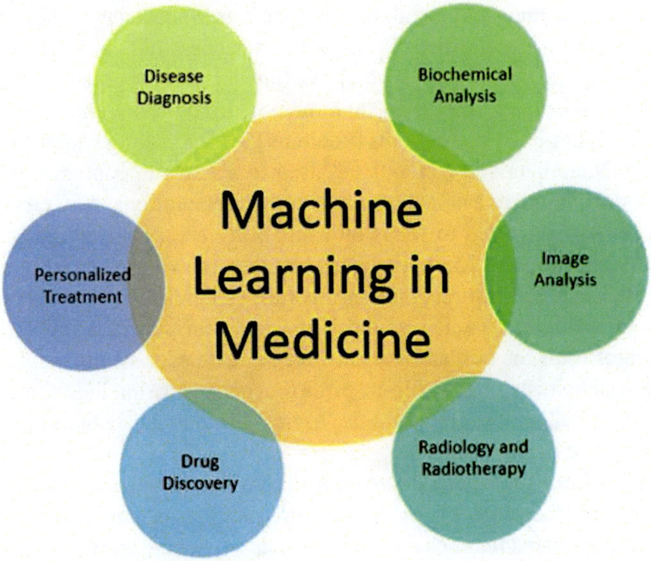

Fig. 6 The main application of ML in medicine

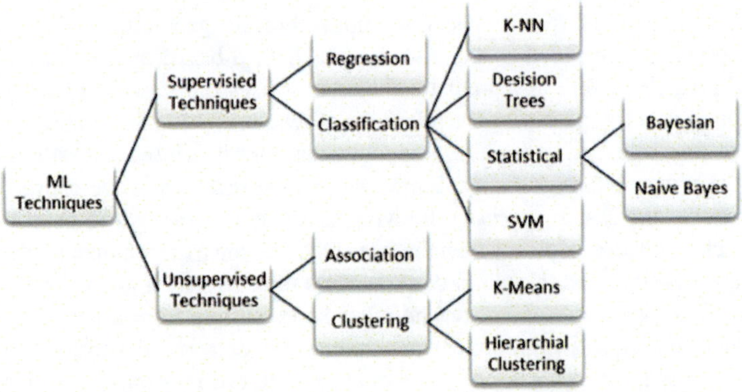

Fig. 7 Machine learning techniques

classification problems. In order to map data in high-dimensional space, the nonlinear model employs a kernel function. Two parameters, Γ, and C dominate the SVM result.

SVM was more useful than other classification techniques. It is difficult to build a linear classifier for separating data classes. SVMs can tackle this challenge by converting the input space to a high-dimensional space and then categorising the input data using a maximum margin hyperplane based on a linear classification decision. Other machine learning algorithms, such as k-nearest neighbour classifiers and neural networks, are deemed slower and less successful than SVMs [29].

For achieving the best results, several machine learning approaches can be combined with meta-heuristic methods, thus meta-heuristic represents an algorithmic framework for high-level autonomous problems [35] for dealing with different types of problems. Meta-heuristic methods are used to achieve maximum results on the limited resources that are available from some of the solutions. Any meta-heuristic algorithm can be suggested to focus on availability areas with a higher value for the diamond find property. Meta-heuristics methods can be used for solving various problems, such as optimization problems, in various fields for example engineering, chemistry, medical, social sciences, business, and transportation problems. Meta-heuristics algorithms can be classified into four basic types as swarm intelligence (SI) based, physics/chemistry-based, bio-inspired (but not swarm intelligence-based), and others. Swarm intelligence algorithms use basic rules to replicate the mass behaviour of the most interacting agents.

Lately, meta-heuristics methods are being implemented to solve several optimization problems. Furthermore, the meta-heuristics methods can be used to remove the repetition and unrelated features from a dataset as a machine learning preprocessing step. This step can be called a feature selection. The process of determining a subset of all feature sets that can maximise the best classification performance is known as feature selection.

Four categories of feature selection approaches are generally used, which are wrapper-based, embedded-based, filter-based, hybrid-based approaches. A filter method doesn't require any predictor, because it uses the statistical properties of the training data for measuring the relevance of features. Using the wrapper method, in the feature selection procedure, a specific classifier is utilised to evaluate a subset of the selected characteristics. Furthermore, The accuracy of the chosen classifier improves the search process. The wrapper method is thought to have a higher median classification accuracy than the filter method, since the wrapper method uses a specific classifier and takes into consideration the correlation among the selected features. The wrapper method, on the other hand, suffers from a long processing time when dealing with high-dimensional microarray datasets. The wrapper method, which is regarded as a black box, is frequently bereft of meaning. The advantages of the wrapper and filter models are combined in the hybrid method. It first selects a subset of features from the input features using the filter method, and then utilises the wrapper method to choose the most informative and relevant characteristics from this subset using the wrapper method. The calculation cost for the wrapper model becomes acceptable as the number of characteristics employed in the wrapper model is reduced. The hybrid method's main flaw is that the filter and wrapper methods cannot be merged because this would result in a worse categorization. The embedded method combines an initial feature set with a training model for building up a criteria in order to determine gene ranking estimations.

3.2 Meta-Heuristics Optimization Algorithms

The meta-heuristic technique uses the optimization algorithm based on swarm intelligence in order to find, select, or generate the optimal solution for the optimization problem particularly with limited computation capacity or with incomplete information. By using meta-heuristic algorithms, Meta-heuristic optimization can handle problems of single-objective and multi-objective optimization. Optimization is taken place everywhere, from holiday planning to internet routing and from engineering design to economics. Since time and financial resources are finite, making the greatest use of what is available is critical. Most of the real-world improvements are considered non-linear and extremely multimodal, within different complex qualifications. Many objectives are oftentimes conflicting. Even for a single objective, there may at times not be perfect solutions at all. Finding optimal or sub-optimal solutions is not an easy process in general. For solving optimization problems, the meta-heuristic algorithms simulate nature. Realistic optimization problems are extremely hard for solving such problems, performance optimization tools must be used, though not sure that the optimal solution will be found. Actually, there are absolutely no efficient algorithms. Additionally, new optimization algorithms have been suggested in order to see if they can handle these hard optimization problems. The next subsections present the five major categories of all existing meta-heuristic algorithms: Evolutionary, swarm intelligence (SI) based, physics/chemistry-based, bio-inspired based, and others [36].

The using of meta-heuristic algorithms is a more effective thing for handling such types of problems. Different meta-heuristic algorithms are presented in order to solve optimization problems and challenges are known as optimization algorithms [37–41]. For instance, Microarray Cancer Classification [42], heartbeat classification [43–47], Deep learning [48, 49], feature selection [50–52], Energy [53–58], fuel cell [59], photovoltaic (pv) [60–62], global optimization problems [63, 64], Image segmentation [65–69], Wireless networks [70], cloud computing [71], Bioinformatics [72], and drug design [73, 74].

The taxonomy of meta-heuristic algorithms is shown in Fig. 8.

Evolutionary Algorithms The generic population-based optimization method is an evolutionary algorithm (EA). Mutation, selection, reproduction, and recombination are all biologically inspired methods used in EA. Individuals in a population play a part in the candidate solutions for the optimization problem. In addition, The fitness

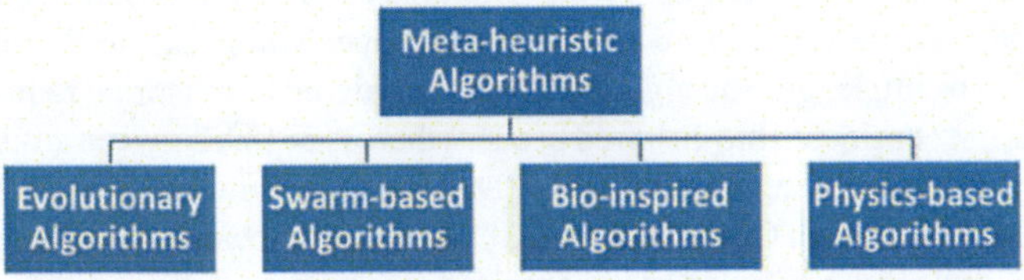

Fig. 8 Taxonomy of meta-heuristic algorithms

function determines the quality of the solutions. After then, population development occurs as a result of repeated application of the foregoing operators [75]. Because, EAs do not assume the underlying fitness scene, they often result good approximations of all kinds of optimization problems. Computational complexity, which is considered a prohibited factor in the majority of real-world EA applications, is resulted from the evaluation of the fitness function. For overcoming this difficulty, physical fitness is one solution. However, the complex problems can be solved by a simple EA, so there may be no relationship between the difficulty of the problem and the difficulty of the method. The known application areas of the evolutionary algorithm are identification, control, simulation, planning, design, and classification. Furthermore, the EAs have solved several optimization problems like the traveling salesman problem, and infinite monkey theorem [76].

Swarm Intelligence-Based Algorithms Swarm intelligence (SI), a sort of artificial intelligence, is defined as organisms' intelligent activity that aids them in completing their objective, such as monitoring prey or locating food resources. SI heavily participates in creating multi-agent systems. Until now, There's no need to explain the mechanisms behind the creation of collective intelligent behavior in the swarm [77]. SI can generally be used in human swarming, crowd simulation, swarm grammars, swarm art, and ant-based routing. SI-based algorithms match the evolving collective behavior of multiple interactive agents who follow a few simple rules. Some self-organizing behavior, which may be showed by the complete system of many agents, can serve as a kind of collective intelligence.

The SI-based algorithm, which is applicable for solving optimization problems, has the characteristics of simplicity, coordination, durability, self-organization, and distribution. These algorithms are characterized by co-evolution, self-organized, learning during iterations, and information sharing among multiple agents for performing efficient searches. Various agents are also parallelized in SI-based algorithms so that optimization on a huge scale becomes more realistic than the opinion on implementation. Some examples of SI-based meta-heuristic algorithms are whale optimization algorithm (WOA) [78], Ant colony optimization (ACO) [79], Artificial Bee Colony (ABC) [15], Particle Swarm Optimization (PSO) [12], Grey Wolf Optimization (GWO) [80], Black widow optimization (BWO) [81], Harris hawks optimization (HHO) [82], and Tunicate Swarm Algorithm (TSA) [14]

Bio-Inspired Algorithms Because they are independent to the established collective behaviour of interacting organisms, bio-inspired algorithms differ from SI-based algorithms. In this form of algorithm, there are no agents. These algorithms simulate various biological inspiration behaviors in animals and humans. Biological connections found in nature can be used as revelation sources for bio-inspired algorithms in general. For example, genetic algorithms are considered bio-inspired rather than SI-based. It is not considered bio-inspired because there is no link between any biological behaviour and the differential evolution (DE) process. Human-Inspired Algorithm (HIA) [83], Brain Storm Optimization (BSO) [84], and Flower pollination algorithm (FPA) [85] are some examples of bio-inspired algorithms.

Physics and Chemistry-Based Algorithms Not all meta-heuristic algorithms are Swarm Intelligence-based or bio-inspired, as there are other algorithms inspired by chemistry and/or physics. The majority of the algorithms that aren't biologically inspired were created by modelling some chemical and/or physical laws, including river systems, electrical charges, gravity, etc. Simulated annealing (SA) [86], ions motion algorithm (IMA) [87], and atom search optimization (ASO) [88] are examples on this type of metaheuristics.

Other Algorithms When new algorithms are developed, researchers may seek inspiration outside of nature. As a result, certain algorithms are not inspired by physics and chemistry, nor are they motivated by biology. Because these algorithms were built utilising distinct characteristics from multiple sources, like emotional, social, it can be difficult to categorise them into the four categories stated above.

Examples of meta-heuristic algorithms, that may be named in others category, are differential search algorithm (DSA) [89], backtracking search optimization algorithm (BSA) [90], and artificial cooperative search (ACS) [91].

4 Application Areas of Intelligent Computational in Cancer Diagnosis

The most recent and related studies that have been offered for tackling gene selection issues are discussed in this area, and their merits and downsides are evaluated. This collection contains four categories of studies: [3, 92–94]. The most common gen selection methods are categorised into four categories: filter, wrapper, hybrid, and embedding. The study evaluated is summarised in Table 1, which includes the feature selection strategy used, the category assigned to each of these investigations, and the classifier used to assess the accuracy of these procedures.

4.1 Filter-Based Studies

The statistical features of the training data are employed to quantify the importance of genes in filter-based studies, therefore no predictor is required. Filter-based studies generally use one of two ways to assess the significance of genes: univariate or multivariate strategies [17, 23, 95]. The given genes are first assessed and arranged according to a specific criterion in the univariate strategy. The best subset of the genes is then chosen based on their fitness. In a univariate method, there are numerous criteria that can be applied, such as Signal-to-noise ratio [23], Laplacian score (LS) [96, 97], mutual information [98], and information gain [99]. The key advantages of univariate approaches are their speed and efficiency. However, because these techniques neglect the association between selected genes, they result in lesser accuracy. On the other hand, because methods based on a multivariate strategy take into consideration

Table 1 A fast review of recent studies on the gene selection issue, including the feature selection strategies used and the categories they fall into

Algorithm	Feature selection technique	Classifier technique	Feature selection category
[144]	mRMR with MRFO	SVM	Hybrid
[42]	IG with BMO	SVM	Hybrid
[108]	Information Gain	SVM, ANN and R.Forest	Filter
[127]	X^2 statistic, information gain, symmetrical uncertainty and ReliefF	Decision trees, Naive Bayes and SVM	Filter and wrapper
[136]	mRMR with SVM-RFE	Decision tree	Hybrid
[143]	SVM-RFE	SVM	Embedded
[142]	MSVM-RFE	SVM	Embeded
[137]	mRMR with PSO	SVM	Hybrid
[124]	Fuzzy preference based rough set (FPRS)	Transductive SVM	Wrapper
[105]	mRMR	SVM, Linear discriminant analysis, Naive Bayes, and logistic regression	Filter
[126]	1-norm SVM with squared loss (SVMSL)	SVM and nearest neighbor (NN)	Wrapper
[133]	mRMR with ABC	SVM	Hybrid
[17]	mRMR	SVM, linear discriminate analysis, and Naive Bayes	Filter
[121]	GA and PSO	SVM	Wrapper
[131]	Binary bat algorithm (BBA)	Extreme learning machine (ELM)	Wrapper
[142]	Spider Monkey optimization	SVM	Wrapper
[138]	IG with SVM	LIBSVM	Hybrid
[135]	mRMR with genetic algorithm	SVM	Hybrid
[134]	mRMR with genetic bee colony (GBC)	SVM	Hybrid
[130]	Particle swarm optimization (PSO)	Functional link artificial neural network (FLANN)	Wrapper
[128]	Gene selection programming (GSP)	SVM	Wrapper
[120]	Artificial bee colony (ABC)	SVM	Wrapper
[123]	Fire fly (FF)	SVM	Wrapper
[129]	Particle swarm optimization (PSO)	K-nearest neighborhood (KNN)	Wrapper
[25]	Particle swarm optimization (PSO)	Decision tree	Wrapper

the interdependencies between selected genes, they have a higher classification accuracy than techniques based on a univariate strategy. Random subspace method (RSM) [100–102], mutual correlation (MC) [103, 104], minimal-redundancy- maximal-relevance method (MRMR) [17, 105], and Relevance- redundancy feature selection (RRFS) [106, 107] are examples of a multivariate method. Multivariate strategy approaches may belong to a local optimum because they constantly hunt for the optimal subset in only one iteration.

The mRMR introduced in [17] is a heuristic technique for analysing redundancy and relevancy of features and determining predicted features in discrete and continuous datasets. The results demonstrate that using mRMR to improve feature selection performance is a good idea. The mRMR was also used for gene expression profiling by the authors of [105]. The mRMR-selected genes considerably improved classification accuracy.

The authors of [108] employed the singular method of Singular Value Decomposition (SVD) [109] to reduce redundant information and choose informative genes. Furthermore, the authors applied the information gain (IG) technique to pick the useful data aspects that result in higher cancer classification performance. However, the usage of SVD, which is slow and computationally expensive, is a key drawback of this work [110].

4.2 Wrapper-Based Studies

The gene selection procedure in the wrapper model employs a specific classifier to assess the subset of selected genes. Furthermore, the accuracy of the chosen classifier can be used to improve the search process. The wrapper model employs two search techniques: stochastic and greedy search strategies [92, 111]. There are a variety of approaches based on the stochastic search strategy, such as the firefly algorithm [112], practical swarm optimization (PSO) [113, 114], genetic algorithm (GA), ant colony optimization (ACO) [93, 115, 116]. The greedy search strategy [117, 118] employs two methods: sequential backward selection and sequential forward selection.

The wrapper model is said to have a higher average classification accuracy than the filter model since it uses a specific classifier and examines the association between selected genes [119]. The wrapper model, on the other hand, has a high computational cost, which is compounded by cancer microarray profiles' high dimensionality. In addition, the wrapper model is seen as a black box with a distinct lack of interpretation [119]. The following paragraphs summarise the most relevant wrapper-based gene selection research.

The authors proposed using the Artificial Bee Colony (ABC) approach in combination with SVM as a classifier for microarray gene expression profiles in [120]. When compared to other studies, the study's main disadvantages are its lower performance and larger number of selected genes.

The authors examined the use of a Genetic Algorithm (GA) and a Particle Swarm Optimization (PSO) for the categorization of high dimensional microarray data in [121], where both are integrated with SVM.

The authors proposed merging the Spider Monkey Optimization Technique with the SVM classification algorithm in [122]. This combination is used to eliminate unneeded and redundant genes and choose the most informative genes, which are then utilised to detect cancer types using the selected genes from the microarray gene expression data.

Authors proposed a wrapper feature selection technique to categorise cancer microarray gene expression in [123], which combined the FireFly algorithm with the SVM classifier (FF-SVM).

Because the identical gene selection process was employed in the previous four studies, they all had the same disadvantages, namely the high computing cost produced by the high dimensionality of cancer microarray data.

By integrating fuzzy preference-based rough set (FPRS) with transductive SVM, the authors of [124] developed a prediction technique for discovering more relevant genes from microarray gene expression data and enhancing classification performance. When compared to the classical SVM, the transductive SVM is more resilient and can generate superior results. However, one of the most significant disadvantages is its computing expenses [125].

The authors employed a 1−norm support vector machine with squared loss (1−norm SVMSL) coupled with a subsequent classifier, such as a nearest neighbour (NN), or support vector machine (SVM), to accomplish rapid gene selection for cancer classification in [126]. To achieve the best performance with rapid test and storage, the 1−norm SVMSL requires a limited number of genes. The number of genes to be picked, on the other hand, was not considered.

On diffuse large B-cell lymphoma and acute leukaemia microarray datasets, the authors of [127] used wrapper, filter, and a correlation-based feature selector (CFS) algorithms, which were all coupled with different machine learning methods such as naive Bayes, decision tree, and SVM.

In [128], the authors suggested a novel Gene Selection Programming (GSP) strategy for choosing genes relevant to efficient and effective cancer categorization in [128]. GSP is based on the Gene Expression Programming (GEP) approach, but with enhanced mutation, a new fitness function formulation, a new population initiation algorithm, and recombination operators. As a classifier for the GSP, the authors utilised SVM with a linear kernel.

In [25], the authors created a new technique that incorporates the Particle Swarm Optimization (PSO) algorithm as well as a decision tree as a classifier. The performance of the suggested method was compared to several well-known classification methods (self-organizing map, Naive Bayes, artificial immune recognition system, CART decision tree, backpropagation neural network, C4.5, support vector machine, and decision tree). Furthermore, the tests are carried out on eleven gene expression cancer datasets.

In [129], the authors suggested a new approach for selecting a limited subset of beneficial genes that are relevant to the desired classification goal by combining the

Particle Swarm Optimization (PSO) algorithm with adaptive K-nearest neighbourhood (KNN) as a classifier. The suggested method is tested on three benchmark gene expression cancer datasets to determine the smallest set of genes that are significant.

Authors, in [130], have presented a new integrated approach utilizing particle swarm optimization (PSO) and functional link artificial intelligence (FLANN) for building a more reliable Classifier. The proposed approach is compared with two different classification techniques: BPN and FLANN, and the experimental results proved the superiority of the proposed approach is better predicting the disease as compared to other methods.

Authors, in [131], have proposed a novel method using binary bat algorithm (BBA) combined with an extreme learning machine as the classifier. Also, they suggested a new fitness function to optimize the feature selection process by using a binary bat algorithm. This study conducts the experiments on eight microarray cancer datasets and is compared with the original fitness function found in the literature.

4.3 Hybrid-Based Studies

The hybrid technique combines the benefits of the filter and wrapper models. It uses the filter method to choose a subset of genes from the input genes, and then uses the wrapper strategy to select the most relevant and informative genes from that subset. Because the wrapper technique uses a small number of genes, the computing time becomes tolerable. There are a variety of methods based on the hybrid method, including the multiple-filter-multiple-wrapper (MFMW) method [94], Fisher score with a GA and PSO [132], an the multiple-filter-multiple-wrapper (MFMW) method [94]. The fundamental flaw in the hybrid method is that the wrapper and filter models are not fully integrated, which could result in poor categorization [119].

The study proposed in [133], which employed (mRMR) paired with the artificial bee colony (ABC) algorithm to pick meaningful genes from microarray datasets, is an example of hybrid-based gene selection for cancer microarray gene expression. The authors of this work employed SVM as a classifier to assess the accuracy of the selected genes' classification. Another paper, [134], described a novel hybrid gene selection approach, the Genetic Bee Colony (GBC) algorithm, which combined the Genetic Algorithm (GA) with the Artificial Bee Colony (ABC) algorithm to take advantage of the benefits of both algorithms. In addition, this method integrated minimum redundancy maximum relevance (mRMR) with the GBC algorithm to pick informative genes, which was then supplemented with SVM to assess classification accuracy.

Although the two research were able to reap the benefits of combining the filter and wrapper models, they were unable to attain the optimum results with such a limited number of genes.

In [135], the authors suggested a unique gene selection technique that combines the mRMR filter method with the GA wrapper method. The experimental findings demonstrated that this combination was effective, with the chosen gene set being more representative of the specified class. In addition, [136] proposes a new strategy that combines an mRMR filter method with SVM-RFE to eliminate redundancy in the selected gene set. The results suggest that the mRMR filter approach is more efficient when paired with the SVM-RFE in this experiment, which was done on four benchmark cancer microarray datasets. In [137], a new hybrid gene selection technique based on the mRMR filter method and the partial swarm optimization (PSO) algorithm was proposed. The experimental results showed that the mRMR-PSO hybrid algorithm achieves greater classification accuracy than the [17, 105, 135, 136] algorithms for colon cancer and leukaemia microarray datasets.

In [138], the authors merged Information Gain with SVM to achieve the best cancer classification performance. SVM was used to filter the redundant genes after IG had selected the most related and informative genes from the initial datasets. Finally, the collected informative genes are evaluated using the LIBSVM classifier [139]. The authors of this study were concerned with selecting the smallest number of genes while ignoring the maximum accuracy.

4.4 Embedded-Based Studies

The embedded model combines a training model with a starting feature set to create a criterion for quantifying gene rank estimations. The first order inductive learner (FOIL) rule-based feature subset selection algorithm [140], Random forest (RF) [141], and support vector machine based on recursive feature elimination (SVM-RFE) [16] are only a few examples of embedded-based approaches. Although the embedded model has the advantage of interacting with the learning model, training a classifier with all genes takes time, especially given the complexity of gene expression microarray datasets.

Multiple linear SVMs were used in a backward elimination strategy for gene selection in [142]. Statistical analysis of the weight vectors of multiple linear SVMs trained on subsamples of the original training data is used to obtain the feature ranking scores at each phase of this procedure.

Authors in [143] introduced a new feature selection approach for multiclass classification based on feature ranking scores obtained from weight vectors of multiple linear binary SVM classifiers from One-versus-one (OVO) or one-versus-all (OVA) (OVA). However, one of the study's key drawbacks is that it is computationally intensive because it runs over all features one by one and does not account for any association between them [16].

The two investigations before this can be classified as embedded-based studies. As a result, they both suffer from the same flaws in this category. Because of the large complexity of gene expression microarray datasets, training a classifier with all genes takes a long time.

5 Open Issues and Challenges

There are several challenges facing the work in this research area including:

1. When using gene expression patterns to classify or predict cancer, the problem is that genes (features) outweigh samples (instances), resulting in low prediction efficiency or effect if the model is not chosen logically. To get over this problem, feature selection is used.
2. When biologists and physicians are interested in your research, another concern is the predictability of the prediction model. In this case, decision rules are used to deal with the problem.
3. The intricacy of microarray datasets, as well as the degree of noise and irrelevant genes, and the small number of samples, all make categorising a specific sample even more challenging.
4. Because most genes are directly or indirectly related to one another, microarray data analysis is complicated. For example, a gene with a high expression level can be easily activated by a gene with a high regulation level.
5. It's worth noting that when applied to complicated and high-dimensional data like a microarray dataset, evolutionary techniques confront several difficulties, mostly in terms of processing performance. This prompts us to propose a hybrid technique that combines one of the evolutionary approaches with one of the filter-based gene selection approaches to address these issues.
6. Small round blue cell tumours (SRBCTs), which are a type of multi-class microarray dataset encompassing four different children tumours, are so termed because they have a similar look on standard histology, making clinical diagnosis challenging.

6 Conclusions and Future Research Issues

This chapter discusses a number of meta-heuristics algorithms and machine learning approaches used in the field of bioinformatics for cancer diagnosis. These techniques are used to diagnose cancer based on biological data. Some of these methods did not do as well as others in terms of cancer categorization, and some of them had flaws. With the smallest number of genes, BMO and MRFO have the highest accuracy. More experimental results on more real and benchmark datasets can be undertaken in the future to verify and broaden this research topic. Furthermore, the chosen genes can be investigated further to better understand their relationships with other genes and disease-specific indicators. Furthermore, the proposed methodology can be used to solve large-scale problems in healthcare that are more difficult to solve.

Conflict of Interest The authors declare that there is no conflict of interest.

References

1. Z. Ji, K. Yan, W. Li, H. Hu, X. Zhu, Mathematical and computational modeling in complex biological systems. BioMed. Res. Int. **2017** (2017)
2. S. Namasudra, Data access control in the cloud computing environment for bioinformatics. Int. J. Appl. Res. Bioinform. (IJARB) **11**(1), 40–50 (2020)
3. V. Bolón-Canedo, N. Sánchez-Marono, A. Alonso-Betanzos, J.M. Benítez, F. Herrera, A review of microarray datasets and applied feature selection methods. Inf. Sci. **282**, 111–135 (2014)
4. W. Dubitzky, M. Granzow, C.S. Downes, D. Berrar, Introduction to microarray data analysis, in *A Practical Approach to Microarray Data Analysis* (Springer, 2003), pp. 1–46
5. A. Benso, S. Di Carlo, G. Politano, A. Savino, GPU acceleration for statistical gene classification, in *2010 IEEE International Conference on Automation, Quality and Testing, Robotics (AQTR)*, vol. 2 (IEEE, 2010), pp. 1–6
6. S.-B. Guo, M.R. Lyu, T.-M. Lok, Gene selection based on mutual information for the classification of multi-class cancer, in *International Conference on Intelligent Computing* (Springer, 2006), pp. 454–463
7. P.M. Narendra, K. Fukunaga, A branch and bound algorithm for feature subset selection. IEEE Trans. Comput. **9**, 917–922 (1977)
8. E.H. Houssein, M. Younan, A.E. Hassanien, Nature-inspired algorithms: a comprehensive review, in *Hybrid Computational Intelligence: Research and Applications* (2019), p. 1
9. E.H. Houssein, A.A. Ewees, M.A. ElAziz, Improving twin support vector machine based on hybrid swarm optimizer for heartbeat classification. Pattern Recognit. Image Anal. **28**(2), 243–253 (2018)
10. M.H. Sulaiman, Z. Mustaffa, M.M. Saari, H. Daniyal, Barnacles mating optimizer: a new bio-inspired algorithm for solving engineering optimization problems. Eng. Appl. Artif. Intell. **87**, 103330 (2020)
11. W. Zhao, Z. Zhang, L. Wang, Manta ray foraging optimization: an effective bio-inspired optimizer for engineering applications. Eng. Appl. Artif. Intell.**87**, 103300 (2020)
12. J. Kennedy, R. Eberhart, Particle swarm optimization, in *Proceedings of ICNN'95-International Conference on Neural Networks*, vol. 4 (IEEE, 1995), pp. 1942–1948
13. D. Whitley, A genetic algorithm tutorial. Stat. Comput. **4**(2), 65–85 (1994)
14. S. Kaur, L.K. Awasthi, A. Sangal, G. Dhiman, Tunicate swarm algorithm: a new bio-inspired based metaheuristic paradigm for global optimization. Eng. Appl. Artif. Intell. **90**, 103541 (2020)
15. D. Karaboga, An idea based on honey bee swarm for numerical optimization, Technical report-tr06, Erciyes University, Engineering Faculty, Computer ... (Technical Report, 2005)
16. I. Guyon, J. Weston, S. Barnhill, V. Vapnik, Gene selection for cancer classification using support vector machines. Mach. Learn. **46**(1–3), 389–422 (2002)
17. H. Peng, F. Long, C. Ding, Feature selection based on mutual information criteria of max-dependency, max-relevance, and min-redundancy. IEEE Trans. Pattern Anal. Mach. Intell. **27**(8), 1226–1238 (2005)
18. H. Enderling, K.A. Rejniak, Simulating cancer: computational models in oncology. Front. Oncol. **3**, 233 (2013)
19. P. Macklin, H.B. Frieboes, J.L. Sparks, A. Ghaffarizadeh, S.H. Friedman, E.F. Juarez, E. Jonckheere, S.M. Mumenthaler, Progress towards computational 3-d multicellular systems biology, in *Systems Biology of Tumor Microenvironment* (Springer, 2016), pp. 225–246
20. D. Hanahan, R.A. Weinberg, Hallmarks of cancer: the next generation. Cell **144**(5), 646–674 (2011)
21. P. Macklin, When seeing isn't believing: how math can guide our interpretation of measurements and experiments. Cell Syst. **5**(2), 92–94 (2017)
22. S. Guo, D. Guo, L. Chen, Q. Jiang, A centroid-based gene selection method for microarray data classification. J. Theor. Biol. **400**, 32–41 (2016)

23. T.R. Golub, D.K. Slonim, P. Tamayo, C. Huard, M. Gaasenbeek, J.P. Mesirov, H. Coller, M.L. Loh, J.R. Downing, M.A. Caligiuri et al., Molecular classification of cancer: class discovery and class prediction by gene expression monitoring. Science **286**(5439), 531–537 (1999)
24. A. Jain, D. Zongker, Feature selection: evaluation, application, and small sample performance. IEEE Trans. Pattern Anal. Mach. Intell. **19**(2), 153–158 (1997)
25. K.-H. Chen, K.-J. Wang, M.-L. Tsai, K.-M. Wang, A.M. Adrian, W.-C. Cheng, T.-S. Yang, N.-C. Teng, K.-P. Tan, K.-S. Chang, Gene selection for cancer identification: a decision tree model empowered by particle swarm optimization algorithm. BMC Bioinform. **15**(1), 49 (2014)
26. I. Guyon, S. Gunn, M. Nikravesh, L.A. Zadeh, *Feature Extraction: Foundations and Applications*, vol. 207 (Springer, 2008)
27. E.P. Xing, M.I. Jordan, R.M. Karp et al., Feature selection for high-dimensional genomic microarray data, in *ICML*, vol. 1 (Citeseer, 2001), pp. 601–608
28. R. Simon, Supervised analysis when the number of candidate features (p) greatly exceeds the number of cases (n). ACM SIGKDD Explor. Newsl. **5**(2), 31–36 (2003)
29. X. Wang, O. Gotoh, Microarray-based cancer prediction using soft computing approach. Cancer Inform. **7**, CIN–S2655 (2009)
30. J.L. Pettifor, Book review of encyclopedia of applied ethics. Can. J. Couns. Psychother. **46**(4) (2012)
31. A.D. Baxevanis, G.D. Bader, D.S. Wishart, *Bioinformatics* (Wiley, 2020)
32. M. Schena, D. Shalon, R.W. Davis, P.O. Brown, Quantitative monitoring of gene expression patterns with a complementary DNA microarray. Science **270**(5235), 467–470 (1995)
33. A. Maseleno, N. Sabani, M. Huda, R. Ahmad, K.A. Jasmi, B. Basiron, Demystifying learning analytics in personalised learning. Int. J. Eng. Technol. **7**(3), 1124–1129 (2018)
34. R. Rodríguez-Peéez, M. Vogt, J. Bajorath, Support vector machine classification and regression prioritize different structural features for binary compound activity and potency value prediction. ACS Omega **2**(10), 6371–6379 (2017)
35. K. Hussain, M.N.M. Salleh, S. Cheng, Y. Shi, Metaheuristic research: a comprehensive survey. Artif. Intell. Rev. **52**(4), 2191–2233 (2019)
36. I. Fister Jr, X.-S. Yang, I. Fister, J. Brest, D. Fister, A brief review of nature-inspired algorithms for optimization. arXiv:1307.4186 (2013)
37. E.H. Houssein, A.G. Gad, K. Hussain, P.N. Suganthan, Major advances in particle swarm optimization: theory, analysis, and application. Swarm Evol. Comput. **63**, 100868 (2021)
38. F.A. Hashim, E.H. Houssein, M.S. Mabrouk, W. Al-Atabany, S. Mirjalili, Henry gas solubility optimization: a novel physics-based algorithm. Futur. Gener. Comput. Syst. **101**, 646–667 (2019)
39. E.H. Houssein, M.R. Saad, F.A. Hashim, H. Shaban, M. Hassaballah, Lévy flight distribution: a new metaheuristic algorithm for solving engineering optimization problems. Eng. Appl. Artif. Intell. **94**, 103731 (2020)
40. F.A. Hashim, K. Hussain, E.H. Houssein, M.S. Mabrouk, W. Al-Atabany, Archimedes optimization algorithm: a new metaheuristic algorithm for solving optimization problems. Appl. Intell. **51**(3), 1531–1551 (2021)
41. F.A. Hashim, E.H. Houssein, K. Hussain, M.S. Mabrouk, W. Al-Atabany, Honey badger algorithm: new metaheuristic algorithm for solving optimization problems. Math. Comput. Simul. **192**, 84–110 (2021)
42. E.H. Houssein, D.S. Abdelminaam, H.N. Hassan, M.M. Al-Sayed, E. Nabil, A hybrid barnacles mating optimizer algorithm with support vector machines for gene selection of microarray cancer classification. IEEE Access **9**, 64,895–64,905 (2021)
43. E.H. Houssein, I.E. Ibrahim, N. Neggaz, M. Hassaballah, Y.M. Wazery, An efficient ECG arrhythmia classification method based on manta ray foraging optimization. Expert Syst. Appl. **181**, 115131 (2021)
44. E.H. Houssein, D.S. AbdElminaam, I.E. Ibrahim, M. Hassaballah, Y.M. Wazery, A hybrid heartbeats classification approach based on marine predators algorithm and convolution neural networks. IEEE Access (2021)

45. Y.M. Wazery, E. Saber, E.H. Houssein, A.A. Ali, E. Amer, An efficient slime mould algorithm combined with k-nearest neighbor for medical classification tasks. IEEE Access (2021)
46. A. Hamad, E.H. Houssein, A.E. Hassanien, A.A. Fahmy, Hybrid grasshopper optimization algorithm and support vector machines for automatic seizure detection in EEG signals, in *International Conference on Advanced Machine Learning Technologies and Applications* (Springer, 2018), pp. 82–91
47. E.H. Houssein, M. Kilany, A.E. Hassanien, ECG signals classification: a review. Int. J. Intell. Eng. Inform. **5**(4), 376–396 (2017)
48. E.H. Houssein, M. Dirar, K. Hussain, W.M. Mohamed, Assess deep learning models for Egyptian exchange prediction using nonlinear artificial neural networks. Neural Comput. Appl. **33**(11), 5965–5987 (2021)
49. E.H. Houssein, M.M. Emam, A.A. Ali, P.N. Suganthan, Deep and machine learning techniques for medical imaging-based breast cancer: a comprehensive review. Expert Syst. Appl. 114161 (2020)
50. N. Neggaz, E.H. Houssein, K. Hussain, An efficient henry gas solubility optimization for feature selection. Expert Syst. Appl. **152**, 113364 (2020)
51. K. Hussain, N. Neggaz, W. Zhu, E.H. Houssein, An efficient hybrid sine-cosine Harris hawks optimization for low and high-dimensional feature selection. Expert Syst. Appl. **176**, 114778 (2021)
52. D.S. Abd Elminaam, A. Nabil, S.A. Ibraheem, E.H. Houssein, An efficient marine predators algorithm for feature selection. IEEE Access **9**, 60,136–60,153 (2021)
53. E.H. Houssein, M.A. Mahdy, A. Fathy, H. Rezk, A modified marine predator algorithm based on opposition based learning for tracking the global MPP of shaded PV system. Expert Syst. Appl. **183**, 115253 (2021)
54. E.H. Houssein, B.E.-D. Helmy, H. Rezk, A.M. Nassef, An enhanced Archimedes optimization algorithm based on local escaping operator and orthogonal learning for PEM fuel cell parameter identification. Eng. Appl. Artif. Intell. **103**, 104309 (2021)
55. M.H. Hassan, E.H. Houssein, M.A. Mahdy, S. Kamel, An improved manta ray foraging optimizer for cost-effective emission dispatch problems. Eng. Appl. Artif. Intell. **100**, 104155 (2021)
56. A. Korashy, S. Kamel, E.H. Houssein, F. Jurado, F.A. Hashim, Development and application of evaporation rate water cycle algorithm for optimal coordination of directional overcurrent relays. Expert Syst. Appl. **185**, 115538 (2021)
57. S. Deb, E.H. Houssein, M. Said, D.S. Abd Elminaam, Performance of turbulent flow of water optimization on economic load dispatch problem. IEEE Access (2021)
58. S. Deb, D.S. Abdelminaam, M. Said, E.H. Houssein, Recent methodology-based gradient-based optimizer for economic load dispatch problem. IEEE Access **9**, 44,322–44,338 (2021)
59. E.H. Houssein, F.A. Hashim, S. Ferahtia, H. Rezk, An efficient modified artificial electric field algorithm for solving optimization problems and parameter estimation of fuel cell. Int. J. Energy Res. (2021)
60. E.H. Houssein, G.N. Zaki, A.A.Z. Diab, E.M. Younis, An efficient manta ray foraging optimization algorithm for parameter extraction of three-diode photovoltaic model. Comput. Electr. Eng. **94**, 107304 (2021)
61. E.H. Houssein, Machine learning and meta-heuristic algorithms for renewable energy: a systematic review, in *Advanced Control and Optimization Paradigms for Wind Energy Systems* (2019), pp. 165–187
62. A.A. Ismaeel, E.H. Houssein, D. Oliva, M. Said, Gradient-based optimizer for parameter extraction in photovoltaic models. IEEE Access **9**, 13,403–13,416 (2021)
63. E.H. Houssein, M.A. Mahdy, M.J. Blondin, D. Shebl, W.M. Mohamed, Hybrid slime mould algorithm with adaptive guided differential evolution algorithm for combinatorial and global optimization problems. Expert Syst. Appl. **174**, 114689 (2021)
64. E.H. Houssein, M.A. Mahdy, M.G. Eldin, D. Shebl, W.M. Mohamed, M. Abdel-Aty, Optimizing quantum cloning circuit parameters based on adaptive guided differential evolution algorithm. J. Adv. Res. **29**, 147–157 (2021)

65. E.H. Houssein, B.E.-D. Helmy, D. Oliva, A.A. Elngar, H. Shaban, A novel black widow optimization algorithm for multilevel thresholding image segmentation. Expert Syst. Appl. **167**, 114159 (2021)
66. E.H. Houssein, M.M. Emam, A.A. Ali, An efficient multilevel thresholding segmentation method for thermography breast cancer imaging based on improved chimp optimization algorithm. Expert Syst. Appl. 115651 (2021)
67. E.H. Houssein, K. Hussain, L. Abualigah, M. Abd Elaziz, W. Alomoush, G. Dhiman, Y. Djenouri, E. Cuevas, An improved opposition-based marine predators algorithm for global optimization and multilevel thresholding image segmentation. Knowl. Based Syst. 107348 (2021)
68. E.H. Houssein, M.M. Emam, A.A. Ali, Improved manta ray foraging optimization for multilevel thresholding using Covid-19 CT images. Neural Comput. Appl. 1–21 (2021)
69. E.H. Houssein, B.E.-D. Helmy, A.A. Elngar, D.S. Abdelminaam, H. Shaban, An improved tunicate swarm algorithm for global optimization and image segmentation. IEEE Access **9**, 56,066–56,092 (2021)
70. M.M. Ahmed, E.H. Houssein, A.E. Hassanien, A. Taha, E. Hassanien, Maximizing lifetime of large-scale wireless sensor networks using multi-objective whale optimization algorithm. Telecommun. Syst. **72**(2), 243–259 (2019)
71. E.H. Houssein, A.G. Gad, Y.M. Wazery, P.N. Suganthan, Task scheduling in cloud computing based on meta-heuristics: review, taxonomy, open challenges, and future trends. Swarm Evol. Comput. 100841 (2021)
72. F.A. Hashim, E.H. Houssein, K. Hussain, M.S. Mabrouk, W. Al-Atabany, A modified henry gas solubility optimization for solving motif discovery problem. Neural Comput. Appl. **32**(14), 10,759–10,771 (2020)
73. E.H. Houssein, N. Neggaz, M.E. Hosney, W.M. Mohamed, M. Hassaballah, Enhanced Harris hawks optimization with genetic operators for selection chemical descriptors and compounds activities. Neural Comput. Appl. 1–18 (2021)
74. E.H. Houssein, M.E. Hosney, D. Oliva, W.M. Mohamed, M. Hassaballah, A novel hybrid Harris hawks optimization and support vector machines for drug design and discovery. Comput. Chem. Eng. **133**, 106656 (2020)
75. E. Zitzler, M. Laumanns, L. Thiele, SPEA2: improving the strength Pareto evolutionary algorithm. TIK-report **103** (2001)
76. T. Back, *Evolutionary Algorithms in Theory and Practice: Evolution Strategies, Evolutionary Programming, Genetic Algorithms* (Oxford University Press, 1996)
77. X.-S. Yang, S. Deb, Y.-X. Zhao, S. Fong, X. He, Swarm intelligence: past, present and future. Soft. Comput. **22**(18), 5923–5933 (2018)
78. S. Mirjalili, A. Lewis, The whale optimization algorithm. Adv. Eng. Softw. **95**, 51–67 (2016)
79. G.-C. Luh, C.-Y. Lin, Structural topology optimization using ant colony optimization algorithm. Appl. Soft Comput. **9**(4), 1343–1353 (2009)
80. S. Mirjalili, S.M. Mirjalili, A. Lewis, Grey wolf optimizer. Adv. Eng. Softw. **69**, 46–61 (2014)
81. V. Hayyolalam, A.A.P. Kazem, Black widow optimization algorithm: a novel meta-heuristic approach for solving engineering optimization problems. Eng. Appl. Artif. Intell. **87**, 103249 (2020)
82. A.A. Heidari, S. Mirjalili, H. Faris, I. Aljarah, M. Mafarja, H. Chen, Harris hawks optimization: algorithm and applications. Futur. Gener. Comput. Syst. **97**, 849–872 (2019)
83. L.M. Zhang, C. Dahlmann, Y. Zhang, Human-inspired algorithms for continuous function optimization, in *IEEE International Conference on Intelligent Computing and Intelligent Systems*, vol. 1 (IEEE, 2009), pp. 318–321 (2009)
84. Y. Shi, An optimization algorithm based on brainstorming process, in *Emerging Research on Swarm Intelligence and Algorithm Optimization* (IGI Global, 2015), pp. 1–35
85. X.-S. Yang, Flower pollination algorithm for global optimization, in *International Conference on Unconventional Computing and Natural Computation* (Springer, 2012), pp. 240–249
86. S. Kirkpatrick, C.D. Gelatt, M.P. Vecchi, Optimization by simulated annealing. Science **220**(4598), 671–680 (1983)

87. B. Javidy, A. Hatamlou, S. Mirjalili, Ions motion algorithm for solving optimization problems. Appl. Soft Comput. **32**, 72–79 (2015)
88. W. Zhao, L. Wang, Z. Zhang, Atom search optimization and its application to solve a hydrogeologic parameter estimation problem. Knowl. Based Syst. **163**, 283–304 (2019)
89. K. Abaci, V. Yamacli, Differential search algorithm for solving multi-objective optimal power flow problem. Int. J. Electr. Power Energy Syst. **79**, 1–10 (2016)
90. P. Civicioglu, Backtracking search optimization algorithm for numerical optimization problems. Appl. Math. Comput. **219**(15), 8121–8144 (2013)
91. P. Civicioglu, Artificial cooperative search algorithm for numerical optimization problems. Inf. Sci. **229**, 58–76 (2013)
92. Y. Saeys, I. Inza, P. Larrañaga, A review of feature selection techniques in bioinformatics. Bioinformatics **23**(19), 2507–2517 (2007)
93. Y. Li, G. Wang, H. Chen, L. Shi, L. Qin, An ant colony optimization based dimension reduction method for high-dimensional datasets. J. Bionic Eng. **10**(2), 231–241 (2013)
94. Y. Leung, Y. Hung, A multiple-filter-multiple-wrapper approach to gene selection and microarray data classification. IEEE/ACM Trans. Comput. Biol. Bioinform. (TCBB) **7**(1), 108–117 (2010)
95. S. Tabakhi, P. Moradi, F. Akhlaghian, An unsupervised feature selection algorithm based on ant colony optimization. Eng. Appl. Artif. Intell. **32**, 112–123 (2014)
96. B. Liao, Y. Jiang, W. Liang, W. Zhu, L. Cai, Z. Cao, Gene selection using locality sensitive Laplacian score. IEEE/ACM Trans. Comput. Biol. Bioinform. (TCBB) **11**(6), 1146–1156 (2014)
97. X. He, D. Cai, P. Niyogi, Laplacian score for feature selection, in *Advances in Neural Information Processing Systems* (2006), pp. 507–514
98. R. Cai, Z. Hao, X. Yang, W. Wen, An efficient gene selection algorithm based on mutual information. Neurocomputing **72**(4–6), 991–999 (2009)
99. L.E. Raileanu, K. Stoffel, Theoretical comparison between the Gini index and information gain criteria. Ann. Math. Artif. Intell. **41**(1), 77–93 (2004)
100. A. Bertoni, R. Folgieri, G. Valentini, Bio-molecular cancer prediction with random subspace ensembles of support vector machines. Neurocomputing **63**, 535–539 (2005)
101. C. Lai, M.J. Reinders, L. Wessels, Random subspace method for multivariate feature selection. Pattern Recogn. Lett. **27**(10), 1067–1076 (2006)
102. X. Li, H. Zhao, Weighted random subspace method for high dimensional data classification. Stat. Interface **2**(2), 153 (2009)
103. M. Haindl, P. Somol, D. Ververidis, C. Kotropoulos, Feature selection based on mutual correlation, in *Iberoamerican Congress on Pattern Recognition* (Springer, 2006), pp. 569–577
104. S.N. Ghazavi, T.W. Liao, Medical data mining by fuzzy modeling with selected features. Artif. Intell. Med. **43**(3), 195–206 (2008)
105. C. Ding, H. Peng, Minimum redundancy feature selection from microarray gene expression data. J. Bioinform. Comput. Biol. **3**(02), 185–205 (2005)
106. A.J. Ferreira, M.A. Figueiredo, An unsupervised approach to feature discretization and selection. Pattern Recogn. **45**(9), 3048–3060 (2012)
107. A.J. Ferreira, M.A. Figueiredo, Efficient feature selection filters for high-dimensional data. Pattern Recogn. Lett. **33**(13), 1794–1804 (2012)
108. H. Vural, A. Subaşı, Data-mining techniques to classify microarray gene expression data using gene selection by SVD and information gain. Model. Artif. Intell. **2**, 171–182 (2015)
109. H. Abdi, Singular value decomposition (SVD) and generalized singular value decomposition, in *Encyclopedia of Measurement and Statistics* (2007), pp. 907–912
110. W. Ahmed, Fast orthogonal search for training radial basis function neural networks. Ph.D. Dissertation, University of Maine (1994)
111. I.A. Gheyas, L.S. Smith, Feature subset selection in large dimensionality domains. Pattern Recogn. **43**(1), 5–13 (2010)

112. A. Srivastava, S. Chakrabarti, S. Das, S. Ghosh, V.K. Jayaraman, Hybrid firefly based simultaneous gene selection and cancer classification using support vector machines and random forests, in *Proceedings of Seventh International Conference on Bio-inspired Computing: Theories and Applications (BIC-TA 2012)* (Springer, 2013), pp. 485–494
113. B. Sahu, D. Mishra, A novel feature selection algorithm using particle swarm optimization for cancer microarray data. Procedia Eng. **38**, 27–31 (2012)
114. E. Martinez, M.M. Alvarez, V. Trevino, Compact cancer biomarkers discovery using a swarm intelligence feature selection algorithm. Comput. Biol. Chem. **34**(4), 244–250 (2010)
115. M.M. Kabir, M. Shahjahan, K. Murase, A new hybrid ant colony optimization algorithm for feature selection. Expert Syst. Appl. **39**(3), 3747–3763 (2012)
116. H. Yu, G. Gu, H. Liu, J. Shen, J. Zhao, A modified ant colony optimization algorithm for tumor marker gene selection. Genomics, Proteomics Bioinform. **7**(4), 200–208 (2009)
117. I. Inza, P. Larrañaga, R. Blanco, A.J. Cerrolaza, Filter versus wrapper gene selection approaches in DNA microarray domains. Artif. Intell. Med. **31**(2), 91–103 (2004)
118. I. Inza, B. Sierra, R. Blanco, P. Larrañaga, Gene selection by sequential search wrapper approaches in microarray cancer class prediction. J. Intell. Fuzzy Syst. **12**(1), 25–33 (2002)
119. M. Ghoneimy, E. Nabil, A. Badr, S.F. El-Khamisy, Bioscience Research
120. H.M. Alshamlan, G.H. Badr, Y.A. Alohali, ABC-SVM: artificial bee colony and SVM method for microarray gene selection and multi class cancer classification. Int. J. Mach. Learn. Comput **6**(3), 184 (2016)
121. E. Alba, J. Garcia-Nieto, L. Jourdan, E.-G. Talbi, Gene selection in cancer classification using PSO, SVM and GA, SVM hybrid algorithms, in *IEEE Congress on Evolutionary Computation* (IEEE, 2007), pp. 284–290
122. R.R. Rani, D. Ramyachitra, Microarray cancer gene feature selection using spider monkey optimization algorithm and cancer classification using SVM. Procedia Comput. Sci. **143**, 108–116 (2018)
123. N. Almugren, H. Alshamlan, FF-SVM: new firefly-based gene selection algorithm for microarray cancer classification, in *2019 IEEE Conference on Computational Intelligence in Bioinformatics and Computational Biology (CIBCB)* (IEEE, 2019), pp. 1–6
124. U. Maulik, D. Chakraborty, Fuzzy preference based feature selection and semisupervised SVM for cancer classification. IEEE Trans. Nanobiosci. **13**(2), 152–160 (2014)
125. M.-S. Chen, T.-Y. Ho, D.-Y. Huang, Online transductive support vector machines for classification, in *2012 International Conference on Information Security and Intelligent Control* (IEEE, 2012), pp. 258–261
126. L. Zhang, W. Zhou, B. Wang, Z. Zhang, F. Li, Applying 1-norm SVM with squared loss to gene selection for cancer classification. Appl. Intell. **48**(7), 1878–1890 (2018)
127. Y. Wang, I.V. Tetko, M.A. Hall, E. Frank, A. Facius, K.F. Mayer, H.W. Mewes, Gene selection from microarray data for cancer classification–a machine learning approach. Comput. Biol. Chem. **29**(1), 37–46 (2005)
128. R. Alanni, J. Hou, H. Azzawi, Y. Xiang, A novel gene selection algorithm for cancer classification using microarray datasets. BMC Med. Genomics **12**(1), 1–12 (2019)
129. S. Kar, K.D. Sharma, M. Maitra, Gene selection from microarray gene expression data for classification of cancer subgroups employing PSO and adaptive k-nearest neighborhood technique. Expert Syst. Appl. **42**(1), 612–627 (2015)
130. J. Dev, S.K. Dash, S. Dash, M. Swain, A classification technique for microarray gene expression data using PSO-FLANN. Int. J. Comput. Sci. Eng. **4**(9), 1534 (2012)
131. K. Chatra, V. Kuppili, D.R. Edla, A.K. Verma, Cancer data classification using binary bat optimization and extreme learning machine with a novel fitness function. Med. Biol. Eng. Comput. **57**(12), 2673–2682 (2019)
132. W. Zhao, G. Wang, H. Wang, H. Chen, H. Dong, Z. Zhao, A novel framework for gene selection. Int. J. Adv. Comput. Technol. **3**(3), 184–191 (2011)
133. H. Alshamlan, G. Badr, Y. Alohali, mRMR-ABC: a hybrid gene selection algorithm for cancer classification using microarray gene expression profiling. Biomed. Res. Int. **2015** (2015)

134. H.M. Alshamlan, G.H. Badr, Y.A. Alohali, Genetic bee colony (GBC) algorithm: a new gene selection method for microarray cancer classification. Comput. Biol. Chem. **56**, 49–60 (2015)
135. A. El Akadi, A. Amine, A. El Ouardighi, D. Aboutajdine, A new gene selection approach based on minimum redundancy-maximum relevance (MRMR) and genetic algorithm (GA), in *2009 IEEE/ACS International Conference on Computer Systems and Applications* (IEEE, 2009), pp. 69–75
136. H. Liu, L. Liu, H. Zhang, Ensemble gene selection by grouping for microarray data classification. J. Biomed. Inform. **43**(1), 81–87 (2010)
137. M.J. Abdi, S.M. Hosseini, M. Rezghi, A novel weighted support vector machine based on particle swarm optimization for gene selection and tumor classification. Comput. Math. Methods Med. **2012** (2012)
138. L. Gao, M. Ye, X. Lu, D. Huang, Hybrid method based on information gain and support vector machine for gene selection in cancer classification. Genomics, Proteomics Bioinform. **15**(6), 389–395 (2017)
139. C.-C. Chang, C.-J. Lin, LIBSVM: a library for support vector machines. ACM Trans. Intell. Syst. Technol. (TIST) **2**(3), 1–27 (2011)
140. G. Wang, Q. Song, B. Xu, Y. Zhou, Selecting feature subset for high dimensional data via the propositional foil rules. Pattern Recogn. **46**(1), 199–214 (2013)
141. R. Díaz-Uriarte, S.A. De Andres, Gene selection and classification of microarray data using random forest. BMC Bioinform. **7**(1), 3 (2006)
142. K.-B. Duan, J.C. Rajapakse, H. Wang, F. Azuaje, Multiple SVM-RFE for gene selection in cancer classification with expression data. IEEE Trans. Nanobiosci. **4**(3), 228–234 (2005)
143. K.-B. Duan, J.C. Rajapakse, M.N. Nguyen, One-versus-one and one-versus-all multiclass SVM-RFE for gene selection in cancer classification, in *European Conference on Evolutionary Computation, Machine Learning and Data Mining in Bioinformatics* (Springer, 2007), pp. 47–56
144. E.H. Houssein, H.N. Hassan, M.M. Al-Sayed, E. Nabil, Gene selection for microarray cancer classification based on manta rays foraging optimization and support vector machines (2021)

Elitist-Ant System Metaheuristic for ITC 2021—Sports Timetabling

Ghaith M. Jaradat

Abstract Timetabling problem is one of the most difficult operational tasks and an important step in raising the industrial productivity, capability, and capacity. Such tasks are usually tackled using metaheuristics techniques that provide an intelligent way of suggesting solutions or decision making. Swarm intelligence techniques including ant systems have proved to be effective examples. Different recent experiments showed that ant system algorithms are reliable for timetabling in many discrete-events applications such as educational and personnel timetabling, job and machine scheduling, and other similar aspects. However, obtaining an optimal solution is extremely difficult but obtaining a near-optimal solution using metaheuristic algorithms is possible. This research paper seeks the enhancement of ant system algorithm for an efficient timetabling task. This algorithm aims to generate feasible and high-quality timetables via minimizing constraints violations in a reasonable execution time. This enhanced version is a hybrid elitist-ant system metaheuristic which is tested on a round-robin tournament known as the international timetabling competition-2021 dedicated for sports timetabling. The competition includes several hard and soft constraints to be satisfied to build a feasible high-quality timetable. It consists of three categories of complexities or difficulties, namely early, test and middle instances. Results showed that the proposed elitist-ant system metaheuristic has obtained competitive timetables for almost all the instances in terms of feasibility and quality. The feasibility is measured by minimizing the violation of hard constraints to zero, while minimizing soft constraints towards zero violations as much as possible. The performance of the elitist-ant system is evaluated by the consumed computational time to produce a timetable, consistency, and robustness. The elitist-ant system showed a robust and consistent performance in producing a diversity of timetables in a reasonable computational time.

Keywords Sports timetabling · Elitist-ant system · International timetabling competition 2021 · Round-robin tournament

G. M. Jaradat (✉)
Department of Computer Science, Faculty of Computer Sciences and Informatics, Amman Arab University, Amman 2234-11953, Jordan
e-mail: g.jaradat@aau.edu.jo

E. H. Houssein et al. (eds.), *Integrating Meta-Heuristics and Machine Learning for Real-World Optimization Problems*, Studies in Computational Intelligence 1038,
https://doi.org/10.1007/978-3-030-99079-4_3

1 Introduction

Timetabling is the distribution of resources such as players and games over a fixed period, a timeslot. This task can be difficult and very time-consuming. If the process of generating timetables is automated using algorithms, then this helps saving time and money for the athlete's/sports federations as in the case of sports timetabling.

Among the metaheuristic algorithms is the Swarm Intelligence (SI), which mimics the collective behavior of decentralized, self-organized systems, natural or artificial. The concept is employed in work on artificial intelligence. Inspired by biological systems, SI consists typically of a population of simple agents interacting locally with one another and with their environment. The agents follow simple rules for interacting between each other leading to an intelligent global behavior. SI refers to a general set of algorithms, for example, SI in natural systems include ant colonies, bird flocking, animal herding, bacterial growth, and microbial intelligence.

Ant Colony Optimization (ACO) is a family of population-based foraging behavior metaheuristics, first proposed by Dorigo et al. [1], namely as the Ant Systems (AS). AS is inspired by foraging behavior of real ants communicating indirectly via the distribution and dynamically changing information known as pheromone trails. These trails' weight reflects the collective search experience, which will be exploited by successive ants attempting to solve a given problem instance more effectively. Examples on successful works that inspired this study include [2, 3] who developed a powerful extension of the AS and elitist-AS to solve course timetabling, vehicle routing, knapsack and travelling salesman problems.

Technically, ant systems are computational methods that iteratively optimize a problem via improving a candidate solution given a quality measure. It solves a problem by a population of candidate solutions (*aka.* Ants) and moving these ants around in search-space according to simple mathematical formulations over the ant's path. Each ant's movement is influenced by its local best-known path but is also guided toward the best-known path in the search space, which are updated as better path found by other ants. This is expected to move the colony of ants toward best solutions.

The constraints that appear in real-life problem instances are extremely diverse, where each competition has its own requirements. In the ITC 2021 competition, two types of constraints are considered for optimization: (i) hard constraints representing fundamental properties of the timetable that can never be violated, and (ii) soft constraints which represent preferences that should be satisfied whenever possible. While many optimization objectives in the literature, the ITC2021 considers problem instances only where the objective is to minimize the penalties of violating soft constraints. This was assumed by Van Bulck et al. [4] which makes the problem formulation more attractive for a wider timetabling community, while retaining empirical complexity of the problems.

1.1 Problem Statement

Scheduling or timetabling Sports events is considered a big business in hundreds of billion dollars industry, many times as big as automobile and movie industries. For example, early in the twenty-first century a globally famous team such as Manchester United football club had a market cap of £400 million; and TV networks such as the US national TV paid billions of dollars per year for NFL games and others paid more than $500 million per year for football English premier league. So, rights holders wanted profitable and beneficial feasible schedules. Feasible timetables are also important for extensive players traveling, weekends ticket sales are better, and better to play division teams at the end of season. In addition, sports timetabling has a wide range of problem types that to be considered. Thus, the idea is simply "which pairs of teams play each other and when?". This presents a challenging task for algorithms in terms of easiness or hardness when dealing with small instances. Finally, a significant theoretical background of the sports timetabling problems provides a fertile environment to test new optimization algorithms or enhance existing ones. Therefore, all of this provide a significant theoretical contribution to both sports timetabling problems as well as optimization algorithms design and enhancement.

In this complex computational task, along with new technologies and scheduling needs, there is a big need to have scheduling schemes in which they are fast and fully utilizing the available resources in sports timetabling tasks. The growth of sports timetabling computational complexities makes it necessary to optimize timetabling methods to better solve sports timetabling problems and to fully benefit from available resources. This work tries to utilize a version of ant systems known as the elitist-ant system to improve and optimize sports timetabling problems.

1.2 Objectives

The main goal of this work is to better utilize the elitist-AS in sports timetabling problems (e.g., round-robin tournament). To achieve this goal, the following objectives must be met:

1. Explore recent literature reviews concerning sports timetabling problems based on elitist-AS or similar algorithms.
2. Propose an elitist-AS algorithm to solve sports timetabling problems.
3. Evaluate the proposed algorithm in international timetabling competition (ITC-2021) sports timetabling.

1.3 Scope

This study focuses on the use of elitist-AS algorithm for ITC-2021 sports timetabling problem.

1.4 Hypothesis

The enhanced elitist-AS algorithm could improve the performance of sports timetabling techniques in ITC-2021.

1.5 Contribution

In this paper, the Elitist-Ant System (EAS), an enhanced version of the original AS algorithm, has been implemented to optimize the round-robin sports games timetabling by completely satisfying hard constraints and minimizing the violations of soft constraints. This intends to produce a feasible high-quality timetable of three categories of datasets dedicated for the sports timetabling problem (provided by ITC-2021) in a 2-round-robin fashion. The proposed algorithm is considered as a hybridization of a dynamic and adaptive version of EAS algorithm and the iterated local search (ILS); named as EAS-ISL. It is expected to outperform the traditional AS and other ACO versions, as well as similar approaches in terms of feasibility, optimality, and speed. Technically, the main considered issue is to strike a balance between diversity and quality of the search process around the neighborhood of feasible timetables produced by the proposed EAS-ILS. In other words, not to diversify the exploration around a feasible timetable too much, while not to intensify the exploitation of a better-quality timetable too much. So, this helps the search process to escape local optima and relief the stagnation.

2 Literature Review

Scheduling or timetabling task is one of the most important steps in managing and manipulating resources for improving capabilities of a computational system. For demonstrating the difficulty of obtaining an optimal solution, Jaradat et al. [3] compared three population-based metaheuristics with each other for three classes of scheduling, namely, travelling salesman, vehicle routing, and knapsack problems. These metaheuristics are one ACO-based, and two memetic-based algorithms. The goal was to generate an optimal schedules or assignments.

Until recently, sports schedules mostly constructed by hand which are time consuming (with 10 teams, there are numerous possible schedules), many constraints (including television networks, teams, cities), and no new constraints can be added.

Sports timetabling is an old traditional task and has widely spread in the field of computing, research, and industry specially in the last decade years. Many heuristics have been implemented to resolve the sports scheduling/timetabling problem such as: linear, constraints (CP), and integer programming (IP), particle swarm optimization, genetic algorithm (GA), ACO. It is known that the use of optimization techniques (including metaheuristics) can easily adapt when new constraints are added or changing the structure of the problem's formulation. So, all professional sports agencies and some institutes construct their sports timetables using optimization.

2.1 Studies Based on Round-Robin Tournaments for Sport Timetabling Problems

A variety of studies and implementations of sport timetabling problems have been conducted in the first decade of the millennium as well as the last two decades of the twentieth century. Concerning several constraints, several studies mainly focused on the formulation and settings of an efficient generation of a feasible timetable.

Trick [5] discussed the impact of premature sets from completing a round-robin schedule and how they determine the minimum size of a premature set. The author constructed a round-robin schedule via two methods, circle method and greedy algorithm. The circle method continues rotating to get all the slots depending on an initial set of games that have all differences. In the other hand the greedy algorithm sets games in lexicographic order, where slots are in cyclic order. It then repeatedly assigns each game to either current slot or next slot it can feasibly generates. Their study also considered some additional requirements such as carry-over effect and how to balance it, venues, fixed or prohibited games, and the objective function. They ended up using IP and CP algorithms for properly handling the additional requirements and improving the problem's formulation. Similar studies presented by [6, 7]. Others proposed GAs for solving the sports timetabling problem such as [8], while [9, 10] tackled multi-objective versions of the problem. Other worth mentioning studies are conducted by [11, 12] who developed a framework for the problem solver and comprehensive revision of the potential problem formulations, respectively. Best results in the literature are obtained by hybrid methods of IP and CP, and metaheuristics such as [5]. Devising optimal tournament timetables is crucial to players, teams, fans, cities, security force, and sponsors. Fair and balanced timetables for all teams, satisfying many hard and soft constraints, is the major issue for attractiveness and confidence in the outcome of professional league tournaments. Therefore, this study is motivated to propose a hybridization between two metaheuristics.

2.2 *Time-Constrained Double Round-Robin Tournaments ITC 2021*[1]

Since 2002, a highly reputed optimization community (namely EURO) including two working groups such as PATAT-automated timetabling and OR in sports have organized a series of educational timetabling competitions. This is an attempt to open opportunities for any aspiring timetable (or solver), in which to encourage conducting research on automated timetabling methods. This time the organizers focus on sport timetabling namely the international timetabling competition (ITC-2021: sports timetabling). Since the literature is extensive of only case studies and their methods have been developed for a specific problem. Add to that, problem instances are rarely shared and without a benchmark. Determining which algorithm works best for solving a specific problem lacks a sufficient insight. All of which have motivated the organizers of the competition to benchmark instances and settings up a proper environment for testing a potential algorithm for sports timetabling.

The aim of the ITC-2021 competition is to stimulate the development of solvers for the construction of round-robin timetables, where teams play against each other in a fixed number of times. Besides the round-robin timetabling format, many other formats exist such as the knock-out format, where teams are tied in pairs and the loser of each game is eliminated. Nevertheless, round-robin tournaments are the most researched format (see [13]), and are very common in practice (see [14]). Most sports competitions organize a double round-robin tournament (2RR) where teams meet twice but single, triple, and even quadruple times. According to [15], categorized round-robin tournaments into two: (i) time-constrained and (ii) time-relaxed. A timetable is time-constrained (e.g., compact) if it uses the minimal number of timeslots needed and is time-relaxed otherwise. The ITC-2021 only considers time-constrained double round-robin tournaments with an even number of teams. Under this setting, the total number of timeslots is exactly equal to the total number of games per team, and hence each team plays exactly one game per timeslot. For more information, please refer to [16, 17]. For a comprehensive survey on round-robin implementations for scheduling tasks, please refer to Rasmussen and Trick (2007).

Unlike elimination and king of the hill tournaments, Round-robin tournament considers the following scenario: if there are n teams then each team plays exactly k times against all other teams. Simply, let's say for solving single round-robin tournaments ($k = 1$), a complete graph is used for graph construction and graph coloring to find a solution (see Fig. 1 for illustration). Where, vertices represent teams while edges represent a game between two teams. The formulation of the problem is as follows:

- How many games can be played is formulated as $n\ (n - 1)/2$ representing graph edges.
- What will be the chromatic number is formulated as $n - 1$.

[1] https://www.sportscheduling.ugent.be/ITC2021/index.php.

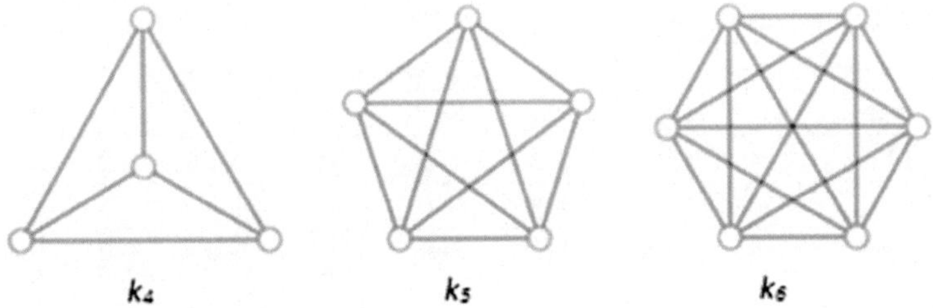

Fig. 1 Single round-robin tournaments

Table 1 A timetable for 6 team round-robin tournament

	Week 1	Week 2	Week 3	Week 4	Week 5
Game 1	6–1	5–1	4–1	3–1	2–1
Game 2	5–2	4–6	3–5	2–4	6–3
Game 3	4–3	3–2	2–6	6–5	5–4

- How many teams can play in the same week is formulated as *n*/2 representing same color comes under same week.
- How many weeks can be conducted for the games is formulated as the amount of games/number of plays per week.

Thus, for timetabling 6 teams (even number of teams) in a round-robin tournament of 5 weeks is conducted as follows: First week with numbering teams from 1 to *n* and fixing the position of a team in the timetable, for example, team 1 in position (1, 1). For the following 4 weeks, iteratively rotating the remaining teams (2, 3, 4, 5, 6) by 1. Then, the resulting solution is presented in Table 1.

2.2.1 ITC-2021—Description and File Format

Van Bulck et al. [4] provided a comprehensive description of the ITC-2021—sports timetabling problem and its instance file format. In details, they explained regulation and procedures for the international timetabling competition (ITC-2021). They illustrated the meta data of the problem and its instances, its resources, its structure, and most importantly its constraints. It is well-known that such an optimization problem it is comprised of two constraint types, namely, hard, and soft constraints. Hard constraints are fundamental properties of the timetable that must not be violated, while soft constraints are preferences that should be satisfied as much as possible. The hard constraints include capacity, game, break, fairness, and separation constraints. All of which are subjected to an objective function penalized/weighted for soft

Table 2 A 2RR-based solution

First half of the season					Second half of the season				
S1	S2	S3	S4	S5	S6	S7	S8	S9	S10
1–2	2–5	2–4	2–3	6–2	4–2	5–2	2–1	3–2	2–6
3–4	4–1	1–6	5–1	4–5	6–1	1–4	4–3	1–5	5–4
5–6	6–3	5–3	6–4	1–3	3–5	3–6	6–5	4–6	3–1

constraints. For more details, please refer to [4] and the official website of the competition[2] including problem instances' three categories. It is also worth checking the work of [4] that describes the benchmark and file format of a round-robin based sports timetabling problem. An excellent implementation of round-robin classification and format for the sports timetabling was conducted by [18].

To simplify the process of timetabling sports games, consider four components of the problem: input, scope, constraints, and output. The input fetches a set of teams and a set of timeslots; while the output generates a timetable consisting of fixtures that determine for each game on which timeslot it is to be played. The scope is either based on 2RR, time-constrained timetables, or phased timetables. See Table 2 to examine an output of a feasible timetable where at the first half of the season each team plays against each other team once.

The most challenging component of the problem is the constraints. Instances are subject to several constraints which are hard and soft constraints. They mainly concerning with: (i) all hard constraints should be satisfied to achieve feasibility; (ii) while minimizing the sum of penalties of the violated soft constraints to achieve optimality.

2.2.2 Problem Instances

The tournament is structured as 2RR with 18, 20, or 22 teams; time-constrained; and phased or no symmetry. Constraints are of 9 constraint types, and 5 constraint groups. These types are:

a. Capacity constraints:
   - CA1 Place constraints: no home or away game in each timeslot.
   - CA2 Top team and bottom team constraints.
   - CA3 Limit length of home or away sequences.
   - CA4 Enforce complementary home-away patterns (shared venues).
b. Break constraints: (a team has a break when it plays two consecutive games with the same home-away status):

[2] https://www.sportscheduling.ugent.be/ITC2021/.

- BR1 No breaks at the beginning or end of season. (See team no. 2 in 2nd and 3rd slots in Table 2).
- BR2 Limit on the total number of breaks in the timetable.

c. Fairness and Attractiveness constraints: (considering only one constraint in the competition):

- FA2 Ranking balancedness: limit on the maximal difference in home games played between any two teams at any point in the timetable.

d. Game constraints:

- GA1 Fixed and forbidden game assignments.

e. Separation constraints:

- SE1 No repeaters: two games with the same opponents are separated by at least a given number of timeslots.

Van Bulck et al. [18] have presented constraints violation penalties which are formulated as each constraint $c \in C$ has its own XML notation and precise description, and description of the deviation vector (for hard constraints) is denoted as sum the overall elements in the deviation vector, where $D_c = [d_1, d_2 \dots d_q]$, $\alpha_c = \sum_{k=1}^{q} d_k$. The same description is meant for the violations of soft constraint with the p_c penalty. The objective is to minimize $\sum_{c \in Csoft} p_c \alpha_c$, while $\alpha_c = 0$ for all $c \in C_{hard}$.

Provided by [18], an example of a deviation vector, consider the xml presentation of the SE1 constraint, <SE1 teams = '0;1;2;3' min = '1' model = 'SLOTS' type = 'HARD'>. Each pair of teams in 'teams' has at least 'min' timeslots between two consecutive mutual games. Each pair of teams in 'teams' triggers a deviation equal to the sum of the number of timeslots less than 'min' for all consecutive mutual games. So, let us say there is at least one timeslot, a penalty p_c of 10, $D_c = [0, 1, 0, 0, 1, 0]$, and then $\alpha_c = 10(0 + 1 + 0 + 0 + 1 + 0) = 20$. For illustration, see Table 3.

Due to their milestone regarding submissions and the availability of their instances (early, middle, and late groups of instances), they consider in the first place the solution value as the only criteria that matters to them in the competition. The rules set is drawn by the organizers in the simplest form as follows:

- There are no computation time or technology restrictions (including commercial or open-source solvers).
- They do not expect the code to run on their hardware, but they may ask to see the source code to check that the rules have been followed.
- Finalists are expected to write a short paper, describing their approach taken in sufficient details.

Table 3 A deviation vectors

S1	S2	S3	S4
1–2	**1–3**	**3–1**	2–1
3–4	**2–4**	**4–2**	4–3

- Only solutions in their accredited XML format will be considered.
- The same version of the algorithm must be used for all instances.

Some worth reading publications regarding sports timetabling are: [19–24].

3 The ESA-ILS Algorithm

To overcome the problems of the typical population-based metaheuristic such as premature convergence and low accuracy, the ESA-ILS initializes the process by a pseudo-random sequence, which is used to balance the diversity of ants. Then, an effective diagnosis mechanism of premature is adopted to determine local convergence, and algorithm correction is performed by random mutation, which could activate the ants in the stagnation state to let them escape from a local optimum. Simulation experiments are conducted to show the feasibility and effectiveness of the ESA-ILS. The ESA-ILS aims at solving the ITC 2021-sports timetabling problem by utilizing its hybridization architecture including elitism and diversity-quality balancing mechanism. The improvement is mainly focusing as much as possible on avoiding falling into a local optimum while diversifying the search around elite solutions.

One of the challenging issues is to completely satisfy hard constraints (utterly minimizing penalties), while satisfying as much as possible soft constraints (e.g., minimizing penalties). Many versions of enhanced ant algorithms are among of the best scheduling algorithm in timetabling (e.g., course and exam timetabling) such as the hybrid ant colony system proposed by [25]. However, ant algorithms randomly generate the first population. As randomness decreases the probability of the algorithm to converge to the best solution. The proposed EAS-ILS aims to overcome the population diversity trade-off. Some existing ant system and swarm intelligence versions for solving different sports timetabling problems, rather than ITC-2021, such as: ant based hyper-heuristic by [26–28] and by [29–31].

Sports timetabling is a very important part of the timetabling problems in a variety of industries. Aiming at the characteristics of sports timetabling and considering all stakeholders. The ESA-ILS algorithm is based on adaptive weights. It uses adaptive weights to make weight change with the increase of the number of iterations, and introduces random weights in the later stage, which avoids search stagnation meaning being trapped in local optimum. Applying the ESA-ILS to sports timetabling can achieve a better timetabling plan.

3.1 ESA-ILS

In the basic ant system, each ant is considered as a potential solution to the numerical optimization problem in a dimensional space. Every ant has a pheromone value in this search space and a weight assigned to it. Also, each ant has a local memory which keeps the best pheromone value that is experienced by the ant so far. A globally shared memory (elite pool) keeps the best global pheromone found so far. This information contributes to the path of each ant presenting the relative influences of the foraging experiences. Defining an upper bound for the pheromone increases the performance of the approach giving the ants pheromone values update. The pheromone improves the performance, since it adjusts the convergence over time and improves the search precision of the ants with a uniformly distributed random number in [0, 1]. In addition, the pheromone evaporation rate is also considered for more efficient control and constraining of convergence. This prevents the exhaustive exploration of the search space, which can occur when ant pheromones vary without control. It is known that the successes of approach are problem dependent. Therefore, in this study the authors considered implementing the ESA-ILS with the best solution found so far.

The ESA-ILS algorithm is based on the path of the ant to its globally closest best and globally closest worst timetables. Decreasing the weight means that during convergence, the algorithm increases the importance of the foraging behavior and pay less attention to the actual pheromone value of the ant. In cases where the ant is stuck in local optima, using a linearly decreasing weight leaves the ant with a pheromone value less than what the ant requires to escape those optima. Therefore, the algorithm decreases the penalty value linearly until a certain threshold is reached. At such point, the weight is reset to its upper bound. Here, a lower and upper bound value for the weight is defined. To achieve a balanced diversification, some randomness is added to the upper and lower bounds as well as the pheromone evaporation rate.

Hence, in this work, the ESA-ILS provides a guidance of ants away from the closest worst timetable or rather path. It works by correcting their paths towards which they are supposed to be, prior to updating their paths. This can be achieved by updating the value of weight randomly to prevent the ants from being stuck in their local or global search. This approach also facilitates global search and rapid local search; where the ants (of the ESA) explore the search space in the beginning of the run and turns into local search (ILS) at the ending of the run. The foraging behavior presents choosing the closest best and worst timetables from the elite pool with respect to the ant, then the similarity of the ant to each of these timetables is measured. See Fig. 2 which demonstrates a generic pseudocode of the ESA-ILS.

The penalty value is decreased linearly until a certain threshold is reached. At such a point, the penalty is reset to its upper bound. Here a lower and upper bound value is defined for the penalty. To achieve diversification, some randomness is added to the upper and lower bounds and the pheromone evaporation rate. For further information on the mathematical formulations of the ESA and the ILS please refer to [3, 32].

3.2 *EAS-ILS Implementation*

In general, the proposed algorithm starts with generating a random population of feasible solutions (timetables), where their hard constraints have zero violations.

Algorithm: ESA-ILS.
Step 1: Initialization phase
Randomly **initialize** *ants* and set initial *pheromone* values
Define external memory (elite pool) for best *pheromone_trails* (***Bests***).
While *StoppingCriterion* is not met **do**
Step 2: Construction phase
    For each *ant*   //solution construction
        Apply Nearest Neighbor construction heuristic;
    End for
Step 3: Improvement phase
    While *non-improvement_stoppingCriterion* is not met **do**
        Locally **improve** each constructed solution;   // employ ILS
        Update *size* & *content* of elite pool.
    End While
If the *best_solution* is updated **Then**
Step 4: Intensification phase
    Randomly **explore** the neighbors of the *best_solution* found so far (elite solution);
Step 5: Global Pheromone update phase
    Update *pheromone_trails* route appearing in solution;   // diversity pool
Else
Step 6: Diversification phase
    Pheromone evaporation;   // diversity control
    Reinitialize pheromone trails;
    Generate new population of ants using solutions from elite pool by performing perturbations;
End If
End While
Step 7: Return *best_ant*   // best solution

Fig. 2 A generic pseudocode of ESA-ILS [3]

Then it selects several solutions based on their fitness values (objective function cost) from those feasible ones, where their soft constraints are less violated towards zero if existed. Those selected solutions are then permutated locally to generate new feasible solutions with probably lesser soft constraints violations.

The proposed algorithm considers permutations made to the order of teams and the determination of the round each team plays. It keeps an elite portion of the population in a steady state reproduction, where a group of elite solutions (feasible and low soft constraints violations) are the local optimum to start with the permutations towards optimality (zero hard and zero soft violations).

The proposed algorithm implements direct and indirect encodings for the problem-solving representation. The direct encoding is only used to represent a timetable as a $n \times n$ array with some complex operator. In the other hand, the indirect encoding is used to perform a permutation of teams in a $n \times n$ array and orders the teams in a 2RR fashion (home and away) which influences the produced timetable towards optimality.

For example, let us say, there are 9 rounds (weeks) of games, so each team (e.g., out of 6 teams) plays each other once. Then it assigns each game a location (timeslot), based on a series of constraints (hard and soft). Each generated timetable is then evaluated based on the fitness value. Permutations are performed based on the index of the round (season) assignment for each team that violates a constraint. The whole process is iteratively repeated by generating a population of timetables and reproduce them until a better quality is found. For choosing the home teams, first the algorithm iterates through unscheduled games generated in the population initialization step. Initially, using a degree heuristic to generate feasible timetables but not necessarily optimal, meaning that by choosing higher degree first by assigning a value to the variable that is involved in the largest number of constraints on the other unassigned variables. Here, variables are the sequences of games. Based on that, the proposed algorithm applies a series of common prioritized rules as follows:

1. If a game has been already assigned into a timeslot due to a triplet (e.g., a group of 3 teams), continue to next game.
2. If the game is between two teams in a triplet, then check the following:
   a. If one of the other triplet games has been assigned a timeslot, then set the home team of this game and the other game in the triplet.
   b. If none of the games in the triplet have been scheduled yet, continue to next rule.
3. If team 1 requires a home game on a certain date, make it the home team.
4. If team 2 requires a home game on a certain date, make it the home team.
5. Let $c1$ and $c2$ be the number of consecutive home games for team 1 and team 2 respectively:
   a. If $c1 > c2$ and $c1 \geq 2$, then make team 1 the home team.
   b. If $c2 > c1$ and $c2 \geq 2$, then make team 1 the home team.
6. Set the team that has few home games scheduled to be the home team.

If both have an equal number of home games, arbitrarily assign team one to play at home. The fitness evaluation of generated timetable is calculated based on one of three ratings:

- Travel time rating rates the average time spent traveling by each team.
- Consecutive rating: rates number of times consecutive away games exceeds 2.
- Location balance rating: rates number of teams not scheduled with e.g., 4/5 home games.

For generating better quality solutions (feasible timetables with minimum soft violations), permutations and reproductions are based on either order of games or mutation (perturbation) of teams. A mutation is simply the neighborhood (neighbors of feasible solutions) around a local optimum (a feasible timetable with minimum soft violations). These neighbors again are based on either of the following neighborhood structures:

- Position-based: remove an element from one position and insert into another.
- Order-based: swap two random elements in a timetable.
- Scramble sub-list: pick a sub-list of consecutive elements in the timetable and randomly shuffle them.

However, it is known that most swapping does not lead to feasible double round-robin timetables, where the following possible permutations need to be controlled:

- Swap locations for a pair of games.
- Swap 2 slots of games.
- Swap timetable of 2 teams.
- Partially swap 2 slots: swap 1 game then minimum number to preserve feasibility.
- Partially swap 2 teams: swap games in single slot then minimum number to preserve feasibility.

The criteria of searching either neighborhood structure based on the permutation rate and weight which is it turn depends on the soft constraints and sometimes on the size of the dataset itself. Where, for example, given a round-robin schedule then assign home/away to minimize breaks. Therefore, a hybrid metaheuristic (EAS-ILS) is proposed with a dynamic and adaptive characteristic which could solve this issue so easily via providing a multiple phase approach for solving each subproblem. The following simplified phases are considered:

Phase 1: find home/away feasible patterns, where one sequence per team is conducted (Table 4).

Phase 2: assign games consisting only a feasible home/away pattern (Table 5).

Phase 3: assign teams to entries (Table 6).

Table 4 Feasible home/away pattern for one team

Game	Location				
1	H	A	H	A	H
2	A	H	A	H	A
3	H	H	A	A	H
4	H	A	H	H	A
5	A	A	H	H	A
6	A	H	A	A	H

Table 5 Feasible games assignment

Game	Pattern				
1	+2	−3	+6	−4	+5
2	−1	+4	−5	+6	−3
3	+6	+1	−4	−5	+2
4	+5	−2	+3	+1	−6
5	−4	−6	+2	+3	−1
6	−3	+5	−1	−2	+4

* (+ denotes Home; – denotes Away)

Table 6 Feasible teams' assignment

Game	Pattern				
F	+E	−A	+B	−D	+C
E	−F	+D	−C	+B	−A
A	+B	+F	−D	−C	+E
D	+C	−E	+A	+F	−B
C	−D	−B	+E	+A	−F
B	−A	+C	−F	−E	+D

* (+ denotes Home; − denotes Away)

This phase represents dividing the problem into subproblems for faster solving (including faster enumeration of timeslots assigned to each game) or satisfying each constraint. This phase may guarantee flexibility and robustness.

For a proper selection of neighborhood structures, the EAS-ILS employs an iterated local search for selecting a structure based on the soft constraint at hand for converging the exploitation of a solution space. This local search is chosen based on a preliminary experiment conducted by the authors, where the iterated local search heuristic proved reliable over similar heuristics.

4 Results and Discussion

The EAS-ILS have been tested over 54 instances (15 *Early*, 15 *Middle*, 15 *Late*, 8 *Test*, and 1 *Demo*) from ITC-2021.[3] The EAS-ILS also comprises experimenting the impact of using an elite pool beside implicit solution recombination. It is also compared to the original version of AS algorithm that employs an explicit recombination. As such, the aim of this study is to investigate the effect of some parameters including elite pool on the performance of EAS-ILS and how it maintained a balance between quality and diversity. Using an elite pool, the performance is examined by testing EAS-ILS on sports timetabling problem, in terms of consistency, efficiency, effectiveness as well as generality.

4.1 Experimental Setup

Parameters shown in Table 7 are determined experimentally (e.g., elite pool size) and some are based on the literature. For example, the population size in a generic foraging optimization metaheuristic is preferred to be relatively small [3]. There are many instances provided for the sports timetabling problem. Therefore, it is

[3] https://www.sportscheduling.ugent.be/ITC2021/instances.php.

Table 7 Parameter's settings of EAS-ILS

Parameter	Value
Population size (no. of ants dispatched per iteration)	25
Max. no. of iterations	100,000
Max. no. of non-improvement iterations	100
Initial pheromone	0.01
Pheromone Evaporation rate	0.25, $\rho \in [0, 1]$
Initial Pheromone matrices of *teams* and *timeslots*	0.5
Controlling ratio (exploration versus exploitation)	1.0, $\alpha \in [1.0, 1.1]$
Importance of soft constraints (penalty/weight)	0.3
Initial elite pool size	5 (elite solutions)
Local search	Iterated local search
Elitism (search update per iteration)	Best ant updates global pheromone
Max. no. of employed neighborhood structures per solution	3
Max. no. of possible permutations performed per solution	5
Elite percentage	0.2
Permutation weight	0.2
Permutation rate	0.2
Similarity measurement	Least similar = Best diverse
Selection mechanism in *Step 6*	Roulette wheel

determined to test the EAS-ILS on new instances provided by ITC-2021 (e.g., 53 instances).

Tuning parameters of EAS-ILS is a combinatorial problem to find an approximate optimum, but parameters are all correlated which makes it complex and highly non-linear. Thus, the focus is on testing different general improvements for EAS rather than intensively fine-tuning each parameter to get a slight improvement for a particular instance. It is possible that better solutions would be found using a set of instance-dependent parameters. However, the aim is to design a robust solver that can efficiently solve a variety of instances for the sports timetabling problem.

The consistency test of EAS-ILS is implemented 25 times for every instance from each benchmark dataset with several iterations as termination condition, or a predefined running time in seconds (depends on instance's size), or once an optimal solution is found. Using NetBeans IDE 8.2, a Java code is developed to represent the

EAS-ILS Algorithm on a core i7 machine with 16 GB of RAM. Parameters are tuned in a way that first calibrating the elite pool size, pheromone evaporation rate, and non-improvement iterations number. Then the perturbation operator is fixed. Based on preliminary testing, it is observed that these parameter settings give satisfying results. Refer to Table 7. The size of the memory structures (e.g., elite pool) in the EAS-ILS metaheuristic was intentionally fixed in this study depending on the instance's characteristics. As for the update strategy, it was maintained. The purpose of weight/penalty is to minimize the cost (soft constraints violation) as much as possible. When varying the value of weight, it is intended to expand search space and span the search space to increase the convergence. Please refer to the official website of the competition[4] to overview the problem instances that are available by the ITC-2021 showing the number of teams, slots, and constraints classification for each instance. Table 8 overviews the problem instances that are available by the ITC-2021 showing the number of teams, slots, and constraints classification for each instance.

4.2 Experimental Results

Results obtained for testing EAS-ILS on the sports timetabling problem ITC-2021 are shown in Table 8. This table also shows descriptive statistics for EAS-ILS:

- *Lower bound* (*Best LB*): is the lower bound quality timetables determined by the organizers of ITC-2021. They are presented in the form of pairs, e.g., test_5.xml (0, 2), where 0 means no hard constraints violation (a feasible timetable), while 2 means the soft constraints violation which is close to zero.
- *Best UB*: the best quality timetables found so far that are obtained by the competitors.
- Feasibility (f): a feasible timetable that has the hard constraints violation = 0.
- Quality (q): the best timetable found so far that has the minimal soft constraints violation possible (lower bound).

See Appendix for detailed parametric statistics:

- first quartile ($Q1$): 25%
- standard deviation (σ)
- median (m): 50%
- Median
- third quartile ($Q3$): 75%
- Minimum: the best-found timetable—lowest soft constraints violation
- Maximum: the worst-found timetable—highest soft constraints violation.

For comparison purposes at the outset, the original EAS have been applied and compared against EAS-ILS. They are applied 25 times for each instance using the

[4] https://www.sportscheduling.ugent.be/ITC2021/instances.php.

Table 8 Instances' specifications

Instance	Teams	Slots	Constraints[5]
Demo	4	6	2RR, C, P I SE1 I SC
Test_1	6	10	2RR, C, P I BR2, CA1, CA3, GA1, SE1 I SC
Test_2	6	10	2RR, C, NULL I BR1, CA1, CA2, FA2 I SC
Test_3	6	10	2RR, C, NULL I CA1, CA2, CA3, CA4 I SC
Test_4	6	10	2RR, C, P I BR1, BR2, CA1, CA2, CA3, CA4, FA2, GA1, SE1 I SC
Test_5	16	30	2RR, C, P I BR1, CA1, CA2, CA4, GA1 I SC
Test_6	18	34	2RR, C, NULL I BR1, BR2, CA1, CA2, CA3, CA4, GA1 I SC
Test_7	20	38	2RR, C, P I BR1, BR2, CA1, CA2, CA3, CA4, GA1, SE1 I SC
Test_8	20	38	2RR, C, P I BR2, CA1, CA2, CA3, CA4, GA1, SE1 I SC
Early_1	16	30	2RR, C, P I BR1, BR2, CA1, CA2, CA4, FA2, GA1, SE1 I SC
Early_2	16	30	2RR, C, P I BR1, BR2, CA1, CA3, FA2, GA1 I SC
Early_3	16	30	2RR, C, P I BR1, BR2, CA1, CA2, CA3, FA2, GA1 I SC
Early_4	18	34	2RR, C, P I BR1, BR2, CA1, CA2, CA4, GA1, SE1 I SC
Early_5	18	34	2RR, C, P I BR1, BR2, CA1, CA2, CA3, CA4, GA1, SE1 I SC
Early_6	18	34	2RR, C, P I BR2, CA1, CA2, CA3, CA4, FA2, GA1, SE1 I SC
Early_7	18	34	2RR, C, NULL I BR1, BR2, CA1, CA2, CA3, CA4, GA1, SE1 I SC
Early_8	18	34	2RR, C, NULL I BR1, CA1, CA2, CA3, CA4, FA2, GA1 I SC
Early_9	18	34	2RR, C, NULL I BR1, BR2, CA1, CA2, CA3, FA2, GA1 I SC
Early_10	20	38	2RR, C, P I BR1, BR2, CA1, CA2, CA3, CA4, SE1 I SC
Early_11	20	38	2RR, C, NULL I BR1, BR2, CA1, CA2, CA3, CA4, GA1, SE1 I SC
Early_12	20	38	2RR, C, P I BR1, BR2, CA1, CA2, CA3, CA4, GA1 I SC
Early_13	20	38	2RR, C, NULL I BR1, BR2, CA1, CA2, CA3, GA1 I SC
Early_14	20	38	2RR, C, NULL I BR1, BR2, CA1, FA2, GA1 I SC
Early_15	20	38	2RR, C, NULL I BR1, BR2, CA1, CA2, CA3, CA4, FA2, GA1 I SC
Middle_1	16	30	2RR, C, P I BR1, BR2, CA1, CA2, CA4, SE1 I SC
Middle_2	16	30	2RR, C, P I BR1, BR2, CA1, CA2, CA3, CA4, GA1, SE1 I SC
Middle_3	16	30	2RR, C, NULL I BR1, BR2, CA1, CA2, CA3, CA4, FA2, GA1, SE1 I SC
Middle_4	18	34	2RR, C, P I BR1, CA1, CA2, CA3, CA4, GA1 I SC
Middle_5	18	34	2RR, C, P I BR1, BR2, CA1, CA2, CA3, FA2, GA1 I SC
Middle_6	18	34	2RR, C, P I BR1, BR2, CA1, CA2, CA3, CA4, GA1, SE1 I SC
Middle_7	18	34	2RR, C, NULL I BR1, BR2, CA1, CA2, CA3, CA4, GA1, SE1 I SC
Middle_8	18	34	2RR, C, NULL I BR1, CA1, CA2, CA3, CA4, GA1 I SC
Middle_9	18	34	2RR, C, NULL I BR1, BR2, CA1, CA2, CA3, CA4, FA2, GA1 I SC
Middle_10	20	38	2RR, C, P I BR1, BR2, CA1, CA2, CA4, GA1 I SC
Middle_11	20	38	2RR, C, P I BR1, CA1, CA2, CA3, CA4, FA2, GA1 I SC
Middle_12	20	38	2RR, C, P I BR1, BR2, CA1, CA2, CA3, FA2, GA1, SE1 I SC

(continued)

[5] Detailed formulations are available in: https://www.sportscheduling.ugent.be/RobinX/threeField.php.

Table 8 (continued)

Instance	Teams	Slots	Constraints
Middle_13	20	38	2RR, C, NULL | BR1, CA1, CA2, CA3, CA4, GA1, SE1 | SC
Middle_14	20	38	2RR, C, NULL | BR1, BR2, CA1, CA2, CA3, CA4, FA2, GA1 | SC
Middle_15	20	38	2RR, C, NULL | BR1, BR2, CA1, CA2, CA3, GA1, SE1 | SC
Late_1	16	30	2RR, C, NULL | BR1, CA1, CA2, CA3, CA4, FA2, GA1 | SC
Late_2	16	30	2RR, C, NULL | BR1, BR2, CA1, CA2, CA3, CA4, GA1 | SC
Late_3	16	30	2RR, C, NULL | BR1, BR2, CA1, CA2, CA3, CA4, FA2, GA1, SE1 | SC
Late_4	18	34	2RR, C, P | BR1, CA1, CA4, GA1, SE1 | SC
Late_5	18	34	2RR, C, P | BR2, CA1, CA2, CA3, CA4, FA2, GA1 | SC
Late_6	18	34	2RR, C, P | BR1, BR2, CA1, CA2, CA4, GA1, SE1 | SC
Late_7	18	34	2RR, C, NULL | BR1, BR2, CA1, CA2, CA3, GA1, SE1 | SC
Late_8	18	34	2RR, C, P | BR1, BR2, CA1, CA2, CA3, GA1, SE1 | SC
Late_9	18	34	2RR, C, NULL | BR1, BR2, CA1, CA2, CA3, FA2, GA1 | SC
Late_10	20	38	2RR, C, P | BR1, BR2, CA1, CA2, CA3, CA4, GA1, SE1 | SC
Late_11	20	38	2RR, C, P | BR1, BR2, CA1, CA2, CA3, FA2, GA1 | SC
Late_12	20	38	2RR, C, NULL | BR1, BR2, CA1, CA2, CA3, CA4, SE1 | SC
Late_13	20	38	2RR, C, NULL | BR2, CA1, CA2, CA3, CA4, FA2, GA1, SE1 | SC
Late_14	20	38	2RR, C, NULL | BR1, CA1, CA2, CA3, CA4, FA2, GA1 | SC
Late_15	20	38	2RR, C, NULL | BR1, BR2, CA1, CA3, FA2, GA1 | SC

same parameters settings with a relaxed computational time. Notice that the original EAS does not have an external memory and diversification and intensification mechanisms.

Results shown in Table 9 indicate that the original EAS is unable to produce feasible timetables for most instances and incomparable results to the ones obtained by the organizers (Van Bulk, *aka* RobinX framework[6]) of the ITC-2021 labeled as (*lower bound*). Therefore, an enhancement of the structure of the original version is required, which is achieved by developing the EAS-ILS metaheuristic outperforming the original EAS across all instances. In which the quality of the solution produced for most instances are considered poor compared to the Van Bulk's and other participants' approaches, noticing that the results of the Van Bulk's shown are the best values among the whole number of trials. Although the organizers and participants presented their lower and upper bounds of their obtained results for each instance, they didn't demonstrate the number of trials implemented for each instance. The competition well be concluded in late August 2021 and will announce the winners including their algorithms and methods developed to obtain the best results. In addition, participants did not yet publish their results and statistics of their implementations. Hence, it is unable to fully compare the proposed EAS-ILS metaheuristic against those participants in the competition.

[6] https://www.sportscheduling.ugent.be/RobinX/index.php.

Table 9 The best objective function values obtained by the competitors compared to the best upper bound of the EAS-ILS vs mean of upper bound of the original EAS

Instance*	*Best LB*	*Best UB*	EAS-ILS$ *Best UB* (f, q)	Original EAS& *Median UB* (f, q)
Early_1	(0, 0)	(0, 362)[d]	0, 362	170, 3467
Early_2	(0, 0)	(0, 145)[e]	0, 145	234, 2349
Early_3	(0, 0)	(0, 992)[e]	0, 992	50, 4385
Early_4	(0, 0)	(0, 507)[e]	0, 507	295, 743
Early_5	(0, 0)	(0, 3127)[d]	0, 3127	426, 3133
Early_6	(0, 0)	(0, 3325)[e]	0, 3325	229, 7423
Early_7	(0, 0)	(0, 4763)[d]	0, 4763	354, 4738
Early_8	(0, 0)	(0, 1064)[f]	0, 1064	27, 5470
Early_9	(0, 0)	(0, 108)[d]	0, 108	47, 4987
Early_10	(0, 0)	(0, 3400)[d]	0, 3400	376, 12,627
Early_11	(0, 0)	(0, 4426)[e]	0, 4426	115, 12,826
Early_12	(0, 0)	(0, 380)[g]	0, 380	317, 255
Early_13	(0, 0)	(0, 121)[d]	0, 121	25, 1566
Early_14	(0, 0)	(0, 4)[h]	0, 4	21, 160
Early_15	(0, 0)	(0, 3362)[e]	0, 3362	7, 3422
Middle_1	(0, 0)	(0, 5177)[d]	0, 5177	284, 6215
Middle_2	(0, 0)	(0, 7381)[d]	0, 7381	390, 7349
Middle_3	(0, 0)	(0, 9701)[i]	0, 9701	245, 10,878
Middle_4	(0, 7)	(0, 7)[d]	0, 7	82, 331
Middle_5	(0, 0)	(0, 413)[i]	0, 413	67, 4512
Middle_6	(0, 0)	(0, 1120)[e]	0, 1120	192, 3365
Middle_7	(0, 0)	(0, 1783)[e]	0, 1783	207, 3938
Middle_8	(0, 0)	(0, 129)[d]	0, 129	125, 1165
Middle_9	(0, 0)	(0, 450)[d]	0, 450	99, 5540
Middle_10	(0, 0)	(0, 1250)[d]	0, 1250	336, 1908
Middle_11	(0, 0)	(0, 2446)[e]	0, 2446	292, 8843
Middle_12	(0, 0)	(0, 911)[e]	0, 911	245, 9578
Middle_13	(0, 0)	(0, 252)[e]	0, 252	144, 9830
Middle_14	(0, 0)	(0, 1172)[j]	0, 1172	13, 3488
Middle_15	(0, 0)	(0, 485)[e]	0, 485	30, 9417
Late_1	(0, 0)	(0, 1922)[e]	0, 1922	179, 4709
Late_2	(0, 0)	(0, 5400)[d]	0, 5400	368, 6119
Late_3	(0, 0)	(0, 2369)[d]	0, 2369	100, 6829
Late_4	(0, 0)	(0, 0)[d]	0, 0	51, 1311

(continued)

Table 9 (continued)

Instance*	*Best LB*	*Best UB*	EAS-ILS$ *Best UB* (*f*, *q*)	Original EAS& *Median UB* (*f*, *q*)
Late_5	(0, 0)	(0, 1923)[e]	0, 1923	395, 4438
Late_6	(0, 0)	(0, 923)[d]	0, 923	145, 2839
Late_7	(0, 0)	(0, 1558)[g]	0, 1558	271, 2249
Late_8	(0, 0)	(0, 934)[d]	0, 934	66, 3267
Late_9	(0, 0)	(0, 563)[d]	0, 563	135, 6024
Late_10	(0, 0)	(0, 1945)[e]	0, 1945	342, 11,203
Late_11	(0, 0)	(0, 202)[e]	0, 202	248, 1221
Late_12	(0, 0)	(0, 3428)[e]	0, 3428	120, 12,683
Late_13	(0, 0)	(0, 1820)[f]	0, 1820	115, 9494
Late_14	(0, 0)	(0, 1203)[i]	0, 1203	43, 3341
Late_15	(0, 0)	(0, 20)[g]	0, 20	18, 275
Test_1	(0, 1066)	(0, 1066)[a]	0, 1066	4, 1122
Test_2	(0, 176)	(0, 176)[a]	0, 176	3, 206
Test_3	(0, 1253)	(0, 1253)[a]	0, 1253	33, 1255
Test_4	(0, 4535)	(0, 4535)[a]	0, 4535	37, 4481
Test_5	(0, 2)	(0, 2)[a]	0, 2	506, 3
Test_6	(0, 0)	(0, 3144)[b]	0, 3144	166, 4987
Test_7	(0, 0)	(0, 4421)[c]	0, 4421	314, 13,216
Test_8	(0, 0)	(0, 3165)[b]	0, 3165	264, 11,678
Demo	(0, 0)	(0, 0)[a]	0, 0	0, 2

Note Execution time is not reported yet by the organizers as well as competitors
[a] Obtained by David Van Bulk using Integer Programming model
[b] Obtained by Team DITUoI@Arta—Angelos Dimitsas, Christos Valouxis and Christos Gogos
[c] Obtained by Team TU/e using Integer Programming
[d] Obtained by Team UoS (the winner team in the first round of the competition)
[e] Obtained by Team DES
[f] Obtained by George Fonseca and Tulio A. M. Toffolo using descent heuristic method
[g] Obtained by Saturn
[h] Obtained by Team Gionar
[i] Obtained by Team MODAL using integer programming
[j] Obtained by Team Udine
* Feasibility is guaranteed for all instances with lower bound of (0, 0) except middle_4, test_1, test_2, test_3, test_4, and test_5
$ Statistics only for feasible timetables
& Statistics including both feasible and infeasible timetables

Both EAS-ILS and original EAS have used the same local search within. It is obvious that the hybridization in EAS-ILS including the local search routine and the selection mechanism have a greater effect on enhancing the quality of solutions (see q in Table 9) while maintaining a search diversity (see $\boldsymbol{\sigma}$ and w in Table 9). Notice that the EAS-ILS is an enhanced version of the original Elitist-AS. On the other hand, the original EAS performs only an opened global ant's update. This is the main drawback of the algorithm which hinders the search process from being balanced between quality and diversity for a smooth and controlled convergence. For this exact reason, a hybrid enhanced EAS version is developed in this work, EAS-ILS, to control the convergence by balancing space exploration and solution exploitation.

Since the competition is not closed yet, the author could not compare the EAS-ILS's results against those competitor optimization approaches, they are not reported or published yet. So, it is limited to compare this study's results to the original EAS and Van Bulk's reported results. Van Bulk et al. [18] have proposed and presented three types of heuristics that handle fairness issues. First, a relax-and-fix based heuristics using team- and time-based variable groupings. Second, an adaptive large neighborhood method that repeatedly rectifies a timetable. Third, a memetic algorithm that makes use of local search to reschedule all home games of a team. For numerous artificial and real-life instances, these heuristics generate high-quality timetables in terms of the overall unfairness using considerably less computational resources compared to IP models. Finally, they looked at the trade-off between a timetable that is as fair as possible for the league overall, versus a timetable that equitably splits its unfair aspects over the teams.

Table 9 shows the results obtained by the EAS-ILS based on the parameters presented in Table 7 and compared to the original EAS and the Van Bulk's approach (including the best-known results) applied over 54 ITC-2021 instances. Best results obtained by EAS-ILS are presented in bold type which are outperforming the original EAS and are the same as the results obtained by Van Bulk's approach.

It is shown that EAS-ILS can produce feasible timetables for all instances on most runs (trials). From Table 9, EAS-ILS algorithm has obtained feasible timetables across all instances w.r.t. hard constraints satisfaction (violations = 0), while in term of quality, the algorithm has obtained similar or competitive timetables to the ones produced by RobinX. The experiments ran 25 times across all 54 instances as follows:

- 15 s for instances of the size (Test: $4 \times 6, 6 \times 10, 16 \times 30, 18 \times 34, 20 \times 38$)—as an experimental setup,
- 90 s for instances of the size (early, middle, late: 16×30),
- 900 s for instances of the size (early, middle, late: 18×34), and
- 9000 s for instances of the size (early, middle, late: 20×38).

These time settings are predetermined experimentally with least possible and fair ratio. Of course, if the algorithm ran for a longer time, it might be able to obtain better quality timetables.

4.3 Discussion

It is shown that EAS-ILS algorithm is relatively consistent in producing feasible and high-quality results for all instances (indicated by σ, $Q1$ and $Q3$, and by the small differences between mean and lower and upper bounds), see Box and Whisker plots in Appendix for details. The closer the value of the median to the best obtained result, the more consistent the approach.

The effectiveness of EAS-ILS maybe due to an efficient search space exploration and effective solution space exploitation. This may present the ability of EAS-ILS to maintain a balance between the diversity and quality of the search. For clarification of the analysis, this study relies on the fact that the standard deviation tells us how tightly a set of values is clustered around the average of those same values. In which, a standard deviation is a measure of dispersal, or variation, in a group of numbers. Hence, it gives us some indication about the consistency of the algorithm.

For translating the consistency of the algorithm, Table 10 presents the statistical tests and descriptions and factors that indicate the performance of the EAS-ILS for all instances across 25 trial runs. Results of the *t*-test statistical comparison of EAS-ILS against the original EAS. Generally, the *t*-test statistical comparison is carried out with 24 degrees of freedom at a 0.05 level of significance.

Note the mean (mean difference) for each instance group shows the average value of the differences between the best results obtained across the 25 runs for both presented algorithms. To find the results of the for the difference in the means (hypothesis test) of instances, the *p*-value is found for the test. The *p*-value is labelled as "Sig. (2-tailed)" value in the output (where "Sig." stands for significance level) which refers to the *t*-test for Equality of Means. For example, the *p*-value is 0.000 for the Early_14 instance in Table 10, which implies that the difference in means is statistically highly significant at *p*-value ≤ 0.001, significant at *p*-value ≤ 0.01, and marginally significant at *p*-value ≤ 0.05 levels. This is applicable to all instances presented in the table.

The exploration in the algorithm is performed by utilizing an external memory (e.g., elite pool) that contains a collection of elite solutions found during the search to prevent premature convergence of the search. The elite solutions in the memory are utilized by the diversification mechanism to regenerate new solutions. The diversification mechanism reinitializes the generation of ants to divert the search perhaps toward the global solution when it stagnates. The memory provides good solutions that may leads to the global solution. Intensive solution exploitation is performed by further explore the elite solution's (obtained from the iterated local search routine) neighbors to significantly enhance the quality of the solution.

Table 10 Samples t-test for EAS-ILS across 25 run/instance

	Test value = 0					
	t	df	Sig. (2-tailed)	Mean difference	95% confidence interval of the difference	
					Lower	Upper
Test8	31.666	24	0.000	3580.440	3347.08	3813.80
Early1	55.989	24	0.000	380.720	366.69	394.75
Early2	30.215	24	0.000	151.800	141.43	162.17
Early3	127.614	24	0.000	1019.840	1003.35	1036.33
Early4	78.808	24	0.000	516.240	502.72	529.76
Early5	533.375	24	0.000	3154.840	3142.63	3167.05
Early6	46.538	24	0.000	3560.600	3402.69	3718.51
Early7	235.955	24	0.000	4877.240	4834.58	4919.90
Early8	167.212	24	0.000	1103.600	1089.98	1117.22
Early9	22.755	24	0.000	119.200	108.39	130.01
Early11	294.116	24	0.000	4453.920	4422.67	4485.17
Early12	29.946	24	0.000	499.000	464.61	533.39
Early13	141.550	24	0.000	123.400	121.60	125.20
Early14	5.185	24	0.000	43.360	26.10	60.62
Early15	70.050	24	0.000	3629.160	3522.23	3736.09
Middle6	2216.244	24	0.000	1122.800	1121.75	1123.85
Middle7	26,881.545	24	0.000	1783.120	1782.98	1783.26
Middle11	382.308	24	0.000	2485.000	2471.58	2498.42
Middle13	2476.162	24	0.000	252.520	252.31	252.73
Middle15	559.086	24	0.000	487.400	485.60	489.20
Late1	1023.340	24	0.000	1923.880	1920.00	1927.76
Late5	1617.298	24	0.000	1936.440	1933.97	1938.91
Late10	683.716	24	0.000	1950.160	1944.27	1956.05
Late11	472.138	24	0.000	205.800	204.90	206.70
Late12	199.813	24	0.000	3459.320	3423.59	3495.05
Late14	4913.676	24	0.000	1203.600	1203.09	1204.11

Note Some instances are missing from the table, due to their standard deviation equals to zero, their t-test can be computed (*t cannot be computed because $\sigma = 0$*)

The diversification mechanism assists the penalty boundaries to further diversify the ant search by exploring different regions of the search space, whilst the intensification mechanism assists the updating the pheromone to further intensify the search around the elite solution space. This may eventually maintain a balance between diversity and quality of the search. These mechanisms proved to be effective and significant in producing good quality results after all. In contrast, the original EAS has no effective diversity control mechanism, and it totally relies on the local search routine to improve the quality of solutions, which is not enough to maintain a balanced convergence. In the EAS, elite solutions permutations are performed implicitly based on the changes of the pheromone. Specifically, the premature convergence occurred in the intensification phase, where the neighborhoods of an elite solution are randomly explored. Therefore, the solution recombination relies on limited information about the elite solution. This prevents it from estimating the effectiveness of performing a permutation to the solution. In addition, the population size in the EAS is maintained relatively small. Hence, some information about the current elite solution is lost during the search in successive iterations where it does not guarantee an effective convergence toward optimality.

In general, EAS-ILS, besides diversification and intensification mechanisms, elitism is used in to manage the pheromone updates, where only best ant in each iteration is used to update the pheromone. Therefore, they are significantly crucial guidance of the search. The diversification mechanism avoids premature convergence of the search, which is accomplished by the penalty/weight and generating new solutions from those in the elite pool, whilst the intensification mechanism intensifies the search around an elite solution, which is accomplished by the pheromone update and adding high quality solutions into the elite pool.

Finally, this research work has contributed the following:

A balance between diversification and intensification of the search is maintained by hybridizing original EAS with a dynamic update strategy; diversification and intensification mechanisms; an elite pool containing elite solutions; and the iterated local search to explore and exploit neighborhoods of an elite solution.

An effective exploitation of a memory is achieved by selecting only elite solutions (local optima) and stores them in the elite pool to generate a new high-quality population (ants). The pool is dynamically updated once there is a much diverse solution than the worst diverse solution stored in the pool, where the worst will be replaced by the better new solution. Hence, the EAS-ILS has produced feasible and high-quality timetables for all instances.

5 Conclusions

An enhanced version of the EAS named as EAS-ILS is proposed, which is a hybrid version containing a dynamic and adaptive diversification mechanisms utilizing elite pool and iterated local search. It is observed when utilizing the elite pool alongside the local search routine, the search converges in stability and very quickly towards better quality solutions. This suggests that once the local search is employed a smaller population size can effectively explore the search space, which may lead to better performance than generating numerous neighborhood structures from randomly constructed solutions. Therefore, generating new solutions from elite pool rather than from scratch. In addition, for each iteration, randomly updating the value of weight has a significant role to prevent the ants from being stuck in their local or global search.

Briefly stated, results demonstrate the superiority of EAS-ILS over the original EAS in terms of consistency, efficiency, and generality. This is mainly factored by using elite pool in generating diverse and high-quality timetables that are also consistent as opposed to the original EAS. The observations prove the capacity of EAS-ILS in generating feasible and high-quality timetables across all instances. So, EAS-ILS can be used for further research and practice in sports timetabling and other complex scheduling problems.

Although, the performance of EAS-ILS is sufficient regarding a guided convergence, it had some drawbacks. Such as, a clear and meaningful representation of pheromones regarding the team assignments. That is, it is hard to estimate the effects of improvements made to a global optimum. This is most unlikely to be achieved due to the implicit solution recombination operators and the indirect solution representation used in the ACOs. These operators are very useful in a guided sampling of the search space using only information about the structure of the global optimum. Then, it can be only applied to indirect representation, which is not suitable for measuring the diversity of the search in term of measuring similarities between solutions in the population within a Hamming or Euclidean space (e.g., as in GAs). In the perspective of this study, any ACO considers the neighborhood space since pheromones are indirectly representing the team-timeslot assignments in a timetable using floating numbers to indicate feasibility or cost. Therefore, to utilize the collection of elite solutions effectively and manipulate a dynamic population. In future work the author might introduce a new approach concerning the drawbacks of EAS-ILS, such as the diversity measurements of the population to determine the right degree of the search diversity. This can be achieved by measuring similarities within a Euclidean space to perform an explicit recombination of elite solutions. In addition, for fair and more critical scientific research study, comparing the ESA-ILS against algorithms participated in the competition will reconsidered once they are announced and published. The author would also adopt two newly developed optimization metaheuristics (arithmetic optimization by [33, 34]) and test them on the datasets of ITC2021.

Appendix

(See Figs. 3, 4, 5, 6 and 7).

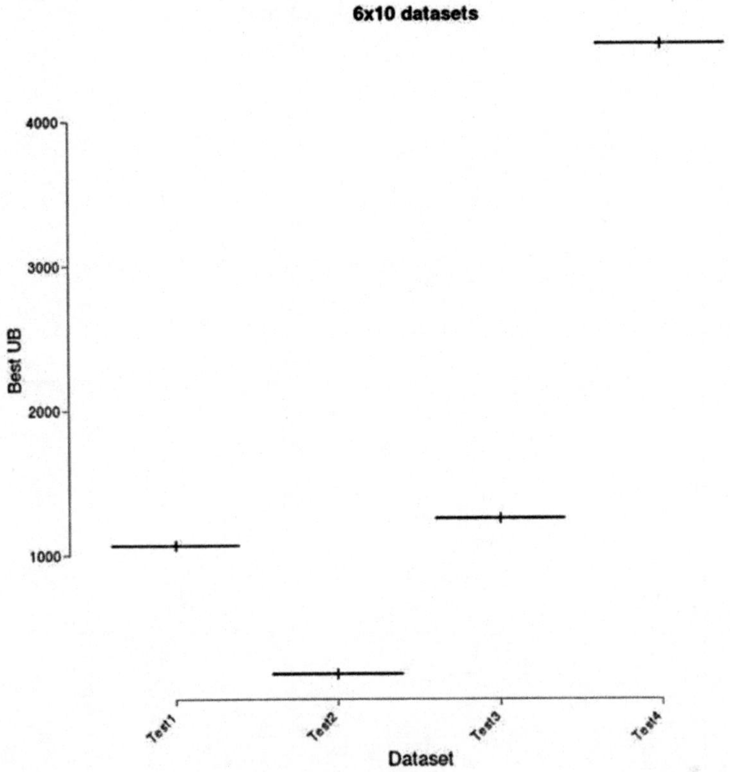

Fig. 3 Statistics and Box-Whisker Plot for 6 × 10 Datasets

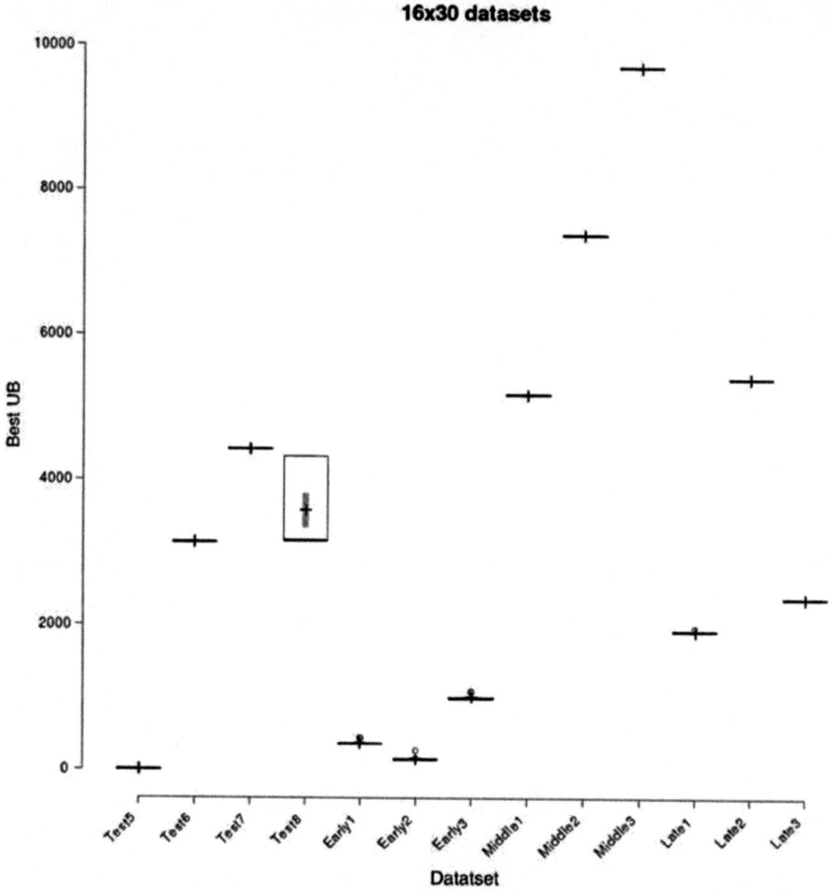

Fig. 4 Statistics and Box-Whisker Plot for 16 × 30 Datasets

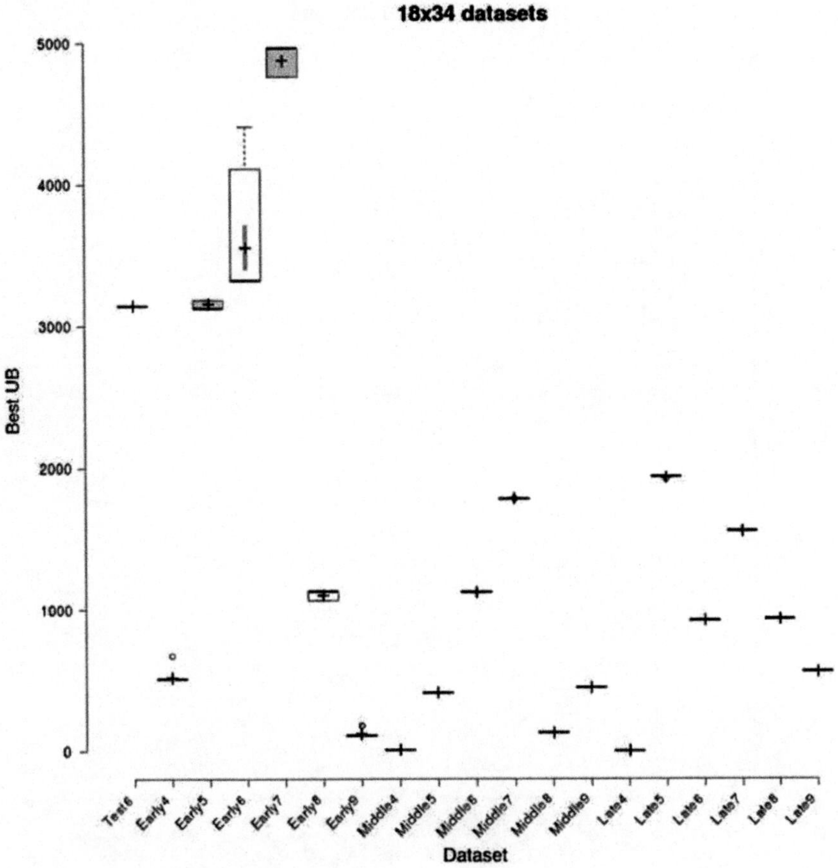

Fig. 5 Statistics and Box-Whisker Plot for 18 × 34 Datasets

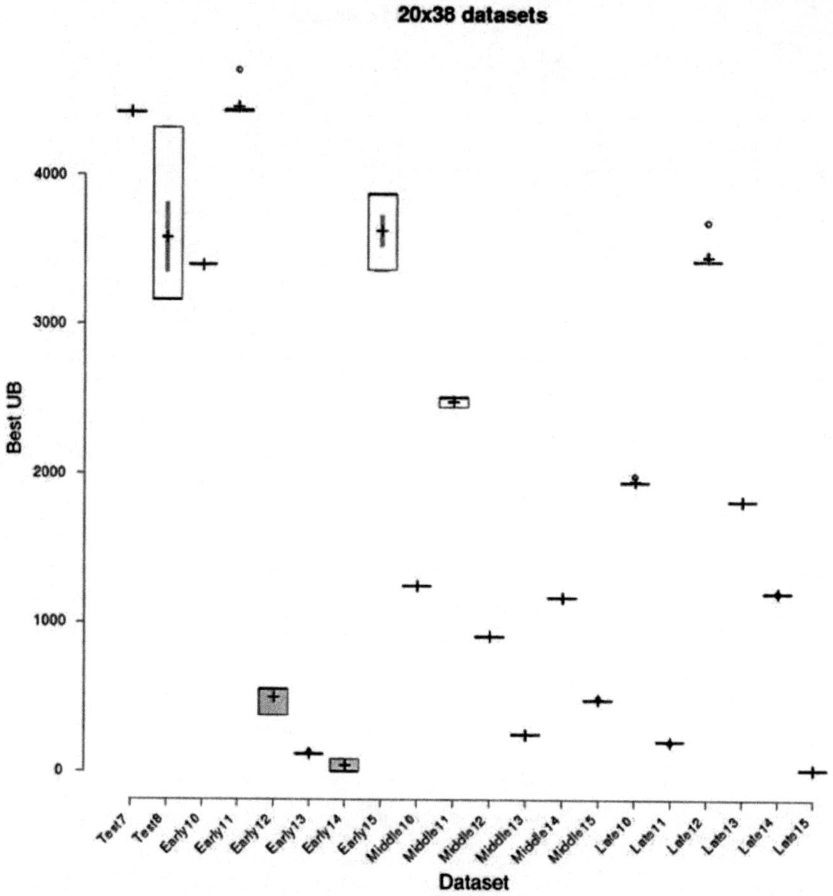

Fig. 6 Statistics and Box-Whisker Plot for 20 × 38 Datasets

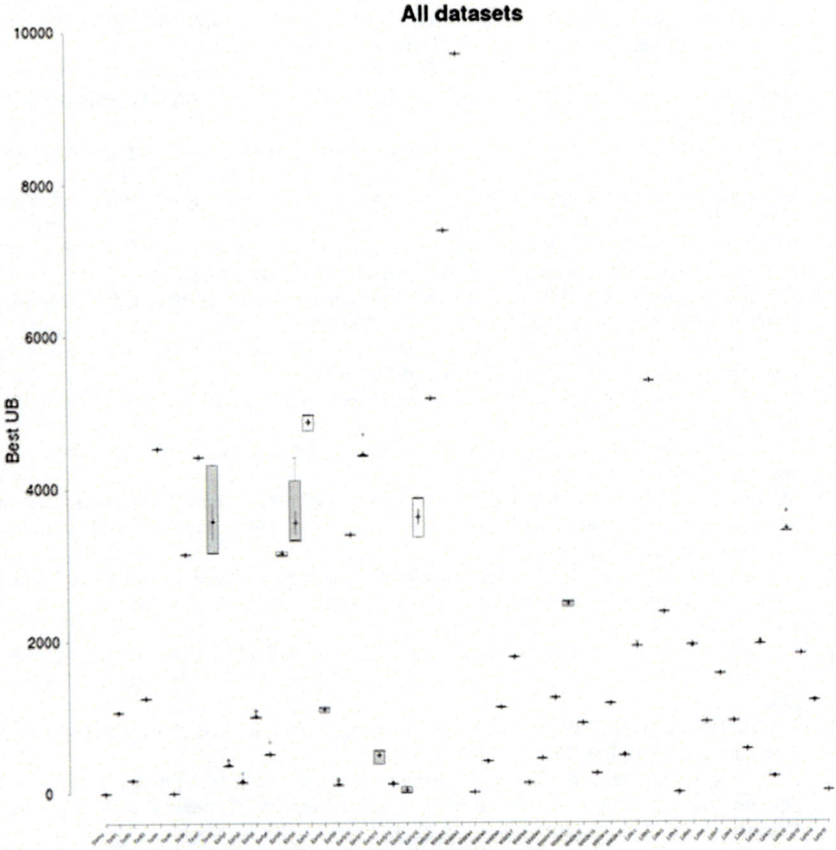

Fig. 7 Statistics and Box-Whisker Plot for all Datasets

References

1. M. Dorigo, V. Maniezzo, A. Colorini, Positive feed-back as a search strategy. Tech. Rept. 91-016 (Dipartmento di Elettronica, Politecnico di Milano, Italy, 1991)
2. G. Jaradat, M. Ayob, An elitist-ant system for solving the post-enrolment course timetabling problem, in *The 2010 International Conference on Database Theory and Application (DTA 2010)* (2010), pp. 167–176
3. G. Jaradat, M. Ayob, I. Almarashdeh, The effect of elite pool in hybrid population-based meta-heuristics for solving combinatorial optimization problems. Appl. Soft Comput. **44**, 45–56 (2016). https://doi.org/10.1016/j.asoc.2016.01.002
4. D. Van Bulck, D. Goossens, J. Beliën, M. Davari, International timetabling competition 2021: sports timetabling. itc2021.ugent.be
5. M.A. Trick, Integer and constraint programming approaches for round robin tournament scheduling (2004)
6. A. Aggoun, A. Vazacopoulos, Solving sports scheduling and timetabling problems with constraint programming, in *Economics, Management and Optimization in Sports*, ed. S. Butenko et al. (Springer-Verlag, Berlin, Heidelberg, 2004)

7. M.A. Trick, A schedule-then-break approach to sports timetabling, in *PATAT 2000. LNCS*, vol. 2079, ed. E. Burke, W. Erben (Springer, Heidelberg 2001), pp. 242–253
8. J. Schönberger, D.C. Mattfeld, H. Kopfer, Memetic algorithm timetabling for non-commercial sport leagues. Eur. J. Oper. Res. **153**(1), 102–116 (2004). ISSN 0377-2217. https://doi.org/10.1016/S0377-2217(03)00102-4
9. A. Duarte, C.C. Ribeiro, S. Urrutia, A hybrid ILS heuristic to the referee assignment problem with an embedded MIP strategy. Lect. Notes Comput. Sci. **4771**, 82–95 (2007)
10. G. Durán, Sports scheduling and other topics in sports analytics: a survey with special reference to Latin America. TOP **29**, 125–155 (2021). https://doi.org/10.1007/s11750-020-00576-9
11. M. Grobner, P. Wilke, S. Buttcher, A standard framework for timetabling problems, in *PATAT 2002, LNCS 2740*, ed. E. Burke, P. De Causmaecker (2003), pp. 24–38
12. G. Kendall, S. Knust, C.C. Ribeiro, S. Urrutia, Scheduling in sports: an annotated bibliography. Comput. Oper. Res. **37**(1), 1–19 (2010). ISSN 0305-0548
13. S. Knust, Classification of Literature on Sports Scheduling (2020), http://www2.inf.uos.de/knust/sportssched/sportlit_class/. Accessed 25 March 2021
14. D.R. Goossens, F.C.R. Spieksma, Soccer schedules in Europe: an overview. J. Sched. **15**, 641–651 (2011)
15. G.L. Nemhauser, M.A. Trick, Scheduling a major college basketball conference. Oper. Res. **46**, 1–8 (1998)
16. D. Van Bulck, D. Goossens, J. Schönberger, M. Davari, ITC2021—Sports Timetabling Problem Description and File Format (2020), https://www.sportscheduling.ugent.be/ITC2021/images/OrganizationITC2021_V7.pdf
17. D. Van Bulck, D. Goossens, J. Schönberger, M. Guajardo, An instance data repository for the round-robin sports timetabling problem. Manag. Labour Stud. **45**(2), 184–200 (2020). https://doi.org/10.1177/0258042X20912108
18. D. Van Bulck, D. Goossens, J. Schönberger, M. Guajardo, RobinX: a three-field classification and unified data format for round-robin sports timetabling. Eur. J. Oper. Res. **280**, 568–580 (2019)
19. T.A.M. Toffolo, J. Christiaens, F.C.R. Spieksma, G.V. Berghe, The sport teams grouping problem. Ann. Oper. Res. **275**, 223–243 (2019)
20. R. Linfati, G. Gatica, J.W. Escobar, A flexible mathematical model for the planning and designing of a sporting fixture by considering the assignment of referees. Int. J. Ind. Eng. Comput. **10**(2), 281–294 (2019). https://doi.org/10.5267/j.ijiec.2018.6.004
21. T. Wauters, S. Van Malderen, Decomposition and local search-based methods for the traveling umpire problem. Eur. J. Oper. Res. **238**(3), (2014)
22. M. Triska, N. Musliu, An improved SAT formulation for the social golfer problem. Ann. Oper. Res. **194**, 427–438 (2012)
23. C.C. Ribeiro, Sports scheduling: problems and applications. Int. Trans. Oper. Res. **19**, 201–226 (2012). https://doi.org/10.1111/j.1475-3995.2011.00819.x
24. M. Goerigk, S. Westphal, A combined local search and integer programming approach to the traveling tournament problem. Ann. Oper. Res. **239**, 343–354 (2016)
25. M. Ayob, G. Jaradat, Hybrid ant colony systems for course timetabling problems, in *2009 2nd Conference on Data Mining and Optimization* (2009), pp. 120–126. https://doi.org/10.1109/DMO.2009.5341898.
26. H. Crauwels, D. Oudheusden, Ant Colony Optimization and Local Improvement (2003), https://www.researchgate.net/publication/228716201_Ant_colony_optimization_and_local_improvement
27. P. Chen, G. Kendall, G.V. Berghe, An ant based hyper-heuristic for the travelling tournament problem, in *2007 IEEE Symposium on Computational Intelligence in Scheduling* (2007), pp. 19–26. https://doi.org/10.1109/SCIS.2007.367665
28. D.C. Uthus, P.J. Riddle, H.W. Guesgen, Ant colony optimization and the single round robin maximum value problem, in *Ant Colony Optimization and Swarm Intelligence. ANTS 2008*, ed. M. Dorigo, M. Birattari, C. Blum, M. Clerc, T. Stützle, A.F.T. Winfield. Lecture Notes in Computer Science, vol. 5217 (Springer, Berlin, Heidelberg 2008). https://doi.org/10.1007/978-3-540-87527-7_23

29. N. Kumyaito, P. Yupapin, K. Tamee, Planning a sports training program using adaptive particle swarm optimization with emphasis on physiological constraints. BMC Res. Notes **11**, 9 (2018). https://doi.org/10.1186/s13104-017-3120-9
30. A. Madureira, N. Sousa, I. Pereira, Swarm Intelligence for Scheduling (2011)
31. R.V. Rasmussen, M.A. Trick, Round robin scheduling—a survey. Eur. J. Oper. Res. **188**(3), 617–636 (2007)
32. G.M. Jaradat, Hybrid elitist-ant system for a symmetric traveling salesman problem: case of Jordan. Neural Comput. Appl. **29**, 565–578 (2018). https://doi.org/10.1007/s00521-016-2469-3
33. L. Abualigah, A. Diabat, S. Mirjalili, M. Abd Elaziz, A.H. Gandomi, The arithmetic optimization algorithm. Comput. Methods Appl. Mech. Eng. **376**, 113609 (2021). ISSN 0045-7825. https://doi.org/10.1016/j.cma.2020.113609
34. L. Abualigah, D. Yousri, M. Abd Elaziz, A.A. Ewees, M.A.A. Al-qaness, A.H. Gandomi, Aquila optimizer: a novel meta-heuristic optimization algorithm. Comput. Ind. Eng. **157**, 107250 (2021). ISSN 0360-8352. https://doi.org/10.1016/j.cie.2021.107250

Swarm Intelligence Algorithms-Based Machine Learning Framework for Medical Diagnosis: A Comprehensive Review

Essam Halim Houssein, Eman Saber, Yaser M. Wazery, and Abdelmgeid A. Ali

Abstract When building medical diagnosis software, one of the most difficult challenges in disease prediction. Machine Learning (ML) approaches have proven to be effective in a range of applications, including medical diagnostics. ML applications that require a high level of speed and accuracy are becoming more frequent. With these applications, the curse of dimensionality, in which the number of features exceeds the number of patterns, is a serious issue. One of the dimensionality reduction strategies that might increase task accuracy while reducing computational complexity is Feature Selection (FS). The goal of the FS method is to locate a subset of features that have the least inner similarity and are most relevant to the target class. By building classifier systems that can help clinicians forecast and detect diseases at an early stage, ML algorithms can aid in the solution of health-related problems. Swarm Intelligence (SI) algorithms are utilized in the detection and treatment of diseases. We can increase the reliability, performance, accuracy, and speed of diagnosing disease on the existing system by merging ML and SI methods. The goal of this paper is to use ML and SI to diagnose diseases. This paper discusses various recently discovered SI and ML algorithms that have been effectively applied to a range of optimization problems, focusing on their implementation areas, strengths, and weaknesses.

Keywords Classification · Machine learning (ML) · Feature selection (FS) · Medical diagnostic · Swarm Intelligence Algorithms (SI)

1 Introduction

When doing analysis or attempting to construct relationships between many features, people may make mistakes. It becomes more difficult for them to discover answers to difficulties as a result of this. In any data set, Machine Learning (ML) [1] methods

E. H. Houssein (✉) · E. Saber · Y. M. Wazery · A. A. Ali
Faculty of Computers and Information, Minia University, Minia, Egypt
e-mail: essam.halim@mu.edu.eg

E. H. Houssein et al. (eds.), *Integrating Meta-Heuristics and Machine Learning for Real-World Optimization Problems*, Studies in Computational Intelligence 1038,
https://doi.org/10.1007/978-3-030-99079-4_4

employ the same set of features to describe an instance. Choosing which learning algorithm to use is an important step.

Algorithms for solving optimization problems and challenges are known as optimization algorithms [2–6]. For instance, Microarray Cancer Classification [7], Heartbeat classification [8–11], Human emotion recognition [12], Deep learning [13, 14], Feature Selection [15–17], Energy [18–21], Fuel cell [22], Photovoltaic (PV) [23–25], Global optimization problems [26, 27], Image segmentation [28–32], Wireless networks [33], Cloud computing [34], Bioinformatics [35], and Drug design [36, 37]. There are two types of optimization algorithms: deterministic and stochastic. Deterministic optimization algorithms often find the same solution for a given problem, while stochastic optimization algorithms may find different solutions due to the randomness factor. Stochastic algorithms can prevent local optima by using randomness, but they are less reliable. However, by balancing exploration and exploitation, as well as increasing the number of runs, they can be enhanced [38]. Individual-based and collective-based stochastic optimization algorithms are two more types of stochastic optimization algorithms. The collective-based algorithm produces several random solutions and develops them over time, while the individual-based algorithm only operates for a single solution.

The Swarm Intelligence (SI) algorithm is one of the categories of optimization techniques. Beni [39] developed the term SI to describe a cellular robotic system in which simple agents organize themselves through neighborhood interactions, and it was later adopted by [40]. In SI, it's common to have a set of feasible solutions that can be improved over time with the help of previous solutions. Exploration and exploitation should be supported by the process to get deep into a global optimum without being trapped by local optima. SI is an Artificial Intelligence (AI) technique that has the collective actions of a decentralized, self-organized machine. SI has more benefits such as scalability, adaptability, collective robustness, human simplicity, and also can solve complex problems. SI systems are typically made up of a population of simple agents that communicate with one another and their surroundings on a local level. Nature, particularly biological systems, is a common source of inspiration [41]. SI algorithms are effective in solving difficult optimization problems in stationary environments, including Ant Colony Optimization (ACO) [42], Particle Swarm Optimization (PSO) [43], Firefly Algorithms (FA) [44], Slime Mould Algorithm (SMA) [45], Sine Cosine Algorithm (SCA) [46], Moth-Flame Optimization Algorithm (MFO) [47], Whale Optimization Algorithm (WOA) [48], Genetic Algorithm (GA) [49], Search and Rescue optimization algorithm (SAR) [50], and many others.

Various computer researchers have tried to employ various strategies in the medical field to increase the accuracy of data categorization for given data, classification techniques that would better generate adequate information to identify possible patients and therefore improve diagnosis accuracy.

In summary, the main contribution of this paper is the combination of ML classification with SI to increase disease diagnosis speed and accuracy.

The paper is structured as follows; Sect. 2 provides a summary of the relevant work. Section 3 illustrates SI algorithms and ML fundamentals and background. Application areas of SI algorithms and ML in medical diagnosis are illustrated in Sect. 4. The Conclusion and future work are demonstrated in Sect. 6.

2 Literature Reviews

Several studies have used ML, and evolutionary algorithms to diagnose a variety of diseases. While the efficiency of the used algorithms improved in diagnosing a specific disease, no generalized model that can improve the accuracy of diagnosing a broad range of datasets has been created. Many researchers have used SI algorithms and ML in medical datasets in recent years, increase the reliability of disease prediction. The authors of [51] present a new model that optimizes the values of Extreme Learning Machine's (ELM) input weights and hidden neurons using the Competitive Swarm Optimizer (CSO). The classical ELM and the regularized form of ELM are also considered. The model aims to reduce the number of neurons in the hidden layer to improve generalization, stabilize the classifier, and create more compact networks. Based on 15 medical classification issues, the proposed model is tested. Experimental findings show that the proposed model can achieve better generalization efficiency and stability with a smaller number of hidden neurons.

In the same context, [52], Wavelets transform and singular value decomposition incorporating SVM, the authors proposed approach has been tested using five benchmark medical data sets, each representing a different disease. The proposed method's expected outcome is to reduce a subset of features to achieve a satisfactory disease diagnosis with the highest precision, sensitivity, and specificity across a wide range of diseases. The authors of [53], proposed the Directed Bee Colony (DBC) algorithm that is used to optimize a high-performance Artificial Neural Network (ANN). Cancer, diabetes, and heart disease are all diagnosed using methodological research. For diagnosis, the performance review is based on three main factors: running time, uniqueness, and classification precision. On 16 other algorithms, a systematic comparison of different meta-analytic studies is made. DBC was ranked second and first in two separate criteria systems. In terms of uniqueness of answer, DBC has been scored first both times. Also, by GA and PSO, the running time is roughly 101% and 21% more than DBC.

In [54], the authors proposed a hybrid algorithmic approach for heart disease prediction. Integrating the K-means clustering algorithm and ANN, this paper provides an effective prediction technique for determining and extracting unknown information about heart disease. The k-means algorithm is used to group different attributes and the back propagation technique in neural networks is used to predict. The primary purpose of this study is to develop a prototype for better predicting heart problems. To prevent CNNs from slipping into the local minimum and to train efficiently, an Exponentially Decreasing Learning Rate (EDLR) architecture is used. The accuracy of the proposed model is quite competitive, according to the study's findings. The authors proposed an indigenous diabetes detection equipment in [55]. There are two parts to the proposed methodology. The Lung Cancer data set is acquired in Phase I, and the data is interpreted using two different ways. Principal component analysis and PSO are used in Phase-II to execute the FS (PSO). C4.5 DT, NB, ID3 DT, logistic regression, and kNN are then used to achieve classification. The proposed method uses less CPU time and has a high degree of accuracy. It could also be used to detect medical conditions other than diabetes.

More related works are summarized in Table 1.

3 Basics and Background

The high-dimensionality problem of a data set with a feature subset size much larger than the pattern size is a critical issue with ML techniques. Supervised and unsupervised techniques [64] are the two most common kinds of feature selection methods. In supervised methods, a collection of training data is available, each of which is determined by taking features values and the classmark, however, in unsupervised methods, training data does not include a class tag. In general, FS algorithms perform better and more consistently in the supervised mode. As a result, selecting a function in the unsupervised mode is more difficult, and several studies have focused on this field [65]. To achieve these goals, several different search techniques are used. Single-objective and multi-objective approaches are the two types of methods that exist. The population is optimized in single-objective models by using only one objective function. Therefore, the optimization algorithm's accuracy is significantly influenced by the choice of goal and specification of the fitness function. Furthermore, several optimization problems have several goals, and defining a objective function with just one goal decreases optimization efficiency. Consider a variety of goals in the fitness role of the FS issue as one solution to solve these challenges [66]. Filter, wrapper, embedding, and hybrid models are four types of feature sets to choose from. The subsections that follow clarify each of these categories.

1. *Filter model:* Calculates feature relevance using probabilistic data properties and performs the FS process independently of ML algorithms. The filter model is divided into two groups depending on how features are evaluated: univariate and multivariate. These univariate approaches assume that features are independent of one another, and potential connections between features are not considered. While multivariate techniques evaluate the suitability of features in terms of their interaction.
2. *Wrapper model:* The efficiency of the selected function subset has been measured using a classifier in the wrapper-based methods [67]. In this approach, a search method is utilized to discover the best feature subset and a classifier calculates a subset of features at each step of the search strategy. Finally, as the final feature subset, the best-produced feature subset is chosen.

Table 1 Related works based on machine learning and optimization algorithms

Refs.	Dataset used	Models developed	Results obtained
[56]	Four well-known medical datasets	The Fruit Fly optimization Algorithm (FOA) with SVM called (FOA-SVM)	FOA-SVM is a computationally efficient and reliable approach to solving medical data classification problems
[57]	Cleveland heart disease database	SVM with PSO called (PSO-SVM) based features selection method to predict the heart disease	SVM's Accuracy = 79.35%, (PSO-SVM)'s Accuracy = 84.36%
[58]	Heart disease dataset	NB with Particle Swarm Optimization (PSO) called (NB+PSO) approach	NB's Accuracy = 79.12%, (NB+PSO)'s Accuracy = 87.91%
[59]	Diabetes type-II dataset from UCI	The "Diabetes Diagnoser" expert system is driven by four AI-based algorithms	Accuracy = 97.34%
[60]	Eighteen benchmark datasets	Active learning (AL) and PSO algorithms (AL–PSO)	Enhanced the performance and reduce the cost of labeling while improving the efficiency of the classifier
[61]	Five benchmarked medical datasets from UCI	Integrates GA with kNN classifier	Achieve good results with a p value of 0.0011 in the t-Test
[62]	Iraqi cancer patients in 2010–2012 dataset	Grasshopper Optimization Algorithm (GOA) with SVM called GOA-SVM	Accuracy up to 100%
[63]	Three medical datasets	Integrates Salp Swarm Algorithm (SSA) with SVM called SSA-SVM	SSA-SVM outperformed another competitors algorithms and enhanced results in performance evaluation metrics
[15]	Eighteen diseases datasets	Henry Gas Solubility Optimization (HGSO) is combined with SVM called HGSO-SVM	HGSO-SVM accuracy = 88.88%
[55]	Pima indian diabetes dataset	Principal Dimensionality Reduction (PCA), PSO as feature selection, C4.5 Decision Tree (C4.5 DT) as classification methods	PCA_C4.5 DT accuracy = 95.58%, PSO_C4.5 DT accuracy = 94.01%

3.1 Swarm Intelligence Algorithms Overview

SI algorithms are computational and behavioral metaphors for addressing search and optimization problems that use collective biological patterns provided by social insects (moths, ants, bees, and others) as stimulus to model algorithmic solutions. In recent years, several algorithms inspired by natural events have been proposed, and particular meta-heuristic search algorithms with population-based frameworks are capable of tackling high-dimensional optimization problems [68]. Figure 1 shows research utilizing SI compared to the whole volume of research over the last decade. Figure 1a is based on a Scopus search using the terms "Swarm Intelligence". Figure 1b emphasizes the importance of SI algorithms study in a variety of fields. This section introduces a number of SI algorithms, emphasizing important variants and implementations.

Some of the algorithms accessible are detailed as follows:

1. *Genetic Algorithm (GA):* The GA [49] is a search optimization algorithm that is based on natural selection mechanics. The basic idea behind this algorithm is to replicate the principle of 'survival of the fittest'; simulating the processes that occur in natural systems, where the strong adapt and survive while the weak perish. In GA, basic genetic operators such as crossover, replication, and mutation are used to create a new population [69]. A collection of strings may be used to describe a population (also known as chromosomes) and each of these chromosomes will be given a fitness score using objective objective function. A chromosome's objective function determines its ability to survive and reproduce offspring. A random number generator, a crossover process, a reproduction process, a fitness evaluation unit, and a mutation procedure are the five major components of a simple GA [70]. GA has the advantage of requiring fewer parameter sets and initializing itself from a pool of probable solutions rather than a single answer. Because the crossover and mutation processes are random, one of the major drawbacks of GA is that it consumes a long time to get the best results.

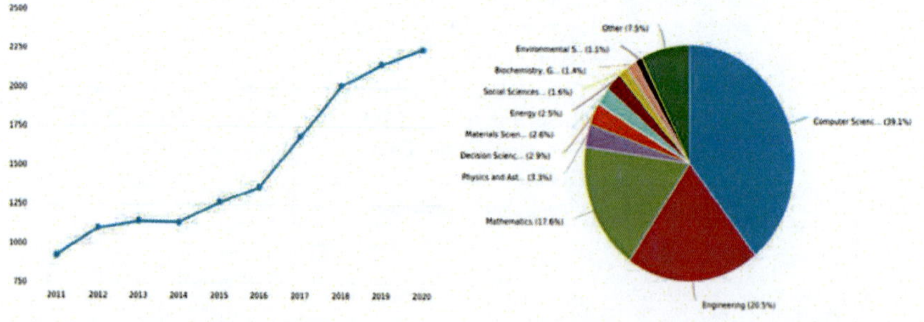

(a) Swarm intelligence applications (b) Swarm intelligence research area

Fig. 1 Swarm intelligence research performed in the last decade (2011–2020)

GA has a wide range of uses, including scheduling [71], machine learning [72], robotics [73], and many others.

2. *Particle Swarm Optimization (PSO):* The PSO [43] is an optimization technique that guides particles to find globally optimal solutions using a simple mechanism that mimics swarm behavior in birds flocking and fish schooling. Separation, alignment, and cohesion are three simple behaviors of PSO. Separation refers to avoiding crowded local flockmates, whereas alignment refers to moving in the general direction of the typical local flockmates. Cohesion is defined as a movement toward the average location of the local flockmates. PSO was demonstrated to be an effective optimization approach by scanning an entire high-dimensional problem space. It uses the principle of social interaction to solve problems and does not use the gradient of the problem to be solved, because it does not require a different optimization problem, as traditional optimization methods do [74]. The number of particles, the location of the agent in the solution space, the velocity, and the agents' position are all PSO parameters [75]. The convergence can be accomplished by attracting all particles to the particle that has the best solution [76]. Networking [77], machine learning [78], image processing [79], and several other applications use PSO.
3. *Grey Wolf Optimizer (GWO):* From top to bottom, The GWO [80] separates the agents (grey wolves) into multiple hierarchy divisions known as alpha, beta, delta, and omega. To find the solutions, each hierarchy has different functions, which in this case are prey. Grey wolves in the wild have a hunting mechanism and a leadership structure that the GWO version mimics. Alpha is the most powerful agent in the GWO hierarchy. Beta and delta are the other alpha subordinates who help manage the bulk of omega wolves in the hierarchy. In addition, three key stages of hunting are introduced for optimization: searching for prey, encircling prey, and attacking prey. Because of its simple formula and a small number of parameters, the GWO algorithm is simple to implement. The GWO algorithm has a few drawbacks, including low solving accuracy, poor local searching ability, and a slow convergence rate [81].
4. *Whale Optimization Algorithm (WOA):* The WOA [48] is a type of SI meta-heuristic algorithm that is new. The concept is inspired by the humpback whale's unique predatory activity in the ocean. Encircling prey, looking for prey, and attacking prey are the three stages that the WOA algorithm optimizes through. It has a simple principle and implementation, has fewer parameters to modify, and is very stable. However, due to the random search mechanism, the simple WOA, like other SI optimization algorithms, often falls into local optimal solutions and obtains low precision solutions. It performs badly for high-dimensional problems. WOA has a wide range of uses in recent years, including image processing [82], civil engineering [83], and electrical engineering [84].
5. *Sine Cosine Algorithm (SCA):* The SCA [46], which is based on the Sine and Cosine functions, is used in global optimization functions for exploitation and exploration phases. Using a mathematical model based on sine and cosine functions, SCA provides several initial random agent solutions and allows them to fluctuate outwards or towards the best possible solution. SCA, like other population-

based optimization algorithms, suffers from low diversity, local optima stagnation, and true solution skipping [85]. SCA has a wide range of uses, including image processing [86], robotics [87], and machine learning [88].

6. *Firefly Algorithm (FA):* The FA [44] was inspired by the behavior of fireflies, which use flashing lights to attract each other. In terms of inspiration, FA is very similar to the Glowworm Swarm Optimization (GSO) algorithm. The brightness of the fireflies' flickering will be determined by their health. Over distance, the brightness of the object decreases. If no brighter firefly is present, the less bright firefly will migrate at random.
7. *Moth-Flame Optimization Algorithm (MFO):* The MFO [47] represents the best solutions found so far by using the location of each moth as a solution representation and a matrix of flames as a solution representation. Each variable in an optimization problem is represented by the dimensions of a location. The upper and lower bounds are used to assign a population of moths to a random location at the start of the algorithm. The flame matrix is updated after calculating its fitness value. Every iteration, the location of each moth to the corresponding flame will be updated. The moth location was updated with the spiral feature. While the main benefit of these methods is their fast search times, the main disadvantage is their tendency to focus on exploitation rather than exploration, which increases the risk of them being trapped in local optima [89]. It has recently a wide range of uses, including medical [90], image processing [91], and economics [92].
8. *Slime Mould Algorithm (SMA):* The SMA [45] is inspired by the natural oscillation characteristic of slime mould. A unique mathematical model that uses adaptive weights to simulate the process of producing positive and negative feedback of a slime mould propagation wave based on a bio-oscillator to form the best path for connecting food with excellent exploratory ability and exploitation propensity is one of the novel features of the proposed SMA. To demonstrate its effectiveness, the proposed SMA is compared to current metaheuristics using a large collection of benchmarks. In addition, four classic engineering challenges are used to assess the algorithm's effectiveness in solving limited situations. The findings show that the suggested SMA performs well in a variety of search environments, with competitive, if not exceptional, outcomes. SMA has recently a wide range of uses, including medical [93], image processing [94], and economics [95].
9. *Search and Rescue Optimization Algorithm (SAR):* The SAR [50] is a technique for solving single-objective continuous optimization problems with a single objective. Human explorations during search and rescue operations inspired SAR. Humans, like other living things, seek out various goals in groups. Searching can be done for a variety of reasons, including hunting, locating food supplies, and locating missing persons. "Search and rescue operations" are one sort of group search. A search is a methodical effort that uses available personnel and resources to locate those who are in need. Rescue is a procedure for rescuing people who are in danger and transporting them to a safe location. SAR has recently a wide range of uses, including medical [96], Agriculture [97], and Engineering [98].

SI-based FS approaches are evaluated and explained in Table 2 in the remainder of this section.

Table 2 Swarm intelligence-based feature selection approaches for medical applications

Refs.	Year	SI method	No. of Objectives	Type
[99]	2015	PSO	Multi Objective	Wrapper
[100]	2016	PSO	Single Objective	Wrapper
[101]	2018	PSO	Single Objective	Wrapper
[102]	2018	PSO	Single Objective	Wrapper
[103]	2019	PSO	Single Objective	Wrapper
[104]	2014	ACO	Single Objective	Filter
[105]	2015	ACO	Single Objective	Filter
[106]	2016	ACO	Single Objective	Filter
[107]	2015	ABC	Single Objective	Wrapper
[108]	2017	ABC	Single Objective	Wrapper
[109]	2014	GSA	Single Objective	Wrapper
[110]	2015	GSA	Single Objective	Wrapper
[111]	2019	GSA	Single Objective	Wrapper
[112]	2018	DE	Multi Objective	Filter
[113]	2020	DE	Multi Objective	Wrapper
[114]	2020	DE	Single Objective	Wrapper
[115]	2018	BA	Single Objective	Wrapper
[116]	2020	BA	Single Objective	Hybrid
[117]	2017	WOA	Single Objective	Wrapper
[118]	2018	WOA	Single Objective	Wrapper
[119]	2019	WOA	Single Objective	Filter
[120]	2015	GWO	Multi Objective	Hybrid
[121]	2019	GWO	Single Objective	Wrapper
[122]	2020	GWO	Single Objective	Wrapper

3.2 *Machine Learning Techniques Overview*

A classification technique divides data samples into target classes and expects the same for each data point, such as categorizing a patient as "higher risk" or "lower risk" based on their diseases. Let us identify three types of machine learning: supervised, unsupervised, and reinforcement learning [64]. A collection of training data is accessible in supervised approaches, each of which is defined by the features values and the classmark. The Naive Bayes (NB) Classifier and SVM are examples of supervised learning solutions for classification and regression issues. In unsupervised methods' train data does not contain a class tag. The train data in unsupervised methods does not contain a class tag. Unsupervised learning is used to tackle two sorts of problems: association and clustering (K–Means clustering and Neural Networks). However, reinforcement systems, such as Q–Learning and SARSA, interact with the environment and solve two types of trouble: Exploitation and Exploration. Figure 2

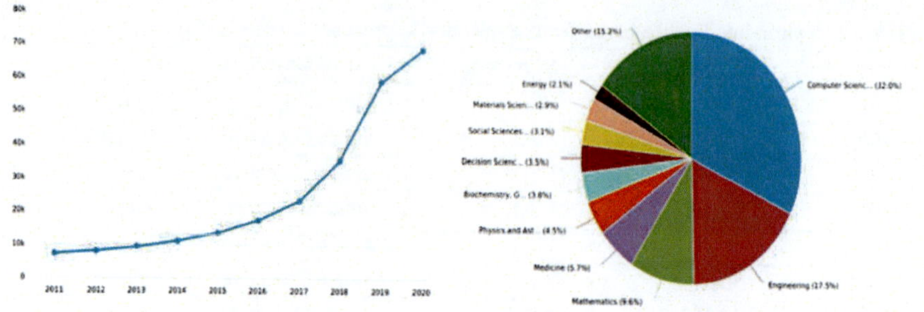

(a) Machine learning applications (b) Machine learning research area

Fig. 2 Machine learning research performed in the last decade (2011–2020)

shows research utilizing ML compared to the whole volume of research over the last ten years. Figure 2a is based on a Scopus search using the term "Machine learning". Figure 2b emphasizes the importance of ML techniques study in a variety of fields.

The following are some of the most used techniques.

1. *Decision Tree (DT):* Based on parameters, a tree- or graph-like shape is built in the DT technique [123], which includes a predefined target variable. The decision is made by traversing from root to leaf, and the traversing continues until the conditions are met. The DT is distinguished by the fact that it does not require domain knowledge to create, that it is simple to interpret, and as well as the fact that it can deal with both numerical and categorical data. However, the DT is limited to one output property, and it is an unstable classifier.
2. *Naïve Bayes (NB):* The supervised learning technique nave Bayes [1] is based on the Bayes' theorem. It is one of the most promising known algorithms for learning to categorize text documents because of its improved outcomes in multi-class problems and independence laws. NB is distinguished by its simplicity, ease of implementation, and increased accuracy in results due to a higher probability value. However, NB makes a strong assumption about the form of the data distribution and suffers from a loss of accuracy.
3. *Support Vector Machine (SVM):* The SVM [124] is a supervised learning technique in which a dataset is represented as points in an n-dimensional space, where n denotes the number of features in the dataset. The purpose of SVM is to create a hyperplane that divides the datasets into distinct categories and is as far away from the different types as possible. For maximum distances between the closest data points of each class, the hyperplane should be chosen with a large margin. SVM is characterized by higher accuracy, the ability to manage complex nonlinear data points, and the removal of sample overfitting. It's tough to use with huge datasets, and it takes a long time to run.
4. *k-Nearest Neighbors (kNN):* The kNN is a widely used machine learning algorithm for categorization on benchmark data sets [37, 125]. The rising availability of data presented in new forms has generated recent interest in kNN. However,

the performance of kNN is affected by some factors, most notably the distance measure chosen, and the k parameter chosen. The anonymous data points are discovered using the well-known data points known as nearest neighbors in this classification technique. The kNN [126] has a simple definition, and it is also known as lazy learning, where "k" stands for nearest neighbor. The goal of the kNN algorithm is to find k samples from the training data set that are like a new sample. The ease with which kNN can be implemented and the speed with which it can train set it apart. However, determining the nearest neighbor in the massive training data takes a long time and requires a lot of storage space.

5. *Artificial Neural Network (ANN):* The ANN [127] is inspired by the human brain's neural network, and it is made up of three layers: input, secret layer, and output layer, also known as MLP (Multilayer Perceptron). Each hidden layer is built up of probabilistic behavior, like a neuron. The ability of the ANN to learn and model nonlinear and complicated relationships, as well as its ability to generalize the model and forecast unknown data, distinguishes it. However, because there are so many parameters to configure, tuning the network can be challenging, and huge neural networks take a long time to process.
6. *Logistic Regression:* For dichotomic output variables, logistic regression, also known as the logit model [128], was developed for disease classification prediction. It is a way of evaluating data using statistics. In this case, the outcomes are determined by one or more independent variables. Logistic regression is distinguished by its ease of implementation and training performance, as well as its ability to manage nonlinear and interaction effects. However, it is impossible to predict continuous outcomes, and consistent results require a large sample size.

4 Application Areas of SI Algorithms and ML in Medical Diagnosis

This section summarizes the important applications of SI algorithms and ML in medicine and public health, such as improved diagnosis, patient death prevention, and decision support. In terms of approaches and results, this paper explores a variety of applications in healthcare that combine SI with ML techniques.

4.1 Swarm Intelligence Algorithms in Medical Diagnosis

The results of important diseases like cancer, heart disease, and cardiology have all been predicted using SI algorithms. Diseases have been diagnosed and treated using SI approaches. SI algorithms have been used to diagnose a variety of disorders. Many medical domain problems collect data sets with very large feature dimensions. To deal with the problem of high dimensionality, research has demonstrated that applying FS to diverse medical sector data sets has a favorable impact [129]. Some application areas of SI algorithms in medical diagnosis such as GA has a wide range of uses, including scheduling [71], machine learning [72], robotics [73], PSO has used in several applications like Networking [77], machine learning [78], image processing [79]. MFO has recently a wide range of uses, including medical [90], image processing [91], and SMA has recently used in several applications like medical [93], image processing [94].

4.2 Machine Learning in Medical Diagnosis

Furthermore, as the health sector becomes more computerized, ML techniques are critical to assisting physicians in identifying and treating abnormalities at an early level. ML algorithms improve the diagnostic process's accuracy, reliability, and speed [52]. For the diagnosis of various diseases, ML approaches have been applied.

1. *Artificial Neural Network (ANN):* The structure of the human brain, which uses neurons for processing, is the inspiration for ANN [130]. The ANN can be used to handle hard mathematical issues, large-scale signal processing challenges, and even parallel computations. Because detecting the symptoms of a Urinary Tract Infection (UTI) can be difficult, [131] created an ANN-based model to help with UTI diagnosis. Using clinically available data, this device may distinguish between cystitis and urethritis. Furthermore, they mentioned that ML methods can be used to prevent intrusive and expensive methods. [132] proposed a model to forecast the occurrence of diarrhoea, which is one of the leading causes of death worldwide. The proposed approach is based on ANN and may be useful in preventing diarrhoea with a 95.63% accuracy rating.
2. *Decision Tree (DT):* DT [133] is a method of prediction that draws inferences (in the form of leaves) from observations (in branches). Although DT in its current form is incapable of dealing with incomplete, inappropriate, or even uncertain features, it can be extended to do so. The fact that thyroid treatment is a long-term procedure is well-known. [134] investigated the efficacy of ML approaches for thyroid diagnosis and classification. On clinical results, DT had a 97.35% accuracy rate. When the authors compared the results to those of other ML approaches, they found that DT was the most effective.
3. *Support Vector Machine (SVM):* The main goal of SVM [124] is to find a hyperplane with the shortest distance between the current classes' features. Support Vector Clustering could be utilized with unlabeled data because SVM is a supervised method. [135] designed a classifier to predict whether someone has glaucoma. PCA is used to extract features from retinal images, and SVM is used to classify them. The suggested classifier might reach a higher accuracy than current state-of-the-art approaches for glaucoma detection. In the later stages of kidney disease, the organ can fail gradually. [136] established a method based on SVM that aid in the early detection of kidney disease in this regard and has the

potential to enhance the decision-making process for assessing the chronic state of kidney disease. [137] suggested a ML-based model for detecting extracranial and intracranial artery stenosis. In comparison to traditional approaches, the proposed SVM model achieved higher accuracy and sensitivity. The high rate of false positives in influenza tests results in high expenditures. [138] suggested the first influenza prediction model reduce the high costs. The proposed model makes use of the SVM model and achieves a 90% accuracy rate.

4. *k-Nearest Neighbors (kNN):* Using the familiar data points, knn [126] is used to discover anonymous data points. [139] introduces a distance-based ensemble for a kNN approach and demonstrates the implications of its use in heart disease diagnosis. Two different versions of the ensemble have been used. One uses three and the other five distances. With the UCI heart disease Cleveland data collection, our ensemble provided an average accuracy of nearly 85% for all configurations and versions tested.

4.3 *Swarm Intelligence Algorithms and Machine Learning in Medical Diagnosis*

Disease diagnosis is still an open question. A disease diagnostic model's most important aspect is that it helps physicians make quick decisions while reducing diagnostic errors. The current methods are incompatible with all disease datasets. Although they perform well on a particular dataset, they perform poorly on other disease datasets. The main aim of biomedical data analytics is to create various predictive models in medical domains and other complex diseases. Aside from that, there are several ML and SI algorithms that are employed in the diagnosis of diseases. Table 3 lists some of the relevant works.

5 Open Issues and Challenges

There are several challenges to overcome when attempting to solve the optimization problem. For example, constraints in the search space generate areas of infeasible solutions, potentially rendering an optimization algorithm unsuccessful, despite having good performance in the search space without constraints. Furthermore, the existence of local optima is a major barrier to solving the realistic optimization problem. An algorithm for solving this type of problem should provide techniques to avoid mistaking local optima for the global optimum. Having a more intensive exploration phase to prevent local optima, on the other hand, does not result in convergence and, as a result, may not be able to achieve the global optimum.

Collective-based algorithms are very popular because of their high potential for avoiding local optima, and their drawback can be relieved by higher-performance

Table 3 Application areas of SI algorithms and ML techniques in medical diagnosis

Refs.	Dataset used	Models developed	Accuracy
[57]	Cleveland heart disease database	SVM with PSO called (PSO-SVM)	84.36%
[140]	Heart Disease	PSO with Neural Network Feed Forward Back Propagation (PSO+FFBP)	91.94%
[141]	Cancer Dataset	Neuro-fuzzy (NF)	81.2%
[142]	Liver Disease	SVM and NB Classifier (SVM-NB)	79.66%
[143]	Heart Disease	Relief approach with the Rough set method (RFRS)	92.59%
[144]	Breast, Heart, Hepatitis, Diabetes datasets	Fuzzy Support Vector Machine (FSVM)	96.76%, 83.53%, 64.44%, 61.94%
[145]	Heart Disease	Self-Regulated PSO with the Extreme Learning Machine classifier (SRLPSO + ELM)	91.33%
[146]	Heart Disease	Extreme Learning Machine (ELM) algorithm	80%
[147]	Diabetes Dataset	Adaboost, Grading Learning, Logiboost, and CART algorithms	78.64%
[148]	Heart Disease	Artificial Bee Colony with SVM (ABC- SVM)	86.76 %
[149]	Liver Dataset	Artificial Immune System and GA (AIS-GA)	88.7%
[150]	Diabetes dataset	DT and NB (DT-NB)	79.57%
[151]	Heart dataset	Wind-driven Swarm Optimization (WSO)	77.8%
[152]	Heart Disease	Fuzzy logic and GA (FGA)	84.44%
[15]	Eighteen diseases datasets	Henry gas solubility optimization (HGSO) is combined with SVM called HGSO-SVM	88.88%
[58]	Heart Disease	NB with PSO (NB-PSO)	87.91%

computers or concurrent computing. Their benefits also include algorithm simplicity and the ability to operate on a variety of problems. In addition, medical databases may contain redundant and incorrect attributes, slowing down the processing phase and affecting accuracy. Thus, the integration of SI algorithms and ML techniques addressed our challenge of increasing the chances of obtaining a true global optimum and enhancing the performance and accuracy of classification models with FS approaches.

6 Conclusions and Future Research Issues

The healthcare sector is rapidly changing all over the world. The healthcare industry generates a large volume of heterogeneous data. It is critical for the healthcare industry to effectively obtain, collect, and mine data. Medical data is complex, making it difficult to interpret. The consistency and efficiency of data must be maintained to make appropriate decisions. Furthermore, medical databases can contain redundant and inappropriate attributes, which can slow down processing and compromise accuracy. To improve the performance and accuracy of classification models, FS methods are needed. SI and ML can be useful tools in this situation. We reviewed some implementations of SI with ML in healthcare. SI algorithms have been successfully applied to healthcare datasets, and the findings show that these algorithms produce reliable and useful results. Due to many features in healthcare datasets, some of them may be redundant or irrelevant. To increase the performance of the classifier, SI algorithms can be used to choose a useful subset of features, which can lead to more effective decision-making in medical datasets. This will help in the early detection of diseases, enabling effective treatment and prevention. Future research in this area could include analyzing actual hospital data to uncover additional characteristics that influence early disease prediction. Data from different diseases that have not been researched by other researchers can also be incorporated into SI algorithms for timely prognosis.

Conflict of Interest The authors declare that there is no conflict of interest.

References

1. S.B. Kotsiantis, I. Zaharakis, P. Pintelas, Supervised machine learning: a review of classification techniques. Emerg. Artif. Intell. Appl. Comput. Eng. **160**(1), 3–24 (2007)
2. E.H. Houssein, M. Younan, A.E. Hassanien, Nature-inspired algorithms: a comprehensive review. Hybrid Comput. Intell. Res. Appl. 1 (2019)
3. E.H. Houssein, A.G. Gad, K. Hussain, P.N. Suganthan, Major advances in particle swarm optimization: theory, analysis, and application. Swarm Evolut. Comput. **63**, 100868 (2021)
4. F.A. Hashim, E.H. Houssein, M.S. Mabrouk, W. Al-Atabany, S. Mirjalili, Henry gas solubility optimization: a novel physics-based algorithm. Future Gener. Comput. Syst. **101**, 646–667 (2019)
5. F.A. Hashim, K. Hussain, E.H. Houssein, M.S. Mabrouk, W. Al-Atabany, Archimedes optimization algorithm: a new metaheuristic algorithm for solving optimization problems. Appl. Intell. **51**(3), 1531–1551 (2021)
6. F.A. Hashim, E.H. Houssein, K. Hussain, M.S. Mabrouk, W. Al-Atabany, Honey badger algorithm: new metaheuristic algorithm for solving optimization problems. Math. Comput. Simul. (2021)
7. E.H. Houssein, D.S. Abdelminaam, H.N. Hassan, M.M. Al-Sayed, E. Nabil, A hybrid barnacles mating optimizer algorithm with support vector machines for gene selection of microarray cancer classification. IEEE Access **9**, 64895–64905 (2021)

8. E.H. Houssein, A.A. Ewees, M.A. ElAziz, Improving twin support vector machine based on hybrid swarm optimizer for heartbeat classification. Pattern Recognit. Image Anal. **28**(2), 243–253 (2018)
9. E.H. Houssein, I.E. Ibrahim, N. Neggaz, M. Hassaballah, Y.M. Wazery, An efficient ecg arrhythmia classification method based on manta ray foraging optimization. Expert Syst. Appl. **181**, 115131 (2021)
10. E.H. Houssein, D.S. AbdElminaam, I.E. Ibrahim, M. Hassaballah, Y.M. Wazery, A hybrid heartbeats classification approach based on marine predators algorithm and convolution neural networks. IEEE Access (2021)
11. E.H. Houssein, M. Kilany, A.E. Hassanien, Ecg signals classification: a review. Int. J. Intell. Eng. Inf. **5**(4), 376–396 (2017)
12. A.E. Hassanien, M. Kilany, E.H. Houssein, H. AlQaheri, Intelligent human emotion recognition based on elephant herding optimization tuned support vector regression. Biomed. Signal Process. Control **45**, 182–191 (2018)
13. E.H. Houssein, M. Dirar, K. Hussain, W.M. Mohamed, Assess deep learning models for Egyptian exchange prediction using nonlinear artificial neural networks. Neural Comput. Appl. **33**(11), 5965–5987 (2021)
14. E.H. Houssein, M.M. Emam, A.A. Ali, P.N. Suganthan, Deep and machine learning techniques for medical imaging-based breast cancer: a comprehensive review. Expert Syst. Appl. 114161 (2020)
15. N. Neggaz, E.H. Houssein, K. Hussain, An efficient henry gas solubility optimization for feature selection. Expert Syst. Appl. **152**, 113364 (2020)
16. K. Hussain, N. Neggaz, W. Zhu, E.H. Houssein, An efficient hybrid sine-cosine harris hawks optimization for low and high-dimensional feature selection. Expert Syst. Appl. **176**, 114778 (2021)
17. D.S. Abd Elminaam, A. Nabil, S.A. Ibraheem, E.H. Houssein, An efficient marine predators algorithm for feature selection. IEEE Access **9**, 60136–60153 (2021)
18. E.H. Houssein, M.A. Mahdy, A. Fathy, H. Rezk, A modified marine predator algorithm based on opposition based learning for tracking the global mpp of shaded pv system. Expert Syst. Appl. **183**, 115253 (2021)
19. E.H. Houssein, B.E.D. Helmy, H. Rezk, A.M. Nassef, An enhanced archimedes optimization algorithm based on local escaping operator and orthogonal learning for pem fuel cell parameter identification. Eng. Appl. Artif. Intell. **103**, 104309 (2021)
20. M.H. Hassan, E.H. Houssein, M.A. Mahdy, S. Kamel, An improved manta ray foraging optimizer for cost-effective emission dispatch problems. Eng. Appl. Artif. Intelli. **100**, 104155 (2021)
21. A. Korashy, S. Kamel, E.H. Houssein, F. Jurado, F.A. Hashim, Development and application of evaporation rate water cycle algorithm for optimal coordination of directional overcurrent relays. Expert Syst. Appl. **185**, 115538 (2021)
22. E.H. Houssein, F.A. Hashim, S. Ferahtia, H. Rezk, An efficient modified artificial electric field algorithm for solving optimization problems and parameter estimation of fuel cell. Int. J. Energy Res. (2021)
23. E.H. Houssein, G.N. Zaki, A.A.Z. Diab, E.M. Younis, An efficient manta ray foraging optimization algorithm for parameter extraction of three-diode photovoltaic model. Comput. Electr. Eng. **94**, 107304 (2021)
24. E.H. Houssein, Machine learning and meta-heuristic algorithms for renewable energy: a systematic review. Adv. Control Optim. Paradigms Wind Energy Syst. 165–187 (2019)
25. A.A. Ismaeel, E.H. Houssein, D. Oliva, M. Said, Gradient-based optimizer for parameter extraction in photovoltaic models. IEEE Access **9**, 13403–13416 (2021)
26. E.H. Houssein, M.A. Mahdy, M.J. Blondin, D. Shebl, W.M. Mohamed, Hybrid slime mould algorithm with adaptive guided differential evolution algorithm for combinatorial and global optimization problems. Expert Syst. Appl. **174**, 114689 (2021)
27. E.H. Houssein, M.A. Mahdy, M.G. Eldin, D. Shebl, W.M. Mohamed, M. Abdel-Aty, Optimizing quantum cloning circuit parameters based on adaptive guided differential evolution algorithm. J. Adv. Res. **29**, 147–157 (2021)

28. E.H. Houssein, B.E.D. Helmy, D. Oliva, A.A. Elngar, H. Shaban, A novel black widow optimization algorithm for multilevel thresholding image segmentation. Expert Syst. Appl. **167**, 114159 (2021)
29. E.H. Houssein, M.M. Emam, A.A. Ali, An efficient multilevel thresholding segmentation method for thermography breast cancer imaging based on improved chimp optimization algorithm. Expert Syst. Appl. 115651 (2021)
30. E.H. Houssein, K. Hussain, L. Abualigah, M. Abd Elaziz, W. Alomoush, G. Dhiman, Y. Djenouri, E. Cuevas, An improved opposition-based marine predators algorithm for global optimization and multilevel thresholding image segmentation. Knowl.-Based Syst. 107348 (2021)
31. E.H. Houssein, M.M. Emam, A.A. Ali, Improved manta ray foraging optimization for multilevel thresholding using covid-19 ct images. Neural Comput. Appl. 1–21 (2021)
32. E.H. Houssein, B.E.D. Helmy, A.A. Elngar, D.S. Abdelminaam, H. Shaban, An improved tunicate swarm algorithm for global optimization and image segmentation. IEEE Access **9**, 56066–56092 (2021)
33. M.M. Ahmed, E.H. Houssein, A.E. Hassanien, A. Taha, E. Hassanien, Maximizing lifetime of large-scale wireless sensor networks using multi-objective whale optimization algorithm. Telecommun. Syst. **72**(2), 243–259 (2019)
34. E.H. Houssein, A.G. Gad, Y.M. Wazery, P.N. Suganthan, Task scheduling in cloud computing based on meta-heuristics: Review, taxonomy, open challenges, and future trends. Swarm Evol. Comput. 100841 (2021)
35. F.A. Hashim, E.H. Houssein, K. Hussain, M.S. Mabrouk, W. Al-Atabany, A modified henry gas solubility optimization for solving motif discovery problem. Neural Comput. Appl. **32**(14), 10759–10771 (2020)
36. E.H. Houssein, N. Neggaz, M.E. Hosney, W.M. Mohamed, M. Hassaballah, Enhanced harris hawks optimization with genetic operators for selection chemical descriptors and compounds activities. Neural Comput. Appl. 1–18 (2021)
37. E.H. Houssein, M.E. Hosney, D. Oliva, W.M. Mohamed, M. Hassaballah, A novel hybrid harris hawks optimization and support vector machines for drug design and discovery. Comput. Chem. Eng. **133**, 106656 (2020)
38. O. Udomkasemsub, K. Akkarajitsakul, T. Achalakul, Hybrid moth-flame and salp-swarm optimization algorithm. Int. J. Model. Optim. **9**(4) (2019)
39. G. Beni, The concept of cellular robotic system, in *Proceedings IEEE International Symposium on Intelligent Control 1988*. (IEEE, 1988), pp. 57–62
40. J. Branke, Memory enhanced evolutionary algorithms for changing optimization problems, in *Proceedings of the 1999 Congress on Evolutionary Computation-CEC99 (Cat. No. 99TH8406)*, vol. 3 (IEEE, 1999), pp. 1875–1882
41. V. Kothari, J. Anuradha, S. Shah, P. Mittal, A survey on particle swarm optimization in feature selection, in *International Conference on Computing and Communication Systems* (Springer, 2011), pp. 192–201
42. N.M. Al Salami, Ant colony optimization algorithm. UbiCC J. **4**(3), 823–826 (2009)
43. J. Kennedy, R. Eberhart, Particle swarm optimization, in *Proceedings of ICNN'95-International Conference on Neural Networks*, vol. 4 (IEEE, 1995), pp. 1942–1948
44. X.S. Yang, Firefly algorithms for multimodal optimization, in *International Symposium on Stochastic Algorithms* (Springer, 2009), pp. 169–178
45. S. Li, H. Chen, M. Wang, A.A. Heidari, S. Mirjalili, Slime mould algorithm: a new method for stochastic optimization. Future Gener. Comput. Syst. **111**, 300–323 (2020)
46. S. Mirjalili, Sca: a sine cosine algorithm for solving optimization problems. Knowl.-Based Syst. **96**, 120–133 (2016)
47. S. Mirjalili, Moth-flame optimization algorithm: a novel nature-inspired heuristic paradigm. Knowl.-Based Syst. **89**, 228–249 (2015)
48. S. Mirjalili, A. Lewis, The whale optimization algorithm. Adv. Eng. Softw. **95**, 51–67 (2016)
49. J.H. Holland, Genetic algorithms. Sci. Am. **267**(1), 66–73 (1992)
50. A. Shabani, B. Asgarian, M. Salido, S.A. Gharebaghi, Search and rescue optimization algorithm: a new optimization method for solving constrained engineering optimization problems. Expert Syst. Appl. **161**, 113698 (2020)

51. M. Eshtay, H. Faris, N. Obeid, Improving extreme learning machine by competitive swarm optimization and its application for medical diagnosis problems. Expert Syst. Appl. **104**, 134–152 (2018)
52. Q. Al-Tashi, H. Rais, S.J. Abdulkadir, Hybrid swarm intelligence algorithms with ensemble machine learning for medical diagnosis, in *2018 4th International Conference on Computer and Information Sciences (ICCOINS)* (IEEE, 2018), pp. 1–6
53. S. Agrawal, B. Singh, R. Kumar, N. Dey, Machine learning for medical diagnosis: A neural network classifier optimized via the directed bee colony optimization algorithm, in *U-Healthcare monitoring systems* (Elsevier, 2019), pp. 197–215
54. A. Malav, K. Kadam, P. Kamat, Prediction of heart disease using k-means and artificial neural network as hybrid approach to improve accuracy. Int. J. Eng. Technol. **9**(4), 3081–3085 (2017)
55. D.K. Choubey, P. Kumar, S. Tripathi, S. Kumar, Performance evaluation of classification methods with pca and pso for diabetes. Network Model. Anal. Health Inf. Bioinform. **9**(1), 1–30 (2020)
56. L. Shen, H. Chen, Z. Yu, W. Kang, B. Zhang, H. Li, B. Yang, D. Liu, Evolving support vector machines using fruit fly optimization for medical data classification. Knowl.-Based Syst. **96**, 61–75 (2016)
57. J. Vijayashree, H.P. Sultana, A machine learning framework for feature selection in heart disease classification using improved particle swarm optimization with support vector machine classifier. Program. Comput. Softw. **44**(6), 388–397 (2018)
58. U.N. Dulhare, Prediction system for heart disease using naive bayes and particle swarm optimization. Biomed. Res. **29** (2018)
59. A. Sarwar, M. Ali, J. Manhas, V. Sharma, Diagnosis of diabetes type-ii using hybrid machine learning based ensemble model. Int. J. Inf. Technol. **12**(2), 419–428 (2020)
60. N. Zemmal, N. Azizi, M. Sellami, S. Cheriguene, A. Ziani, M. AlDwairi, N. Dendani, Particle swarm optimization based swarm intelligence for active learning improvement: application on medical data classification. Cogn. Comput. **12**(5), 991–1010 (2020)
61. R.T. Prasetio, Genetic algorithm to optimize k-nearest neighbor parameter for benchmarked medical datasets classification. J. Online Inf. **5**(2) (2020)
62. H.T. Ibrahim, W.J. Mazher, O.N. Ucan, O. Bayat, A grasshopper optimizer approach for feature selection and optimizing svm parameters utilizing real biomedical data sets. Neural Comput. Appl. **31**(10), 5965–5974 (2019)
63. A.Z. Ala'M, A.A. Heidari, M. Habib, H. Faris, I. Aljarah, M.A. Hassonah, Salp chain-based optimization of support vector machines and feature weighting for medical diagnostic information systems, in *Evolutionary machine learning techniques* (Springer, 2020), pp. 11–34
64. C. Tang, X. Liu, M. Li, P. Wang, J. Chen, L. Wang, W. Li, Robust unsupervised feature selection via dual self-representation and manifold regularization. Knowl.-Based Syst. **145**, 109–120 (2018)
65. D. Ding, X. Yang, F. Xia, T. Ma, H. Liu, C. Tang, Unsupervised feature selection via adaptive hypergraph regularized latent representation learning. Neurocomputing **378**, 79–97 (2020)
66. L. Hu, W. Gao, K. Zhao, P. Zhang, F. Wang, Feature selection considering two types of feature relevancy and feature interdependency. Expert Syst. Appl. **93**, 423–434 (2018)
67. J. Miao, L. Niu, A survey on feature selection. Proc. Comput. Sci. **91**, 919–926 (2016)
68. Z. Beheshti, S.M.H. Shamsuddin, A review of population-based meta-heuristic algorithms. Int. J. Adv. Soft Comput. Appl **5**(1), 1–35 (2013)
69. M. Yong-Jie, Y. Wen-Xia, Research progress of genetic algorithm. Appl. Res. Comput. **4**, 1201–1206 (2012)
70. J. Carr, An introduction to genetic algorithms. Senior Project **1**(40), 7 (2014)
71. L.R. Abreu, J.O. Cunha, B.A. Prata, J.M. Framinan, A genetic algorithm for scheduling open shops with sequence-dependent setup times. Comput. Oper. Res. **113**, 104793 (2020)
72. Y. Li, W. Cheng, L.H. Yu, R. Rainer, Genetic algorithm enhanced by machine learning in dynamic aperture optimization. Phys. Rev. Accel. Beams **21**(5), 054601 (2018)

73. R.M.C. Santiago, A.L. De Ocampo, A.T. Ubando, A.A. Bandala, E.P. Dadios, Path planning for mobile robots using genetic algorithm and probabilistic roadmap, in *2017IEEE 9th International Conference on Humanoid, Nanotechnology, Information Technology, Communication and Control, Environment and Management (HNICEM)* (IEEE, 2017), pp. 1–5
74. Y. Lu, M. Liang, Z. Ye, L. Cao, Improved particle swarm optimization algorithm and its application in text feature selection. Appl. Soft Comput. **35**, 629–636 (2015)
75. H. Gao, S. Kwong, J. Yang, J. Cao, Particle swarm optimization based on intermediate disturbance strategy algorithm and its application in multi-threshold image segmentation. Inf. Sci. **250**, 82–112 (2013)
76. Q. Bai, Analysis of particle swarm optimization algorithm. Comput. Inf. Sci. **3**(1), 180 (2010)
77. S. Tabibi, A. Ghaffari, Energy-efficient routing mechanism for mobile sink in wireless sensor networks using particle swarm optimization algorithm. Wireless Pers. Commun. **104**(1), 199–216 (2019)
78. Y. Khourdifi, M. Bahaj, Heart disease prediction and classification using machine learning algorithms optimized by particle swarm optimization and ant colony optimization. Int. J. Intell. Eng. Syst. **12**(1), 242–252 (2019)
79. H. Lei, T. Lei, T. Yuenian, Sports image detection based on particle swarm optimization algorithm. Microprocess. Microsyst. **80**, 103345 (2021)
80. S. Mirjalili, S.M. Mirjalili, A. Lewis, Grey wolf optimizer. Adv. Eng. Softw. **69**, 46–61 (2014)
81. H. Faris, I. Aljarah, M.A. Al-Betar, S. Mirjalili, Grey wolf optimizer: a review of recent variants and applications. Neural Comput. Appl. **30**(2), 413–435 (2018)
82. S.J. Mousavirad, H. Ebrahimpour-Komleh, Multilevel image thresholding using entropy of histogram and recently developed population-based metaheuristic algorithms. Evol. Intel. **10**(1), 45–75 (2017)
83. Y. Moodi, S.R. Mousavi, A. Ghavidel, M.R. Sohrabi, M. Rashki, Using response surface methodology and providing a modified model using whale algorithm for estimating the compressive strength of columns confined with frp sheets. Constr. Build. Mater. **183**, 163–170 (2018)
84. P.R. Sahu, P.K. Hota, S. Panda, Power system stability enhancement by fractional order multi input sssc based controller employing whale optimization algorithm. J. Electr. Syst. Inf. Technol. **5**(3), 326–336 (2018)
85. S. Gupta, K. Deep, A hybrid self-adaptive sine cosine algorithm with opposition based learning. Expert Syst. Appl. **119**, 210–230 (2019)
86. J. Yang, X. Kang, E.K. Wong, Y.-Q. Shi, Jpeg steganalysis with combined dense connected cnns and sca-gfr. Multimedia Tools Appl. **78**(7), 8481–8495 (2019)
87. H. Paikray, P. Das, S. Panda, Optimal multi-robot path planning using particle swarm optimization algorithm improved by sine and cosine algorithms. Arab. J. Sci. Eng. **46**(4), 3357–3381 (2021)
88. I. Kabin, M. Aftowicz, D. Klann, Y. Varabei, Z. Dyka, P. Langendoerfer, Horizontal sca attack using machine learning algorithms. Crypto Day Matters 30 (2019)
89. M. Shehab, L. Abualigah, H. Al Hamad, H. Alabool, M. Alshinwan, A.M. Khasawneh, Moth–flame optimization algorithm: variants and applications. Neural Comput. Appl. **32**(14), 9859–9884 (2020)
90. M. Wang, H. Chen, B. Yang, X. Zhao, L. Hu, Z. Cai, H. Huang, C. Tong, Toward an optimal kernel extreme learning machine using a chaotic moth-flame optimization strategy with applications in medical diagnoses. Neurocomputing **267**, 69–84 (2017)
91. A.K.M. Khairuzzaman, S. Chaudhury, Moth-flame optimization algorithm based multilevel thresholding for image segmentation. Int. J. Appl. Metaheuristic Comput. (IJAMC) **8**(4), 58–83 (2017)
92. A.A. Elsakaan, R.A.A. El-Sehiemy, S.S. Kaddah, M.I. Elsaid, Economic power dispatch with emission constraint and valve point loading effect using moth flame optimization algorithm. Adv. Eng. Forum **28**. Trans Tech Publ, pp 139–149 (2018)
93. A.M. Anter, D. Oliva, A. Thakare, Z. Zhang, Afcm-lsma: new intelligent model based on lévy slime mould algorithm and adaptive fuzzy c-means for identification of covid-19 infection from chest x-ray images. Adv. Eng. Inf. 101317 (2021)

94. M. Abdel-Basset, V. Chang, R. Mohamed, Hsma_woa: A hybrid novel slime mould algorithm with whale optimization algorithm for tackling the image segmentation problem of chest x-ray images. App. Soft Comput. **95**, 106642 (2020)
95. T.A. Ahmed, M. Ebeed, A. Refai, S. Kamel, Solving combined economic and emission dispatch problem using the slime mould algorithm. Sohag Eng. J. **1**(1), 62–70 (2021)
96. Q. Tian, Y. Wu, X. Ren, N. Razmjooy, A new optimized sequential method for lung tumor diagnosis based on deep learning and converged search and rescue algorithm. Biomed. Signal Process. Control **68**, 102761 (2021)
97. C. Muppala, V. Guruviah, Detection of leaf folder and yellow stemborer moths in the paddy field using deep neural network with search and rescue optimization. Inf. Process. Agric. (2020)
98. L. Cai, Y. Wu, S. Zhu, Z. Tan, W. Yi, Bi-level programming enabled design of an intelligent maritime search and rescue system. Adv. Eng. Inf. **46**, 101194 (2020)
99. H. Banka, S. Dara, A hamming distance based binary particle swarm optimization (hdbpso) algorithm for high dimensional feature selection, classification and validation. Pattern Recogn. Lett. **52**, 94–100 (2015)
100. P. Moradi, M. Gholampour, A hybrid particle swarm optimization for feature subset selection by integrating a novel local search strategy. Appl. Soft Comput. **43**, 117–130 (2016)
101. O.S. Qasim, Z.Y. Algamal, Feature selection using particle swarm optimization-based logistic regression model. Chemom. Intell. Lab. Syst. **182**, 41–46 (2018)
102. S. Gunasundari, S. Janakiraman, S. Meenambal, Multiswarm heterogeneous binary pso using win-win approach for improved feature selection in liver and kidney disease diagnosis. Comput. Med. Imaging Graph. **70**, 135–154 (2018)
103. E. Pashaei, E. Pashaei, N. Aydin, Gene selection using hybrid binary black hole algorithm and modified binary particle swarm optimization. Genomics **111**(4), 669–686 (2019)
104. S. Tabakhi, P. Moradi, F. Akhlaghian, An unsupervised feature selection algorithm based on ant colony optimization. Eng. Appl. Artif. Intell. **32**, 112–123 (2014)
105. P. Moradi, M. Rostami, Integration of graph clustering with ant colony optimization for feature selection. Knowl.-Based Syst. **84**, 144–161 (2015)
106. B.Z. Dadaneh, H.Y. Markid, A. Zakerolhosseini, Unsupervised probabilistic feature selection using ant colony optimization. Expert Syst. Appl. **53**, 27–42 (2016)
107. E. Hancer, B. Xue, D. Karaboga, M. Zhang, A binary abc algorithm based on advanced similarity scheme for feature selection. Appl. Soft Comput. **36**, 334–348 (2015)
108. P. Shunmugapriya, S. Kanmani, A hybrid algorithm using ant and bee colony optimization for feature selection and classification (ac-abc hybrid). Swarm Evol. Comput. **36**, 27–36 (2017)
109. X. Han, X. Chang, L. Quan, X. Xiong, J. Li, Z. Zhang, Y. Liu, Feature subset selection by gravitational search algorithm optimization. Inf. Sci. **281**, 128–146 (2014)
110. J. Xiang, X. Han, F. Duan, Y. Qiang, X. Xiong, Y. Lan, H. Chai, A novel hybrid system for feature selection based on an improved gravitational search algorithm and k-nn method. Appl. Soft Comput. **31**, 293–307 (2015)
111. M. Taradeh, M. Mafarja, A.A. Heidari, H. Faris, I. Aljarah, S. Mirjalili, H. Fujita, An evolutionary gravitational search-based feature selection. Inf. Sci. **497**, 219–239 (2019)
112. E. Hancer, B. Xue, M. Zhang, Differential evolution for filter feature selection based on information theory and feature ranking. Knowl.-Based Syst. **140**, 103–119 (2018). https://www.sciencedirect.com/science/article/pii/S0950705117304987
113. Y. Zhang, D.-W. Gong, X.-Z. Gao, T. Tian, X.-Y. Sun, Binary differential evolution with self-learning for multi-objective feature selection. Inf. Sci. **507**, 67–85 (2020)
114. E. Hancer, A new multi-objective differential evolution approach for simultaneous clustering and feature selection. Eng. Appl. Artif. Intell. **87**, 103307 (2020)
115. M.A. Tawhid, K.B. Dsouza, Hybrid binary bat enhanced particle swarm optimization algorithm for solving feature selection problems. Appl. Comput. Inf. (2018)
116. M.A. Al-Betar, O.A. Alomari, S.M. Abu-Romman, A triz-inspired bat algorithm for gene selection in cancer classification. Genomics **112**(1), 114–126 (2020)

117. M.M. Mafarja, S. Mirjalili, Hybrid whale optimization algorithm with simulated annealing for feature selection. Neurocomputing **260**, 302–312 (2017)
118. M. Mafarja, S. Mirjalili, Whale optimization approaches for wrapper feature selection. Appl. Soft Comput. **62**, 441–453 (2018)
119. H. Nematzadeh, R. Enayatifar, M. Mahmud, E. Akbari, Frequency based feature selection method using whale algorithm. Genomics **111**(6), 1946–1955 (2019)
120. E. Emary, W. Yamany, A.E. Hassanien, V. Snasel, Multi-objective gray-wolf optimization for attribute reduction. Proc. Comput. Sci. **65**, 623–632 (2015)
121. Q. Tu, X. Chen, X. Liu, Multi-strategy ensemble grey wolf optimizer and its application to feature selection. Appl. Soft Comput. **76**, 16–30 (2019)
122. M. Abdel-Basset, D. El-Shahat, I. El-henawy, V.H.C. de Albuquerque, S. Mirjalili, A new fusion of grey wolf optimizer algorithm with a two-phase mutation for feature selection. Expert Syst. Appl. **139**, 112824 (2020)
123. P. Argentiero, R. Chin, P. Beaudet, An automated approach to the design of decision tree classifiers. IEEE Trans. Pattern Anal. Mach. Intell. **PAMI-8**(1), 51–57 (1982)
124. C. Cortes, V. Vapnik, Support-vector networks. Mach. Learn. **20**(3), 273–297 (1995)
125. S. Zhang, X. Li, M. Zong, X. Zhu, R. Wang, Efficient knn classification with different numbers of nearest neighbors. IEEE Trans. Neural Netw. Learn. Syst. **29**(5), 1774–1785 (2017)
126. J.H. Friedman, F. Baskett, L.J. Shustek, An algorithm for finding nearest neighbors. IEEE Trans. Comput. **100**(10), 1000–1006 (1975)
127. G.P. Zhang, Neural networks for classification: a survey. IEEE Trans. Syst. Man Cybern. Part C (Appl. Rev.) **30**(4), 451–462 (2000)
128. D.W. Hosmer Jr., S. Lemeshow, R.X. Sturdivant, *Applied logistic regression*, vol. 398. (John Wiley & Sons, 2013)
129. M. Habib, I. Aljarah, H. Faris, S. Mirjalili, Multi-objective particle swarm optimization: theory, literature review, and application in feature selection for medical diagnosis. Evol. Mach. Learn. Tech. 175–201 (2020)
130. S.E. Dreyfus, Artificial neural networks, back propagation, and the kelley-bryson gradient procedure. J. Guid. Control. Dyn. **13**(5), 926–928 (1990)
131. I.A. Ozkan, M. Koklu, I.U. Sert, Diagnosis of urinary tract infection based on artificial intelligence methods. Comput. Methods Programs Biomed. **166**, 51–59 (2018)
132. I.R. Abubakar, S.O. Olatunji, Computational intelligence-based model for diarrhea prediction using demographic and health survey data. Soft. Comput. **24**(7), 5357–5366 (2020)
133. J.R. Quinlan, Decision trees as probabilistic classifiers, in *Proceedings of the Fourth International Workshop on Machine Learning* (Elsevier, 1987), pp. 31–37
134. I. Ioniţă, L. Ioniţă, Prediction of thyroid disease using data mining techniques. Broad Res. Artif. Intell. Neurosci. (BRAIN) **7**(3), 115–124 (2016)
135. V. Thangaraj, V. Natarajan, Glaucoma diagnosis using support vector machine, in *2017 International Conference on Intelligent Computing and Control Systems (ICICCS)* (IEEE, 2017), pp. 394–399
136. M. Ahmad, V. Tundjungsari, D. Widianti, P. Amalia, U.A. Rachmawati, Diagnostic decision support system of chronic kidney disease using support vector machine, in *Second International Conference on Informatics and Computing (ICIC)*. (IEEE, 2017), pp. 1–4
137. K.C. Hsu, C.H. Lin, K.R. Johnson, C.H. Liu, T.Y. Chang, K.L. Huang, Y.-C. Fann, T.-H. Lee, Autodetect extracranial and intracranial artery stenosis by machine learning using ultrasound. Comput. Biol. Med. **116**, 103569 (2020)
138. E. Marquez, V. Barrón, Artificial intelligence system to support the clinical decision for influenza. in *IEEE International Autumn Meeting on Power, Electronics and Computing (ROPEC)*. (IEEE, 2019), pp. 1–5
139. A.P. Pawlovsky, An ensemble based on distances for a knn method for heart disease diagnosis, in *International Conference on Electronics, Information, and Communication (ICEIC)* (IEEE, 2018), pp. 1–4
140. M.G. Feshki, O.S. Shijani, Improving the heart disease diagnosis by evolutionary algorithm of pso and feed forward neural network, in *Artificial Intelligence and Robotics (IRANOPEN)* (IEEE, 2016), pp. 48–53

141. G. Cosma, G. Acampora, D. Brown, R.C. Rees, M. Khan, A.G. Pockley, Prediction of pathological stage in patients with prostate cancer: a neuro-fuzzy model. PLoS One **11**(6), e0155856 (2016)
142. S. Vijayarani, S. Dhayanand, Liver disease prediction using svm and naïve bayes algorithms. Int. J. Sci. Eng. Technol. Res. (IJSETR) **4**(4), 816–820 (2015)
143. X. Liu, X. Wang, Q. Su, M. Zhang, Y. Zhu, Q. Wang, Q. Wang, A hybrid classification system for heart disease diagnosis based on the rfrs method. Comput. Math. Methods Med. **2017** (2017)
144. X. Gu, T. Ni, H. Wang, New fuzzy support vector machine for the class imbalance problem in medical datasets classification. Sci. World J. **2014** (2014)
145. C. Subbulakshmi, S. Deepa, Medical dataset classification: a machine learning paradigm integrating particle swarm optimization with extreme learning machine classifier. Sci. World J **2015** (2015)
146. S. Ismaeel, A. Miri, D. Chourishi, Using the extreme learning machine (elm) technique for heart disease diagnosis, in *IEEE Canada International Humanitarian Technology Conference (IHTC2015)*. (IEEE, 2015), pp. 1–3
147. S.K. Sen, S. Dash, Application of meta learning algorithms for the prediction of diabetes disease. Int. J. Adv. Res. Comput. Sci. Manag. Stud. **2**(12) (2014)
148. B. Subanya, R. Rajalaxmi, Feature selection using artificial bee colony for cardiovascular disease classification, in *2014 International Conference on Electronics and Communication Systems (ICECS)* (IEEE, 2014), pp. 1–6
149. C. Liang, L. Peng, An automated diagnosis system of liver disease using artificial immune and genetic algorithms. J. Med. Syst. **37**(2), 1–10 (2013)
150. A. Iyer, S. Jeyalatha, R. Sumbaly, Diagnosis of diabetes using classification mining techniques. arXiv preprint arXiv:1502.03774 (2015)
151. J.J. Christopher, H.K. Nehemiah, A. Kannan, A swarm optimization approach for clinical knowledge mining. Comput. Methods Programs Biomed. **121**(3), 137–148 (2015)
152. A. Lahsasna, R.N. Ainon, R. Zainuddin, A. Bulgiba, Design of a fuzzy-based decision support system for coronary heart disease diagnosis. J. Med. Syst. **36**(5), 3293–3306 (2012)

Aggregation of Semantically Similar News Articles with the Help of Embedding Techniques and Unsupervised Machine Learning Algorithms: A Machine Learning Application with Semantic Technologies

Nitesh Tarbani and Kanchan Wadhva

Abstract Business news helps leaders and entrepreneurs in decision-making every day. This involves making corporate strategies, taking marketing decisions, planning operations, investing human capital, etc. This news gives a complete idea to leaders and entrepreneurs about what is happening in the corporate world. They maintain track of all mergers and takeovers and make interested people informed. Today, it is essential for people to keep themselves updated about corporate business. As there are so many news websites and the same news article gets published on each of these websites with a little changed title. As a consequence of which people have to spend far longer trying to find information than the time they have to catch up on the news. So, it would be very helpful for such people, if clusters of semantically similar news articles from different websites could be created, and reading only one news item from one cluster will be sufficient. This chapter will explain a few approaches to aggregate similar news articles. The very first step is to collect the data. Initially, for developing the model, data is collected from sites such as Kaggle, UCI, etc. After the model is developed, real-time data can be collected using the news API. The second step is to preprocess the collected data which involves subtasks such as Tokenization, Stop-Word Removal, Stemming/Lemmatization, Case Transformation, etc. The third step here is Embedding Text to Vectors, using various embedding techniques such as Bag-of-Words, TF-IDF, Word2Vec, etc. The next task here is to make clusters of these embedded vectors or numbers, using various unsupervised learning algorithms such as K-means, agglomerative, etc. Finally, the last step here is to find a comparison of various combinations of embedding techniques and clustering algorithms.

N. Tarbani (✉)
Sipna COET, Amravati, India
e-mail: nmtarbani@sipnaengg.ac.in

K. Wadhva
Great Learning, Hyderabad, India
e-mail: kanchan.wadhava@gds.ey.com

E. H. Houssein et al. (eds.), *Integrating Meta-Heuristics and Machine Learning for Real-World Optimization Problems*, Studies in Computational Intelligence 1038,
https://doi.org/10.1007/978-3-030-99079-4_5

Keywords Semantic · News-article · Clustering · Embedding · Unsupervised learning · K-means · Agglomerative · Bag-of-words · TF-IDF · Word2Vec

1 Introduction

In recent years, there is an unbelievable and massive increase in the rate of news that is published, has been observed. People are living in an era that is full of news, information, and data. So, news has become one of the important parts of society and the public. To keep up with the newest data and inputs, people read the daily news. This news may be about superstars, climate, sports, future technologies, and the nation, or many other areas. As the internet is full of news websites, people need to spend a far longer time finding the required information than the time they have for gathering the desired knowledge. To overcome this drawback, this chapter is going to present a machine learning approach to aggregate similar news articles from different websites. This will be achieved in two phases; the first phase will be to embed text into vectors which can be done using various methods such as Bag-of-Words, TF-IDF, Word2Vec, One Hot Encoding, etc. The second phase will be clustering the similar news articles which can be done by applying various unsupervised machine learning algorithms such as K-means, hierarchical, etc.

Working of Various Embedding Techniques

The first embedding technique is Bag-of-Words. A Bag-of-Words model, many times refers as BoW, is used to extract features from the text so that it can be used in modeling and with machine learning algorithms. This embedding technique is very easy and flexible and can be applied in various ways to extract features from document text. This embedding technique embeds text into vectors by describing the occurrence of words in the document. This technique is called as Bag-of-Words as it only cares about the presence or absence of the word in the document, it does not care about the order or structure of the word in the document.

The second embedding technique is Term Frequency-Inverse Document Frequency (TF-IDF). TF-IDF is an embedding technique that assesses how the particular word is related to a document from the collection of documents. This is achieved by multiplying two values: the frequency of the word in the documents, and the word's inverse document frequency across a collection of documents. There are many ways to calculate the term frequency of a word in a document; the simplest one is to find out the number of occurrences of a word in a document. Then the frequency can be adjusted in different ways, such as total words present in documents, the frequency of the most frequent words in a document. Another term is the inverse document frequency of a word across the collection of documents. This actually calculates how rare and common a word is across the collection of documents. For common words, the value of this term will be closer to 0 whereas if the word is rare then its value will be closer to 1. This term is calculated by taking the logarithm of the ratio of a total number of documents and the number of documents containing the word. Hence, for words with more frequency, this value will be closer to 0 and vice-versa. Multiplication of these frequencies, i.e. term frequency and inverse document frequency will generate the TF-IDF score of the word for the particular document. A more relevant word for the document will have a higher TF-IDF score.

The third embedding technique is Word2Vec. Word2Vec is actually a two-layer neural network and it is trained to produce a vector for a word that retains the linguistic contexts of the word. The input to Word2Vec is a huge amount of words and the output is vector space with large dimensions; each unique word is represented with a corresponding vector. The vector assigned to the word is such that vectors of words having similar meanings will be located in close proximity to each other in the vector space. For generating word embedding from text, Word2Vec is considered to be a computationally-efficient embedding technique.

Working of Various Clustering Techniques

Though there are many ways of clustering techniques, only two of those are highlighted in this chapter. Those are K-means and Hierarchical clustering algorithms.

The k-means clustering algorithm tries to group a given unidentified dataset (a data set with no class identity information) into a k (fixed) number of clusters. In this, initially, k centroids are chosen randomly. The real or imaginary data point at the center of the cluster is called a centroid. In the first round, the k-centroids are randomly picked from existing data points from the given dataset such that all centroids are unique ($ci \neq cj$, for all centroids ci and cj). The distance of each point in the data set is calculated from these centroids. Each data point from the given data set will be then assigned to that centroid which is at the minimum distance from that data point. The mean of points assigned to each centroid are calculated and these means will be considered as new centroids for the dataset. This adjustment of centroids is repeated until the centroids get stabilized. These stabilized centroids are then used to create clusters of the points in the dataset. That means an unidentified dataset with no class identity information is now clustered into data points with class identity (Fig. 1).

The second clustering technique which is highlighted in this chapter is Hierarchical clustering, also known as hierarchical cluster analysis, which is a clustering-techniques that attempts to group similar data points into groups called clusters. The output of this clustering technique is a collection of clusters, in which each cluster is separate from every other cluster, and the data points assigned to the cluster are largely similar to each other. In hierarchical clustering, initially, each data point from the data set is considered as a separate cluster. Then, the following two steps are executed repeatedly: (1) identification of the two clusters that are nearest to each other, and (2) merging of two most-closest clusters. These iterations are continued until all the clusters are merged. This is shown in the following diagram (Fig. 2).

The hierarchical relationships between the clusters are shown by the diagram called dendrogram, which is the main output of hierarchical clustering as shown below (Fig. 3).

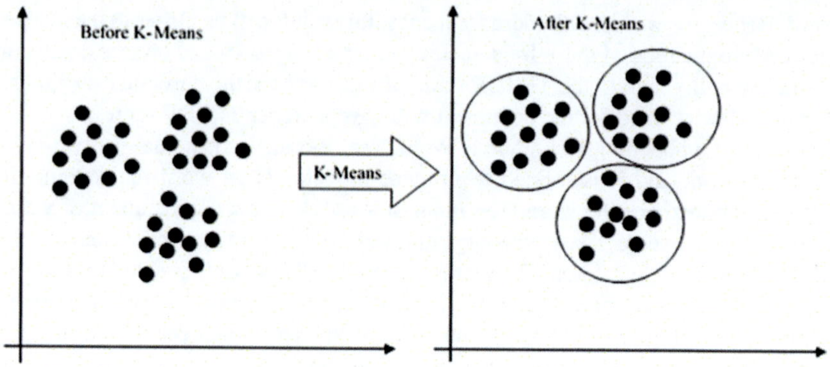

Fig. 1 Working of K-means clustering

2 Review of Literature

Algorithms used in text clustering could be categorized into several groups such as generative algorithms, k-means variations, spectral algorithms, phrase-based methods, dimensionality reduction methods, and vector space models [1]. It has been observed that the vector space model approach gives better results when topics are homogeneous and the number of clusters is already known [2]. Various researches show that Generative algorithms are affected by outliers and they are not found so effective with heterogeneous topics and they also need the number of clusters [1]. From various studies, it can be stated that the most popular methods for hierarchical and partitioned clustering techniques are K-Means and its extension. K-Means and its extension have a few weaknesses such as they are not effective for large data, they are based on random initialization, they are sensitive to outliers, the number of clusters must be already known, etc. [3]. It has been also observed that the accuracy of spectral clustering methods can be high if the vector model of data can be presented as a bipartite graph. The most important benefit of these methods is that there is no need to mention the number of clusters [4–6]. Dimensionality reduction methods which were initially designed for applications related to computer vision are also found effective for document clustering. However, their result may vary across runs on the same data, as they begin with random initialization. Their main advantages include that their performance is high and a few of these methods can also give an ideal number of clusters for given data [7, 8]. Due to the encoding of word order information, the performance of phrase-based methods has been enhanced. The main drawback here is that they don't assure better accuracy as compared to other clustering techniques [9]. To perform news headlines-based clustering, Xia et al. have presented a discriminative bi-term topic model in [10]. In [11], Himelboim et al. present clustering of Twitter network topics for analysis of social networks. To find out short text semantic similarity which is used for query analysis of search engine Sahami et al. have presented a special kernel function [12]. Banerjee et al.

Fig. 2 Working of hierarchical clustering

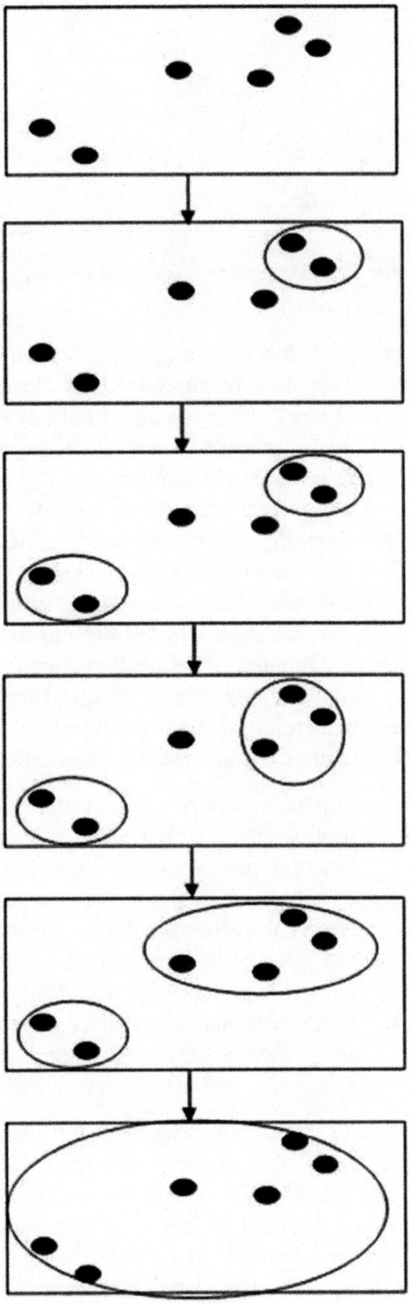

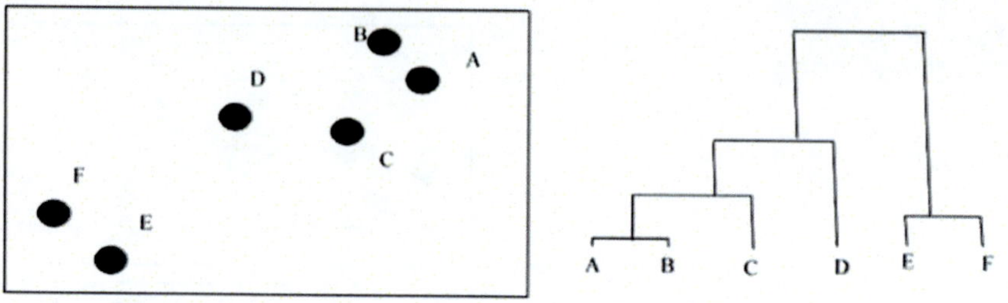

Fig. 3 Dendrogram of hierarchical clustering

proposed that feature generation from Wikipedia could be used to enhance clustering accuracy for short text [13]. Conrad and Bender presented that to design even centric news clustering algorithms, agglomerative clustering methods can be used [14]. Weber proposed a news collecting and clustering method with the help of cosine similarity-based clustering [15].

Sonia Bergamaschi et al. proposed RELEVANTNews, a web feed reader that clusters similar news available in various news websites on different days on its own. This web feed reader tool is based on the previously developed tool RELEVANT, which is used to calculate relevant values for string attributes. This tool clusters the news title related to a query given by the user. This tool could be using syntactic and lexical similarity of the news to identify a set of related news [16].

Various aggregators are designed and developed. Almost all of them are deployed as commercial products and therefore their internal working is not disclosed [16]. They can be categorized into three different categories:

1. Simple readers give only visualization and collection of RSS feeds from various news websites with the help of a graphical interface. They only provide simple functions which support the user in reading (e.g. news association to a map, different ordering, search engine, etc.);
2. News categorizer provides the news categorized depending on criteria occasionally given by the user. Simple categorizations may exploit the classes and/or the keywords given by the websites;
3. Unconventional aggregators deliver advanced functions for accessing, gathering, categorizing, and storing news so that users can read required news easily.

In [17], Velthune proposed a news search engine. This tool used a naive classifier that categorizes the news in a few classes. Unlike this method, RELEVANTNews finds clusters of news related to the same topic depending on their title. Categorizing thousands of news in a small number of classes leads to huge sets containing many numbers of news articles that make news reading inefficient for the user. Li et al. presented an aggregator, known as RSS Clusgator System (RCS), in which a system was implemented for time-based updating of cluster contents [18]. Radev et al. proposed an advanced aggregator called NewsInEssence, that used TF*IDF clustering algorithm for clustering similar news. It also provided users synthesis of news articles [19]. The main advantage of RELEVANTNews is that it is based on a parameterized clustering algorithm with the help of lexical/dominance/syntactic relationships and to improve the accuracy of the clusters this algorithm can be tuned with the help of parameters passed to it. However, this algorithm does not provide synthesis to readers.

Hamborg et al. used matrix-based analysis (MNA) for the collection of news which involved the following five steps: In the first and second steps they collected data, extracted useful articles from news websites, and then these news articles were stored in the database. In the third step clusters of news articles were created. In the fourth and fifth steps, a summary of the articles was created and then articles were presented to the user using various visualization techniques. Before the clustering step, to provide stats of the entities, they used matrix-based analysis. At the beginning of the analysis, users used to provide his/her requirement and for this purpose default values were prepared by MNA. Once this is done, matrix initialization was extended over the two desired selected dimensions and searched in all cells for the cell documents. The summarization step was achieved in the following way: summary of the topic, a summary of cells, and summarization of both using TF-IDF for all cells present in the matrix [20].

Grozea et al. were collecting the content from various websites like articles found, blogs containing news headlines, etc. Their belief was summarized and short data are provided by Rich Site Summary (RSS). Summarized and short data are desirable for news clustering which is still an effective solution for article indexing. A user subscribed to Rich Site Summary feeds can get fast access to the desired news article which eliminates the need of searching for news on a large number of websites. The main step in this application is to create HTTP requests from clients which are to be received by a web server. Once an HTTP request is created and received at a web server, python was used for downloading feeds from Rich Site Summary and for accessing articles from it based on input from the user. Periodically, subscribed users receive some requests from web servers. In case of any updates/upgrades, these will be downloaded and stored on the client-side [21].

Paliouras et al. aimed to retrieve required information from definite sources with the help of Rich Site Summary along with HTML based on wrappers and parsers. After retrieving information, using web-based interface authors adjusted this information to various news classes and personalized views. They also gave a brief about how the content scanner was developed with the help of HTML and Rich Site Summary. In this, wrapping was the first step where recognizing the URL of new articles according to the category of news and then this address of each category was stored in the database along with the matching wrapper. After Wrapping, new items were used to find information for indexing the article [22].

According to [23], Sundaramoorthy et al. were gathering the news from multiple news web-sites, magazines, television, and websites and combining all of them into one concise website. As the content and data are short and summarized, it allows users to have quick access to desired news. Here, the work was to periodically recover Rich Site Summary reports from definite websites. To get more accurate results along with Rich Site Summary they used web crawling also. Web crawling is a technique used to access and retrieve a large amount of information from various websites.

In [24], for text clustering and classification, Amer and Abdalla have proposed set theory based similarity measures (STB-SM) using KNN for classification, the K-means algorithm for grouping, and BoW model for feature selection.

3 Proposed Methodology

Proposed methodology is divided into following five steps.

3.1 Data Collection

Data collection is done in two phases. Initially, for the result analysis, the data is manually collected from various datasets available on the internet. Once the model is developed, the model can be used for real-time news that will be collected from various news APIs.

3.2 Data Preprocessing

In this step, the collected data is cleaned using various techniques of natural language processing such as Tokenization, Removing Stop Words, Stemming, and Case Transformation.

Tokenization: It is the process of segmenting text into words, clauses, or sentences. In this, words are separated out and punctuations will be removed.

Removing Stop Words: It is the removal of commonly used words which are unlikely to be useful for learning. The words such as in, of, are, etc. are removed from the sentences.

Stemming: This process is to reduce related words to a common stem.

Case Transformation: All alphabets in sentences are converted to lowercase.

3.3 Embedding Text to Vectors

As Machine Learning algorithms cannot be applied to text directly, there is a need to embed all sentences into vectors. To achieve this, there are many techniques commonly used such as Bag-of-Words, TF-TDF, Word2Vec, etc. In this chapter, all these techniques are tried to identify which of these techniques is better.

3.4 Clustering

In this step, clusters of similar news articles are created. To achieve this, various unsupervised machine learning algorithms are tried such as K-Means, Agglomerative Clustering, Hierarchical clustering, Hashing, etc.

3.5 Clustering-Validation (Result Analysis)

In this step, validation of created clusters is found out for each clustering algorithm. The algorithm giving better clusters is used for the final model (Fig. 4).

Among the combinations of the above-mentioned embedding techniques and clustering techniques, the combination of TF-IDF and Hierarchical clustering is found to be the most successful combination. Following is the python code for the same. The result analysis of these combinations is shown in the next heading.

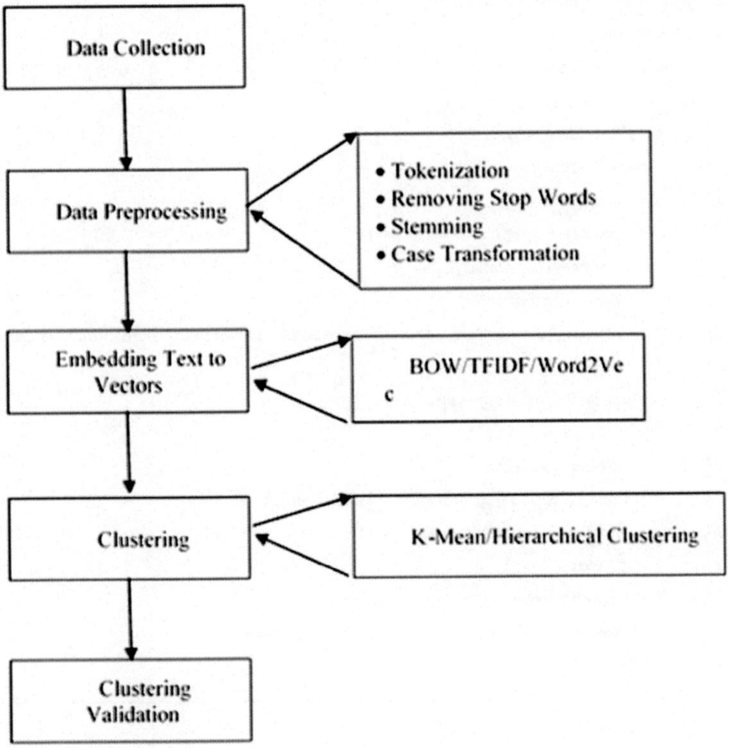

Fig. 4 Proposed methodology

To run the code successfully, various libraries are imported into the python code such as re, nltk, sklearn.feature_extraction.text, scipy. cluster etc. The complete code is written in the function cluster. Here the default value of the threshold is 0.6. A threshold value is required in hierarchical clustering Threshold value of hierarchical clustering states how much dissimilarities between the news articles are allowed in the same cluster. In the above code, the first 8 lines are used to import various libraries. In line no. 13–19, all articles are preprocessed. In line no. 22 and 23, TF-IDF embedding is applied. In line no. 25 and 26, hierarchical clustering is applied. Final clusters which are stored in object C, are returned in line 27.

```
1.  import re
2.  import nltk
3.  nltk.download('stopwords')
4.  from nltk.corpus import stopwords
5.  from nltk.stem.porter import PorterStemmer
6.  from sklearn.feature_extraction.text import CountVectorizer, TfidfTransformer
7.  from scipy.cluster import  hierarchy
8.  import numpy as np
9.  def cluster(ip,threshold=0.6):
10.     print(threshold)
11.     ps=PorterStemmer()
12.     cleanded_titles=[]
13.     for i in range(len(ip)):
14.             review=re.sub('[^A-Za-z]',' ',ip[i])
15.             review=review.lower()
16.             review=review.split()
17.             review=[ps.stem(word) for word in review if not word in stopwords.words('english')]
18.             review = ' '.join(review)
19.             cleanded_titles.append(review)
20.     while("" in cleanded_titles) :
21.             cleanded_titles.remove("")
22.     X = CountVectorizer().fit_transform(cleanded_titles)
23.     X = TfidfTransformer().fit_transform(X)
24.     X = X.todense()
25.     Z = hierarchy.linkage(X,"average", metric="cosine")
26.     C = hierarchy.fcluster(Z, threshold, criterion="distance")
27.     return list(C)
```

Table 1 Accuracy of embedding and clustering techniques

Embedding technique	Clustering technique	Accuracy (%)
Bag-of-Words	K-Means	69.54
TF-IDF	K-Means	79.33
Word2Vec	K-Means	81.82
Bag-of-Words	Hierarchical	73.23
TF-IDF	Hierarchical	91.42
Word2Vec	Hierarchical	89.67

4 Result Analysis

To check the accuracy of the proposed methodology, a dataset of 1.6 million news articles was taken from the Kaggle website. Applying embedding and clustering techniques to all these 1.6 million news articles was not possible because of hardware constraints. So, a subset of 1189 records is taken from the complete dataset. Various embedding techniques were applied to a subset of news articles and then clustering techniques were applied to embedded news articles. The value of K has to be passed to K-Means clustering. Looking into the chosen subset, the value of K was selected as 22. Whereas, the threshold value has to be passed to Hierarchical clustering. The threshold value equal to 0.6 was giving a better result than any other threshold value. The following table shows the observed result for various embedding techniques and clustering techniques (Table 1).

Table 2 shows the actual and predicted number of news items in each cluster, obtained using TF-IDF embedding Technique and Hierarchical Clustering Technique. It can be that for 5 clusters out of 22 clusters, accuracy is 100%. For all other clusters, accuracy is more than 80%. So, out of a total of 1189 news articles, 1087 were correctly clustered, which is equal to 91.42%.

5 Conclusion

As the internet is full of news websites, people need to spend a far longer time finding required information than the time they actually have for gathering the desired information. To overcome this drawback, this chapter has presented a machine learning approach to aggregate similar news articles from different websites. This is achieved in two phases; the first phase is to embed text into vectors which can be done using various methods such as Bag-of-Words, TF-IDF, Word2Vec, One Hot Encoding, etc. The second phase is clustering the similar news articles which can be done by applying various unsupervised machine learning algorithms such as K-means, hierarchical, etc.

As the result analysis shows, among the tried combinations of embedding techniques and clustering techniques, the combination of TF-IDF and Hierarchical

Table 2 TF-IDF embedding technique and hierarchical clustering technique prediction

Cluster number	Number of titles in actual cluster (story)	Number of titles in predicted cluster (story)	Cluster wise accuracy
1	24	24	100.00
2	10	10	100.00
3	30	30	100.00
4	8	8	100.00
5	9	9	100.00
6	31	30	96.77
7	31	30	96.77
8	161	148	91.93
9	121	111	91.74
10	53	48	90.57
11	21	19	90.48
12	126	104	82.54
13	10	9	90.00
14	112	106	94.64
15	123	114	92.68
16	36	32	88.89
17	51	45	88.24
18	61	49	80.33
19	60	58	96.67
20	52	49	94.23
21	32	29	90.63
22	27	25	92.59
Total	**1189**	**1087**	**91.42**

clustering is found to be the most successful combination. The accuracy of this combination is found to be 91.42%.

Futurework in news aggregation can be that various deep learning-based embedding techniques such as BERT can be applied for embedding text into vectors and for clustering purposes Locality Sensitivity Hashing-based clustering technique can be applied. This may increase the accuracy of the system. As K-Means needs the value of K to be provided and hierarchical clustering needs threshold to be provided. There is a need to choose these values based on many trials. Locality Sensitivity Hashing-based clustering technique may be the solution for these drawbacks of K-Means and Hierarchical clustering.

References

1. N.O. Andrews, E.A. Fox, *Recent Developments in Document Clustering* (2007)
2. G. Salton, A. Wong, C.S. Yang, A vector space model for automatic indexing. Commun. ACM **18**(11), 613–620 (1975)
3. M. Steinbach, G. Karypis, V. Kumar, A comparison of document clustering techniques. Technical Report (Department of Computer Science and Engineering, University of Minnesota, 2000)
4. F. Bach, M. Jordan, Learning spectral clustering, in *Advances in Neural Information Processing Systems 16 (NIPS)*. ed. by S. Thrun, L. Saul, B. Schölkopf (MIT Press, Cambridge, 2004), pp. 305–312
5. D. Cheng, S. Vempala, R. Kannan, G. Wang, A divide-and-merge methodology for clustering, in *PODS '05: Proceedings of the Twenty-Fourth ACM SIGMOD-SIGACT-SIGART Symposium on Principles of Database Systems* (ACM Press, New York, NY, USA, 2005), pp. 196–205
6. C.H.Q. Ding, X. He, H. Zha, M. Gu, H.D. Simon, A min–max cut algorithm for graph partitioning and data clustering, in *ICDM '01: Proceedings of the 2001 IEEE International Conference on Data Mining* (IEEE Computer Society, Washington, DC, USA, 2001), pp 107–114
7. S. Osinski, J. Stefanowski, D. Weiss, Lingo: search results clustering algorithm based on singular value decomposition, in ed. M.A. Klopotek, S.T. Wierzchon, K. Trojanowski, *Intelligent Information Systems, Advances in Soft Computing* (Springer, Berlin, 2004), pp 359–368
8. D. Greene, P. Cunningham, Producing accurate interpretable clusters from high-dimensional data, in *9th European Conference on Principles and Practice of Knowledge Discovery in Databases*, vol. 3721 (University of Dublin, Trinity College, Dublin, 2005), pp. 486–494
9. O. Zamir, O. Etzioni, Web document clustering: a feasibility demonstration, in *SIGIR '98: Proceedings of the 21st Annual International ACM SIGIR Conference on Research and Development in Information Retrieval* (ACM Press, New York, NY, USA, 1998), pp. 46–54
10. Y. Xia, N. Tang, A. Hussain, E. Cambria, *Discriminative Bi-Term Topic Model for Headline-Based Social News Clustering in FLAIRS Conference* (2015)
11. I. Himelboim, M.A. Smith, L. Rainie, B. Shneiderman, C. Espina, Classifying Twitter topic-networks using social network analysis. Soc. Media + Soc. **3**(1), (2017)
12. M. Sahami, T.D. Heilman, A web-based kernel function for measuring the similarity of short text snippets, in *WWW* (ACM, New York, NY, USA, 2006), pp. 377–386
13. S. Banerjee, K. Ramanathan, A. Gupta, Clustering short texts using Wikipedia, in *Proceeding SIGIR '07 Proceedings of the 30th Annual International ACM SIGIR Conference on Research and Development in Information Retrieval* (2007), pp. 787–788
14. J.G. Conrad, M. Bender, Semi-supervised events clustering in news retrieval, in *NewsIR@ECIR* (2016)
15. M. Weber, Finding news in a haystack: event based clustering with social media based ranking. Master thesis for the media technology programme, Leiden University, The Netherlands, 2012
16. S. Bergamaschi, F. Guerra, M. Orsini, C. Sartori, M. Vincini, Relevant news: a semantic news feed aggregator, in *Semantic Web Applications and Perspectives*, vol. 314, ed. G. Semeraro, E. Di Sciascio, C. Morbidoni, H. Stoemer (2007), pp. 150–159
17. A. Gulli, The anatomy of a news search engine, in *WWW (Special Interest Tracks and Posters)*, ed. A. Ellis, T. Hagino (ACM, New York, 2005), pp. 880–881
18. X. Li, J. Yan, Z. Deng, L. Ji, W. Fan, B. Zhang, Z. Chen, A novel clustering-based RSS aggregator, in Williamson et al. [11], pp. 1309–1310
19. D.R. Radev, J. Otterbacher, A. Winkel, S. Blair-Goldensohn, Newsinessence: summarizing online news topics. Commun. ACM **48**(10), 95–98 (2005)
20. F. Hamborg, N. Meuschke, B. Gipp, Matrix-based news aggregation: exploring different news perspectives, in *Proceedings of the 17th ACM/IEEE Joint Conference on Digital Libraries* (IEEE Press, 2017), pp. 69–78

21. C. Grozea, D.C. Cercel, C. Onose, S. Trausan-Matu, Atlas: news aggregation service, in *2017 16th RoEduNet Conference: Networking in Education and Research (RoEduNet)* (IEEE, 2017), pp. 1–6
22. G. Paliouras, A. Mouzakidis, V. Moustakas, C. Skourlas, PNS: a personalized news aggregator on the web, vol. 104 (1970), pp. 175–197
23. K. Sundaramoorthy, R. Durga, S. Nagadarshini, Newsone—an aggregation system for news using web scraping method, in *2017 International Conference on Technical Advancements in Computers and Communications (ICTACC)* (IEEE, 2017), pp. 136–140
24. A.A. Amer, H.I. Abdalla, A set theory based similarity measure for text clustering and classification. J. Big Data **7**, 74 (2020). https://doi.org/10.1186/s40537-020-00344-3

Integration of Machine Learning and Optimization Techniques for Cardiac Health Recognition

Essam Halim Houssein, Ibrahim E. Ibrahim, M. Hassaballah, and Yaser M. Wazery

Abstract Cardiovascular disease (CVD) remains the primary reason for illness and death throughout the world despite tremendous progress in diagnosis and treatment. Artificial intelligence (AI) technology can drastically revolutionize the way we perform cardiology to enhance and optimize CVD results. With boosting of information technology and the increased volume and complexity of data, aside from a large number of optimization problems that arise in clinical fields, AI approaches such as machine learning and optimization have become extremely popular. AI also can help improve medical expertise by uncovering clinically important information. Early on, the treatment of vast amounts of medical data was a significant task, leading to adaptations in the biological field of machine learning. Improvements are carried out and tested every day in algorithms for machine learning so that more accurate data may be analyzed and provided. Machine learning has been active in the realm of healthcare, from the extraction of information from medical papers to the prediction and diagnosis of a disease. In this perspective, this chapter provides an overview of how to use meta-heuristic algorithm on CVD's classification process for enhancing feature selection process, and various parameters optimization.

Keywords Feature selection · Metaheuristics algorithms · Cloud · CVD · Engineering design problems

E. H. Houssein (✉) · Y. M. Wazery
Faculty of Computers and Information, Minia University, Minia, Egypt
e-mail: essam.halim@mu.edu.eg

I. E. Ibrahim
Faculty of Computers and Information, Luxor University, Luxor, Egypt

M. Hassaballah
Department of Computer Science, Faculty of Computers and Information, South Valley University, Qena, Egypt

E. H. Houssein et al. (eds.), *Integrating Meta-Heuristics and Machine Learning for Real-World Optimization Problems*, Studies in Computational Intelligence 1038,
https://doi.org/10.1007/978-3-030-99079-4_6

1 Introduction

These days, cardiovascular disorders are increasingly frequent, describing a variety of conditions in which your heart is affected. Global mortality from (Health) CVDs are estimated in the World Health Organization to be 17.9 million [1]. Heart death, the primary reason for which is cardiac arrhythmia, is the death caused by cardiovascular illnesses. When our heartbeat functions abnormally, it is the cardiac arrhythmia which is associated with cardiac disorders [2].

The traditional paradigm of CVD diagnostics is built on medical and clinical assessments of individuals. The results are interpreted according to a set of medical quantitative factors to classify patients on the basis of the medical conditions taxonomy [3]. Often owing to the vast amount of heterogeneous data, the standard rule-based diagnostics paradigm is ineffective and requires extensive analysis and medical know-how to ensure appropriate accuracy in diagnosis.

Our project can benefit anyone who can use their medical history to diagnose cardiac illness. It recognises who are all affected by heart diseases such chest pain or high blood resurrection and can help diagnose diseases with fewer medical tests and efficient therapy to ensure that they can be treated. Diagnostics with computer aid (CAD) is now one of medical diagnostics' leading study subjects. CAD's primary notion is to give doctors a second perspective on computer output. Electrocardiogram (ECG) [4–7] is an electric heart activity measurement that analyses the performance of a person's heart. This test allows the doctor to diagnose the patient's condition. Because the ECG signals are non-steady, they cannot be examined using previous techniques because they are constrained [8].

The ECG provides a realistic record of the direction and size of the electrical turmoil caused by depolarization of the auricles and ventricles and re-polarization. The P-QRS-T waves are a single cardiac cycle in the ECG signal. Due to atrial depolarization, P-wave in the ECG signal arises. It is the activation of the heart's high chambers. Due to ventricular depolarization that leads to major pumping contraction, QRS-complex ensues. T-wave develops because of ventricular re-polarization. It is the arousal of the ventricles [9].

The characteristics for a signal are divided into two divisions (i) Morphologic and (ii) statistical characteristics. The morphological characteristics are QRS, T and P, the amplitudes are R, S, P and T and the intervals of demarcation are QRS, T, P, RR and ST. Statistical characteristics, like variance, mean, skewness, defect, spectral entropy and kurtosis [2]. There were a lot of descriptors to characterise the ECG signal. A new descriptor or a new mix of existing features is provided in every consecutive published paper. The ECG is based on various transforms, e.g. transform wavelet (WT), analysis component principle (PCA), functionality of Hermits and statistic features, for example, mean, differences.

Machine learning (ML) techniques [10] have been used in various sectors for the objectives of classification and prediction. It was recently noted that ML have been strongly developed for a wide variety of medical jobs, with a considerable impact on the accuracy of the categorization. Modern CADS systems employ ML in order to detect arithmetic of the ECG signal, which reduces the cost of continuous cardiac surveillance and improves the quality of the forecasts. But the ECG-based categorization of automatic arrhythmia often presents many major obstacles [11].

Simple classification models like Linear Discriminant or K-nearest neighbor have been utilized to date successfully. More advanced categorizations, such as fractal analysis, chaotic modelling, bi-spectral coherency analysis and artificial neural networks, have been used by others. The most prominent Artificial neural networks (ANNs) family with ECG classification is the multi-layer perception, although the other ANNs paradigms, such the Fuzzy neural grid, convolution neural network and Support Vector Machine have also widely been employed [12].

The meta-heuristic algorithms [13] such as Archimedes optimization algorithm [14], Lévy flight distribution [15], Honey Badger Algorithm [16], Particle Swarm Optimization (PSO) [17], and Henry gas solubility optimization [18] plays important rule in solving several real-world applications such as; Image segmentation [19–23], Heartbeats Classification [5–7, 24], Computer networks [25], drug design and discovery [26], cloud computing [27], Energy [28–31], fuel cell [32–34], photovoltaic (PV) system [35–38], Optimization problems [39, 40], Stock prediction [41], Feature selection [42, 43], ECG signals classification [2, 4], chemical descriptors [44, 45], and detecting covid-19 [46, 47].

The rest of the chapter is structured as follows: Sect. 2 explains the Cardiac health recognition. Section 3 introduces an overview of the Machine Learning Techniques. Section 4 presents the Optimization Techniques. The integration of Machine Learning and Optimization Techniques are discussed in Sect. 4. Some open issues and Challenges are introduced in Sect. 6. Section 7 concludes the chapter.

2 Cardiac Health Recognition

Socially speaking, the prevention of heart disease that is the world's most common cause of mortality is tremendously significant. Millions of individuals around the world are at danger of heart disease, according to official data [48]. This section reviews methodologies to efficiently recognise cardiac diseases based on the electrocardiograms (ECG) signal analysis.

Signal processing of ECG is widely utilised to diagnose a range of heart problems which is the largest cause of global premature mortality. Many algorithms for the automated, computerized and accurate detection of rhythms in an ECG record have been designed in recent years. The ECG is just a recording of the electrical activity of the heart. Measurement of electrodes put on the skin. The pulse rate, rhythm, and indirect indications of blood flow to the heart muscle are all measured. In the rear wall of the right atrium near the superior vena cava, a pacemaker termed the síatrial node (SA) generally determines the rhythm of the heart. The SA node consists of specialised cells, which are spontaneously generated at 100–110 action potential ("beats") a minute. The rhythm of the sinus generally governs the ventricular and atrial rhythms. The SA node's action potential spreads to the atria, depolarizes this tissue and causes atrial contraction. The pulse then goes through the atrioventricular

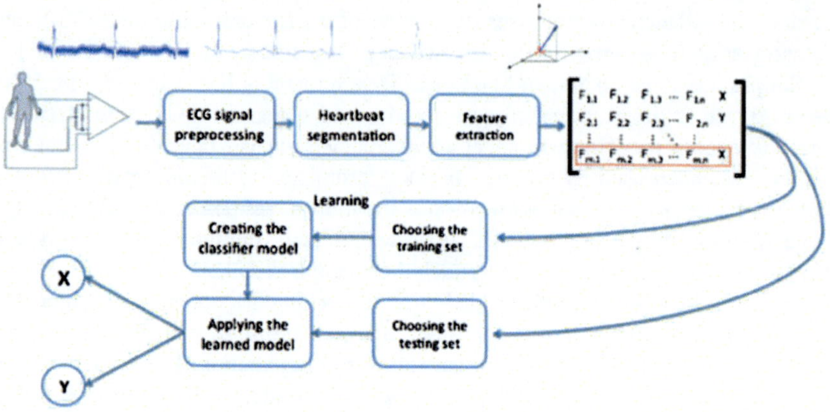

Fig. 1 Overall ECG classification system design

node into the ventricles (AV node). The depolarization pulse throughout the ventricles leads fast to specialised conduction routes (bundling branches and Purkinje fibres) within the ventricle to trigger ventricular contraction. The pacemaker activity of the SA node thereby controls the regular heart rhythm.

Traditional ECG classification system consists of (1) ECG raw data collection. (2) ECG data preprocessing. (3) heartbeats segmentation. (4) Features extraction. (5) Classification model. Figure 1 demonstrate overall ECG classification system design.

Firstly, ten electrodes are required to generate 12 electrical heart images. There are 12 leads: Leads of the Limb (I, II, III), Leads of Augmented Limb (aVR, aVL, aVF), and Leads of Precordial (V1, V2, V3, V4, V5, V6). Each heartbeat is shown as a number of electric waves with peaks and valleys. Two types of information are provided by every ECG. The first is the period of the electrical wave crossing the heart and the second is the electrical activity travelling through the heart muscle which determines whether or not the electrical activity is normal or slower or unregulated. Heart includes three hundred billion cardiovascular myocytes.

Cardio Signal is very cardiovascular, all cardio beat covering P wave, T wave and QRS Complex. PR, QRS, RR, QT and ST are all picks and intervals and segments of cardiac signals are ST and PR. The standard amplitude or interval value is preserved. As features are the above peaks, segments and intervals. The usual ECG wave and intervals are displayed in Fig. 2. Different waves comprising the ECG are the sequence of atria and ventricles depolarized and re-polarized.

The P wave is the depolarization wave that travels over the whole atria from the SA node. The short isoelectric period (null voltage) following the P wave is the time when an impulses are moving into the AV node and into the His bundle. The PR interval refers to the time between the commencement of the P wave and the start of the QRS complex. Ventricular depolarization is represented by the QRS complex. The time interval between QRS complexes can be used to compute the ventricular rate. This short duration implies that ventricular depolarization happens quite quickly in most cases. Following the QRS and finishing at the start of the T wave comes the

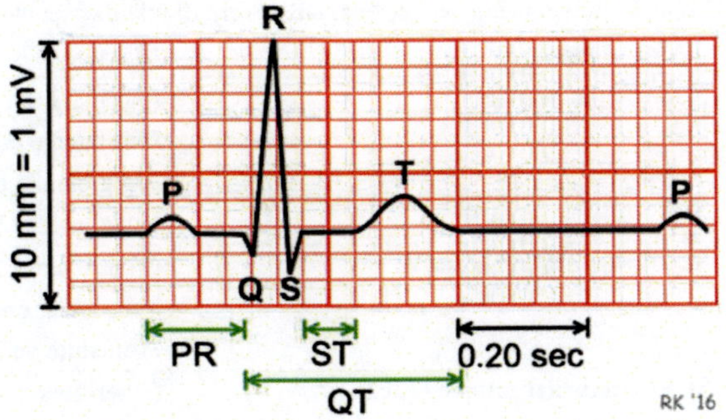

Fig. 2 Normal ECG Waves

Table 1 ECG Waves and intervals description

Wave	Amplitude in mV	Intervals	Period in ms
P	0.25	P and P-R	p = 80 and P-R = 120–200
R	1.60	RR	600–1200
Q	25% of R	QRS, QT, and ST	QRS< 120, QT = 200–400, and ST = 50–150
T	0.1–0.5	T	160

isoelectric phase (ST segment), during which both ventricles are totally depolarized. The ST segment is critical in the diagnosis of ventricular ischemia or hypoxia since it can become either depressed or raised in certain situations.

Ventricular repolarization is represented by the T wave. The T wave has a positive deflection in general. This is because the last cells in the ventricles to depolarize are the first to repolarize. This happens because the last cells to depolarize are found in the subepicardial region of the ventricles, where their action potentials are shorter than those found in the subendocardial sections of the ventricular wall.

The QT interval measures the time it takes for both ventricular depolarization and repolarization to occur, and hence approximates the length of an average ventricular action potential.

Table 1 and Fig. 2 below shows the usual ECG waveform duration and also the ECG waveform amplifier amplitude.

2.1 ECG Data

Most of the research have been based on data obtained from publicly accessible databases for instance the MIT-BIH Arrhythmia dataset (MIT-BIH), the European ST-T (EDB) and the INCART dataset in St. Petersburg (INCART). each dataest comprises information record consists of mixed attributes containing both the numerical and categorical data, in which each heartbeat was labelled in the dataset used for diagnostic pathways by cardiologists [49]. The extracting and processing of the data source records is the first step in ECG analysis (Table 2).

Table 2 Relationship between MIT-BIH arrhythmia dataset classes and AAMI standard classes

AAMI Classes	Classes
Non-ectopic beats(N)	(Normal beats Left bundle branch block beat Right bundle branch block beat Nodal escape beats)
Ventricular ectopic beats(V)	(Ventricular flutter wave Ventricular escape beat Premature ventricular contraction)
Supraventricular ectopic beats(S)	(Aberrated atrial premature beats Supraventricular premature beats Atrial premature beats Nodal (junctional) premature beats Nodal (junctional) escape beats Atrial escape beat)
Fusion beats(F)	(Fusion of ventricular and normal beats)
Unknown beat(Q)	(Paced beats Unclassifiable beats)

2.2 *MIT-BIH Arrhythmia Database*

The MIT-BIH dataset [49] contains forty-eight ECG files of approximately thirty minutes duration, 360 Hz sampling rate, and eleven-bit resolution from forty-seven different patients. The records have two signals: the first (MLII) is the modified lead for all records, and the second (based on the records) is to V 1, 2, 4, or 5. The MIT-BIH database forms are divided into five categories based on the Association for the Advancement of Medical Instrumentation's (AAMI) recommended practise, as shown in Table 3.

It is interesting to note that the dataset is particularly unbalanced, as nearly 90% of the beats belong to category (N), while only 15 samples belong to category (S) (F). In this study, the inter-patient partitioning paradigm proposed by Chazal et al. [50] is used to split the dataset into 2 domains (DS1 and DS2) in order to compare our

Table 3 Classes distribution according the AAMI standard

AAMI classes	MT-BIH classes	Total
Normal (N)	e, j, R, L, N	89695
Supraventricular ectopic beat (S)	A, a, x, J	2946
Ventricular-ectopic beats (VEBs)	V, E, !	7459
Fusion (F)	F	811
Unknown beat (Q)	f, Q	15

Table 4 Division of records into training (DS-1) and evaluation (DS-2) datasets

Dataset	Records number
Train (DS-1)	101, 106, 108–09, 112, 114–16, 118-19, 122, 124, 201, 203, 205, 207–09, 215, 220, 223, 230.
Test (DS-2)	100, 103, 105, 111, 113, 117, 121, 123, 200, 202, 210, 212-14, 219, 221–22, 228, 231-34.
Records not intended in DS-1 or DS-2	102, 104, 107, 217

Table 5 ECG databases were used for classification process in this work

Database	Files	Leads	Sample Rate	Duration
MIT-BIH	48	1	360 HZ	30 min
EDB	90	1	250 HZ	120 min
INCART	75	1	257 HZ	30 min

findings to those of other previous works. Every dataset contains data from 22 beat forms with a similar percentage as listed in Table 4, with the exception of records; 102, 104, 107, and 217, which are not regarded as suggested by the AAMI.

2.2.1 European ST-T Database (EDB)

The EDB was used to implement ST and T-wave architectures. This collection contains 90 annotated samples corresponding to 79 subjects' outpatient ECG records. Every recording lasts 2 h and includes two signals recorded at 250 samples per second.

2.2.2 St. Petersburg INCART Database (INCART)

The INCART dataset is a 12-lead arrhythmia dataset made up of 75 annotated recordings from 32 Holter records. Each record is 30 minutes long and contains 12 regular leads, each sampled 257 Hz. The initial data came from patients.

For the classification task, only one channel is used for each recording. Table 5 summarises the main characteristics used.

2.3 Data Filtering

The elimination of undesirable artefacts and noise is a prevalent difficulty in ECG interpretation. The ECG signal analysis can be influenced considerably by several types of noise such as power lines interference and basic line walking. Different kinds of filtering are used to remove noise sources. In order to reduce interference in electricity and base line wandering from ECG signals, Singhal [51] was applied as a decomposition technique.

For power line interface removal, Padma et al. [52] employed adaptive noise filtering, which is the interference power line, since the ECG signal also contains 50 Hz signal, and the extremely 50 Hz signal in the ECG signal will be lost if the standard band reject filter is used. The frequency of the power line can thus be deleted when 50 Hz signal in the original waveform is retained by means of adaptive noise filtering.

Lee et al. [53] employed the Ecg signal with an adaptive Least mean squares filter, but its convergence and performance create distortion and even poor performance, depending on the surroundings and the condition of the patient. They proposed a sub-band adaptive filtering (DSAF) that performed better in experiments, and as a result, they proposed an LMS adaptive filter.

Dehghani et al. [54] utilized the processing and analysis of computer-based signals. Baseline hiking is usually about 15 percent amplitude, full-scale deflection at frequencies between 0 and 0.15 Hz, which can only be suppressed by a high pass digital filter. To remove the walking baseline, a high pass Kaiser Window FIR filter was applied. They showed that after baseline roaming there are still various noise kinds that remain affect the ECG signal. This noise may be stochastic in a broadband way that typical digital filters are not able to eliminate it. Undecimated transformers, which have a better balance between smoothness and accuracy, are used for the elimination of broadband disturbances than the discrete transforms wavelet.

Morphologic filtering is a strong traditional technique to ECG drift estimation and cancellations (MF). The literature on the baseline estimate for ECG known as 1-stage MF and 2-stage MF algorithms has provided the two principal versions of MF algorithms. For morphologically smooth ECG the one-stage MF deploys a Structure Element (SE). Half the ECG period was used by the SE. Because the ECG estimate is quick and uncomplicated, the 1-stage MF baseline estimate can be prefixed quickly. The 2-stage MF method has a more accurate baseline evaluation. The 2-stage MF contains two phases of ECG morphological smoothing. The length of the ECG is therefore a key need for the establishment of two phases for each stage. While both traditional MF baseline estimates require setup of the MF parameter via ECG features prediction [55].

2.4 Heartbeats Segmentation

Before extracting features from a continuous ECG signal, it should be separated into discrete heartbeats. A heartbeat phase should capture as much information about the current heartbeat as feasible, but not the previous or next heartbeat component. The ECG signal is divided into separated heartbeats on the basis of R peaks identified. In the Literature for Ecg signals segmentation, the annotation positions of the database could be used, as well as any other Specialized QRS detection algorithms [56]. After

R-peak positions are obtained, the heartbeat will be segmented using the RR intervals obtained from R-peak positions.

Different automation techniques used deep learning algorithm for heartbeats segmentation. The model based on LSTM, which converts a series of heartbeats into a series of labels, where context dependence are recorded by the cell-status of the network, was proposed by Moussaw et al. [57]. Hannun et al. [58] suggested 33-layers arythmia detection neural networks, mapping ECG data to label sequences. However, it is not possible to acquire accurate arythmic areas. A modified U-net is utilised by Oh et al. [59] to recognise regions of heartbeat and the raw data, however this methodology requires further procedures to detect annotated arhythmia.

2.5 *Feature Extraction*

After noise removal, extraction of the feature is performed. The most significant aspect of the ECG signal analysis consists of the detection of fiduciary points within heartbeat. An essential objective of signal processing is to extract specific information from a certain heartbeat. For this reason, heartbeats are often converted to various domains, which allows for simpler reading of the needed information [60].

In addition to intervals such RR, PR, QRS, ST, and QT, the so-called wavelet transform (WT), presented in the following, is one of those used transformations. Wavelet transformation especially interesting for signal compression to reduce the number of bits of information required to store or transmit digitized ECG signals without significant loss of signal quality. The bulk of the feature was established using morphological traits that were also previously evaluated as very relevant. Some morphological elements are beats, beats power, beats min, beats max and beats Max-Min. Statistical aspects, including various time indexes of order and histogram variance have been accomplished.

Zhao et al. [61] suggested a method for removing features with wavelet transformers and vector machines. The authors introduced a novel feature extraction method for accurate cardiac rhythm detection. The proposed classification system is made up of three parts: data pre-processing, feature extraction, and ECG signal categorization. In order to achieve the feature vector of ECG data, two different extraction methods are employed jointly. The transformation wavelet is used to extract the transform coefficients as characteristics of each ECG section. At the same time, auto-regressive modelling (AR) is also used to take over ECG waveform temporal structures.

In [62] Mahmoodabadi et al. proposed a method for extracting ECG features that uses the Daubechies Wavelets transform. They created and tested a multi-resolution wavelet transform-based electrocardiogram (ECG) feature extraction method. Modified Lead II (MLII) ECG signals were chosen for processing. Better detection was accomplished using a wavelet filter with a scaling function that was more closely related to the form of the ECG signal. Their results showed that their proposed method for extracting ECG features has a sensitivity of 99.18 percent and a positive predictivity of 98 percent.

A new ECG approach for extracting features and detected corruption has been formulated by Sufi et al. [63]. It presents a new ECG hiding method, using a template-based approach to match cross correlation to identify all ECG features and the addition of noise. Without knowing the templates for feature matching and the noise, it is extremely impossible to reconstruct the obscured features. Three templates and three noises for P wave, QRS Complex and T wave were regarded to be significant components, which are only 0.4% to 0.9% of the original ECG file size.

2.6 Feature Selection

Selection of features (FS) plays a significant part in ECG pattern recognition system model training and testing. The aim of FS approaches is to identify prominent characteristics and remove noisy or dispensable features to provide optimal performance of the design recognition model. Instead of changing the original representation of variables, Fs strategies just choose a subset of the original properties in distinct from the dimensionality reduction methods based on projection or combinations. [64]. The basis for selecting features is to generate an efficient feature pool that includes positive candidates. State-of-the-art FS techniques use time-frequency algorithms to choose characteristics from sub-signals. These approaches translate sensory data into another place which may more vividly demonstrate the characteristics of original data. The statistical-based techniques of determining functional properties require examining the link between each input variable and the destination variable by use of statistics. The selection of statistical measures depends on the data type of the variables of input and output. These methods can be rapid and effective.

Swarm-based function selection strategy enables you to identify the best feature subset for developing the most accurate classifier model from the extreme huge dimensionality of features. In data mining there are still certain kind of studies not done yet. The use of swarm intelligence algorithms in high-dimensional information focused on medical data categorization for the feature selection process is covered in this research. we aim to investigate and explore to use swarm intelligence algorithms performance for features selection and parameters optimization to ECG classification.

2.7 Classification

There are different sorts of arrhythmias and every kind is linked to a pattern, so its type may be identified and classified. The rhythms can be categorised into two main types. The first class comprises of rhythms created by one irregular heartbeat, which is referred to here as morphological arrhythmia. The other category includes rhythms created by a number of irregular heartbeats, hereafter known as rhythmic arrhythms. This survey focuses on the classification of normal heartbeats and those that make up the former group. There are two key types such as raw time-series or functional techniques. Time series-based approaches directly use their original time series and functional approaches indirectly to calculate the required functions using the original time series. Different machine Learning techniques were investigated to classify both raw time-series or functional techniques in the following section. These approach allows for the analysis of larger and more diverse datasets in order to produce more accurate results and guide better decisions in real time without the need for human interference.

3 Machine Learning Techniques

Various important applications for machine learning in medicine include: smart electronic records of the health system, drug discovery, biological signals and detection and diagnosis of diseases. The development of ML systems is thought to emulate medical expertise in the identification of the illnesses, in most cases of disease identification and diagnosis. The use of machine training in the classification of diseases is fairly common and scientists are much more interested in creation of such systems to make diabetes and cardiovascular diseases easier to track and to diagnose.

The study was written in order to provide effective healthcare with a precise diagnosis of CVD at an early stage, by combining the principles of machine learning and swarm optimization algorithms. In recent years, a variety of technology has grown in ML to enable computer programmes to learn from the data, producing a model that recognises common models and can decide on the basis of information, the incompleteness of the medical database is difficult [65].

Machine training methods can be divided into four groups, including supervised learning, unsupervised learning, semi-Supervised learning and reinforcement learning. We employ known or labelled data for the training data in supervised learning. Since the information is known, the learning, i.e. successful execution, is supervised. The input data passes through and is used for the modelling process using the machine learning algorithm. After training on the basis of the known data, unknown data can be used for the model and a new answer can be given.

The training data in unsupervised learning is unknown and unlabeled, implying that no one has ever looked at it before. The input is not oriented towards the algorithm from which the uncontrolled term comes without the aspect of known data. This data is supplied into the algorithm for machine learning and used for model training. The trained model seeks a pattern and provides the desired answer

Semi-supervised machine learning combines the advantages of both supervised and unsupervised machine learning techniques. More frequent supervised machine learning methods involve training a machine learning algorithm on a "labelled" dataset in which each record contains the outcome information. Based on the information it already knows, the algorithm can detect patterns and find links between your target variable and the rest of the dataset. Unsupervised machine learning algorithms, on the other hand, learn from a dataset without an outcome variable.

In semi-supervised learning, an algorithm learns from a dataset that contains both labelled and unlabeled data, with the latter typically being the majority.

In reinforcement learning, the algorithm, like older kinds of data analysis, discovers data through trial and error and then selects which action leads in bigger rewards. Reinforcement learning is made up of three primary components: the agent, the environment, and the actions. The learner or decision-maker is the agent, the environment is everything with which the agent interacts, and the actions are what the agent does. When an agent chooses activities that maximise the expected reward over a set period of time, this is known as reinforcement learning. When the agent is working within a sound policy framework, this is the easiest to do.

The most popular machine learning method in medical applications is a classification because it connects with daily problems, the most commonly used are Support vector machine, Decision Trees, Random forests, K-nearest neighbor and artificial neural networks. Next sections provides more details about structure and classification ability for these models.

3.1 Support Vector Machine (SVM)

The SVM is developed in statistical study and afterwards used in machine learning, statistics and signal processing. A variety of practical problems have been solved via SVM. SVM is a controlled way of learning that produces input-output mapping functions from a set [66]. The map function might be either a classification function or a regression function. SVM works by converting input raw data from its primary domain P into a feature space of higher dimension, then searches for the optimal hyperplane there. SVM aims to minimize the number of misclassification errors directly, by finding a hyper-plane or a function $g(x) = f\left(w_i^T \cdot x + b_i\right)$ that correctly separates two classes with a maximum margin [67] (Fig. 3).

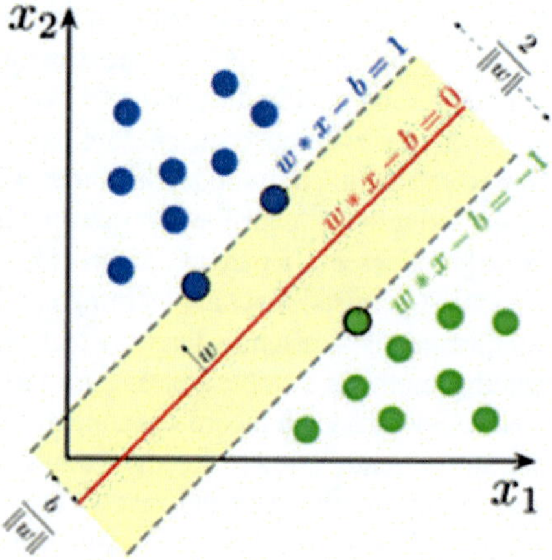

Fig. 3 Block representation of the SVM model

Fig. 4 Block representation of the Decision Tree model

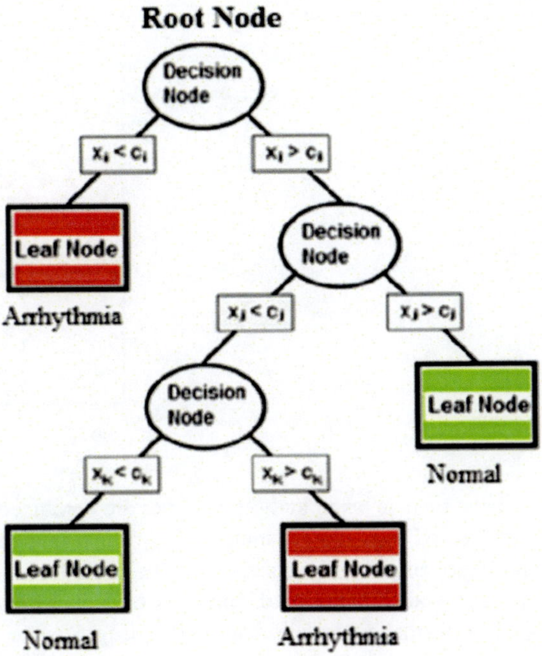

3.2 *Decision Trees*

It is like a flowchart with leaves, nodes and branches. It's like a tree structure. Decision Tree has grown widespread in the finding of knowledge because no specific subject knowledge is needed to construct the decision tree classification. Decision tree is an assembly that helps to separate different record groups into subsequent, smaller record sets by simply applying rules of decision. The members of the subsequent sets become increasingly homogeneous with each consecutive division. Entropy, the gain of information and the Gini index are dividing criteria to divide the samples into subsets. Better performance is achieved before and after trimming [68] (Fig. 4).

3.3 *Random Forests (RF)*

RF is a method similar to a Bayesian procedure used to identify an ensemble based on combination of tree predictors In RF, we grow multiple trees, as opposed to a single tree, where each tree is built upon values of random vectors sampled independently following the whole forest distribution. Thus, it is an ensemble classifier containing several random decision trees. Individual classification output of these

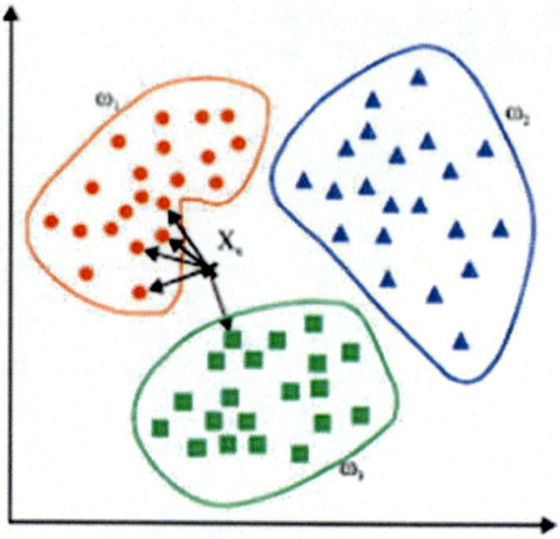

Fig. 5 Block representation of the KNN model

decision trees is taken and these values are combined to produce final outputs of a classifier. RF considers a small subset of dividing characteristics for classification and regression tree-like classifier. Utilizing suitable randomness makes RF precise regresses and classifiers and it does not over-fit. Also, random input features provide good performance in classification [69] (Fig. 5).

3.4 K-Nearest Neighbor (KNN)

KNN is one of the simplest and easiest algorithms for machine learning based on supervised learning. It suggests a similarity between the new class and the instances available and puts the new class in the closest category to the categories available. KNN is a non-parametric method that requires no assumption during data classification the underlying data [70].

3.5 Perceptron

Perceptron is one of artificial neuron's initial models. Rosenblatt first proposed it in 1958. It is a single-layer neural network with weights that may be trained to create the correct target vector when given the appropriate input vector. The Perceptron learning rule was utilised as the training method. Because of its capacity to generalise from its training vectors and work with randomly distributed connections, the Perceptron sparked a lot of curiosity. Perceptrons are particularly well adapted to pattern classification challenges [71]. Figure 6 shows a schematic diagram of a perceptron. The perceptron inputs are indicated by X = {x1, x2, x3 }. W = { w1, w2, w3 } is marked for its synaptic weights. The outward bias is marked with y.

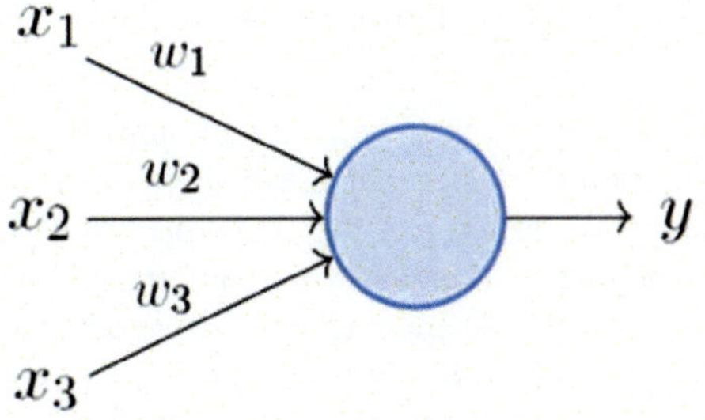

Fig. 6 Block representation of Perceptron model

3.6 *Artificial Neural Network (ANN)*

The human nervous system sets the basis for an artificial neural network. A standard ANN is associated with the input layer, hidden layers and the output layer. Mathematically change the numerical weight between these layers until the mistake has been minimised to its smallest value. The input characteristics, including ageing, sex, obesity, smoking, etc. determine the number of nodes in the input layer, while the output layer has a number of nodes which represents the available categories [72].

3.7 *Summarizing and Comparison*

To investigate the state of the art of ML in ECG analysis, and whether there is an effect of classification on reported ML performance, a comparison was performed using following terms primary problem (Multi class or binary), predictors (Numerical or categorical), classification Power , implementation, inter-pretabilty level, and normalization need detailed in Table 6.

Table 6 Machine learning algorithm illustration

Algorithm	Problem	Predictors type	Power	Implementation	Interpretabilty	Normal form
SVM	binary	Numerical/ categorical	High	Very difficult	Medium	Yes
DT	Multiclass/ binary	Numerical/ categorical	High	Difficult	Good	No
RF	Multiclass/ binary	Numerical/ categorical	High	Difficult	Good	No
KNN	Multiclass/ binary	Numerical	Medium	Easy	Good	Need
ANN	Multiclass/ binary	Numerical/ categorical	very high	Very difficult	Weak	Yes

4 Optimization Techniques

Optimization is an applied and numerical analysis field. Almost every engineering, scientific, business and life challenge can be stated as a search problem or an optimization. While some of the challenges could be easily solved by typical mathematical analytical optimization methods, most of the problems are difficult to overcome with analysis-based approaches. Fortunately, these difficult difficulties with optimization can be solved by nature inspirations as we know that nature is a very complicated system and always produces an almost optimal answer.

The optimization concept tries to obtain the best level in the parameters of the system in modern society. This notion remains a very difficult undertaking because of the increasing intricacy of real problems. Researchers investigated collective behaviour in animal groups in order to address optimization issues to build various clever algorithms [73].

Natural computing deals with nature-inspired computers and with natural processing. Evolutionary calculation, neural calculations, cellular automatics, swarm intelligence, molecular computing, quantum computing, artificial immune systems, and membrane computing are well recognised instances of natural computing. They form together the computer intelligence discipline

Evolutionary calculation is most important among all natural computing paradigms. It is a method of calculation to obtain optimum answers in a vast array of solutions based on Darwin's theory of survival. Evolutionary algorithms are a class of effective global strategies for optimising multiple difficult problems. Many of these problems need to search for a wide range of solutions, for example among a wide variety of features designed for describe heartbeat behaviour select relevant features, for a range of equations to forecast the ups and downs of the financial market, and for a series of rules to control the environment of a robot. Such computer issues frequently require an adaptable system, which in a changing environment can continue to work successfully [74].

In handling today's challenging challenges, most classic optimization approaches are inefficient. Many academics have therefore begun to suggest new solutions to issues of sophisticated optimization, called meta-heuristic algorithms, at an acceptable time and cost. Most meta heuristic algorithms rely on the core concept of nature, animal performance, or physical phenomena, it can be classified in three main classes: evolutionary-based, physics-based, swarm-based techniques and Human-based algorithms [75].

4.1 Evolutionary Algorithms

It is based on notions of natural selection. Here the population seeks to build on their environmental fitness measurements and has made the best possible effort to find the best optimal solution in the search space. This strategy has no knowledge of their fitness and adaptability. Deference evolution, Genetic Algorithm, Evolutionary

Programming, and Genetic Programming are some of the best-known, and recently created evolutionary algorithms.

4.2 Physics-Inspired Algorithms

Physics-inspired algorithms imitate certain physical/chemical laws, such as electric charges, waterways, chemicals, gas pressure, gravity etc. Exemplary examples of literature algorithms that represent such a method of metaheuristic analysis are annealing simulated (SA), optimising radiation (RO), optimising chemical reaction (CRO), an algorithm for motion (IMA, magnetic loaded system search) (MCSS), a local gravitational searching algorithm (GLSA) and Henry gas solubility optimization (HGSO)

4.3 Swarm-Based Algorithms

The latter category meta-heuristic algorithms is a swarm based on insect, particle and social behaviour. Particle Swarm Optimization , Ant Colony Optimization, and artificial Bee Colony are among the best known and popular swarm-based meta-heuristic optimizations [75].

4.4 Human-Based Algorithms

Human beings are considered to be the most intelligent creature and figure out the best way to solve their problems. There are several optimization algorithms in the literature that are affected by social social conduct. Teaching learning based optimization is one of the referred algorithms in this literature category (TLBO) [76] and Political Optimizer (PO) [77]. All these meta-heuristic algorithms are able to make a significant effort to locate the optimum answer (Table 7).

5 Integration of Machine Learning and Optimization Techniques

Smart data processing is needed to improve cardiac disease recognition quality and transition by accelerating research into active hidden factors in clinical information using machine learning algorithms in order to provide actionable information based on gaps about patients for early detection and prevention of constitutional disorders [5].

Table 7 Taxonomy of many different meta-heuristics

Algorithm	Abbreviation	Category	Year	Refs.
Particle Swarm Optimization	PSO	Swarm-based	1995	[17]
Artificial Bee Colony	ABC	Swarm-based	2005	[78]
Genetic Algorithm	GA	Evolutionary-based	1992	[79]
Ant Colony Optimization	ACO	Swarm-based	2006	[80]
Teaching learning-based optimization	TLBO	Human-based	2011	[76]
Water Wave Optimization	WWA	Physics/Chemistry-based	2015	[81]
Grey Wolf Optimizer	GWO	Swarm-based	2015	[82]
Whale Optimization Algorithm	WOA	Evolutionary-based	2016	[83]
Henry gas solubility optimization	HGSO	Physics/Chemistry-based	2019	[18]
Thermal Exchange Optimization	TEO	Physics/Chemistry-based	2019	[84]
Political Optimizer	PO	Human-based	2020	[77]
Heap-Based Optimizer	HBO	Evolutionary-based	2020	[75]
capuchin search algorithm	CapSA	Evolutionary-based	2021	[85]
Chaos Game Optimization	CGO	Evolutionary-based	2021	[86]

Recent efforts to cardiac disease recognition to facilitate predictive diagnosis have resulted in many meta-heuristic and ML-based approaches. In machine learning, feature selection algorithms based on meta-heuristic are used to pick a collection of relevant features from a super-set of original data. They reduce the feature dimension and minimise the complexity of the system, process time and better precision. Also using ML requires a lot of work setting different appropriate parameters, meta-heuristic algorithms are used to select the best optimal parameters setting for ML algorithms [87] in the following section. We analyzed contributions for feature selection and parameter tuning to acquire a complete picture of available scholarly solutions about integration of ML and Optimization Techniques for cardiac health recognition.

The resulting ECG recording has massive data volumes, quickly filling up storage space. Signal transmission over public telephone networks is another use that involves large volumes of data. In both instances, data compression is a critical procedure, and it serves another ECG signal processing purpose. In feature selection, the aim of the overall process is to seek an optimal feature group to improve the classification accuracy of trained classifiers, to indirectly effect the decreasing time and cost of both the computing and data analysis components, as well as the choose relevant features subset. The optimum selection methods are further categorised into filter, wrapper and embedding depending on the search strategy and the evaluation measurement [88].

Regardless of the learning algorithm, filter approaches use metrics to measure the significance of characteristics, such as information theory, distance, and rough setting theory. To choose relevant features, filtering techniques rely on fundamental data qualities [89].

While wrapper feature selection methods choose a subset of features using a search methodology from the original vector space. It also incorporates a machine-learning technology for performance assessment. The outcomes are reviewed when the produced subset interacts with a performance assessment classifier. Higher accuracy or lower error rate is a classification measure for assessment. This is performed until an optimum functional collection is produced.

Embedded techniques of feature selection to solve filter and wrapper problems. In contrast to computational wrapper approaches, embedding methods take into account the classification bias, in opposition to filter methods [90].

Finding an ideal balance of Features set and classification is still an ongoing topic in developing automatic heartbeat classification systems, particularly when applications involving resource-restricted devices are taken into account. By floating sequential searches, Mar et al. [91] created the method of feature selection in which the authors assessed a collection of feature selection options with the comparison between number and exactness. With the linear discrimination strategy, the author reaches the overall accuracy of 84.6%.

Tahir et al. [92] proposed a new genetic algorithm called binary chaotic genetic algorithm and binary physical fitness evaluation criteria (BCGA). The performance of the classic genetic algorithm, addressing their computational time and the selected functions for a classification task, was augmented by ten distinct chaotic maps. To show its value, the proposed BCGA is used to a feature selection task from an effective database namely AMIGOS (A Dataset for Affect, Personality and Mood Research on Individuals and Groups) where data recorded as electroencephalogram (EEG), electrocardiogram (ECG), and galvanic skin reaction (GSR).

In [93], Five distinct meta-heuristic strategies from SI-based methods such as binary Gray-Wolf Optimizer (bGWO), Ant Lion Optimization (ALO) and Butterfly optimization algorithm BOA have been devised to overcome the task of feature selection for the diagnosis of heart rhythm, as well as Dragonfly Algorithms(DA), Satin-B ird Optimization (SBO), and five new chaotic variations of SBO. There have been calculations of several performance measurements like precision, fitness, optimal set of characteristics and run-time. From the trial, the SBO has been shown to outperform other SI algorithms such as bGWO, DA, BOA and ALO in terms of accuracy and fitness of cardiovascular arrhythmia. BOA and ALO also appear to be the best if dimensional size is the focus.

Author in [94] offer a new single-solution meta-heuristic system for the fuzzy grouping of ECG beats, namely the Vortex Search algorithm (VS). Fuzzy training set cluster centres are located. The test set is then categorised by using these cluster centres to evaluate the clustering performance of the process. The findings are compared with the c-means (FCM), the c-means (PSO) and c-means with the colonial artificial bees algorithms (AA).

The two feature choice strategies, i.e. filter and wrapper, were compared by Doquire et al. [95]. For experimentation, the authors employ 200 features. The filter-based FS technique is used to obtain mutual information and to acquire a balance rating of 73%, with a ranking of 82,99%, and a weighted Linear Discrimination (LD) model is utilised for the wrapper selection, using the forward-looking search strategy for a balanced classification rate of 73%.

A new method for categorising the electrocardiogram (ECG) heart beats is proposed by Verma et al. [96]. For the optimal feature set for ECG classifications, the suggested technique uses the Fisher and BAT Opimization algorithms. First features are calculated for each decomposed EWT mode, using empirical transform wavelet (EWT), and then higher-order statistics and symbolic features. Second, the Fisher ratio is used for function selection. The BAT algorithm is optimised to maximise discriminating between classes.

The selection of optimal parameters affects the performance of least-squares support vector machines significantly (LS-SVMs). Based on particle swarm optimization, a novel hyper-parameter selection strategy for LS-SVMs is proposed in this research (PSO) [97]. The suggested method does not require any prior knowledge of the generalisation performance measure's analytic property and may be used to calculate several hyper-parameters at the same time. SRBF and RBF kernel functions are used in the scaling radial basis kernel function (SRBF) and RBF kernel functions, respectively. We offer a novel noninvasive fetal electrocardiogram (FECG) extraction method using an echo state neural network (ESN). The FECG is extracted from abdominal recordings of pregnant women by cancelling the maternal ECG. By extracting clinically interpret-able characteristics from the FECG, it can be used for fetal health monitoring. On the test database, they show that optimising an ESN by random search offers approximately identical performance to an exhaustive grid search, with 85.6 percent vs. 87.9% accuracy.

Swarm-TWSVM is an electrocardiogram (ECG) signals heartbeat classification approach that was proposed by Houssein et al. [24] to aid in the detection of the cardiac heartbeat process. It is a hybrid of the pso with gravitational search algorithm (PSOGSA) that was hybrid-ed with twin svm, where the hybrid PSOGSA algorithm was used to obtain the best parameters of the twin svm.

carrillo et al. [98] offer a metaheuristic optimization strategy for parameter estimations in arrhythmia classification from unbalanced data. To classify eight forms of arrhythmia, we used an imbalanced subset of those databases. To address the unbalanced class problem, we used a combination of under sampling based on the clustering approach (data level) and feature selection method (algorithmic level). they examined two meta-heuristic approaches based on differential evolution and particle swarm optimization to investigate parameter estimates and enhance classification for their model.Even in the presence of unbalanced data, the final findings indicated an accuracy of 99.95 percent, an F1 score of 99.88 percent, a sensitivity of 99.87 percent, a precision of 99.89 percent, and a specificity of 99.99 percent, all of which are good.

In this publication [99], authors offer a novel noninvasive fetal electrocardiogram (FECG) extraction method using an echo state neural network (ESN). The FECG is extracted from abdominal recordings of pregnant women by cancelling the maternal ECG. By extracting clinically interpretable characteristics from the FECG, it can be used for fetal health monitoring. On the test database, we show that optimising an ESN by random search offers approximately identical performance to an exhaustive grid search, with 85.6 percent vs. 87.9% accuracy.

Thorough this work [100], authors present an improved convolutional neural network for automatic arrhythmia classification using an ECG signal in this research. To lower the risk of death from CVD, it is critical to monitor the heart beat arrhythmia on a regular basis. In computer vision issues, the Visual Geometry Group Network (VGGNet) is commonly employed. However, because ECG signals differ from picture signals in terms of dimensionality and inherent characteristics, the same network cannot be utilised to classify ECG beats. As a result, the authors looked at the impact of reducing the depth and width of a convolutional neural network on cardiac arrhythmia classification. The suggested architecture's hyper-parameters were optimised using the sequential model based global optimization (SMBO) technique.

It is evident in the above studies that the accuracy of classification of the above approaches was not superior and the use of meta-heuristic algorithms appears to have been further strengthened Table 8 summarized the explored studies for integration machine learning and optimization techniques for cardiac recognition.

Table 8 Summary of the Comparison between recent papers for ECg classification

Refs.	Dataset	Leads	Beats #	Method	Performance
[91]	MIT-BIH	lead I1I	109 000	- Feature selection using SFFS algorithm. - a multilayer perceptron (MLP) classifer. - Temporal features. - Morphological features. - Statistical features.	Accuracy = 89%.
[92]	AMIGOS	12 leads	-	- Feature selection using BCGA algorithm. - KNN. - Time–frequency features. - Mel frequency cepstral coefficients. - wavelet transform-based features. - item vocal fold features.	Accuracy = 92.7%
[93]	UCI	12 leads	452	- Feature selection using SBO algorithm. - KNN. - Raw data.	Accuracy = 68%
[94]	MIT-BIH	lead II	2,134	- Feature selection using VS algorithm. - Fuzzy cluster. - QRS detection.	Sensitivity = 93.58%
[95]	MIT-BIH	lead II	110000	- Feature selection using forward search strategy. - Forward-backward search strategy. - Weighted LDA. - Weighted SVM. - QRS waves.	Accuracy = 73.00%
[96]	MIT-BIH	lead II	110000	- Feature selection using BAT algorithm. - KNN. - RR interval. - EWT with HOS and symbolic features.	Accuracy = 99.80%
[97]	Heart (HT)	-	-	- Hyper-parameter using PSO algorithm. - SVM. - Raw data.	Accuracy = 84.49%
[24]	MIT-BIH	lead II	110000	- Hyper-parameter using PSO algorithm. - SVM. - Q, P, S, R, and T waves features.	Accuracy = 99.44%

6 Open Issues and Challenges

The current obstacles and problems with profound learning based on ECG data are considered in this section. In addition, potential opportunities are identified in the context of these issues and problems.

Various research have employed a variety of lead numbers, times, sources (subject backgrounds), etc. This makes it difficult to do a fair comparison of results from different databases. Furthermore, because high-quality data and annotations are difficult to come by, many current studies still rely on the MIT-BIH Arrhythmia Database, which was compiled over 40 years ago. A new high-quality long-term ECG data-set with annotations would be welcomed by researchers, and such a data-set will undoubtedly inspire new novel experiments.

Deep learning models are sometimes seen as black box models as they usually feature a lot of model parameters or sophisticated model structures, making it impossible for a person to comprehend why such a model produces a given result. This difficulty in the medical field is significantly more severe because medical specialists cannot accept diagnosis without explanation. There were some works aimed at increasing ECG deep learning systems' inter-pretability

ECG disease labels have highly biased distributions, since most of the serious conditions seldom occur, yet are very significant. There are several model parameters that make use of minimal datasets on disease labels hard to train an efficient deep learning algorithm. This problem is handled by two major techniques. One is the addition of data, for example data preparation by side-cutting technique, or the generation of synthetic training data systems using generative models.

With respect to the ECG compression methods, there is a set and different bit rate of categorisation. The ECG compression bit rate can be fixed with a simple structure, as a consistent amount of bits is used in each frame.

Machine learning is employed in this work to automate the diagnosis, annotation, and detection of ECG readings, which are appraised and labelled precisely and effectively, but due to the volume of ECG data result of signal segmentation, powerful and high computational environment is needed. The Cloud environment allows researcher to use multiple devices to perform the same task at the same time. For example, if you're optimizing a hyper-parameter in a neural network or features selection need to restart the template 10,000 times to get the best parameter selection (which is a common problem), running on a single computer for two weeks is unrealistic. If you can run the identical model on 100 computers, you should be able to complete the project in a few of hours. In this context, Cloud environment as an advanced strategy to reduce computational times significantly.

Finally, some unique multidisciplinary investigations have been carried out, some of which appear below. 1) Evaluation of safe driving intensity and heart response time by ECG signals. 2) ECG-based emotion detection. 3) ECG analysis of mammals. 4) Age and gender estimates based on ECG data

7 Conclusion

This chapter focused on a different topic related to heart disease and its statistics. In attempt to grant new researchers a basic understanding of machine learning, meta-heuristics and guide those to find their starting point in identifying heart disease, the chapter reviews different machine learning and meta-heuristics algorithms and their integrate methodologies into cardiology recognition with balancing between the feature selection and classification accuracy. Moreover, the ability of meta-heuristics algorithms to configure the optimal setting for different classification model which consists of architecture and parameters.

Conflict of interest There is no conflict of interest, according to the authors.

References

1. WHO, *Cardiovascular diseases (CVDs)*. http://www.who.int/news-room/fact-sheets/detail/cardiovascular-diseases-(cvds) (2017)
2. E.H. Houssein, I.E. Ibrahim, N. Neggaz, M. Hassaballah, Y.M. Wazery, An efficient ecg arrhythmia classification method based on manta ray foraging optimization. Expert Syst. Appl. **181**, 115131 (2021)
3. D. Lai, Y. Bu, Y. Su, X. Zhang, C.-S. Ma, Non-standardized patch-based ECG lead together with deep learning based algorithm for automatic screening of atrial fibrillation. IEEE J. Biomed. Health Inform. **24**, 1569–1578 (2020)
4. E.H. Houssein, M. Kilany, A.E. Hassanien, Ecg signals classification: a review. Int. J. Intell. Eng. Inf. **5**(4), 376–396 (2017)
5. E.H. Houssein, D.S. AbdElminaam, I.E. Ibrahim, M. Hassaballah, Y.M. Wazery, A hybrid heartbeats classification approach based on marine predators algorithm and convolution neural networks. IEEE Access (2021)
6. Y.M. Wazery, E. Saber, E.H. Houssein, A.A. Ali, E. Amer, An efficient slime mould algorithm combined with k-nearest neighbor for medical classification tasks. IEEE Access (2021)
7. A. Hamad, E.H. Houssein, A.E. Hassanien, A.A. Fahmy, Hybrid grasshopper optimization algorithm and support vector machines for automatic seizure detection in eeg signals, in *International Conference on Advanced Machine Learning Technologies and Applications* (Springer, 2018), pp. 82–91
8. C. Chen, Z. Hua, R. Zhang, G. Liu, W. Wen, Automated arrhythmia classification based on a combination network of cnn and lstm. Biomed. Signal Process. Control **57**, 101819 (2020)
9. G. Sannino, G. De Pietro, A deep learning approach for ECG-based heartbeat classification for arrhythmia detection. Future Gener. Comput. Syst. **86**, 446–455 (2018)
10. E.H. Houssein, M.M. Emam, A.A. Ali, P.N. Suganthan, Deep and machine learning techniques for medical imaging-based breast cancer: a comprehensive review. Expert Syst. Appl. 114161 (2020)
11. Z. Li, D. Zhou, L. Wan, J. Li, W. Mou, Heartbeat classification using deep residual convolutional neural network from 2-lead electrocardiogram. J. Electrocardiol. **58**, 105–112 (2020)
12. A.I. Abdullah, Facial expression identification system using fisher linear discriminant analysis and k-nearest neighbor methods. ZANCO J. Pure Appl. Sci. **31**(2), 9–13 (2019)

13. E.H. Houssein, Y. Mina, E. Aboul, Nature-inspired algorithms: a comprehensive review, in *Hybrid Computational Intelligence: Research and Applications* (CRC Press, 2019), p. 1
14. F.A. Hashim, K. Hussain, E.H. Houssein, M.S. Mabrouk, W. Al-Atabany, Archimedes optimization algorithm: a new metaheuristic algorithm for solving optimization problems. Appl. Intell. **51**(3), 1531–1551 (2021)
15. E.H. Houssein, M.R. Saad, F.A. Hashim, H. Shaban, M. Hassaballah, Lévy flight distribution: a new metaheuristic algorithm for solving engineering optimization problems. Eng. Appl. Artif. Intell. **94**, 103731 (2020)
16. F.A. Hashim, E.H. Houssein, K. Hussain, M.S. Mabrouk, W. Al-Atabany, Honey badger algorithm: new metaheuristic algorithm for solving optimization problems. Math. Comput. Simul. (2021)
17. E.H. Houssein, A.G. Gad, K. Hussain, P.N. Suganthan, Major advances in particle swarm optimization: theory, analysis, and application. Swarm Evol. Comput. **63**, 100868 (2021)
18. F.A. Hashim, E.H. Houssein, M.S. Mabrouk, W. Al-Atabany, S. Mirjalili, Henry gas solubility optimization: a novel physics-based algorithm. Future Gener. Comput. Syst. **101**, 646–667 (2019)
19. E.H. Houssein, M.M. Emam, A.A. Ali, An efficient multilevel thresholding segmentation method for thermography breast cancer imaging based on improved chimp optimization algorithm. Expert Syst. Appl. 115651 (2021)
20. E.H. Houssein, K. Hussain, L. Abualigah, M. Abd Elaziz, W. Alomoush, G. Dhiman, Y. Djenouri, E. Cuevas, An improved opposition-based marine predators algorithm for global optimization and multilevel thresholding image segmentation. Knowl.-Based Syst. 107348 (2021)
21. E.H. Houssein, M.M. Emam, A.A. Ali, Improved manta ray foraging optimization for multilevel thresholding using covid-19 ct images. Neural Comput. Appl. 1–21 (2021)
22. E.H. Houssein, B.E.-D. Helmy, D. Oliva, A.A. Elngar, H. Shaban, A novel black widow optimization algorithm for multilevel thresholding image segmentation. Expert Syst. Appl. **167**, 114159 (2021)
23. E.H. Houssein, B.E.D. Helmy, A.A. Elngar, D.S. Abdelminaam, H. Shaban, An improved tunicate swarm algorithm for global optimization and image segmentation. IEEE Access **9**, 56066–56092 (2021)
24. E.H. Houssein, A.A. Ewees, M. Abd ElAziz, Improving twin support vector machine based on hybrid swarm optimizer for heartbeat classification. Pattern Recogn. Image Anal. **28**(2), 243–253 (2018)
25. M.M. Ahmed, E.H. Houssein, A.E. Hassanien, A. Taha, E. Hassanien, Maximizing lifetime of large-scale wireless sensor networks using multi-objective whale optimization algorithm. Telecommun. Syst. **72**(2), 243–259 (2019)
26. E.H. Houssein, M.E. Hosney, M. Elhoseny, D. Oliva, W.M. Mohamed, M. Hassaballah, Hybrid harris hawks optimization with cuckoo search for drug design and discovery in chemoinformatics. Sci. Rep. **10**(1), 1–22 (2020)
27. E.H. Houssein, A.G. Gad, Y.M. Wazery, P.N. Suganthan, Task scheduling in cloud computing based on meta-heuristics: review, taxonomy, open challenges, and future trends. Swarm Evol. Comput. 100841 (2021)
28. M.H. Hassan, E.H. Houssein, M.A. Mahdy, S. Kamel, An improved manta ray foraging optimizer for cost-effective emission dispatch problems. Eng. Appl. Artif. Intell. **100**, 104155 (2021)
29. A. Korashy, S. Kamel, E.H. Houssein, F. Jurado, F.A. Hashim, Development and application of evaporation rate water cycle algorithm for optimal coordination of directional overcurrent relays. Expert Syst. Appl. **185**, 115538 (2021)
30. S. Deb, E.H. Houssein, M. Said, D.S. AbdElminaam, Performance of turbulent flow of water optimization on economic load dispatch problem. IEEE Access (2021)
31. S. Deb, D.S. Abdelminaam, M. Said, E.H. Houssein, Recent methodology-based gradient-based optimizer for economic load dispatch problem. IEEE Access **9**, 44322–44338 (2021)
32. E.H. Houssein, F.A. Hashim, S. Ferahtia, H. Rezk, An efficient modified artificial electric field algorithm for solving optimization problems and parameter estimation of fuel cell. Int. J. Energy Res. (2021)

33. E.H. Houssein, B.E.D. Helmy, H. Rezk, A.M. Nassef, An enhanced archimedes optimization algorithm based on local escaping operator and orthogonal learning for pem fuel cell parameter identification. Eng. Appl. Artif. Intell. **103**, 104309 (2021)
34. E.H. Houssein, M.A. Mahdy, A. Fathy, H. Rezk, A modified marine predator algorithm based on opposition based learning for tracking the global mpp of shaded pv system. Expert Syst. Appl. **183**, 115253 (2021)
35. E.H. Houssein, G.N. Zaki, A.A.Z. Diab, E.M. Younis, An efficient manta ray foraging optimization algorithm for parameter extraction of three-diode photovoltaic model. Comput. Electr. Eng. **94**, 107304 (2021)
36. E.H. Houssein, Machine learning and meta-heuristic algorithms for renewable energy: a systematic review. Adv. Control Optim. Paradigms Wind Energy Syst. 165–187 (2019)
37. D.S. Abdelminaam, M. Said, E.H. Houssein, Turbulent flow of water-based optimization using new objective function for parameter extraction of six photovoltaic models. IEEE Access **9**, 35382–35398 (2021)
38. A.A. Ismaeel, E.H. Houssein, D. Oliva, M. Said, Gradient-based optimizer for parameter extraction in photovoltaic models. IEEE Access **9**, 13403–13416 (2021)
39. E.H. Houssein, M.A. Mahdy, M.J. Blondin, D. Shebl, W.M. Mohamed, Hybrid slime mould algorithm with adaptive guided differential evolution algorithm for combinatorial and global optimization problems. Expert Syst. Appl. **174**, 114689 (2021)
40. E.H. Houssein, M.A. Mahdy, M.G. Eldin, D. Shebl, W.M. Mohamed, M. Abdel-Aty, Optimizing quantum cloning circuit parameters based on adaptive guided differential evolution algorithm. J. Adv. Res. **29**, 147–157 (2021)
41. E.H. Houssein, M. Dirar, K. Hussain, W.M. Mohamed, Assess deep learning models for egyptian exchange prediction using nonlinear artificial neural networks. Neural Comput. Appl. **33**(11), 5965–5987 (2021)
42. K. Hussain, N. Neggaz, W. Zhu, E.H. Houssein, An efficient hybrid sine-cosine harris hawks optimization for low and high-dimensional feature selection. Expert Syst. Appl. **176**, 114778 (2021)
43. N. Neggaz, E.H. Houssein, K. Hussain, An efficient henry gas solubility optimization for feature selection. Expert Syst. Appl. **152**, 113364 (2020)
44. E.H. Houssein, M.E. Hosney, D. Oliva, W.M. Mohamed, M. Hassaballah, A novel hybrid harris hawks optimization and support vector machines for drug design and discovery. Comput. Chem. Eng. **133**, 106656 (2020)
45. E.H. Houssein, N. Neggaz, M.E. Hosney, W.M. Mohamed, M. Hassaballah, Enhanced harris hawks optimization with genetic operators for selection chemical descriptors and compounds activities. Neural Comput. Appl. 1–18 (2021)
46. D.S. Abdelminaam, F.H. Ismail, M. Taha, A. Taha, E.H. Houssein, A. Nabil, Coaid-deep: An optimized intelligent framework for automated detecting covid-19 misleading information on twitter. IEEE Access **9**, 27840–27867 (2021)
47. E.H. Houssein, M. Ahmad, M.E. Hosney, M. Mazzara, Classification approach for covid-19 gene based on harris hawks optimization, in *Artificial Intelligence for COVID-19* (Springer, 2021), pp. 575–594
48. Z. Ebrahimi, M. Loni, M. Daneshtalab, A. Gharehbaghi, A review on deep learning methods for ecg arrhythmia classification. Expert Syst. Appl. **X**, 100033 (2020)
49. A.L. Goldberger, L.A. Amaral, L. Glass, J.M. Hausdorff, P.C. Ivanov, R.G. Mark, J.E. Mietus, G.B. Moody, C.-K. Peng, H.E. Stanley, Physiobank, physiotoolkit, and physionet: components of a new research resource for complex physiologic signals. Circulation **101**(23), e215–e220 (2000)
50. P. De Chazal, M. O'Dwyer, R.B. Reilly, Automatic classification of heartbeats using ECG morphology and heartbeat interval features. IEEE Trans. Biomed. Eng. **51**, 1196–1206 (2004)
51. A. Singhal, P. Singh, B. Fatimah, R.B. Pachori, An efficient removal of power-line interference and baseline wander from ecg signals by employing Fourier decomposition technique. Biomed. Signal Process. Control **57**, 101741 (2020)

52. T. Padma, M.M. Latha, A. Ahmed, Ecg compression and labview implementation. J. Biomed. Sci. Eng. **2**(3), 177 (2009)
53. J.-W. Lee, G.-K. Lee, Design of an adaptive filter with a dynamic structure for ecg signal processing. Int. J. Control Autom. Syst. **3**(1), 137–142 (2005)
54. M. Dehghani, A. Shahabinia, A.A. Safavi, Implementation of wireless data transmission based on bluetooth technology for biosignals monitoring. World Appl. Sci. J. **10**(3), 287–293 (2010)
55. M. Khosravy, N. Gupta, N. Patel, T. Senjyu, C.A. Duque, Particle swarm optimization of morphological filters for electrocardiogram baseline drift estimation, in *Applied Nature-Inspired Computing: Algorithms and Case Studies*. (Springer, 2020), pp. 1–21
56. H. Li, X. Wang, Detection of electrocardiogram characteristic points using lifting wavelet transform and hilbert transform. Trans. Inst. Meas. Control. **35**, 574–582 (2013)
57. S. Mousavi, F. Afghah, Inter-and intra-patient ecg heartbeat classification for arrhythmia detection: a sequence to sequence deep learning approach, in *ICASSP 2019-2019 IEEE International Conference on Acoustics, Speech and Signal Processing (ICASSP)* (IEEE, 2019), pp. 1308–1312
58. A.Y. Hannun, P. Rajpurkar, M. Haghpanahi, G.H. Tison, C. Bourn, M.P. Turakhia, A.Y. Ng, Cardiologist-level arrhythmia detection and classification in ambulatory electrocardiograms using a deep neural network. Nat. Med. **25**(1), 65–69 (2019)
59. S.L. Oh, E.Y. Ng, R. San Tan, U.R. Acharya, Automated beat-wise arrhythmia diagnosis using modified u-net on extended electrocardiographic recordings with heterogeneous arrhythmia types. Comput. Biol. Med. **105**, 92–101 (2019)
60. V. Mondéjar-Guerra, J. Novo, J. Rouco, M.G. Penedo, M. Ortega, Heartbeat classification fusing temporal and morphological information of ECGs via ensemble of classifiers. Biomed. Signal Process. Control **47**, 41–48 (2019)
61. Q. Zhao, L. Zhang, Ecg feature extraction and classification using wavelet transform and support vector machines, in *2005 International Conference on Neural Networks and Brain*, vol. 2 (IEEE, 2005), pp. 1089–1092
62. S. Mahamoodabadi, A. Ahmedian, M. Abolhasani, Ecg feature extraction using daubechies wavelet, in *Proceedings of 5th IASTED Inter (IMAGING and IMAGE PROCESSING, Conf. VISUALIZATION* (2005), pp. 7–9
63. F. Sufi, S. Mahmoud, I. Khalil, A new ecg obfuscation method: a joint feature extraction & corruption approach, in *2008 International Conference on Information Technology and Applications in Biomedicine* (IEEE, 2008), pp. 334–337
64. M. Dash, H. Liu, Feature selection for classification. Intell. Data Anal. **1**, 131–156 (1997)
65. A. Pramod, H.S. Naicker, A.K. Tyagi, Machine learning and deep learning: Open issues and future research directions for the next 10 years. Comput. Anal. Deep Learn. Med. Care Princ. Methods Appl. 463 (2021)
66. C.J. Burges, A tutorial on support vector machines for pattern recognition. Data Min. Knowl. Disc. **2**, 121–167 (1998)
67. A. Tharwat, Parameter investigation of support vector machine classifier with kernel functions. Knowl. Inf. Syst. **61**, 1269–1302 (2019)
68. C. Sridhar, U.R. Acharya, H. Fujita, G.M. Bairy, Automated diagnosis of coronary artery disease using nonlinear features extracted from ecg signals, in *2016 IEEE International Conference on Systems, Man, and Cybernetics (SMC)* (IEEE, 2016), pp. 545–549
69. T. Li, M. Zhou, ECG classification using wavelet packet entropy and random forests. Entropy **18**, 285 (2016)
70. L.E. Peterson, K-nearest neighbor. Scholarpedia **4**(2), 1883 (2009)
71. R.J. Martis, U.R. Acharya, K. Mandana, A.K. Ray, C. Chakraborty, Application of principal component analysis to ECG signals for automated diagnosis of cardiac health. Expert Syst. Appl. **39**, 11792–11800 (2012)
72. S. Celin, K. Vasanth, ECG signal classification using various machine learning techniques. J. Med. Syst. **42**, 241 (2018)
73. H. Allioui, M. Sadgal, A. Elfazziki, Optimized control for medical image segmentation: improved multi-agent systems agreements using particle swarm optimization. J. Ambient Intell. Hum. Comput. 1–19 (2021)

74. E.H. Houssein, I.E. Mohamed, A.E. Hassanien, Salp swarm algorithm: modification and application, in *Swarm Intelligence Algorithms* (CRC Press, 2020), pp. 285–299
75. Q. Askari, M. Saeed, I. Younas, Heap-based optimizer inspired by corporate rank hierarchy for global optimization. Expert Syst. Appl. **161**, 113702 (2020)
76. R.V. Rao, V.J. Savsani, D. Vakharia, Teaching-learning-based optimization: a novel method for constrained mechanical design optimization problems. Comput. Aided Des. **43**(3), 303–315 (2011)
77. Q. Askari, I. Younas, M. Saeed, Political optimizer: A novel socio-inspired meta-heuristic for global optimization. Knowl.-Based Syst. **195**, 105709 (2020)
78. D. Karaboga, B. Basturk, A powerful and efficient algorithm for numerical function optimization: artificial bee colony (abc) algorithm. J. Global Optim. **39**(3), 459–471 (2007)
79. D. Whitley, A genetic algorithm tutorial. Stat. Comput. **4**(2), 65–85 (1994)
80. M. Dorigo, M. Birattari, T. Stutzle, Ant colony optimization. IEEE Comput. Intell. Mag. **1**(4), 28–39 (2006)
81. Y.-J. Zheng, Water wave optimization: a new nature-inspired metaheuristic. Comput. Oper. Res. **55**, 1–11 (2015)
82. S. Mirjalili, S.M. Mirjalili, A. Lewis, Grey wolf optimizer. Adv. Eng. Softw. **69**, 46–61 (2014)
83. S. Mirjalili, A. Lewis, The whale optimization algorithm. Adv. Eng. Softw. **95**, 51–67 (2016)
84. A. Kaveh, A. Dadras, A novel meta-heuristic optimization algorithm: thermal exchange optimization. Adv. Eng. Softw. **110**, 69–84 (2017)
85. M. Braik, A. Sheta, H. Al-Hiary, A novel meta-heuristic search algorithm for solving optimization problems: capuchin search algorithm. Neural Comput. Appl. **33**(7), 2515–2547 (2021)
86. S. Talatahari, M. Azizi, Chaos game optimization: a novel metaheuristic algorithm. Artif. Intell. Rev. **54**(2), 917–1004 (2021)
87. E.H. Houssein, I.E. Ibrahim, M. Hassaballah, Y.M. Wazery, Integration of internet of things and cloud computing for cardiac health recognition, in *Metaheuristics in Machine Learning: Theory and Applications* (Springer, 2021), pp. 645–661
88. B. Xue, M. Zhang, W.N. Browne, X. Yao, A survey on evolutionary computation approaches to feature selection. IEEE Trans. Evol. Comput. **20**(4), 606–626 (2015)
89. P. Bermejo, J.A. Gámez, J.M. Puerta, Speeding up incremental wrapper feature subset selection with naive bayes classifier. Knowl.-Based Syst. **55**, 140–147 (2014)
90. G. Khademi, H. Mohammadi, D. Simon, Gradient-based multi-objective feature selection for gait mode recognition of transfemoral amputees. Sensors **19**(2), 253 (2019)
91. T. Mar, S. Zaunseder, J.P. Martínez, M. Llamedo, R. Poll, Optimization of ecg classification by means of feature selection. IEEE Trans. Biomed. Eng. **58**(8), 2168–2177 (2011)
92. M. Tahir, A. Tubaishat, F. Al-Obeidat, B. Shah, Z. Halim, M. Waqas, A novel binary chaotic genetic algorithm for feature selection and its utility in affective computing and healthcare. Neural Comput. Appl. 1–22 (2020)
93. S. Sharma, G. Singh, Diagnosis of cardiac arrhythmia using swarm-intelligence based metaheuristic techniques: a comparative analysis. EAI Endorsed Trans. Pervasive Health Technol. **6**(23) (2020)
94. B. Doğan, T. Ölmez, Fuzzy clustering of ecg beats using a new metaheuristic approach, in *2nd International Work-Conference on Bioinformatics and Biomedical Engineering (iwbbio)*, 7–9 April 2014 (Granada, Spain, 2014)
95. G. Doquire, G. De Lannoy, D. François, M. Verleysen, Feature selection for interpatient supervised heart beat classification. Comput. Intell. Neurosci. **2011** (2011)
96. A.K. Verma, I. Saini, B.S. Saini, A new bat optimization algorithm based feature selection method for electrocardiogram heartbeat classification using empirical wavelet transform and fisher ratio. Int. J. Mach. Learn. Cybern. **11**(11), 2439–2452 (2020)
97. X. Guo, J. Yang, C. Wu, C. Wang, Y. Liang, A novel ls-svms hyper-parameter selection based on particle swarm optimization. Neurocomputing **71**(16–18), 3211–3215 (2008)

98. J.C. Carrillo-Alarcón, L.A. Morales-Rosales, H. Rodríguez-Rángel, M. Lobato-Báez, A. Muñoz, I. Algredo-Badillo, A metaheuristic optimization approach for parameter estimation in arrhythmia classification from unbalanced data. Sensors **20**(11), 3139 (2020)
99. J. Behar, A. Johnson, J. Oster, G. Clifford, An echo state neural network for foetal ecg extraction optimised by random search. Proc. Adv. Neural Inf. Process. Syst. 1–5 (2013)
100. M. Jangra, S.K. Dhull, K.K. Singh, Ecg arrhythmia classification using modified visual geometry group network (mvggnet). J. Intell. Fuzzy Syst. **38**(3), 3151–3165 (2020)

Metaheuristics for Parameter Estimation of Solar Photovoltaic Cells: A Comprehensive Review

Essam Halim Houssein, Gamela Nageh Zaki, Laith Abualigah, and Eman M. G. Younis

Abstract Recently, renewable energy sources have great significance and contracted to be more and more interesting for the numerous reasons. Among these reasons, they are considered environment friendly, green, safe and sustainable power sources. The use of solar radiation is increasing as a clean source of energy. The Photovoltaic (PV) panels that contain solar cells are used to transform the solar energy into electricity. This article discusses and explains the parameter extraction of solar cell using mathematical techniques. Soft computing and analytical approaches are used for parameter. The determination of the mathematical model parameters of cells and photovoltaic (PV) modules is a big challenge. In recent years, various numerical, analytical and hybrid methods have been proposed for the extraction of the parameters of the photovoltaic model from manufacturer datasheets or experimental data. It is complex to quickly and accurately determine highly credible solutions. This review article critically describes and discusses the main problems of the methods presented in the research literature over the past few years. It analyzes the dynamic behavior of solar cells. Moreover, this paper uses two real world models, single and double diode models, and inspects the execution of the photovoltaic parameters for every model and their impact on power–voltage (P–V) and current–voltage (I–V) characteristics.

Keywords Metaheuristics · Photovoltaic (PV) · PV parameter estimation · Single diode (SD) · Double diode (DD) · Three diode (TD)

E. H. Houssein (✉) · G. N. Zaki · E. M. G. Younis
Faculty of Computers and Information, Minia University, Minia, Egypt
e-mail: essam.halim@mu.edu.eg

L. Abualigah
Faculty of Computer Sciences and Informatics, Amman Arab University, Amman 11953, Jordan

School of Computer Sciences, Universiti Sains Malaysia, Pulau Pinang 11800, Malaysia

E. H. Houssein et al. (eds.), *Integrating Meta-Heuristics and Machine Learning for Real-World Optimization Problems*, Studies in Computational Intelligence 1038,
https://doi.org/10.1007/978-3-030-99079-4_7

1 Introduction

Energy production is a great challenge for this century. Technologies for generating electricity from renewable sources will play an important role in this regard, not only for the growing global awareness of the need to protect the environment, but also to reduce dependence on fossil fuels for electricity production [1–6]. Over the past decades, there has been a significant increment in the combined capability of PV system power plants around the world. Because each photovoltaic system is collected mainly of SCs [7, 8].

The PV system is the most significant technique for transforming solar energy into electricity [9, 10]. Therefore, a photovoltaic model that can be carefully predicted the execution of every photovoltaic module will be very substantial in designing such effective photovoltaic systems. A perfect photovoltaic model can, for example, locate the optimal selection of PV modules [11]. But there exist unknown parameters for the photovoltaic system. Therefore, identify these parameters is always desirable not only for evaluating the performance of cell, but also for improving the design of cell, manufacturing process and quality control [12].

To use the electric circuit models, the parameters (Ipv, Io, a, Rs, Rsh) must first be determined separately for each PV device. Dozens of techniques have been developed to determine the SDM and DDM parameters. These techniques can generally be split into analytical methods and numerical methods. Additionally making suitably simple presumptions, methods of analytical largely rely on the accuracy of different important points on the I-V curve, i.e. MPP, open circuit voltage, short circuit current and slope Curves at intersections [13, 14].

In the relevant literature there is a broad reference to estimate parameters. The difference is in the extraction process can be specified to be in the number of data samples that are involved in the extraction process and in the type of Mathematical model methods used. Mathematically, the extraction of solar cell parameters is usually divided into two categories: numerical methods [15] and analytical methods [16]. Numerical methods are based on algorithms that match curves for getting the optimal match between experiential and theoretical I-V characteristics of solar cells. Overall, curve fitting methods are known to have greater confidence in the result of parameter values because most or all of the sample I-V array is used in the identification process. Moreover, the precision of the matching method rely on the method used, the objective function to be minimized and the initial values of the parameters. In addition, due to the wrong choice of the primary values, fit curves based on slope reduction methods tend to converge to local rather than global extremes. Also, a relatively long time can be calculated for these methods. On the contrary, the analysis requires only a limited set of I-V points corresponding to a set of finite equations whose solutions are parameter values.

The meta-heuristic algorithms [17] such as Henry gas solubility optimization [18], Archimedes optimization algorithm [19], Lévy flight distribution [20], Honey Badger Algorithm [21], Particle Swarm Optimization (PSO) [22] plays important rule in solving several real-world applications such as; Image segmentation [23–

27], Heartbeats Classification [28–31], Computer networks [32], drug design and discovery [33], cloud computing [34], Energy [1–4], fuel cell [5–7], photovoltaic (PV) system [8, 13, 14, 35], Optimization problems [36, 37], Stock prediction [38], Feature selection [39, 40], ECG signals classification [41, 42], chemical descriptors [43, 44], and detecting covid-19 [45].

In the same context, different algorithms have been used to extract the values of the parameters. Such as bird mating optimizer [46], artificial immune system [47], repaired adaptive differential evolution [48], pattern search [49, 50], harmony search-based algorithms [51], simulated annealing [52], chaos particle swarm algorithm [53], artificial bee colony [54], mutative-scale parallel chaos optimization algorithm [55], adaptive differential evolution [56], grey wolf optimizer [57], mine blast algorithm [58], improved shuffled complex evolution algorithm [55], direct search optimization algorithm [59], evaporation rate based water cycle algorithm [60], tabu search [61], and chaos particle swarm optimization [62], among others. For avoiding the possibility disadvantages of analytical methods, many methods of numerical have been proposed, inclusive Newton method. So, different heuristic techniques used to estimate the unknown PV parameters. Techniques for generating different parameters have been proposed in the literature, and since parameter estimation is a multidimensional numerical optimization problem, different authors have generated different parameter values for common I-V data sets. Therefore, it can be said that no method guarantees totally reliable results for the estimation of parameters. The main purpose of this article is to study the performance of some methods. The accuracy of the estimated parameters evaluated. Additionally, the impact of the control parameter on the performance has been inspected [35].

The rest of the paper is structured as follows: Sect. 2 introduces an overview of the mathematical model of PV cell/module. Section 3 presents the The objective function. Metaheuristics for parameter estimation of PV cell are discussed in Sect. 4. Section 5 concludes the paper.

2 Mathematical Model of PV Cell/Module

This section discusses the mathematical model of the photovoltaic [63] that can be approaching by SD [64] and DD [65] model are widely utilized in various relevant studies. Recently, the three diode (TDM) [66] is used. Therefore we discusses these models and their usage as an optimization problem. Figures 1, 2 and 3 display the equipollent circuits diagram of an typical photovoltaic models.

In the single diode model we need to estimate five parameters ($I - ph, I_{sd}, a, R_s$ *and* R_{sh}). In Fig. 1 current (I) is determine as the following:

$$I = I_{Ph} - I_{sd} - I_{sh} \tag{1}$$

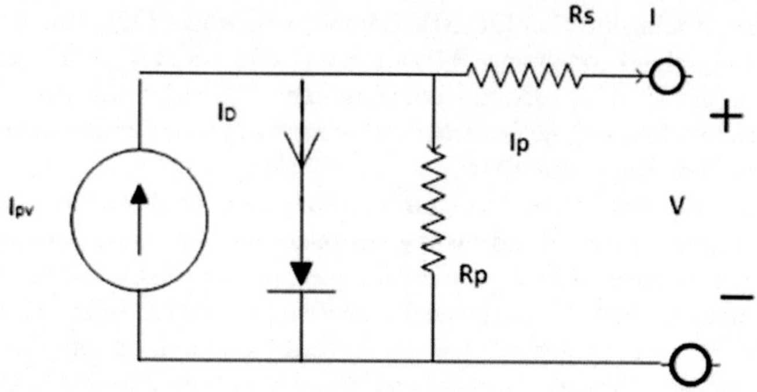

Fig. 1 Circuit of SD model

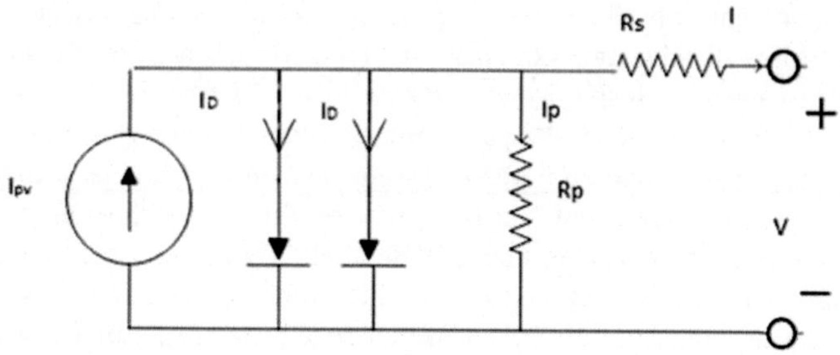

Fig. 2 Circuit of DD model

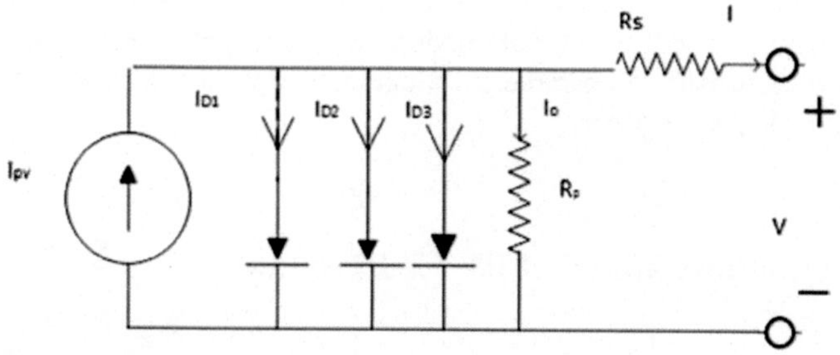

Fig. 3 Circuit of TD model

where (I_{ph}) is the current of photon, I_{sd} is the current of diode and I_{sh} is resistor of shunt. From the equivalent Shockley diode equation, can handle the inside diode parameters fore accurate output. In view of the previous Eq. 1 can write as:

$$I = I_{PV} - I_O[\exp \frac{q(V + R_S I)}{aKN_C T} - 1] - \frac{V + R_S I}{R_{Sh}} \tag{2}$$

So I_{sd} is the current of diode saturation, V is terminal voltage, a is the factor of ideality, R_s and R_{sh} define the series and shunt resistances. In addition some3 constants: constant of Boltzmann $k = 1.380 \times 10^{-23}$, T is temperature of cell and The charge of electron $q = 1.602 \times 10^{-19}$.

The SDM is popular model to use, however, its accuracy is not the best for this reason, the DDM is proposed.

In the DDM in Fig. 2, 7 parameters are needed to obtain ($I_{ph}, I_{sd1}, I_{sd2}, a_1, a_2, R_s, R_{sh}$).

From Fig. 2, Eq. 1 can be written as:

$$I = I_{Ph} - I_{sd1} - I_{sd2} - I_{sh} \tag{3}$$

where (I_{ph}) is the current of photon, I_{sd} and I_{sd2} are the current of first and second diode and I_{sh} is resistor of shunt. From the equivalent Shockley diode equation, can handle the inside diode parameters fore accurate output. In view of the previous Eq. 3 can formulate as:

$$I = I_{Ph} - I_{sd1}\left[\exp\frac{q(V + R_S I)}{a_1 K N_C T} - 1\right] - I_{sd2}\left[\exp\frac{q(V + R_S I)}{a_2 K N_C T} - 1\right] - \frac{V + R_S I}{R_{sh}} \tag{4}$$

Additionally, the tree diode model is proposed as an expansion to SDM and DDM, the important feature of TDM is the capacity to be applied for industrial applications. Figure 3 shows the TDM. With combination of TDM Eq. 1 then is edited as following:

$$I = I_{ph} - I_{sd1} - I_{sd2} - I_{sd3} - I_{sh} \tag{5}$$

As the Shockley equivalence, Eq. 5 write as:

$$\begin{aligned} I = & I_{Ph} - I_{sd1}\left[\exp\left(\frac{q(V + R_S I)}{a_1 K N_c T}\right) - 1\right] - I_{sd2}\left[\exp\left(\frac{q(V + R_s I)}{a_2 K N_c T}\right) - 1\right] \\ & - I_{sd3}\left[\exp\left(\frac{q(V + R_s I)}{a_3 K N_c T}\right) - 1\right] - \frac{V + R_s.I}{R_{sh}} \end{aligned} \tag{6}$$

where I_{sd1}, I_{sd2} and I_{sd3} are diode currents of the first, second and three. n_1, n_2 and n_3 are the diode ideal factors respectively. From Eq. (6) illustrate that exist 9 unknown variables for the TDM ($I_{ph}, I_{sd1}, I_{sd2}, I_{sd3}, a_1, a_2, a_3, R_s$, and R_{sh}) need to estimate.

3 The Objective Function

The major purpose of extracting the SD, DD and TD model parameters is to obtain the parameter values for minimizing the error between the measured and simulated current, which are expressed by an objective function. To facilitate the comparison with the results published in the most recent literature, this article chooses the mean square error (RMSE) as the objective function.

$$RMSE = \sqrt{\frac{1}{K}\sum_{i=1}^{K} f(V_t, I_t, X)^2} \tag{7}$$

$$f_{SDM}(V_t, I_t, X) = I_{PV} - I_O\left[\exp\frac{q(V + R_S I)}{aKN_C T} - 1\right] - \frac{V + R_S I}{R_S} - I \tag{8}$$

$$\begin{aligned} f_{DDM}(V_t, I_t, X) = I_{PV} - I_{o1}\left[\exp\frac{q(V + R_S I)}{a_1 K N_C T} - 1\right] \\ - I_{o2}\left[\exp\frac{q(V + R_S I)}{a_2 K N_C T} - 1\right] - \frac{V + R_S I}{R_S} - I \end{aligned} \tag{9}$$

$$\begin{aligned} f_{TDM}(V_t, I_t, X) = I_{PV} - I_{o1}\left[\exp\frac{q(V + R_s I)}{a_1 K N_c T} - 1\right] - I_{o2}\left[\exp\frac{q(V + R_s I)}{a_2 K N_c T} - 1\right] \\ - I_{o3}\left[\exp\frac{q(V + R_s I)}{a_3 K N_c T} - 1\right] - \frac{V + R_s I}{R_s} - I \end{aligned} \tag{10}$$

4 Metaheuristics for Parameter Estimation of PV Cell

In this section we will give a brief explanation of the latest metaheuristics used for estimating the parameters of the photovoltaic cell. The metaheuristics technique was inspired by nature to solve the optimization problems that were solved by mimicking biological or physical phenomena. The metaheuristics optimization can be classified into main four types, evolutionary algorithms, human-based algorithms, physicals-based algorithms, and swarm-based algorithms.

4.1 Evolutionary Algorithms

Traditional evolutionary algorithms [67–71]. Deterministic techniques are critical to start values and are likely to be perceived in local minimum. Therefore, the evolutionary algorithm offers best precision and higher computational capacity than the deterministic method, whereby the performance depend on the correct setting of the control parameters. Any wrong selection can lead to slow convergence and an early end of iterations. So, the search for an accurate, precise and effective numerical method for solving the problem of extracting the parameters form the solar cells is still ongoing.

Genetic Algorithm (GA) The GA begins by a randomly generated group of individuals in the area of search, called the primary population. For creation the new generation, the algorithm chooses several individuals from the current population as "parents" for participating Some of their genes (the parameters of the photovoltaic in this case) are used for creating "children". These children are members of the new generation [72]. There are three various categories of children: Mutant, Elite and Crossover. The real individuals of the generation with the most basic cost value act as the children of the next generation, those children are the elite. Another set of children (crossover) are create by combination of genes or vectors from two parents. Therefore, every child stores information about the genes of both parents. After all Mutation children are created by randomly changing their parents' genes; in other words, Mutation child creates by changing only one parent. Generally, the GA keeps to create new generations until the stop conditions are met. Some exemplary conditions of stop: desirable fitness interval, number of maximum generations, time end or related modify in fitness.

This technique used 20 times for extracting the variables. The development of the variables for twenty tests is shown in Fig. 4. Generally, the parameters generated from the speculation solution close to the last results gained from the optimization method, which offer that the estimation solution is significant for minimizing time of convergence of GA. Furthermore, Fig. 4 shows that, the GA obtains another result, which is nevertheless close to the approximate solution, and the solutions found in the 20 experiments are lower. These points demonstrate the coherence of the solutions proposed in the GA technique.

In [72] used GA for estimating five parameters for SD in different temperature (25, 45 and 65 °C) and compared the RMSE with other methods. In temperature 25 °C the RMSE of GA was 2.285×10^{-03}% and other methods Quasi-Newton (QN) and Trust-Region (TR) were 2.276×10^{-03}% and 1.804×10^{-03}% respectively. In temperature 45 °C the RMSE of GA, QN and TR were 5.343×10^{-03}%, 7.485×10^{-03}% and 1.211×10^{-03}% respectively. And in temperature 65 °C RMSE of GA, QN and TR were 1.692×10^{-02}%, 2.705×10^{-02}% and 2.541×10^{-3}% respectively.

Recently, [49] improved genetic algorithm in which the traditional GA inserted with a convex collection crossover for determining SDM and DDM. For evaluation of the solutions obtained by the improved GA, the solutions were compared by other algorithms modified GA, GA, HS, CSO, SA and ABSO for SD model were 9.8602×10^{-04}%, 0.01908%, 9.951×10^{-04}%, 9.8744×10^{-04}%, 0.0190% and 9.8602×10^{-04}% respectively, and for DD model were 9.8248×10^{-04}%, 03604%, 0.00126%, 9.8252×10^{-04}%, 0.01664% and 9.8344×10^{-04}% respectively. From the comparison, we can notice that the modified GA can be used as a precise and strong method for estimating SC parameters.

Improved Shuffled Complex Evolution Algorithm The shuffled complex evolution algorithm is a 3-level that includes: population, complex, and simple. The style of the shuffled complex evolution is therefore for treatment global research as top-down population development. To guarantee efficiency and effectiveness, the SCE combines the power of randomized controlled research, competitive development,

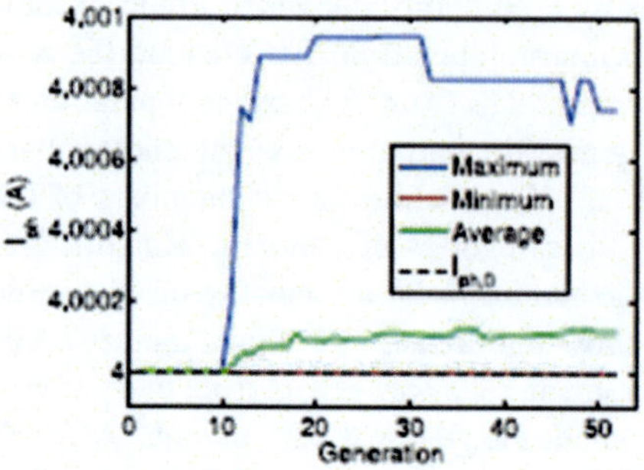

(a) Evolution of I_{ph}.

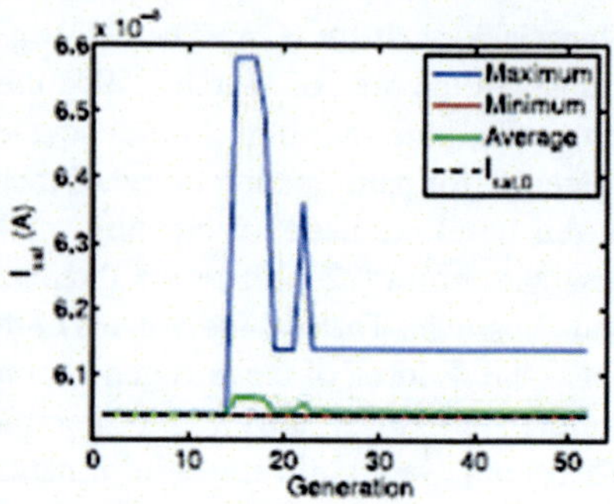

(b) Evolution of I_{sat}.

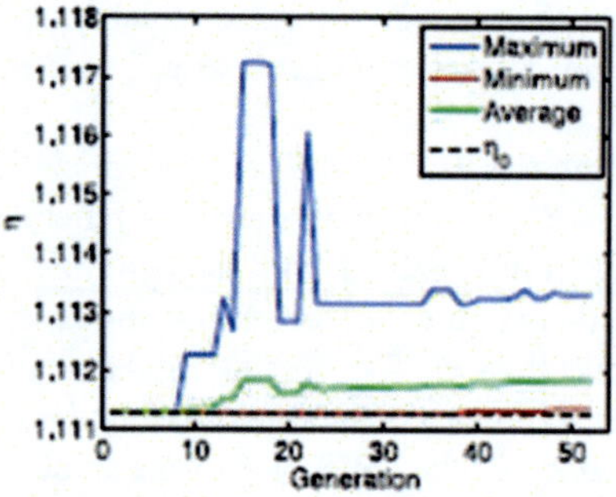

(c) Evolution of η.

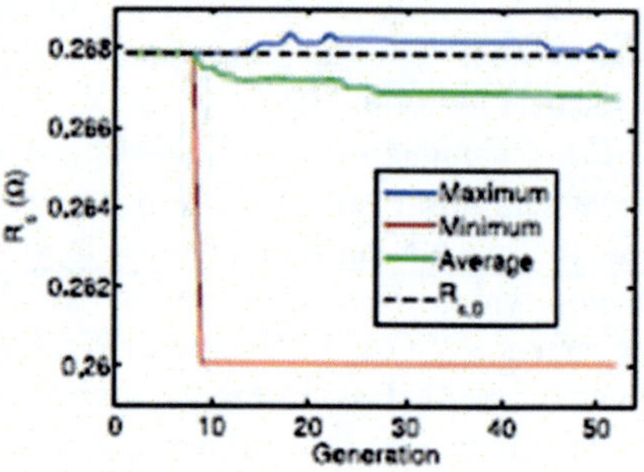

(d) Evolution of R_s.

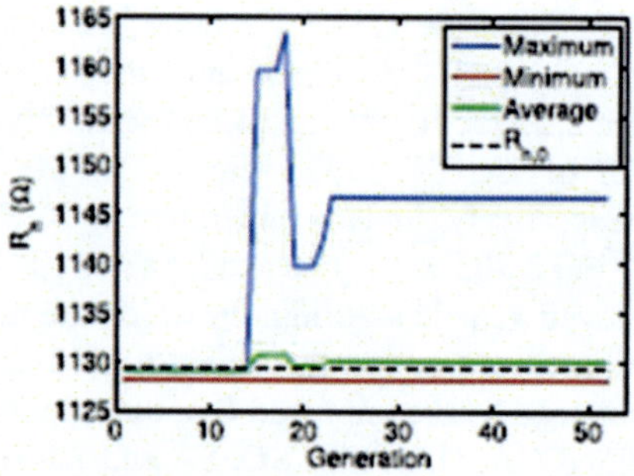

(e) Evolution of R_h.

Fig. 4 Development of the parameters after 20 tests of GA

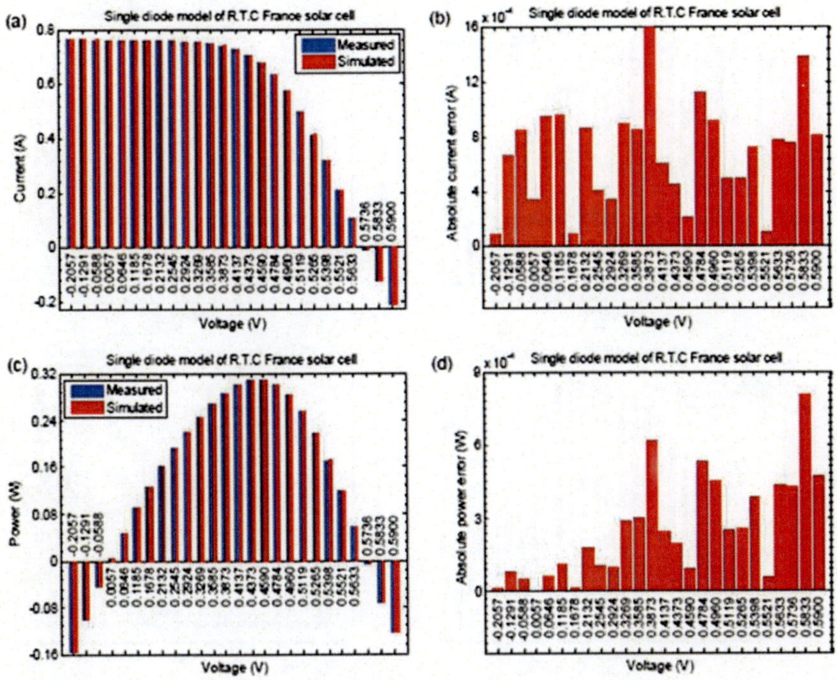

Fig. 5 Comparison between calculated and simulated data of ISCE for SD: **a** I–V data, **b** absolute current error, **c** P–V data and **d** absolute power error

complex mixing and simple adaptability of the Nelder-Mead algorithm simplex [73] to develop the population towards the global optimum. The SCE method starts by a random sample that represents the initial point population that covers the entire possible range of [LB, UB]. Then the initial population points are divided into a number of complexes. Each complex develops independently using a Competitive Complex Development (CCE) strategy for updating the Simplex's worst peak and lead research for improvement. After several generations, the updated abstraction becomes complex again. Eventually, the resulting complexes compelled to mix and new complexes are created through the strategy of complex fusion in order to assure the exchange of information between the various complexes. This evolutionary process reiterated till it meets conformed convergence criterion. The major routine of the ISCE algorithm is the same as that of the traditional SCE algorithm. The major difference is that the ISCE algorithm extends every domain with the extended CCE strategy [62].

The ISCE technique is used to extract the optimal parameters from both the single diode model and the double diode model. Figures 5 and 6 illustrate the comparison between the results of the experiment and those evaluated by ISCE for SCs (single and double diode model).

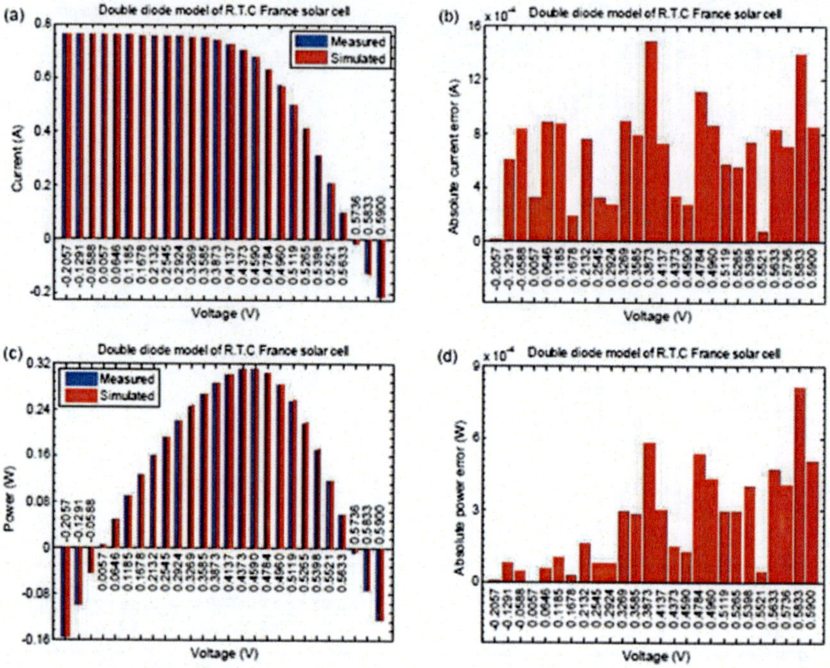

Fig. 6 Comparison between calculated and the simulated data of ISCE for DD: **a** I–V data, **b** absolute current error, **c** P–V data and **d** absolute power error

In [62] used ISCE algorithm to determine SD and DD model parameters. For evaluation the execution of proposed method compared the results with other methods in RMSE. The RMSE obtained by PCE, ABC, CSO, BMO, MABC, GGHS, ABSO, IGHS, CPSO, CWOA and proposed method (ISCE) were $9.86022 \times 10^{-04}\%$, $9.8620 \times 10^{-04}\%$, $9.8602 \times 10^{-04}\%$, $9.8608\times^{-04}\%$, $9.8610\times^{-04}\%$, $9.9097 \times 10^{-04}\%$, $9.9124\times^{-04}\%$, $9.9306\times^{-04}\%$, $1.3900\times^{-03}\%$, $9.8602\times^{-04}\%$ and $9.860219\times^{-04}\%$ respectively for SD model and The RMSE obtained by PCE, ABC, CSO, BMO, MABC, GGHS, ABSO, IGHS, CWOA and proposed method (ISCE) were $9.8252\times^{-04}\%$, $9.861\times^{-04}\%$, $9.8252\times^{-04}\%$, $9.8262\times^{-04}\%$, $9.8276\times^{-04}\%$, $1.0684\times^{-03}\%$, $9.8344\times^{-04}\%$, $9.8635\times^{-04}\%$, $9.8272 \times 10^{-04}\%$ and $9.824849\times^{-04}\%$ respectively for DD model. This comparison illustrate that ISCE has ability to use as robust and precise method for determining SC parameters.

And Other Evolutionary algorithms that are used to estimate unknown parameters are listed in Table 1.

Table 1 Summary of evolutionary metaheuristics algorithms that are used to estimate PV cell model parameters

References	Year	Method	Main result (RMSE)
El-Naggar et al. [47]	2012	PS	SDM: 0.01494% DDM: 0.01518%
Jiang et al. [74]	2013	IADE	SDM: 9.8900×10^{-04}% DDM: 9.8610×10^{-04}%
Niu et al. [75]	2014	BBO-M	SDM: 9.8634×10^{-04}% DDM: 9.8272×10^{-04}%
Zhang et al. [76]	2016	PCE	SDM: 9.86022×10^{-04}% DDM: 9.8252×10^{-04}%
lin et al. [15]	2017	SSO	SDM: 9.8640×10^{-04}% DDM: 9.9129×10^{-04}%
Ramos-Paja et al. [77]	2017	GA	SDM: 3.845×10^{-02}%
lin et al. [15]	2017	MSSO	SDM: 9.8607×10^{-04}% DDM: 9.8281×10^{-04}%
Hamid et al. [49]	2017	GA	SDM: 9.8602×10^{-04}% DDM: 9.8248×10^{-04}% PV: 0.002425%
Louzazni et al. [78]	2018	FA	SDM:5.1382×10^{-04}% DDM: 4.5485×10^{-06}%
Gao et al. [62]	2018	ISCE	SDM: 9.860219×10^{-04}% DDM: 9.8344×10^{-04}%
Kumar et al. [79]	2020	SMA	SDM: 9.8582×10^{-04}% DDM: 9.8148×10^{-04}% TDM: 9.80143×10^{-04}%

4.2 *Human Algorithms*

The Human Algorithm: How Artificial Intelligence Defines Who We Are examines the enormous impact of intelligent technologies on humanity. We must therefore develop and implement laws, guidelines and controls to protect ourselves from the destructive threats posed by technology. Ultimately, a human algorithm is a clear invitation to shape a more human future and push our project to new frontiers.

Harmony Search Algorithm To achieve pleasant harmony, musicians attempt to adapt the height of their machines. In musical improvisation, every musician makes the sounds that are in the imaginable area. The positions together create a harmony. The quaily of the improvised harmony is judged according to aesthetic criteria and the musicians attempt to achieve the best status of harmony by punishing the height of the instruments. This procedure continues till an ideal state of harmony is achieved.

To adjust the pitch of their instruments, musicians utilize one of three principles: (1) play a memory tone, (2) play a tone close to the memory tone, and (3) play a random tone that is in possible range.

In the harmony search algorithm, every solution is called harmony and is defined by a vector x containing d elements corresponding to the dimensions of the problem. At the beginning of the method, a population of harmony vectors in the possible area is randomly generated and stored in the harmony memory (HM). Then a new harmony extemporized. Every parameter for a decision is updated by one of the following three rules: (1) select a value in HM, (2) select a value close to a value in HM, and (3) randomly select a value within a possible area. The worst harmony in HM is deleted and exchanged by the new one if the fineness of the extemporized harmony is better than the worst [80].

GGHS and IGHS [80] improved from HM for better used to identify variables of single and double diode model. The solutions obtained by modified HS compared with HS and other methods. The RMSE fr SD model of GGHS, SA, IGHS, PS, GA, HS and CPSO were $9.9306\times^{-04}$%, 0.01900%, 9.9097×10^{-04}%, 0.01494%, 0.01908%, 9.9510×10^{-04}% and 0.00139% respectively. The RMSE of HS, GGHS. IGHS, SA and PS were 0.00126%, 0.00107%, 9.8635×10^{-04}%, 0.01664%, 0.01518% respectively for DD model.

And Other human algorithms that are used to estimate unknown parameters are listed in Table 2. This Table discusses some of the recently used human metaheuristics to identify parameters of PV solar cell.

4.3 Physicals-Based Algorithms

In recent years, scientists have developed various optimization methods, including meteorological optimization methods. People have used the power of nature to solve problems. Hence, these meteoric methods mimicked the physical and biological processes of nature. In 2007, the Big Crunch optimization algorithm established by the development of the universe was proposed and implemented, and in 2009 the gravitation search algorithm depend on the law of gravity was suggested and implemented to solve complicated problems. However, many physics-based algorithms were later proposed. In this part; Some of the optimization algorithms for physics-based meteorologists have been tested.

Evaporation Rate Based Water Cycle Algorithm ER-WCA is a natural inspired metaheuristic that was proposed in 2015 to optimize the current problems [89]. ER-WCA is a improved WCA that reported in 2012 by [90]. The WCA is depend on the process of navigating on water and downstream of rivers and streams flowing into the sea. ER-WCA modifies in traditional WCA by combining vaporization modes, that has 2 advantages: (1) good balance between exploration and exploitation phases than WCA, and (2) precise results compared to WCA. ER-WCA performed in 4 main stages: initialization, flowing of streams to river or ocean, vaporization and precipitation process, and vaporization rate.

To illustrate the performance of ER-WCA in estimation the value of unknown parameters for PV in Figs. 7 and 8.

Table 2 Summary of human algorithms metaheuristics that are used to estimate PV cell model parameters

References	Year	Method	Main result (RMSE)
Askarzadeh et al. [80]	2012	GGHS	SDM: 9.9097×10^{-04}% DDM: 1.0684×10^{-03}%
Askarzadeh et al. [80]	2012	IGHS	SDM: 9.9306×10^{-04}% DDM: 9.8635×10^{-04}%
Yuan et al. [55]	2014	MPCOA	SDM: 9.4457×10^{-04}% DDM: 9.2163×10^{-04}% PV: 0.002425%
Niu et al. [81]	2014	STBLO	SDM: 9.8602×10^{-04}% DDM: 9.8248×10^{-04}%
Chen et al. [82]	2016	GOTLBO	SDM: 9.8744×10^{-04}% DDM: 9.83177×10^{-04}%
Yu et al. [83]	2017	IJAYA	SDM: 9.8606×10^{-04}% DDM: 9.8380×10^{-04}%
Chen et al. [84]	2018	TLBO-ABC	SDM: 9.8602×10^{-04}% DDM: 9.8415×10^{-04}% PV: 2.4251×10^{-03}%
Yu et al. [85]	2018	LBSA	SDM: 1.0092×10^{-03}% DDM: 1.0165×10^{-03}% PV: 2.4296×10^{-03}%
Kler et al. [86]	2019	HISA	SDM: 2.0166×10^{-03}% DDM: 2.0166×10^{-03}%
Kler et al. [87]	2019	ITLBO	SDM:1.0069×10^{-03}%
Yu et al. [88]	2019	JAYA	PV: 2.4278×10^{-03}%

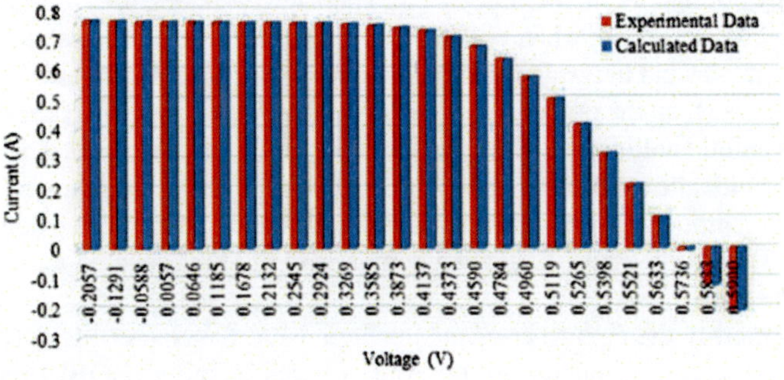

Fig. 7 I-V characteristics of ER-WCA for SD

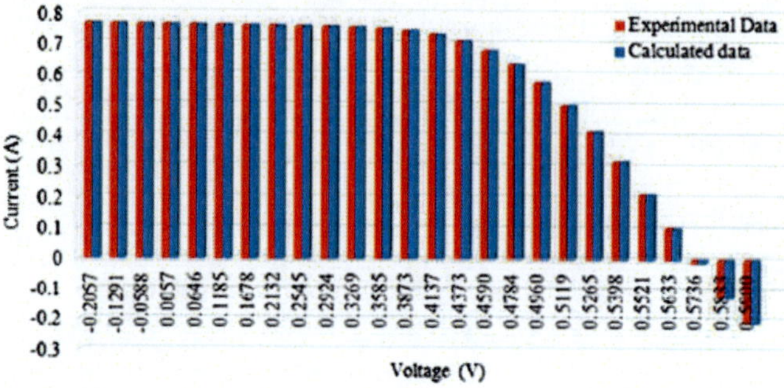

Fig. 8 I-V characteristics of ER-WCA for DD

In [91] ER-WCA used for generation values of parameters for SDM and DDM. The results generated by ER-WCA compared with other method in term RMSE to evaluate the performance of ER-WCA. The RMSE for single diode of ER-WCA, NM-MPSO, GOTLBO, MABC, CSO, BBO-M, ABC and IADE were $9.8602 \times 10^{-04}\%$, $9.8602\times^{-04}\%$, $9.8744\times^{-04}\%$, $9.861\times^{-04}\%$, $9.8602\times^{-04}\%$, $9.8634\times^{-04}\%$, $9.8629\times^{-04}\%$ and $9.89\times^{-04}\%$ respectively and $9.824849 \times 10^{-04}\%$, $9.8250 \times 10^{-04}\%$, $9.83177 \times 10^{-04}\%$, $9.8276 \times 10^{-04}\%$, $9.8252 \times 10^{-04}\%$, $9.8272 \times 10^{-04}\%$ and for double diode were $9.8610 \times 10^{-04}\%$ respectively. The comparison illustrate that ER-WCA obtained the best result than other methods.

Flower Pollination Algorithm The FPA is a new population-based optimization method advanced by Xin-She Yang [92]. It Imitates the flower pollination behavior. Pollination is a natural physiological process in plants that involves the transmission of pollen by pollinators such as insects etc. There are two categories of powder: self-pollination, which occurs when pollen from the same flower fertilizes and the other cross-pollinated what happens when pollen grains transported from another plant to a flower. Flowers, on the other hand, try to expand their pollen in different ways. One of these is abiotic pollination, where the pollen is carried by the wind. The second is biological pollination, which is done with the help of insects and other animals.

The FPA method utilized for identifying the parameters of single and double diode models [93]. Figures 9 and 10 show a comparison of the simulated and calculated FPA results for SD and DD.

In [94] modify the FPA to combine the basic FPA, the Nelder-Mead (NM) simplex method, and the Generalized opposition-based learning (GOBL) mechanism together. The Flower Pollination Algorithm is perfect at global exploration because of the Lévy flight, the NM simplex method is good at local optimum but robustly oversensitive to the start points, the GOBL technique assist metaheuristic algorithms to jumpy out of local search.

The results compared with other algorithms to evaluate the execution of the suggested method. For single diode the RMSE of the basic FPA (BFPA), the FPA collective with the OBL (OFPA), the FPA collective with the NM simplex method (BFPANM), the FPA together with the GOBL (GOFPA), the FPA together com-

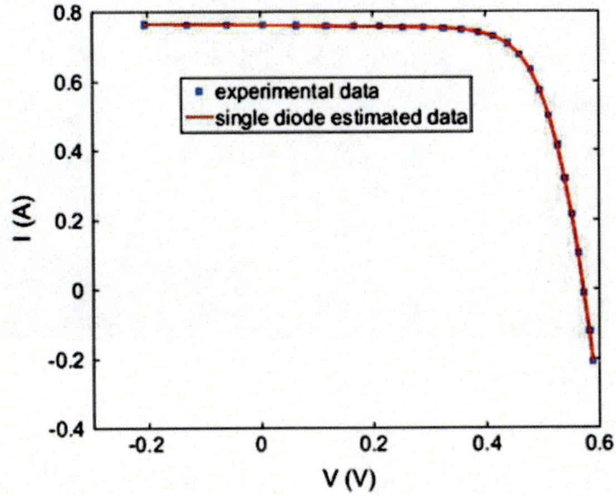

Fig. 9 Comparison between the experimental data and simulated data results by FPA for SD

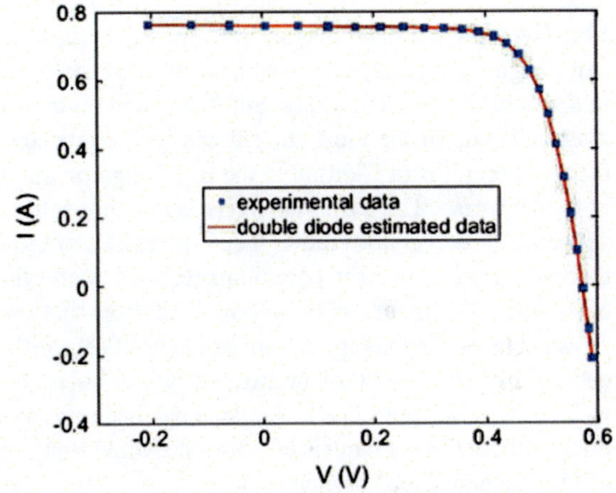

Fig. 10 Comparison between the experimental data and simulated data results by FPA for DD

bined with the OBL and the NM simplex method (OFPANM) and the FPA collective together with the GOBL and the NM simplex method (GOFPANM) together were $9.860219 \times 10^{-04}\%$, $9.864331 \times ^{-04}\%$, $9.860219 \times 10^{-04}\%$, $9.860219 \times 10^{-04}\%$, $9.860219 \times 10^{-04}\%$ and $9.860219 \times 10^{-04}\%$ respectively and for double diode were $9.835164 \times 10^{-04}\%$, $9.824880 \times 10^{-04}\%$, $9.824849 \times 10^{-04}\%$, $9.828786 \times 10^{-04}\%$, $9.824849 \times 10^{-04}\%$ and $9.824849 \times 10^{-04}\%$ respectively. The RMSE of other method for single diode model that compared with the proposed method CPSO, PS, LMSA, IGHS, ABC, ABSO, GOTLBO, BBO-M and GOFPANM were $1.3900 \times 10^{-03}\%$, $1.4940 \times 10^{-02}\%$, $9.8640 \times 10^{-04}\%$, $9.9306 \times 10^{-04}\%$, $9.862 \times 10^{-04}\%$, $9.9124 \times 10^{-04}\%$, $9.87442 \times 10^{-04}\%$, $9.8634 \times 10^{-04}\%$ and $9.8602 \times 10^{-04}\%$ respectively and the RMSE of PS, SA, IGHS, ABC ABSO, STLBO, GOTLBO, BBO-M and GOFPANM for double diode model were $1.5180 \times 10^{-02}\%$, $1.6640 \times$

Fig. 11 White hole, black hole, and wormhole

$10^{-02}\%$, $9.8635 \times 10^{-04}\%$, $9.861 \times 10^{-04}\%$, $9.8344 \times 10^{-04}\%$, $9.8248 \times 10^{-04}\%$, $9.83177 \times 10^{-04}\%$, $9.8272 \times 10^{-04}\%$ and $9.8248 \times 10^{-04}\%$ respectively.

Multi-verse Optimizer Algorithm The big bang theory [95] explains that our universe begins with a huge explosion. According to this theory, the source of everything in this world is the Big Bang, and there was nothing before it. The theory of multiverses is one of the most current and well connected [96] among physicists. This theory is that it is in addition to the Big Bang and that the Big Bang causes the birth of the Universe. The term multiverse is in contrast to universities, which refer to the existence of universities rather than universities of the same lives [96]. In multiverse theory, several universes communicate with each other and can even collide with each other. The theory of polynomials dictates that the universe may have different physical laws. As the inspiration for the MVO algorithm we have chosen three main concepts of the theory of multiverse: white holes, black holes and wormholes as in Fig. 11. A white troupe in Jamaica has been observed in the universities our, but doctors think the big bang is considered a white troupe and could in the main element of the university renaissance [97].

As discussed in the previous section, population-based algorithms divide the research process into two stages: research versus exploitation. We use the concept of white holes and black holes to determine MVO's search placements. You can see, for revenge, help MVO to use the research stations. We assume that the jacket solution is compatible with universities and that the variable version of the solution.

There is an object in this universe. Its attributions are necessarily a solution rate of inflation at each which is linked to the value of the corresponding fitness function of the solution. We also use the term time instead of coming back in this article because it is a current term in multiverse theory and [98] cosmology.

In [99] used Multi-Verse Optimizer algorithm to identify SD model parameters and compared with other methods. The RMSE of MVO, LI, RMF 1A, RMF 1B, RMF 1C/1D, SA and PS were $2.0771 \times 10^{-03}\%$, $2.4777 \times 10^{-03}\%$, $2.1176 \times 10^{-03}\%$, $2.1547 \times 10^{-03}\%$, $2.0465 \times 10^{-03}\%$, $2.6600 \times 10^{-03}\%$ and $1.1800 \times 10^{-02}\%$ respectively for SD model.

Table 3 Summary of physicals-based algorithms that are used to estimate PV cell model parameters

References	Year	Method	Main result (RMSE)
Chegaar et al. [100]	2001	CM	SDM: 0.0013%
Brest et al. [101]	2006	jDE	SDM: 1.0115×10^{-03}% DDM: 1.0703×10^{-03}% PV: 2.4251×10^{-03}%
Zhang et al. [102]	2011	LW	SDM: 9.6964×10^{-03}%
Wang et al. [103]	2011	CoDE	SDM: 1.1019×10^{-03}% DDM: 1.2259×10^{-03}% PV: 2.4376×10^{-03}%
El-Nagga et al. [47]	2012	SA	SDM: 1.7000×10^{-03}% DDM: 1.9000×10^{-02}% PV: 2.7000×10^{-03}%
Gong et al. [52]	2013	$R_{cr} - IJADE$	SDM:9.8602×10^{-04}% DDM: 9.8248×10^{-04}%
Alam et al. [104]	2015	FPA	PV: 2.7425×10^{-03}%
Tong et al. [105]	2016	LSP	SDM: 1.5051×10^{-03}% PV: 0.00218%
Mirjalili and Seyedali [106]	2016	SCA	SDM: 5.8058×10^{-03}% DDM: 9.2482×10^{-03}%
Chellaswamy et al. [59]	2016	DET	SDM: 9.3×10^{-04}% DDM: 9.4×10^{-04}% PV: 0.002131%
Derick et al. [107]	2017	WDO	SDM: 0.00084% DDM: 0.00106%
Gotmare et al. [91]	2017	ER-WCA	SDM: 9.8602×10^{-04}% DDM: 9.8248×10^{-04}%
Ram et al. [108]	2017	BPFPA	SDM: 7.2700×10^{-04}% DDM: 7.2300×10^{-04}%
Turgut [109]	2017	GBEST	SDM: 9.860×10^{-04}% DDM: 9.82485×10^{-04}% PV: 0.00242507%
Xu et al. [94]	2017	GOFPANM	SDM: 9.8602×10^{-04}% DDM: 9.8248×10^{-04}%

And Other Physicals-based algorithms that used to estimate unknown parameters are listed in Table 3.

4.4 *Swarm-Based Algorithms*

This section provides an overview of the optimization techniques, followed by a brief overview of some of the naturally inspired swarm-based algorithms introduced over the last decade. This technique is inspired by the natural processes of plants, the feeding behavior of insects and the social behavior of animals.

Particle Swarm Optimization The best known algorithm is particle swarm optimization, originally developed by Kennedy and Eberhart [110]. PSO imitated by the public behavior of flocks of birds. It utilizes a series of particles (candidate solutions) that hover in the search area for obtaining the optimal value (i.e. the optimal position). Meanwhile, everyone is looking for the best location on their way. In other words, the particles are looking for their optimal results and the best value the swarm has received so far.

Figures 12 and 13 illustrate the performance of PSO in estimation unknown parameters of PV.

In [111] applied PSO algorithm for estimating the SC parameters of SDM by finding five parameters: I_{ph}, I_{sd}, R_s, R_{sh}, n in different temperature. For every value of temperature, PSO is executed and obtains finished after 1000 generations.

The optimization iterated for 100 times so new sets of population to obtain the average of optimized creates. Moreover early convergence was avoided. The obtained solution were compared with other method artificial PSO, chaos PSO and Simulated annealing the RMSE for PSO was $9.86024 \times 10^{-04}\%$ while $9.9124 \times 10^{-04}\%$, 0.00139% and 0.0017% for artificial PSO, chaos PSO and simulated annealing respectively. In [112] performed proposed technique to determine the variables of PV cells and modules depend on ensured convergence particle swarm optimization. The RMSE of the proposed technique for SD was $2.046535 \times 10^{-03}\%$, minimum iteration number 274 and minimum time was 6 s. For double diode, the RMSE was $2.046535 \times 10^{-03}\%$, minimum number iteration was 3712 and minimum time was 87 s.

In [113] displays the extraction parameters for various PV models found from 30 runs of the parallel particle swarm optimization algorithm, in which the size of swarm and number of iteration are 2048 and 80,000 respectively. The RMSE acquired by the proposed technique compared with other method LSO, PS and SA for SDM

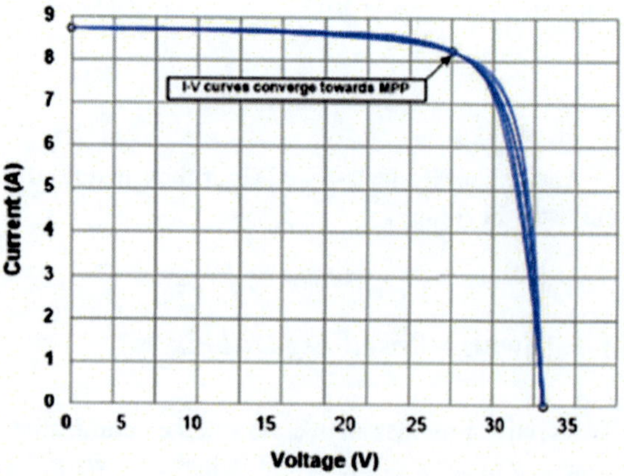

Fig. 12 I-V curves of PSO algorithm

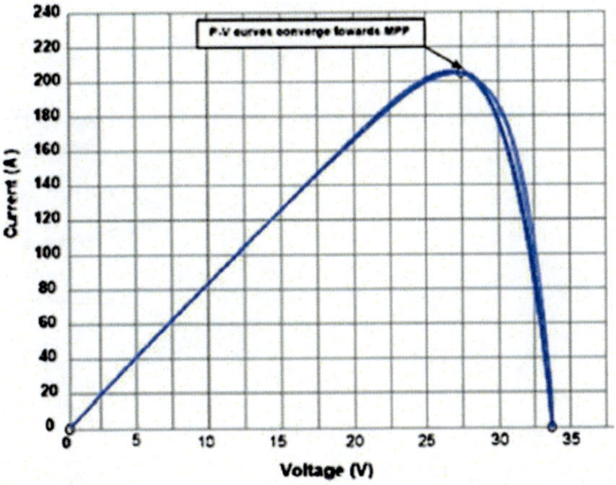

Fig. 13 P-V curves of PSO algorithm

was $1.6715 \times 10^{-03}\%$, $4.6245 \times 10^{-03}\%$, $2.2753 \times 10^{-03}\%$ and $2.0288 \times 10^{-03}\%$ respectively, and for DDM $1.6716 \times 10^{-03}\%$.

The Whale Optimization Algorithm Whale Optimization Algorithm (WOA), which imitates the behavior of humpback whales. Whales are imaginary creatures. They are considered the biggest mammals in the world. This giant mammal has 7 different main species, namely assassin, mink, six species, humpback whale, Wright, Finn and blue. Whales are mainly considered predators [114].

The most interesting thing about humpback whales is their specialized hunting methods. This foraging behavior is called bubble net foraging. He discovered two actions involving bubbles and called them 'upward spiral' and 'double loop'. In the first maneuver, the humpback whale dived 12 m downstream, and then began to form spiral bubbles around the prey and float to the surface as in Figs. 14 and 15 [114].

Recently Abd Elaziz and Oliva [115] showed new improvement to the WOA, that improving the exploration by stratifying an opposition based learning for estimating the SD, DD and TD parameters. The RMSE of modified WOA, HS, GGHS, IGHS, LMSA, ABSO, CWOA, OFPANM, ABC, CSO and CIABC For Single diode model was $9.8602 \times 10^{-04}\%$, $9.9510 \times 10^{-04}\%$, $9.9097 \times 10^{-04}\%$, $9.9306 \times 10^{-04}\%$, $9.8640 \times 10^{-04}\%$, $9.9124 \times 10^{-04}\%$, $9.8602 \times 10^{-04}\%$, $9.8602 \times 10^{-04}\%$, $9.862 \times 10^{-04}\%$, $9.8602 \times 10^{-04}\%$ and $9.8602 \times 10^{-04}\%$ respectively. For the double diode model, the RMSE of improved WOA, GOTLBO, ABC, CWOA, IGHS, ABSO, CSO and BMO was $9.83177 \times 10^{-04}\%$, $9.861 \times 10^{-04}\%$, $9.8272 \times 10^{-04}\%$, $9.8635 \times 10^{-04}\%$, $9.8344 \times 10^{-04}\%$, $9.8252 \times 10^{-04}\%$ and $9.8262 \times 10^{-04}\%$ respectively. The RMSE of improved WOA and ABC for three diode model was $9.8249 \times 10^{-04}\%$ and $9.8466 \times 10^{-04}\%$ respectively.

Moreover, the proposed modified WOA to generate the parameters of SCs. The proposed introduced the best performance and obtaining the optimal solution to determine SC parameters at different irradiance levels (200, 400, 600, 800 and 1000 W/M^2) for Single diode the RMSE was 0.0017%, 0.0091%, 0.0205%, 0.0209%

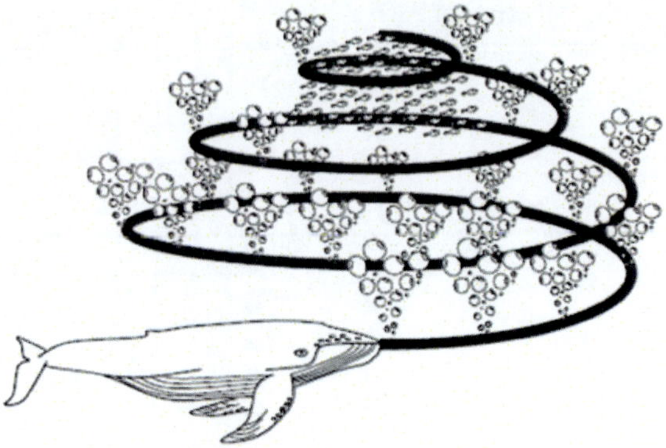

Fig. 14 Bubble-net feeding behavior of whales

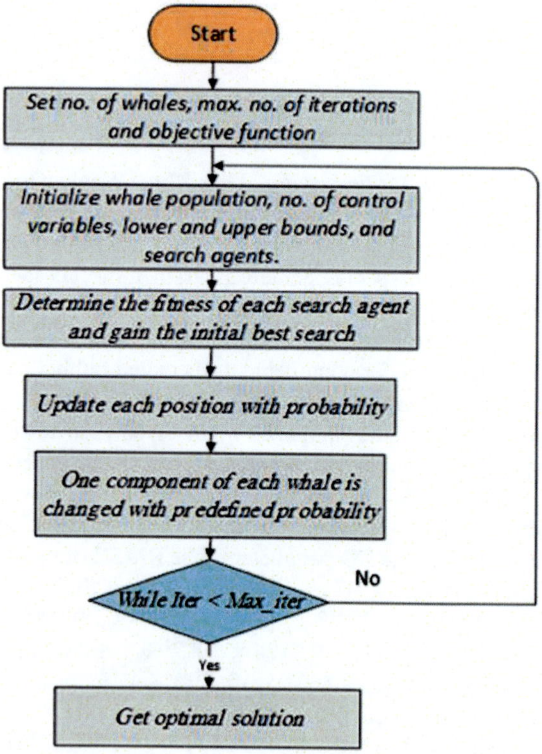

Fig. 15 Flowchart of the WOA

and 0.0247% respectively. The RMSE for double diode was 0.00148%, 0.0090%, 0.0204%, 0.0199% and 0.0243% and for Three diode was 0.00145%, 0.0089%, 0.0203%, 0.0195% and 0.0231% respectively.

The WOA has obtained high quality for estimating the SC parameters because it has high balance between the exploration and exploitation and few parameters can be changed automatically in the process updating the solutions.

Grey Wolf Optimizer This algorithm mimicked the behavior of the gray wolf [116]. The leaders are male and a female, called Alpha. The alpha is primarily accountable for taking decisions about prey, where to sleep, when to wake up, etc. The second level of the gray wolf hierarchy is beta. Betas are subordinate wolves, which help the Alpha with decision-taking or other activities. A beta is male or female and is likely the good candidate for an alpha if one of the alpha wolves dies or gets too old. The lowest ranking grey wolf is omega. Omega plays the damn goat. Omega Wolves must always serve everyone Other big wolves. They are the last wolves they will eat. It may appear that Omega is not an important person in the pack, but dealing with internal conflicts and problems during Omega loss has been observed throughout the pack. If the wolf is not an alpha, beta, or omega, then it is a subordinate machine (delta). The Delta bow to the Alpha and express themselves, but they control the Omega.

In [117] A new hybrid combined GWO and cuckoo search to parameter estimate of PV models, the result compared with other method in term RMSE. For single diode model the RMSE of HISA, MADE, CS, ABC, BLPSO, BMO, LBSA, CWOA, ISCA, IGHS, SA and the proposed method were $2.0166 \times 10^{-03}\%$, $9.8602 \times 10^{-04}\%$, $2.0119 \times 10^{-03}\%$, $9.8620 \times 10^{-04}\%$, $1.0272 \times 10^{-03}\%$, $1.0272 \times 10^{-03}\%$, $9.8602 \times 10^{-04}\%$, $1.0092 \times 10^{-03}\%$, $9.8602 \times 10^{-04}\%$, $9.8602 \times 10^{-04}\%$, $9.9306 \times 10^{-04}\%$, $1.7000 \times 10^{-03}\%$ and $9.8607 \times 10^{-04}\%$ respectively and For double diode model ere $2.0166 \times 10^{-03}\%$, $9.8261 \times 10^{-04}\%$, $2.4440 \times 10^{-03}\%$, $9.8956 \times 10^{-04}\%$, $1.1042 \times 10^{-03}\%$, $9.8262 \times 10^{-04}\%$, $1.0165 \times 10^{-03}\%$, $9.8272 \times 10^{-04}\%$, $9.8237 \times 10^{-04}\%$, $9.8635 \times 10^{-04}\%$, $1.9000 \times 10^{-02}\%$ and $9.8334 \times 10^{-04}\%$ respectively.

Recently IGWO is a developed variant of GWO [118]. To increase the search capacity two modifications have been applied for superior exploration and exploitation by using opposition based learning to provide better exploration. And the Multi-Group Grey Wolf Optimizer algorithm (MG-GWO) is derived in [119] it is used on a single-diode solar model to extract its parameters.

Moth-Flame Optimization Algorithm Moths are rebellious insects that resemble the butterfly family. Exist more than 160,000 different types of this insect that originate in nature. There are two important stages in its life: the larva and the adult. The larvae turn into butterflies inside the tuber. The most important thing about kites is their special night navigation. They are designed to fly at night in the moonlight. It used a technique named transversal steering for navigation. In this technique, a butterfly flies at an angle to the moon, a very efficient method to travel long space in a straight line.

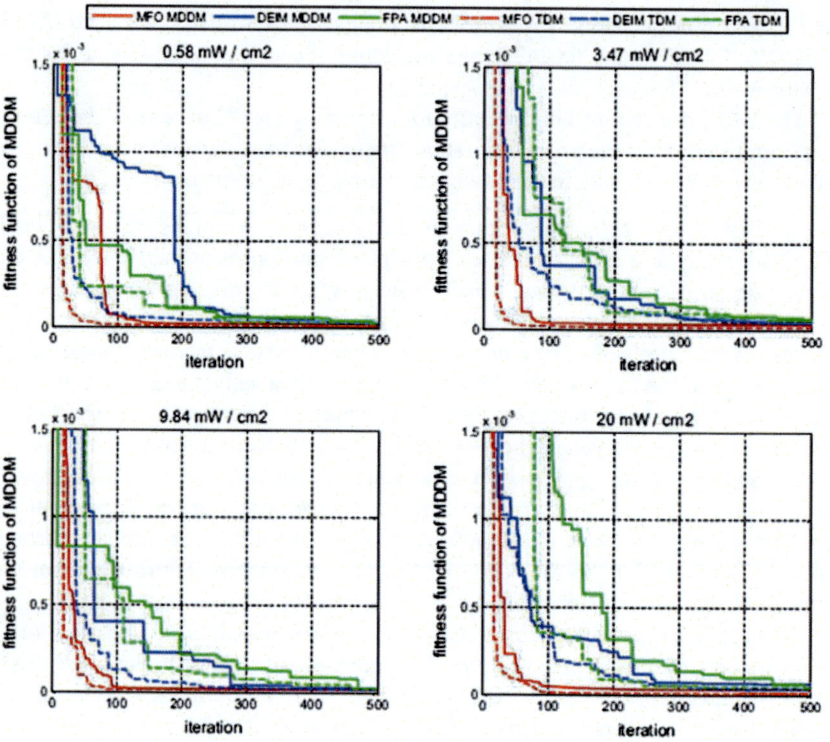

Fig. 16 The fitness function of MFO for MDD and TD Model of SC

Despite the efficiency of the lateral direction, we often see moths moving round the lights. In fact, the artificial light caused the moths to behave this way. This is due to the ineffectiveness of lateral orientation, where it is useful to go straight only when source of light is too far away. When moths view synthetic light, they attempt to preserve an equal angle with the light for flying in a straight streak [120].

In Fig. 16 shows the fitness function with estimate to number of iteration of MFO for MDD and TD Model of solar cell.

For ensuring that the proposed method can be extracted the PV parameters precisely, the unknown parameters and the RMSE for DD, MDD and TD model is carried out and the results compared with two recent algorithms, namely DEIM and FPA algorithms in various irradiance and temperature [93]. For double diode the RMSE of MVO, DEIM and FPA were $9.2898\times^{-05}\%$, $9.4866\times^{-05}\%$ and $1.2661\times^{-04}\%$ respectively, $2.0648\times^{-05}\%$, $2.7693\times^{-05}\%$ and $3.3465\times^{-05}\%$ respectively for MDD model and $2.0144\times^{-05}\%$, $2.0918\times^{-05}\%$ and $3.0115\times^{-05}$ respectively for TD model.

Table 4 Summary of swarm-based metaheuristics algorithms that are used to estimate PV cell model parameters

References	Year	Method	Main result (RMSE)
Timurkutluk et al. [113]	2011	IPSO	SDM: 1.6715×10^{-03}% DDM: 1.6716×10^{-03}%
Wei et al. [57]	2011	CPSO	SDM: 1.3900×10^{-03}%
Askarzadeh et al. [50]	2013	BMO	SDM: 9.8608×10^{-04}% DDM: 9.8262×10^{-04}%
Askarzadeh et al. [122]	2013	ABSO	SDM: 9.9124×10^{-04}% DDM: 9.8344×10^{-04}%
Jamadi et al. [123]	2016	Modify ABC	SDM: 9.861×10^{-04}% DDM: 9.8276×10^{-04}%
Guo et al. [124]	2016	CSO	SDM: 9.8602×10^{-04}% DDM: 9.8252×10^{-04}%
Chen et al. [82]	2016	EHA-NMS	SDM: 9.8602×10^{-04}% DDM: 9.8248×10^{-04}%
Jamadi et al. [123]	2016	MABC	SDM: 9.8610×10^{-04}% DDM: 9.8276×10^{-04}%
Hamid et al. [125]	2016	NM-MPSO	SDM: 9.8602×10^{-04}% DDM: 9.8250×10^{-04}%
Vickers and Neil [126]	2017	ABC	SDM: 9.8620×10^{-04}% DDM: 9.861×10^{-04}%
Oliva et al. [67]	2017	CWOA	SDM: 9.8602×10^{-04}% DDM: 9.8272×10^{-04}%
Xiong et al. [127]	2018	PSO-WOA	SDM: 9.9097×10^{-04}% DDM: 1.6700×10^{-03}% PV: 2.62422×10^{-03}%
Kang et al. [128]	2018	ImCSA	SDM: 9.8602×10^{-04}% DDM: 9.8249×10^{-04}% PV: 2.425×10^{-03}%
Diego and Abd Elaziz [115]	2018	OBWOA	SDM: 9.8602×10^{-04}% DDM: 9.8251×10^{-04}%
Jordehi and A Rezaee [129]	2018	ELPSO	SDM: 7.7301×10^{-04}% DDM: 7.4240×10^{-04}%
Nunes et al. [112]	2018	GCPSO	SDM: 7.7301×10^{-04}% DDM: 7.1827×10^{-04}% PV: 2.0465×10^{-03}%
Chen et al. [130]	2019	CS	SDM: 2.0119×10^{-03}% DDM: 2.4440×10^{-03}%
Diab et al. [131]	2020	COA	SDM: 7.7547×10^{-04}% DDM: 7.6480×10^{-04}%
Rezk et al. [132]	2021	SFS	SDM: 7.931×10^{-04}% DDM: 7.7827×10^{-04}%
Houssein et al. [35]	2021	MRFO	SDM: 7.7301×10^{-04}% DDM:$7.6046E^{-04}$% TDM:$7.6083E^{-04}$%

Recently in [121] improved Multi-Verse Optimizer algorithm by proposing a double-flames generation (DFG) strategy to create two types of target flames for leading moths flying to avoid trap into local optimum when meeting complex multi-modal problems. The results obtained by the improved Multi-Verse Optimizer algorithm (IMFO) compared with other method to evaluate the performance of IMFO. For single diode the RMSE of hybrid water cycle–moth-flame optimization method (WCMFO), the MFO, opposition-based MFO method (OMFO), improved Multi-verse optimizer (IMFO), brain storm optimization method (BSO), artificial bee colony (ABC), comprehensive learning PSO (CLPSO) method, sine cosine algorithm (SCA) and improved JAYA (IJAYA) were $9.8602 \times 10^{-04}\%$, $9.9496 \times 10^{-04}\%$, $1.1927 \times 10^{-03}\%$, $9.8602 \times 10^{-04}\%$, $2.4551 \times 10^{-03}\%$, $9.9049 \times 10^{-04}\%$, $9.9455 \times 10^{-04}\%$, $5.8058 \times 10^{-03}\%$ and $9.8606 \times 10^{-04}\%$ respectively and for double diode $9.8371 \times 10^{-04}\%$, $1.0102 \times 10^{-03}\%$, $9.8652 \times 10^{-04}\%$, $9.8252 \times 10^{-04}\%$, $2.4636 \times 10^{-03}\%$, $1.0001 \times 10^{-03}\%$, $9.9224 \times 10^{-04}\%$, 9.2482×10^{-03} and $9.8380 \times 10^{-04}\%$ respectively. It can be seen that IMFO generate the minimal RMSE among the nine techniques.

And Other Swarm-based algorithms that have been used to estimate unknown parameters are listed in Table 4. This Table discusses some of the recently used meta-heuristic algorithms to identify variables of PV solar cell and confirms that PV cell model parameter extraction is an important research subject in renewable energy.

5 Conclusion

In this article the techniques of metaheuristics for determining SCs parameters have been reviewed. Moreover, the significant research on the modelling and parameter extraction of PV cells have been reviewed. In addition, the extraction techniques were categorized into four main groups: Evolutionary algorithms (GA, ISCE), human algorithms (HS), Physicals-based algorithms (ER-WCA, FPA, MVO) and swarm-based algorithms (PSO, WOA, GWO, MFO). Ten methods were developed over the past 15 years by several researchers to identify the parameters of solar cells. they have been evaluated and critically discussed in this review paper. In addition, we summarized other metaheuristics algorithms that were used to determine the SCs parameters. This study confirms that meta-heuristic algorithms are the best way to find SCs parameters, because of their advantages, which are high accuracy, short computing time, search in global space and avoiding mathematical complexity.

Conflict of Interest The authors declare that there is no conflict of interest.

References

1. M.H. Hassan, E.H. Houssein, M.A. Mahdy, S. Kamel, An improved manta ray foraging optimizer for cost-effective emission dispatch problems. Eng. Appl. Artif. Intell. **100**, 104155 (2021)
2. A. Korashy, S. Kamel, E.H. Houssein, F. Jurado, F.A. Hashim, Development and application of evaporation rate water cycle algorithm for optimal coordination of directional overcurrent relays. Expert Syst. Appl. **185**, 115538 (2021)
3. S. Deb, E.H. Houssein, M. Said, D.S. Abd Elminaam, Performance of turbulent flow of water optimization on economic load dispatch problem. IEEE Access (2021)
4. S. Deb, D.S. Abdelminaam, M. Said, E.H. Houssein, Recent methodology-based gradient-based optimizer for economic load dispatch problem. IEEE Access **9**, 44,322–44,338 (2021)
5. E.H. Houssein, F.A. Hashim, S. Ferahtia, H. Rezk, An efficient modified artificial electric field algorithm for solving optimization problems and parameter estimation of fuel cell. Int. J. Energy Res. (2021)
6. E.H. Houssein, B.E.-D. Helmy, H. Rezk, A.M. Nassef, An enhanced Archimedes optimization algorithm based on local escaping operator and orthogonal learning for PEM fuel cell parameter identification. Eng. Appl. Artif. Intell. **103**, 104309 (2021)
7. E.H. Houssein, M.A. Mahdy, A. Fathy, H. Rezk, A modified marine predator algorithm based on opposition based learning for tracking the global MPP of shaded PV system. Expert Syst. Appl. **183**, 115253 (2021)
8. E.H. Houssein, Machine learning and meta-heuristic algorithms for renewable energy: a systematic review, in *Advanced Control and Optimization Paradigms for Wind Energy Systems* (2019), pp. 165–187
9. S.-X. Lun, C.-J. Du, J.-S. Sang, T.-T. Guo, S. Wang, G.-H. Yang, An improved explicit i–v model of a solar cell based on symbolic function and manufacturer's datasheet. Sol. Energy **110**, 603–614 (2014)
10. J. Bai, S. Liu, Y. Hao, Z. Zhang, M. Jiang, Y. Zhang, Development of a new compound method to extract the five parameters of PV modules. Energy Convers. Manag. **79**, 294–303 (2014)
11. A. Mellit, M. Benghanem, S.A. Kalogirou, Modeling and simulation of a stand-alone photovoltaic system using an adaptive artificial neural network: proposition for a new sizing procedure. Renew. Energy **32**(2), 285–313 (2007)
12. M. AlRashidi, M. AlHajri, K. El-Naggar, A. Al-Othman, A new estimation approach for determining the i–v characteristics of solar cells. Sol. Energy **85**(7), 1543–1550 (2011)
13. D.S. Abdelminaam, M. Said, E.H. Houssein, Turbulent flow of water-based optimization using new objective function for parameter extraction of six photovoltaic models. IEEE Access **9**, 35,382–35,398 (2021)
14. A.A. Ismaeel, E.H. Houssein, D. Oliva, M. Said, Gradient-based optimizer for parameter extraction in photovoltaic models. IEEE Access **9**, 13,403–13,416 (2021)
15. P. Lin, S. Cheng, W. Yeh, Z. Chen, L. Wu, Parameters extraction of solar cell models using a modified simplified swarm optimization algorithm. Sol. Energy **144**, 594–603 (2017)
16. A. Chouder, S. Silvestre, N. Sadaoui, L. Rahmani, Modeling and simulation of a grid connected PV system based on the evaluation of main PV module parameters. Simul. Model. Pract. Theory **20**(1), 46–58 (2012)
17. E.H. Houssein, Y. Mina, E. Aboul, Nature-inspired algorithms: a comprehensive review," in *Hybrid Computational Intelligence: Research and Applications* (CRC Press, 2019), p. 1
18. F.A. Hashim, E.H. Houssein, M.S. Mabrouk, W. Al-Atabany, S. Mirjalili, Henry gas solubility optimization: a novel physics-based algorithm. Futur. Gener. Comput. Syst. **101**, 646–667 (2019)
19. F.A. Hashim, K. Hussain, E.H. Houssein, M.S. Mabrouk, W. Al-Atabany, Archimedes optimization algorithm: a new metaheuristic algorithm for solving optimization problems. Appl. Intell. **51**(3), 1531–1551 (2021)

20. E.H. Houssein, M.R. Saad, F.A. Hashim, H. Shaban, M. Hassaballah, Lévy flight distribution: a new metaheuristic algorithm for solving engineering optimization problems. Eng. Appl. Artif. Intell. **94**, 103731 (2020)
21. F.A. Hashim, E.H. Houssein, K. Hussain, M.S. Mabrouk, W. Al-Atabany, Honey badger algorithm: new metaheuristic algorithm for solving optimization problems. Math. Comput. Simul. **192**, 84–110 (2021)
22. E.H. Houssein, A.G. Gad, K. Hussain, P.N. Suganthan, Major advances in particle swarm optimization: theory, analysis, and application. Swarm Evol. Comput. **63**, 100868 (2021)
23. E.H. Houssein, M.M. Emam, A.A. Ali, An efficient multilevel thresholding segmentation method for thermography breast cancer imaging based on improved chimp optimization algorithm. Expert Syst. Appl. 115651 (2021)
24. E.H. Houssein, K. Hussain, L. Abualigah, M. Abd Elaziz, W. Alomoush, G. Dhiman, Y. Djenouri, E. Cuevas, An improved opposition-based marine predators algorithm for global optimization and multilevel thresholding image segmentation. Knowl. Based Syst. 107348 (2021)
25. E.H. Houssein, M.M. Emam, A.A. Ali, Improved manta ray foraging optimization for multilevel thresholding using Covid-19 CT images. Neural Comput. Appl. 1–21 (2021)
26. E.H. Houssein, B.E.-D. Helmy, D. Oliva, A.A. Elngar, H. Shaban, A novel black widow optimization algorithm for multilevel thresholding image segmentation. Expert Syst. Appl. **167**, 114159 (2021)
27. E.H. Houssein, B.E.-D Helmy, A.A. Elngar, D.S. Abdelminaam, H. Shaban, An improved tunicate swarm algorithm for global optimization and image segmentation. IEEE Access **9**, 56,066–56,092 (2021)
28. E.H. Houssein, D.S. AbdElminaam, I.E. Ibrahim, M. Hassaballah, Y.M. Wazery, A hybrid heartbeats classification approach based on marine predators algorithm and convolution neural networks. IEEE Access (2021)
29. Y.M. Wazery, E. Saber, E.H. Houssein, A.A. Ali, E. Amer, An efficient slime mould algorithm combined with k-nearest neighbor for medical classification tasks. IEEE Access (2021)
30. A. Hamad, E.H. Houssein, A.E. Hassanien, A.A. Fahmy, Hybrid grasshopper optimization algorithm and support vector machines for automatic seizure detection in EEG signals, in *International Conference on Advanced Machine Learning Technologies and Applications* (Springer, 2018), pp. 82–91
31. E.H. Houssein, A.A. Ewees, M. Abd ElAziz, Improving twin support vector machine based on hybrid swarm optimizer for heartbeat classification. Pattern Recognit. Image Anal. **28**(2), 243–253 (2018)
32. M.M. Ahmed, E.H. Houssein, A.E. Hassanien, A. Taha, E. Hassanien, Maximizing lifetime of large-scale wireless sensor networks using multi-objective whale optimization algorithm. Telecommun. Syst. **72**(2), 243–259 (2019)
33. E.H. Houssein, M.E. Hosney, M. Elhoseny, D. Oliva, W.M. Mohamed, M. Hassaballah, Hybrid Harris hawks optimization with cuckoo search for drug design and discovery in chemoinformatics. Sci. Rep. **10**(1), 1–22 (2020)
34. E.H. Houssein, A.G. Gad, Y.M. Wazery, PN. Suganthan, Task scheduling in cloud computing based on meta-heuristics: review, taxonomy, open challenges, and future trends. Swarm Evol. Comput. 100841 (2021)
35. E.H. Houssein, G.N. Zaki, A.A.Z. Diab, E.M. Younis, An efficient manta ray foraging optimization algorithm for parameter extraction of three-diode photovoltaic model. Comput. Electr. Eng. **94**, 107304 (2021)
36. E.H. Houssein, M.A. Mahdy, M.J. Blondin, D. Shebl, W.M. Mohamed, Hybrid slime mould algorithm with adaptive guided differential evolution algorithm for combinatorial and global optimization problems. Expert Syst. Appl. **174**, 114689 (2021)
37. E.H. Houssein, M.A. Mahdy, M.G. Eldin, D. Shebl, W.M. Mohamed, M. Abdel-Aty, Optimizing quantum cloning circuit parameters based on adaptive guided differential evolution algorithm. J. Adv. Res. **29**, 147–157 (2021)
38. E.H. Houssein, M. Dirar, K. Hussain, W.M. Mohamed, Assess deep learning models for Egyptian exchange prediction using nonlinear artificial neural networks. Neural Comput. Appl. **33**(11), 5965–5987 (2021)

39. K. Hussain, N. Neggaz, W. Zhu, E.H. Houssein, An efficient hybrid sine-cosine Harris hawks optimization for low and high-dimensional feature selection. Expert Syst. Appl. **176**, 114778 (2021)
40. N. Neggaz, E.H. Houssein, K. Hussain, An efficient henry gas solubility optimization for feature selection. Expert Syst. Appl. **152**, 113364 (2020)
41. E.H. Houssein, I.E. Ibrahim, N. Neggaz, M. Hassaballah, Y.M. Wazery, An efficient ECG arrhythmia classification method based on manta ray foraging optimization. Expert Syst. Appl. **181**, 115131 (2021)
42. E.H. Houssein, M. Kilany, A.E. Hassanien, ECG signals classification: a review. Int. J. Intell. Eng. Inform. **5**(4), 376–396 (2017)
43. E.H. Houssein, M.E. Hosney, D. Oliva, W.M. Mohamed, M. Hassaballah, A novel hybrid Harris hawks optimization and support vector machines for drug design and discovery. Comput. Chem. Eng. **133**, 106656 (2020)
44. E.H. Houssein, N. Neggaz, M.E. Hosney, W.M. Mohamed, M. Hassaballah, Enhanced Harris hawks optimization with genetic operators for selection chemical descriptors and compounds activities. Neural Comput. Appl. 1–18 (2021)
45. D.S. Abdelminaam, F.H. Ismail, M. Taha, A. Taha, E.H. Houssein, A. Nabil, Coaid-deep: an optimized intelligent framework for automated detecting Covid-19 misleading information on twitter. IEEE Access **9**, 27,840–27,867 (2021)
46. A. Askarzadeh, L. dos Santos Coelho, Determination of photovoltaic modules parameters at different operating conditions using a novel bird mating optimizer approach. Energy Convers. Manag. **89**, 608–614 (2015)
47. K.M. El-Naggar, M. AlRashidi, M. AlHajri, A. Al-Othman, Simulated annealing algorithm for photovoltaic parameters identification. Sol. Energy **86**(1), 266–274 (2012)
48. C. Dai, W. Chen, Y. Zhu, Seeker optimization algorithm for digital IIR filter design. IEEE Trans. Industr. Electron. **57**(5), 1710–1718 (2009)
49. D. Oliva, M. Abd Elaziz, A.H. Elsheikh, A.A. Ewees, A review on meta-heuristics methods for estimating parameters of solar cells. J. Power Sour. **435**, 126683 (2019)
50. A. Askarzadeh, A. Rezazadeh, Extraction of maximum power point in solar cells using bird mating optimizer-based parameters identification approach. Sol. Energy **90**, 123–133 (2013)
51. B. Jacob, K. Balasubramanian, S.M. Azharuddin, N. Rajasekar et al., Solar PV modelling and parameter extraction using artificial immune system. Energy Procedia **75**, 331–336 (2015)
52. W. Gong, Z. Cai, Parameter extraction of solar cell models using repaired adaptive differential evolution. Sol. Energy **94**, 209–220 (2013)
53. M. AlHajri, K. El-Naggar, M. AlRashidi, A. Al-Othman, Optimal extraction of solar cell parameters using pattern search. Renew. Energy **44**, 238–245 (2012)
54. M. AlRashidi, K. El-Naggar, M. AlHajri, Parameters estimation of double diode solar cell model. Int. J. Electr. Comput. Eng. **7**(2), 118–121 (2013)
55. X. Yuan, Y. Xiang, Y. He, Parameter extraction of solar cell models using mutative-scale parallel chaos optimization algorithm. Sol. Energy **108**, 238–251 (2014)
56. J. Tvrdík, Adaptation in differential evolution: a numerical comparison. Appl. Soft Comput. **9**(3), 1149–1155 (2009)
57. H. Wei, J. Cong, X. Lingyun, S. Deyun, Extracting solar cell model parameters based on chaos particle swarm algorithm, in *International Conference on Electric Information and Control Engineering* (IEEE, 2011), pp. 398–402
58. A. Askarzadeh, A. Rezazadeh, Artificial bee swarm optimization algorithm for parameters identification of solar cell models. Appl. Energy **102**, 943–949 (2013)
59. C. Chellaswamy, R. Ramesh, Parameter extraction of solar cell models based on adaptive differential evolution algorithm. Renew. Energy **97**, 823–837 (2016)
60. A.S. Rodríguez, E.C. Murillo, Automatic parametrization of support vector machines for short texts polarity detection. CLEI Electron. J. **20**(1), 6–1 (2017)
61. A. El-Fergany, Efficient tool to characterize photovoltaic generating systems using mine blast algorithm. Electr. Power Compon. Syst. **43**(8–10), 890–901 (2015)

62. X. Gao, Y. Cui, J. Hu, G. Xu, Z. Wang, J. Qu, H. Wang, Parameter extraction of solar cell models using improved shuffled complex evolution algorithm. Energy Convers. Manag. **157**, 460–479 (2018)
63. D.S. Osheba, H.Z. Azazi, S. Shokralla, Parameter estimation of a photovoltaic array using direct search optimization algorithm. J. Renew. Sustain. Energy **9**(4), 043501 (2017)
64. M.M. El-Arini, A.M. Othman, A. Fathy, A new optimization approach for maximizing the photovoltaic panel power based on genetic algorithm and Lagrange multiplier algorithm. Int. J. Photoenergy **2013** (2013)
65. A.M. Humada, M. Hojabri, S. Mekhilef, H.M. Hamada, Solar cell parameters extraction based on single and double-diode models: a review. Renew. Sustain. Energy Rev. **56**, 494–509 (2016)
66. K. Nishioka, N. Sakitani, Y. Uraoka, T. Fuyuki, Analysis of multicrystalline silicon solar cells by modified 3-diode equivalent circuit model taking leakage current through periphery into consideration. Sol. Energy Mater. Sol. Cells **91**(13), 1222–1227 (2007)
67. D. Oliva, M. Abd El Aziz, A.E. Hassanien, Parameter estimation of photovoltaic cells using an improved chaotic whale optimization algorithm. Appl. Energy **200**, 141–154 (2017)
68. J. Ma, K.L. Man, S.-U. Guan, T. Ting, P.W. Wong, Parameter estimation of photovoltaic model via parallel particle swarm optimization algorithm. Int. J. Energy Res. **40**(3), 343–352 (2016)
69. X. Yuan, Y. He, L. Liu, Parameter extraction of solar cell models using chaotic asexual reproduction optimization. Neural Comput. Appl. **26**(5), 1227–1239 (2015)
70. M. Ye, X. Wang, Y. Xu, Parameter extraction of solar cells using particle swarm optimization. J. Appl. Phys. **105**(9), 094502 (2009)
71. O. Hachana, K. Hemsas, G. Tina, C. Ventura, Comparison of different metaheuristic algorithms for parameter identification of photovoltaic cell/module. J. Renew. Sustain. Energy **5**(5), 053122 (2013)
72. C.A. Ramos-Paja, J.D. Bastidas-Rodríguez, D. Gonz ález, S. Acevedo, J. Pel áez Restrepo, Design and control of a buck–boost charger-discharger for dc-bus regulation in microgrids. Energies **10**(11), 1847 (2017)
73. S. Singer, Nelder-mead algorithm. Scholarpedia **4**(7), 2928 (2009)
74. L.L. Jiang, D.L. Maskell, J.C. Patra, Parameter estimation of solar cells and modules using an improved adaptive differential evolution algorithm. Appl. Energy **112**, 185–193 (2013)
75. Q. Niu, L. Zhang, K. Li, A biogeography-based optimization algorithm with mutation strategies for model parameter estimation of solar and fuel cells. Energy Convers. Manag. **86**, 1173–1185 (2014)
76. Y. Zhang, P. Lin, Z. Chen, S. Cheng, A population classification evolution algorithm for the parameter extraction of solar cell models. Int. J. Photoenergy **2016** (2016)
77. J.D. Bastidas-Rodriguez, G. Petrone, C.A. Ramos-Paja, G. Spagnuolo, A genetic algorithm for identifying the single diode model parameters of a photovoltaic panel. Math. Comput. Simul. **131**, 38–54 (2017)
78. M. Louzazni, A. Khouya, K. Amechnoue, A. Gandelli, M. Mussetta, A. Crăciunescu, Metaheuristic algorithm for photovoltaic parameters: comparative study and prediction with a firefly algorithm. Appl. Sci. **8**(3), 339 (2018)
79. C. Kumar, T.D. Raj, M. Premkumar, T.D. Raj, A new stochastic slime mould optimization algorithm for the estimation of solar photovoltaic cell parameters. Optik **223**, 165277 (2020)
80. A. Askarzadeh, A. Rezazadeh, Parameter identification for solar cell models using harmony search-based algorithms. Sol. Energy **86**(11), 3241–3249 (2012)
81. Q. Niu, H. Zhang, K. Li, An improved TLBO with elite strategy for parameters identification of PEM fuel cell and solar cell models. Int. J. Hydrogen Energy **39**(8), 3837–3854 (2014)
82. X. Chen, K. Yu, W. Du, W. Zhao, G. Liu, Parameters identification of solar cell models using generalized oppositional teaching learning based optimization. Energy **99**, 170–180 (2016)
83. K. Yu, J. Liang, B. Qu, X. Chen, H. Wang, Parameters identification of photovoltaic models using an improved JAYA optimization algorithm. Energy Convers. Manag. **150**, 742–753 (2017)
84. X. Chen, B. Xu, C. Mei, Y. Ding, K. Li, Teaching-learning-based artificial bee colony for solar photovoltaic parameter estimation. Appl. Energy **212**, 1578–1588 (2018)

85. K. Yu, J. Liang, B. Qu, Z. Cheng, H. Wang, Multiple learning backtracking search algorithm for estimating parameters of photovoltaic models. Appl. Energy **226**, 408–422 (2018)
86. D. Kler, Y. Goswami, K. Rana, V. Kumar, A novel approach to parameter estimation of photovoltaic systems using hybridized optimizer. Energy Convers. Manag. **187**, 486–511 (2019)
87. S. Li, W. Gong, X. Yan, C. Hu, D. Bai, L. Wang, L. Gao, Parameter extraction of photovoltaic models using an improved teaching-learning-based optimization. Energy Convers. Manag. **186**, 293–305 (2019)
88. K. Yu, B. Qu, C. Yue, S. Ge, X. Chen, J. Liang, A performance-guided JAYA algorithm for parameters identification of photovoltaic cell and module. Appl. Energy **237**, 241–257 (2019)
89. A. Sadollah, H. Eskandar, A. Bahreininejad, J.H. Kim, Water cycle algorithm with evaporation rate for solving constrained and unconstrained optimization problems. Appl. Soft Comput. **30**, 58–71 (2015)
90. H. Eskandar, A. Sadollah, A. Bahreininejad, M. Hamdi, Water cycle algorithm-a novel metaheuristic optimization method for solving constrained engineering optimization problems. Comput. Struct. **110**, 151–166 (2012)
91. A. Gotmare, S.S. Bhattacharjee, R. Patidar, N.V. George, Swarm and evolutionary computing algorithms for system identification and filter design: a comprehensive review. Swarm Evol. Comput. **32**, 68–84 (2017)
92. X.-S. Yang, Flower pollination algorithm for global optimization, in *International Conference on Unconventional Computing and Natural Computation* (Springer, 2012), pp. 240–249
93. D. Allam, D. Yousri, M. Eteiba, Parameters extraction of the three diode model for the multi-crystalline solar cell/module using moth-flame optimization algorithm. Energy Convers. Manag. **123**, 535–548 (2016)
94. S. Xu, Y. Wang, Parameter estimation of photovoltaic modules using a hybrid flower pollination algorithm. Energy Convers. Manag. **144**, 53–68 (2017)
95. J. Khoury, B.A. Ovrut, N. Seiberg, P.J. Steinhardt, N. Turok, From big crunch to big bang. Phys. Rev. D **65**(8), 086007 (2002)
96. M. Tegmark, *Science and Ultimate Reality: Quantum Theory, Cosmology, and Complexity*, ed. by J.D. Barrow, P.C. Davies, C.L. Harper Jr. (Cambridge University Press, Cambridge, 2004)
97. D.M. Eardley, Death of white holes in the early universe. Phys. Rev. Lett. **33**(7), 442 (1974)
98. S. Mirjalili, S.M. Mirjalili, A. Hatamlou, Multi-verse optimizer: a nature-inspired algorithm for global optimization. Neural Comput. Appl. **27**(2), 495–513 (2016)
99. E. Ali, M. El-Hameed, A. El-Fergany, M. El-Arini, Parameter extraction of photovoltaic generating units using multi-verse optimizer. Sustain. Energy Technol. Assess. **17**, 68–76 (2016)
100. M. Chegaar, Z. Ouennoughi, A. Hoffmann, A new method for evaluating illuminated solar cell parameters. Solid-State Electron. **45**(2), 293–296 (2001)
101. J. Brest, S. Greiner, B. Boskovic, M. Mernik, V. Zumer, Self-adapting control parameters in differential evolution: a comparative study on numerical benchmark problems. IEEE Trans. Evol. Comput. **10**(6), 646–657 (2006)
102. C. Zhang, J. Zhang, Y. Hao, Z. Lin, C. Zhu, A simple and efficient solar cell parameter extraction method from a single current-voltage curve. J. Appl. Phys. **110**(6), 064504 (2011)
103. Y. Wang, Z. Cai, Q. Zhang, Differential evolution with composite trial vector generation strategies and control parameters. IEEE Trans. Evol. Comput. **15**(1), 55–66 (2011)
104. D. Alam, D. Yousri, M. Eteiba, Flower pollination algorithm based solar PV parameter estimation. Energy Convers. Manag. **101**, 410–422 (2015)
105. N.T. Tong, W. Pora, A parameter extraction technique exploiting intrinsic properties of solar cells. Appl. Energy **176**, 104–115 (2016)
106. S. Mirjalili, SCA: a sine cosine algorithm for solving optimization problems. Knowl. Based Syst. **96**, 120–133 (2016)
107. M. Derick, C. Rani, M. Rajesh, M. Farrag, Y. Wang, K. Busawon, An improved optimization technique for estimation of solar photovoltaic parameters. Sol. Energy **157**, 116–124 (2017)

108. J.P. Ram, T.S. Babu, T. Dragicevic, N. Rajasekar, A new hybrid bee pollinator flower pollination algorithm for solar PV parameter estimation. Energy Convers. Manag. **135**, 463–476 (2017)
109. O. Turgut, Global best algorithm based parameter identification of solar cell models. Int. J. Intell. Syst. Appl. Eng. **5**(4), 189–205 (2017)
110. O.K. Erol, I. Eksin, A new optimization method: big bang-big crunch. Adv. Eng. Softw. **37**(2), 106–111 (2006)
111. S. Bana, R. Saini, Identification of unknown parameters of a single diode photovoltaic model using particle swarm optimization with binary constraints. Renew. Energy **101**, 1299–1310 (2017)
112. H. Nunes, J. Pombo, S. Mariano, M. Calado, J.F. De Souza, A new high performance method for determining the parameters of PV cells and modules based on guaranteed convergence particle swarm optimization. Appl. Energy **211**, 774–791 (2018)
113. B. Timurkutluk, C. Timurkutluk, M.D. Mat, Y. Kaplan, Anode-supported solid oxide fuel cells with ion conductor infiltration. Int. J. Energy Res. **35**(12), 1048–1055 (2011)
114. S. Mirjalili, A. Lewis, The whale optimization algorithm. Adv. Eng. Softw. **95**, 51–67 (2016)
115. M. Abd Elaziz, D. Oliva, Parameter estimation of solar cells diode models by an improved opposition-based whale optimization algorithm. Energy Convers. Manag. **171**, 1843–1859 (2018)
116. S. Mirjalili, S.M. Mirjalili, A. Lewis, Grey wolf optimizer. Adv. Eng. Softw. **69**, 46–61 (2014)
117. W. Long, S. Cai, J. Jiao, M. Xu, T. Wu, A new hybrid algorithm based on grey wolf optimizer and cuckoo search for parameter extraction of solar photovoltaic models. Energy Convers. Manag. **203**, 112243 (2020)
118. A. Saxena, A. Sharma, S. Shekhawat, Parameter extraction of solar cell using intelligent grey wolf optimizer. Evol. Intell. 1–17 (2020)
119. M. AlShabi, C. Ghenai, M. Bettayeb, F.F. Ahmad, M.E.H. Assad, Multi-group grey wolf optimizer (MG-GWO) for estimating photovoltaic solar cell model. J. Therm. Anal. Calorim. **144**(5), 1655–1670 (2021)
120. S. Mirjalili, Moth-flame optimization algorithm: a novel nature-inspired heuristic paradigm. Knowl. Based Syst. **89**, 228–249 (2015)
121. H. Sheng, C. Li, H. Wang, Z. Yan, Y. Xiong, Z. Cao, Q. Kuang, Parameters extraction of photovoltaic models using an improved moth-flame optimization. Energies **12**(18), 3527 (2019)
122. A. Zadeh, A. Rezazadeh, Artificial bee swarm optimization algorithm for parameters identifications of solar cell modules. Appl. Energy **102**, 943–949 (2013)
123. M. Jamadi, F. Merrikh-Bayat, M. Bigdeli, Very accurate parameter estimation of single-and double-diode solar cell models using a modified artificial bee colony algorithm. Int. J. Energy Environ. Eng. **7**(1), 13–25 (2016)
124. L. Guo, Z. Meng, Y. Sun, L. Wang, Parameter identification and sensitivity analysis of solar cell models with cat swarm optimization algorithm. Energy Convers. Manag. **108**, 520–528 (2016)
125. N.F.A. Hamid, N.A. Rahim, J. Selvaraj, Solar cell parameters identification using hybrid Nelder-Mead and modified particle swarm optimization. J. Renew. Sustain. Energy **8**(1), 015502 (2016)
126. N.J. Vickers, Animal communication: when i'm calling you, will you answer too? Curr. Biol. **27**(14), R713–R715 (2017)
127. G. Xiong, J. Zhang, X. Yuan, D. Shi, Y. He, G. Yao, Parameter extraction of solar photovoltaic models by means of a hybrid differential evolution with whale optimization algorithm. Sol. Energy **176**, 742–761 (2018)
128. T. Kang, J. Yao, M. Jin, S. Yang, T. Duong, A novel improved cuckoo search algorithm for parameter estimation of photovoltaic (PV) models. Energies **11**(5), 1060 (2018)
129. A.R. Jordehi, Enhanced leader particle swarm optimisation (ELPSO): an efficient algorithm for parameter estimation of photovoltaic (PV) cells and modules. Sol. Energy **159**, 78–87 (2018)

130. X. Chen, K. Yu, Hybridizing cuckoo search algorithm with biogeography-based optimization for estimating photovoltaic model parameters. Sol. Energy **180**, 192–206 (2019)
131. A.A.Z. Diab, H.M. Sultan, T.D. Do, O.M. Kamel, M.A. Mossa, Coyote optimization algorithm for parameters estimation of various models of solar cells and PV modules. IEEE Access **8**, 111,102–111,140 (2020)
132. H. Rezk, T.S. Babu, M. Al-Dhaifallah, H.A. Ziedan, A robust parameter estimation approach based on stochastic fractal search optimization algorithm applied to solar PV parameters. Energy Rep. **7**, 620–640 (2021)

Big Data Analysis Using Hybrid Meta-Heuristic Optimization Algorithm and MapReduce Framework

Mohammad Qassem Bashabsheh, Laith Abualigah, and Mohammad Alshinwan

Abstract Clustering large data is a recent and popular challenge that is used in various applications, including social networking, bioinformatics, and many others. In order to manage the rapidly growing data sizes, traditional clustering algorithms must be improved. In this research, a hybrid Harris Hawks Optimizer (HHHO) with K-mean clustering and MapReduce framework is proposed to solve the various data clustering problem. The proposed scheme uses the K-means' ability to solve the various clustering problems. More specifically, the K-means are utilized as initial solutions to the traditional Harris Hawks Optimizer (HHO). In general, HHO tries to refine the candidate solutions to find the best one. MapReduce is a distributed processing computing paradigm that produces datasets using a parallel program on a cluster. In particular, it is adopted in the developed HHHO for parallelization since it offers fault tolerance, load balancing, and data locality. The performance of the presented methodology has been evaluated by means of numerical comparisons which proved the efficiency of the proposed HHHO, where it produces better results than other existing computation methods. Moreover, it has a very good ability in improving and finding optimal and converging sets of data. In addition, the accuracy and error rate of the obtained results are assessed. The proposed method is implemented and evaluated using PYTHON simulation settings.

Keywords Big data analysis · Hybrid Harris hawks optimizer · MapReduce framework

M. Q. Bashabsheh · L. Abualigah (✉) · M. Alshinwan
Faculty of Computer Sciences and Informatics, Amman Arab University, Amman 11953, Jordan
e-mail: laythdyabat@aau.edu.jo

L. Abualigah
School of Computer Sciences, Universiti Sains Malaysia, 11800 Pulau Pinang, Malaysia

E. H. Houssein et al. (eds.), *Integrating Meta-Heuristics and Machine Learning for Real-World Optimization Problems*, Studies in Computational Intelligence 1038,
https://doi.org/10.1007/978-3-030-99079-4_8

1 Introduction

Big Data analytics is a "technology-enabled strategy" for acquiring a rich, detailed, and precise view of customers and market while still gaining a competitive advantage. Companies can make quicker decisions than ever before by receiving a continuous stream of real-time results, as well as keeping an eye on changing developments and discovering new market opportunities [1].

Mobile networks have evolved both the carriers and generators of enormous data. Big data analytics can enhance the execution of mobile networks and maximize operators' revenue [2–4].

Clustering data is a significant concern in many areas of computer science and related fields, such as machine learning, data mining, and pattern recognition [5, 6]. Singh & researchers have been focusing more on developing new data analysis approaches for Big Data, resulting in the ongoing growth of a wide range of algorithms and frameworks [7]. Aggregation is an effective technique for data mining and Big Data analysis. However, it is difficult to apply the clustering methods to Big Data to solve the new problems associated with it. More specifically, clustering algorithms have high processing costs due to the terabytes and petabytes range of the Big Data size. A study aimed to investigate the trend and progress in clustering algorithms is performed to deal with Big Data problems, from the first proposed algorithms to new solutions [8]. As a consequence, a tool to gather and preserve data is required, as well as a system in which the proof is separated into groups that are more comparable in characteristics than the others. However, speed, scale, range, variation, and uncertainty are some of the obstacles that Big Data face when collecting data. One of the most common applications in data analysis is data processing by groups that are homogeneous in some properties. A widely used data collection algorithm is the k-means algorithm which is easy to use and takes no time to complete. Nevertheless, it has some flaws [9]. MapReduce has become a popular method of leveraging the power of large computer clusters that allows to handle the specifics of distributed execution, network access, and fault tolerance, allowing programmers to think in a data-centric way that focuses on applying transformations to data record sets [10, 11].

Recently, high-dimensional data is being viewed thanks to the recent internet advances and modern applications [12]. The development of efficient methods that can reliably predict future findings, as well as looking into the relationship between features and response for scientific purposes, are two fundamental goals of "high-dimensional" data analysis. However, since the sample size is so high, Big Data has two additional objectives: to comprehend heterogeneity and similarity among various subpopulations [13].

In this research, the k-mean outputs were used as inputs for Harris Hawk's algorithm, which uses MapReduce to solve the clustering efficiency problem. Since it is vulnerable to random values, the K-mean algorithm has limitations when dealing with large quantities of data. The benefit of MapReduce is that it is a cost-effective solution to implementing scalable parallel algorithms and Big Data analysis that can help millions of users with a petabyte of data to deal with data mining and machine learning applications that would otherwise cause major issues when analyzing their Big Data [14].

A significant need for developing the current traditional data clustering algorithms arises in order to handle the massive data in the various applications. Therefore, the main challenge is dividing a large volume of data into similar clusters. As the researchers [15] point out, the key to addressing artificial intelligence (AI) problems is successful Big Data analysis, which is becoming increasingly obvious with the proliferation of Big Data. Big Data clustering is undeniably expanding rapidly through all fields of science and engineering. While the potential of such vast data is undeniable, overcoming the myriad obstacles that come with it, necessitates new ways of thinking and learning techniques. Machine learning connections with signal processing techniques are needed for Big Data analysis [16]. The vast majority of data stored in businesses is unstructured. In semantic web areas, it is critical to collect and extract data. Many of these criteria are based on unstructured data processing. Unstructured data accounts for more than 80% of all potentially valuable business data. An approach to organizing complex unstructured data and retrieving important information [1].

The main goal of this research is to identify a new path in the field of Big Data analytics to provide customers, businesses and any other party who needs it with more fertile, deep, and accurate perspectives, enabling enterprises to make smarter decisions in less time. Thus, a new hybrid optimization algorithm, Harris Hawks Optimizer with K-means, is developed for large data clustering to discover the best clustering solution. As a result, a new Hybrid Harris Hawks Optimizer that efficiently utilizes the search space to achieve optimized results is proposed with adopting the MapReduce to improve its capacity. Various evaluation criteria and scenarios are implemented to test the proposed scheduling process.

The paper organization is as follows: Sect. 1 summarizes Big Data gathering, the issues that organizations and people confront in dealing with it, the study's goals, and the recommended solutions to these issues. Section 2 explains literature reviews and related work that are related to the research topics. The proposed technique's approach is presented in Sect. 3. Findings and discussions are presented in Sects. 4, and 5 outlines the research key conclusions and makes recommendations for the future.

2 Literature Review

2.1 *Overview of Big Data Clustering*

To provide a clear understanding of Big Data analysis, this section offers an overview of the Big Data concept, resources, methods, relevance, and limitations. Big Data, according to [17], is a collection of big procedure that is difficult to process and store. In addition, it is challenging to generate new and useful results since it includes a variety of data types. Almost everything around us create Big Data, e.g., digital processes and social media activities produce Big Data for us, which is constantly distributed through networks, sensors, and mobile devices [18, 19]. The pace, volume, and diversity of Big Data come from several places. To extract the ethical value from Big Data, researchers have stated that they need exceptional processing, analytical abilities, and expertise [20, 21]. Big Data analytics is a "technology-enabled technique" for gaining richer, more reliable, and more fertile insights for clients and companies, as well as achieving competitive advantages. "By storing a continuous stream of real-time data", companies can make faster decisions, track increasing patterns, and discover new market opportunities than ever before [1]. Clustering data is considered a major issue in several areas of computer science and related fields, such as machine learning, data mining, and pattern recognition [5].

The well-known k-means clustering algorithm has an algorithm that contains various aspects of complexity, such as being iterative, numerical, and having the potential to parallelize some of the computation's recognition [7]. When conventional data processing systems are fundamentally improved by Big Data clustering, drastic changes occur [22]. For example, to perform any type of analysis on such complex and massive data, it is critical to scale up the hardware platforms and select the necessary hardware-software platforms to complete the process on time [7, 23]. Data mining applications, machine learning, and similarities can all be triggered by Big Data analysis. To approach such data-intensive applications, researchers [8] saw the MapReduce method as a cost-effective solution to implement scalable parallel algorithms for Big Data processing that can handle petabytes of data for millions of users. The resources of Big Data clustering are depicted in Fig. 1.

The resources of Big Data are depicted in Fig. 1. As a simple example of Big Data implementations, the number of websites is immense, necessitating the collection of

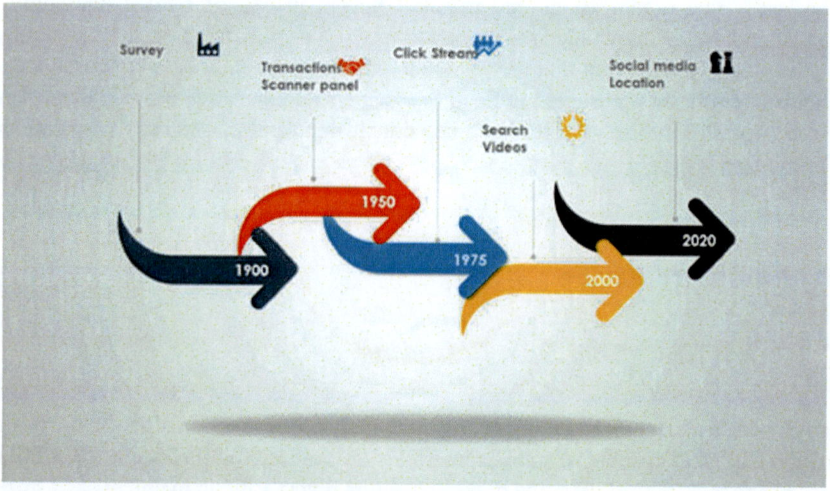

Fig. 1 Recourses of big data clustering

massive data and the production of even more massive data [8]. As a result, a tool to discover the information is required to handle this data. Data collection is a process in which evidence is divided into groups with more similar characteristics than other groups [24]. However, speed, scale, variety, variance, and complexity are all issues that Big Data faces when it comes to data collection. Recently, we have been dealing with high-dimensional data as a result of the internet's recent evolution and modern applications [12]. The development of efficient methods that can reliably predict future findings, as well as the investigation of the relationship between features and response for scientific purposes, are the two main goals of "high-dimensional" data analysis. Despite the large sample size, Big Data has two additional goals: to consider variability and commonality through various subpopulations [13].

2.1.1 Big Data Services

Big Data analyses are now being seen as the foundation for science's newest, paradigm. Vast amounts of data are becoming available in an increasing number of application fields. Infrastructures that enable us to continuously store these data for sharing and processing are taking shape. This allows experts from the previous three paradigms of scientific research to collaborate. The importance of knowledge has long been acknowledged. It can be used for a variety of purposes, including increasing device performance, guiding decision-making, risk management, cost reduction, and sales boost, among others [25].

As a result, it's critical to have a single infrastructure with popular Big Data management features that's still flexible enough to handle a variety of Big Data and Big Data analytics [1]. Customers may use Big Data-as-a-Service to get common services connected to Big Data to improve efficiency and cut costs. As depicted in Fig. 2.

It is typically made up of three layers:

- **Technology-as-a-Service** for Big Data: Infrastructure scale-out offers the required processing and storage capacity for Big Data. A modern Internet and

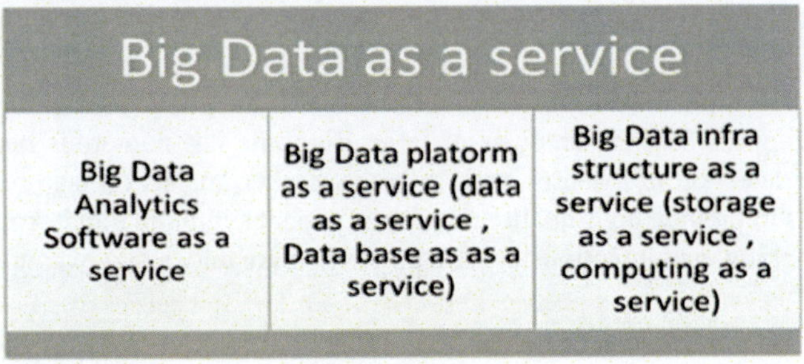

Fig. 2 Big data services

science research initiative produces a vast volume of data with complex interrelationships. A new type of Big Data infrastructure, with the performance to provide quick data access and processing to meet users' just-in-time needs, is needed to assist these Big Data users [26].

- **Big Data Platform-as-a-Service:** A Big Data platform allows users to access, analyze, and develop analytical applications based on broad data sets [27]. Google's BigQuery1 is an example of a Big Data Platform-as-a-Service, which allows users to take advantage of Google's massive computing and storage ability to analyze Big Data and obtain real-time business insights in seconds using the REST GUI. Big Data systems usually include several modules, such as research task definition, data storage and management, data process and integration, discovery and visualization, and so on. Different steps in a Big Data platform are typically needed for Big Data analysis [27]. The Big Data Platform supports APIs that can be used to perform basic data processing. For other measures, users must define their specific processing and analysis rules, allowing the Big Data system to be sufficiently flexible in communicating various types of Big Data analysis tasks [28]. Software-as-a-Service (SaaS) is a form of cloud computing. Big Data analytics is the process of analysing large amounts of data of various types to uncover hidden patterns, unknown connections, and other useful information [29].
- **Big Data analytics algorithms** are difficult and well beyond the capabilities of most organizations' IT departments. Furthermore, both businesses have an insufficient number of qualified Big Data practitioners. As a result, an increasing number of companies are turning to Big Data Analytics Software-as-a-Service to obtain the Business Intelligence (BI) service that transforms their unstructured data into a better understanding [30, 31].

2.1.2 The Importance of Clustering Big Data

According to (Balachandran), the goal of Big Data analytics is to extract useful information from a large data set and transform it into a comprehensible format for further use. The key processes for Big Data are capture, curation, storage, search, exchange, transfer, analysis, and visualization. This field has recently gained a lot of attention because it provides companies with useful knowledge and deeper insight into both structured and unstructured data, which can help them make better decisions [32].

According to [33], the analysis of large Big Data has numerous health benefits. Early detection of diseases such as breast cancer, medical image recognition, and signals to provide high-quality diagnosis, patient symptom monitoring, chronic disease tracking, and infectious disease prevention are only a few examples.

Fig. 3 Big data criteria

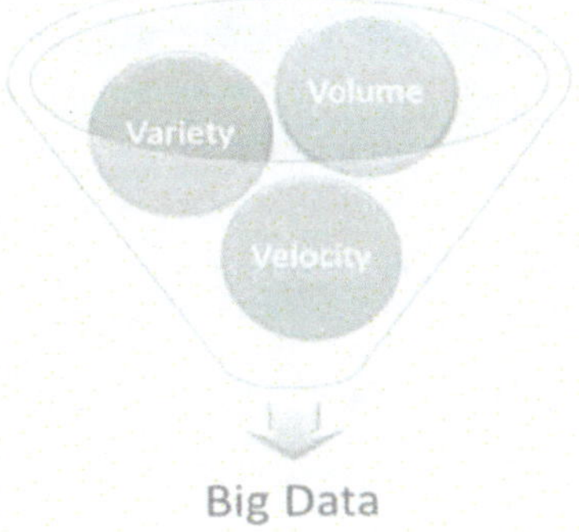

2.1.3 Technologies for Big Data

Balachandran and Prasad [34] explains that a computing platform should meet the following three requirements, as shown in Fig. 3, to support Big Data analytics.

Variety: The platform embraces a wide variety of data and enables companies to manage it in its native format while also converting it to other formats using robust transformation tools.

Velocity: The platform can handle data at any scale, whether it comes from low-latency sources like a sensor or stock data, or huge batch data volumes.

Volume: The platform can handle large volumes of data, both at rest and in real-time.

2.1.4 Limitations of Big Data

Apart from the importance of Big Data, Fig. 4 will answer some of the more common issues.

Data Storage—An extensive and complicated hardware infrastructure is needed to support and analyze large amounts of data required for an enterprise to perform. With the constant evolution of data, data storage appliances are becoming increasingly essential, and multiple cloud enterprises are looking for ample storage areas to stay competitive.

Data quality—Consistency and convenient availability of data are necessary for decision-making. When used in junction with an information management approach, Big Data may help ensure data integrity [35].

Security and privacy are two issues that occur when dealing with large amounts of data. Organizations will have to start incorporating confidential data into the more critical data to use Big Data. Companies will need to start implementing self-configurable security policies that exploit existing trust relationships and allow data and resource sharing within organizations. Data analytics are simplified and not determined because of those policies.

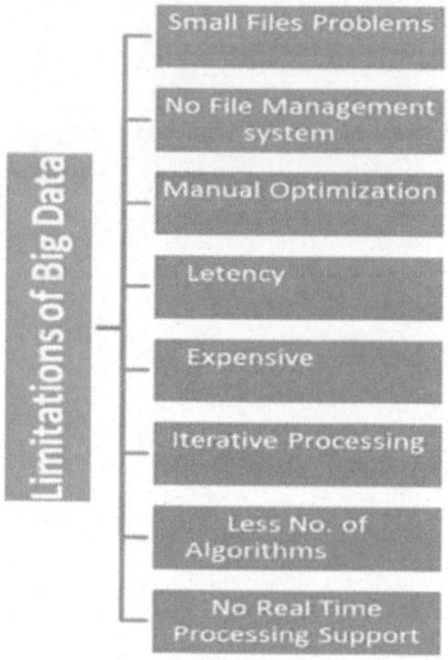

Fig. 4 Limitations of big data

Hacking and multiple cloud computing attacks will impact multiple users, even though only one site is compromised. These threats can be reduced by using monitoring programs, protected file systems, data loss applications, and security hardware to detect unusual behaviour across servers.

Billing and Service Delivery: The prices of the facilities are challenging to calculate due to their on-demand nature. If the provider has excellent and equivalent measurements to present, budgeting and calculating the expense would be complicated. The retailer's service-level agreements (SLAs) are insufficient to ensure availability and scalability. Companies will be reluctant to migrate to the cloud without a robust service level guarantee.

Businesses with interoperability and portability should change services and transfer in and out of the cloud anytime they want, with no lock-in period. Cloud computing services should be easy to incorporate with on-premises IT.

Cloud providers are also unreliable and unavailable around the clock, resulting in frequent outages. It's essential to keep track of the support you're doing, whether you're using internal or third-party tools. These programs must have plans in place to monitor their use, SLAs, execution, robustness, and business dependability [36].

Businesses Businesses will save money on computers for reliability and bandwidth, but bandwidth costs more. This can be a low cost for smaller systems, but it can be costly for data-intensive applications. Sufficient bandwidth is needed to relay intensive and complex data over the network. These issues should not be seen as impediments to cloud computing adoption.

2.2 Tasks Involving Big Data

According to research, the use of modern Big Data analysis tools, as well as an increase in the number and shared use of heterogeneous data sources, help in identifying new patterns and in improving diagnosis accuracy [37]. Big Data has become relevant as both public and private organizations have accumulated large amounts of domain-specific data that can contain useful knowledge about topics such as national intelligence, cyber security, fraud detection, marketing, and medical informatics. Google and Microsoft, for example, analyze large amounts of data for business analysis and decision-making, shaping existing and future technologies [38].

2.3 Analyzing Big Data

The advent of Big Data has been prompted by increased data storage capabilities, increased computer processing power, and the availability of increased data volumes, which provides organizations with greater expertise than they already have to handle computational resources and technology. In addition to the apparent large amounts of data, Big Data is synonymous with several peculiar complexities: length, variety, velocity, and veracity [39].

Traditional computing environments are immediately challenged by the unmanageable vast volume of data, which necessitates flexible data query and analysis storage, as well as a distribution strategy. On the other hand, Big Data's huge amount of information is a major benefit. Many companies, such as Facebook, Yahoo, and Google, still have large amounts of data and have just recently started to reap the rewards [38].

A general trend in Big Data systems is that raw data is becoming more diverse and dynamic, comprising mostly of uncategorized/unmonitored data with a small amount of categorized/monitored data [40]. Working with a large number of distinct data representations in a single repository poses unique Big Data challenges, requiring Big Data to preprocess unstructured data to provide structured data representations. In today's data-intensive technological era, the increasing rate at which data is captured and acquired is as important as Big Data's volume and variety features [30]. While streaming data is prone to data loss if not stored and analyzed immediately, there is the possibility of batch processing fast-moving data into bulk storage later. The speed of the feedback loop, i.e. the method of transforming data input into useful material, is the practical importance of interacting with Big Data-related velocity. This is especially critical in the case of time-sensitive data processing. Some companies, such as Twitter, Yahoo, and IBM, have created streaming data processing products [39].

Big Data Analytics is plagued with a slew of issues. The below are some of the key areas of concern [39]:

- high-dimensionality and data reduction
- data cleansing
- data quality and validation
- data sampling
- parallel and distributed data processing
- real-time analysis and decision making
- tracing and analyzing data provenance
- feature engineering
- data representations and distributed data sources,
- crowdsourcing and semantic input for improved data analysis
- integrating heterogeneous data
- scalability of algorithms,
- parallel and distributed computing
- data visualization,
- exploratory data analysis and interpretation
- data discovery and integration
- developing new models for massive data computation.

2.4 Previous Studies

Several studies have been conducted to solve clustering issues and deal with large amounts of data. [5] claim that the repetition of the K-means algorithm is the most significant factor affecting clustering implementation and propose a new, more efficient parallel clustering model. Experiments on massive real-world and fictitious datasets reveal that the improved algorithm is competitive and outperforms parallel K-means, K-Means, and stand-alone K-means algorithms. Clustering validation also reveals that K-means and accuracy of clustering techniques are significant.

Bahmani proposes the k-means algorithm, which parallelizes the first stage of K-means to increase performance and run it using MapReduce, according to [41]. The first K-means algorithm uses an oversampling factor and then performs consistent repeats, testing each point x with the probability of lpx, while px is the same as K-means, assigning the weight of x to the number of points in the key D dataset that are closer to x than other points, and regrouping the weighted points into k clusters to obtain k primary centers. As a result, the time it takes MapReduce to cluster large-scale datasets may increase. The researcher uses acceleration, extension, and scale to evaluate the efficiency of our proposed algorithm. The results show that the proposed algorithm can effectively accommodate massive data sets on common devices.

The various data processing systems currently available are also studied, in addition to the advantages and disadvantages of each model. Aside from some popular software architectures (such as Hadoop and Spark), there is also some extensive information about these hardware platforms. There are also comprehensive relations between various networks based on certain essential features based on stars (such as scalability and real-time processing). The widely used k-means clustering algorithm is used as a case study to show the benefits and drawbacks of various platforms. The k-means algorithm is suitable for further understanding different Big Data platforms because of its iterative nature, computationally expensive simulations, and the accumulation of local results in a parallel environment [7]. Nevertheless, since many methodical algorithms share similar characteristics, [7] propose a comprehensive summary of the different platforms for readers, allowing them to choose the most suitable platforms based on their data/computation requirements.

Aggregation is one of the primary tasks of data mining, according to [8], but it requires more sophisticated techniques than ever before to help data analysts extract information from terabytes and petabytes of data. The goal of his research is to improve data collection algorithms, however, while traditional sampling and dimensional reduction algorithms are still useful, they are unable to process large amounts of data. Even though parallel clustering is very useful for clustering, the complexity of implementing such algorithms remains a problem. On the other hand, MapReduce design is an excellent basis for compiling algorithms, and the results reveal that MapReduce-based algorithms, compared to Map, can provide superior scalability and speed while maintaining accuracy. Clustering algorithms can be designed on the MapReduce GPU-based architecture to achieve improved scalability and speed, even though "GPUs will be more powerful than CPUs in future business. The researchers present a process analysis that uses the mobilization community learning method to address the inconsistency and sensitivity of k implies, as well as the MapReduce distributed computing paradigm to address inefficiency issues in aggregating broad data sets [9] which is proved to be successful according to numerous comprehensive trials.

Through empirical research conducted on twenty-four cases as an unbalanced review of Big Data, the researchers support the good performance of this method, for which the results are competitive in both the model classification performance and the time required for measurement. MapReduce is used in this form. To tailor the fuzzy model's equations while integrating cost-sensitive learning methods into its design to address data shortcomings. Chi-FRBCS-Bigdata is used as an obscure rule-based classification method which is capable of tackling the complexity of an underrepresented class with large quantities of data while avoiding learning.

Big Data, as described by [42], is "data generated from a variety of sources with enormous volumes. Current systems are not suitable for Big Data processing because they were not designed with Big Data requirements in mind, and most of them are built on a centralized architecture. As a result, they result in high processing costs, low efficiency, poor-quality processing, large speeds, and different data structures," as well as data with these characteristics. To manage these massive data sets easily and effectively, the MapReduce architecture is designed as a parallel distributed programming model. They demonstrate six common Big Data software analysis systems based on the MapReduce method, demonstrating the properties of their databases and how to integrate them. Table 1 presented previous related studies briefly.

Table 1 Previous related studies

Papers	Year	Evaluation	Dataset
[5]	2014	He demonstrated how iteration of the K-mean algorithm influenced assembly efficiency and proposed an effective new parallel assembly model. The optimized algorithm is more effective and works better than the parallel K approaches, according to experimental results on massive real-world and synthetic datasets. Kmeans ǁ The Kmeans ++ algorithms are stand-alone The validation of the collections reveals that the grouping methods' consistency is indeed K-mean	Gauss Distribution Set, UC Irvine Machine Learning repository
[41]	2009	The first K-means algorithm uses an oversampling factor and proceeds to perform consistent repeats, and checks each point x with the likelihood of lpx, thus px is the same as K-means, then sets the number of points in the main D dataset closer to x than other points as the weight of x, and re-groups the weighted points into k clusters to get k primary centers. This can increase the time of clustering large-scale datasets using MapReduce	
[7]	2015	The method clustering algorithm was chosen as a case study to illustrate the advantages and disadvantages of the various platforms. Because of its iterative nature, computation-intensive nature, and aggregation of local results in a parallel setup, a k-mean algorithm is an excellent option for better understanding various Big Data platforms	

(continued)

Table 1 (continued)

Papers	Year	Evaluation	Dataset
[43]	2015	Although parallel clustering is very useful for clustering, the difficulty of implementing such algorithms is still problematic. However, the MapReduce architecture is a very ideal basis for clustering algorithms. The research results show that, compared with MapMap, MapReduce-based algorithms can provide excellent scalability and speed while maintaining the same consistency. Regarding the fact that "GPUs will be more efficient than CPUs in future work", clustering algorithms can be built on the GPU-based MapReduce framework to achieve better scalability and speed	The findings show that, when compared to MapMap, MapReduce-based algorithms can provide superior performance
[8]	2014	The development of data clustering algorithms is the subject of this study. However, while traditional sampling and dimensionality reduction algorithms are still useful, they are not capable of processing large quantities of data because even after sampling PBs of data, the data is still large and clustering algorithms cannot obtain it. As a result, clusters have much to do with distribution	The progress trend of data clustering algorithms was addressed in this report. To summarize, conventional sampling and dimension reduction algorithms are still useful, but they lack the computational capacity to handle large amounts of data
[9]	2011	In this paper, we propose a method for overcoming the uncertainty and vulnerability to outliers of k-means while solving the inefficiency problem in clustering large data sets using an ensemble learning method bagging and a distributed computing framework MapReduce. Extensive testing has been carried out to demonstrate the efficacy of our method	

(continued)

Table 1 (continued)

Papers	Year	Evaluation	Dataset
[44]	2015	Empirical research performed on twenty-four cases as an unbalanced analysis of Big Data using the Chi-FRBCS-BigDataCS algorithm, an obscure rule-based classification system capable of managing the difficulty faced by the underrepresented class of large volumes of data without ignoring learning. López backed up the method's good efficiency, claiming that the findings were comparable in terms of model classification performance and measurement time	

2.5 Research Gap

Based on the business reviewed for aggregation of Big Data, the data should be categorized and produced high-value, structured, and detailed data that organizations can use easily and as quickly as possible. The research did not include the use of an effective optimization method for Big Data aggregation that can help categorize data and create huge data volumes. The value is arranged and described in such a way that companies may use it quickly and simply. The goal of this research to improve existing clustering methods in order to close the search gap. To tackle the data aggregation problem, HHHO, a hybrid Harris Hawks optimizer (HHO) with K-mean clustering and the MapReduce architecture, is presented. To gather and classify data, the KNN method will be utilized, and MapReduce will be used to speed up the processing and classification process.

3 Implementation and Experiment Work

Clustering is often used in machine learning or data mining as a classification or grouping tool. K-nearest neighbor assignment is the primary application of a kNN joiner. Some data points for training are given and some new unlabeled data for research are given. The aim is to find the latest points of the class mark. A kNN query is carried out in the training set to estimate their group membership for any unlabeled data. This method can be seen as a kNN link to the test set with the training framework. Similar images can be identified by the kNN operation. For this reason, descriptive functionalities (points in 128-dimension data space) are first extracted by an Extractor technique from pictures. The kNN operation is then used to detect near points to indicate related images. We consider this sort of information for kNN calculation later in this paper. KNN joint can be applied in a broad variety of areas, including digital, social network, time series analysis, bioinformatics, and medical imaging, along with other approaches.

For each element in R and each element in S, the basic concept is to measure a kNN relation by pairing distance calculations. Data volume (2) Data Dimensionality are the two principal aspects of this problem. Suppose we are in a space in d dimension, this fairway calculation is computationally complex by O (d lRl Olll). Finding k closest neighbors in S for each r in R results in the minimum k distance and leads to minimal lSl à la log lSl complexity. As the quantity of data or its complexity increases (the number of dimensions), this is not possible. This is why a great deal of work has been done to reduce the machine sophistication of the memory. Two main areas of focus are:

Indexes must be measured to reduce the number of distances. These indexes cannot however be scaled to high-dimensional data.

Projections are used to minimize data dimensionality. However, preserving precision is another matter. Despite these efforts, as the amount of data increases, kNN is still considerably restricted on a single computer. Only distributed and parallel solutions are sufficiently efficient for large data sets. MapReduce is a parallel and distributed paradigm that is versatile and scalable and optimized for data-intense processing in particular. First implemented by Google, it became popular with the open-source implementation Hadoop framework. Figure 5 shows the proposed method flowchart and the methodology that the research will be based on.

For more information, Fig. 6 shows the proposed model breakdown. The proposed model starts by feeding it with the dataset, the dataset will be split into two main parts, training and testing. The kNN machine learning after feeding by the data will search for the optimal parameters and results resulting in the update it the best and most optimal parameters the HHO and Maprudes with initials the agents will optimize the data and refeed the classifier for better and best model results.

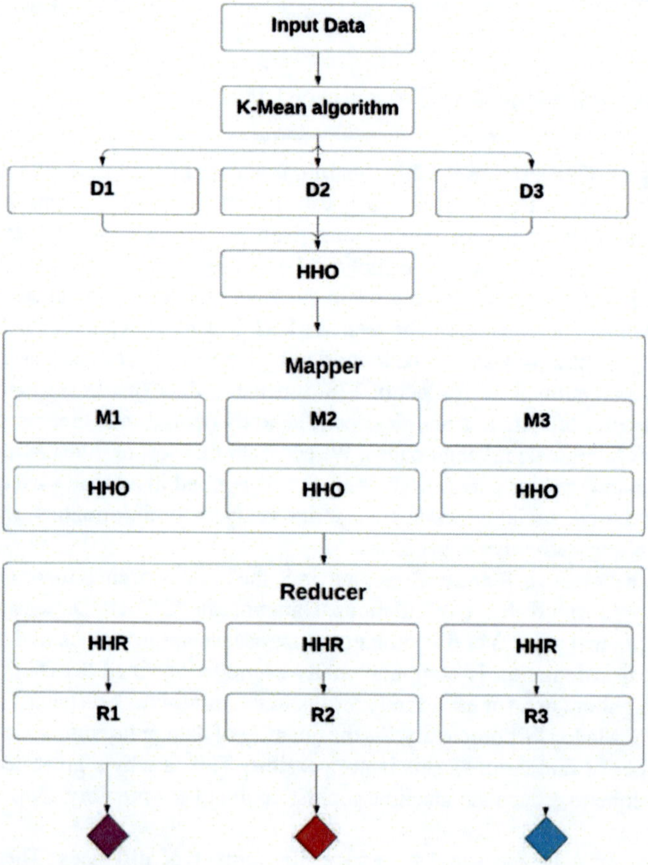

Fig. 5 Proposed method flowchart

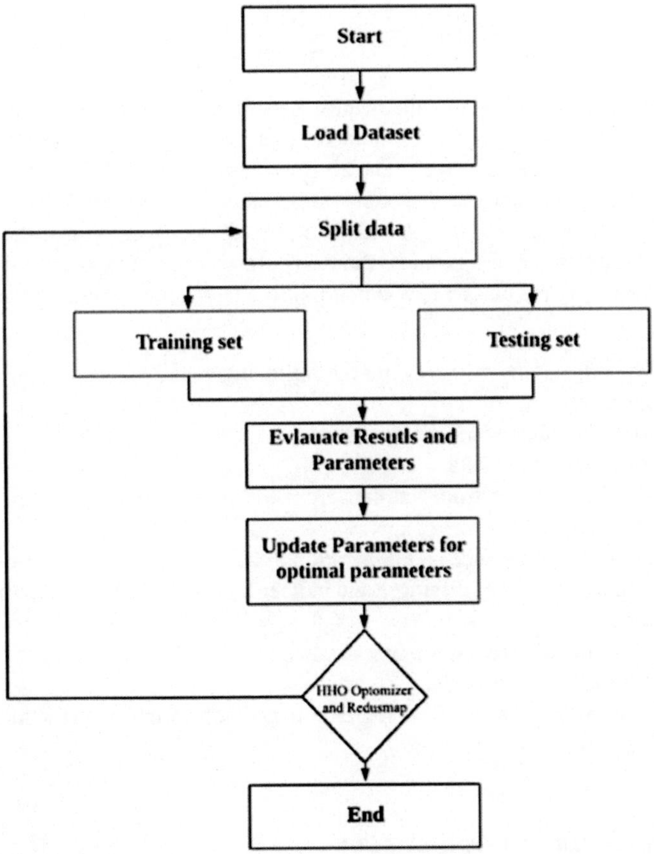

Fig. 6 The proposed model breakdown

3.1 K-means Clustering

K-means classifier is one of the unsupervised machine learning algorithms that is applied when we have information that is unbaled. This calculation aims to determine bunches of the information, with the number of assemblies spoken to by the k variable. The method of calculation is iterative to reach the appointment that every data can point to one of the K bunches depending on the highlights that are given. Figure 6 shows an iteration of k-mean convergence.

k-Nearest Neighbor's Classification is non-parametric, determined by the plurality of its neighbors for a given data point. In two stages the KNN algorithm finishes, the first step determines the number of nearest neighbors, as well as the second, classifies the data point. To find the neighbor is given in the equation:

$$\text{Distance}(\text{x}, \text{y}) = \sqrt{\sum i(xi - yi)} \quad (1)$$

The training method selects the nearest k samples and takes a majority vote of its class, where k should be an odd number to prevent confusion. The design of the KNN classifier is shown in Fig. 7. The classes Class 1 and Class 2 are two. The red asterisks are of class 1 and the blue circles are of class 2. The selected K shall be 5 and 3 samples shall be classes 1 and 2 shall be classes 2 among the 5 nearest neighbors. With a majority vote in the given K, the KNN classification works according to the principle of giving the class a new function. So, class 1 is allocated to the new test example.

- The KNN Algorithm workflow has the following steps:
- Load the dataset
- Set K to the number of neighbors picked
- For an example of the data
- Calculate the distance from the data from the present example to the question example.
- Fill a list with the distance and illustration index
- Sort the organized set of distances and indices by size from smallest to maximum (ascending)
- Select from the filtered list the first K entries
- Get the selected K labels Get
- Return the average K mark if regression and return the K mark mode when classified.

3.2 *Harris Hawk's Optimization*

The co-operative conduct and chasing style of Harris' hawks in the wild is the key inspiration of HHO called surprise punch, see Fig. 8. In this clever tactic, many hawks combine to punch beasts from multiple angles and aim to ambush them. Harris hawks will display varying patterns of chasing depending on the complex design of the situations and the escape patterns. This work imitates such complex patterns and conducts mathematically to use an optimization algorithm to allow kNN to be clustered and shows good results. Exploration, the transition phase from exploration to exploitation, and the exploitation phase:

$$X(t)^{n} = \frac{1}{N}\sum_{t=1}^{N} Xi(t) \quad (2)$$

where Xi(t) specifies each hawk's location in iteration t and N is the total number of hawks The average location may be calculated in a variety of methods, but we used the most basic technique.

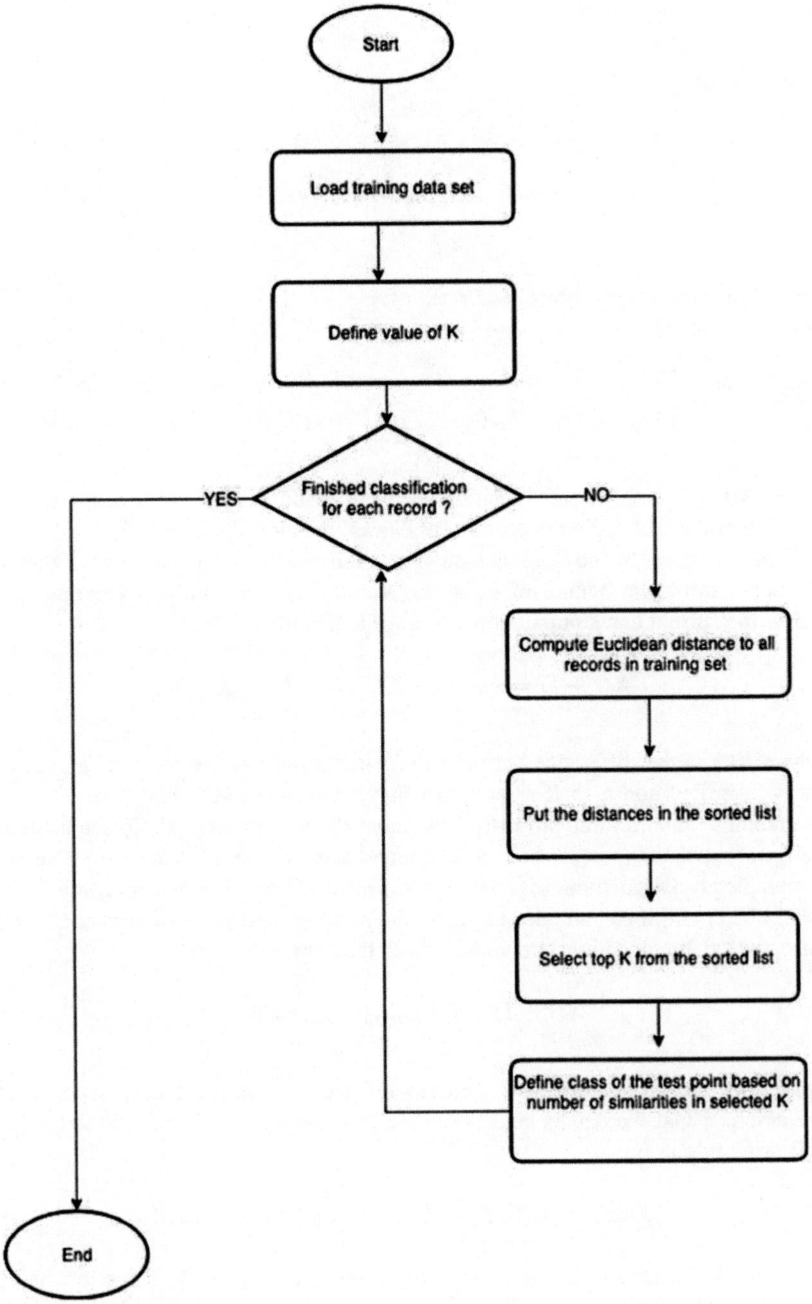

Fig. 7 KNN algorithm

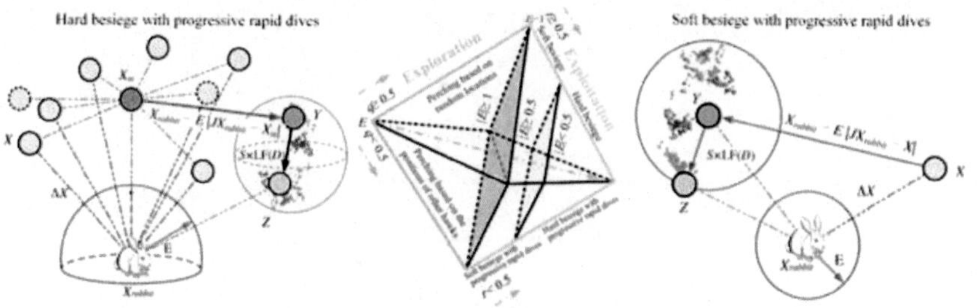

Fig. 8 Harris Hawks optimizer (Aljarah et al. 2019)

$$X(t+1) = \begin{cases} X_{rand}(t) - r1|X_{rand}(t) - 2r_2X(t)| & q \geq 0.5 \\ (X_{rabbit}(t) - X_m(t)) - r_3(LB + r_4(UB - LB)) & q < 0.5 \end{cases} \quad (3)$$

where X(t + 1) is the hawks' position vector in the next iteration t, Xrabbit(t) is the position of the rabbit, X(t) is the current hawks' position vector, r1, r2, r3, r4, and q are random values inside (0,1) that are updated in each iteration, LB and UB display the upper and lower bounds of variables, Xrand(t) is a randomly picked hawk,and Xm is the current hawk population's average location.

$$X(t+1) = \Delta X(t) - E|JX_{rabbit}(t) - X(t)| \quad (4)$$

where X(t) is the difference between the rabbit's position vector and the current location in iteration t, r5 is a random value between 0 and 1, and J = 2(1 r5) is the rabbit's random jump strength throughout the fleeing method. To replicate the nature of rabbit movement, the J value changes at random in each iteration. The prey is completely fatigued and has very little fleeing energy. Furthermore, the Harris hawks barely orbit the intended target before performing the surprise pounce. In this case, the existing locations are updated with the help of:

$$X(t+1) = X_{rabbit}(t) - E|\Delta X(t)| \quad (5)$$

while they are in a competitive scenario and want to catch the prey. As a result, we assumed that the hawks could evaluate (decide) their next move based on the following rule in Eq:

$$\text{JXrabbit(t)X(t)}|\text{Y} = \text{Xrabbit(t)E}|\text{JXrabbit(t)X(t)}| \quad (6)$$

Then they compare the likely effect of such a movement to the prior dive to determine whether or not it will be a good dive. If it wasn't reasonable (when they realized the price was too much).

They begin to conduct erratic, sudden, and quick dives when approaching the rabbit if it was not logical (when they detect the prey exhibiting more deceptive moves). We assumed they'd dive based on LF-based patterns, following the following rule:

$$Y = X_{rabbit}(t) - E|JX_{rabbit}(t) - X(t)| \quad (7)$$

Hence, the final strategy for updating the positions of hawks in the soft besiege phase can be performed by Eq:

$$X(t+1) = \begin{cases} Y\ if\ \ F(Y) < F(X(t)) \\ Z\ if\ \ (F(Z) < F(X(t)) \end{cases} \quad (8)$$

where N is the total number of hawks and Xt is the location of each hawk in iteration t.

3.3 MapReduce

Figure 9 shows the MapReduce, which is a basic but efficient scheduling model to divorce a task and to work parallel around the cluster in a humiliating way. Google has popularized the method and is commonly used by businesses that process vast volumes of data [45].

On data structures represented as key/value pairs, MapReduce algorithms run. The knowledge is separated into blocks; each block is displayed as a key and value. The key is usually a descriptive block data structure when the value is the real block data. On input key/value pairs, MapReduce methods perform separate parallel operations, and their performance is often key/value pairs. Two steps of the MapReduce model: map and reduction, which work as follows:

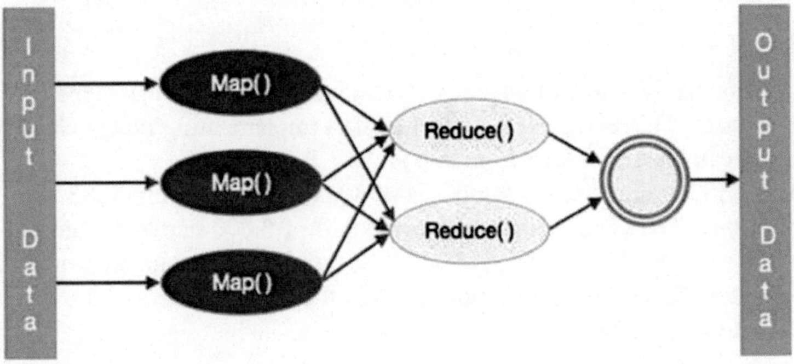

Fig. 9 Map and reduce algorithm

Map: Each input key/value pair has a Map Feature that performs some user-specified processing and sends new key/value pairs to be used as the intermediate storage to minimize.

Shuffle/Sort: Any input key/value pair can use a map feature, which allows some user-defined processing and issues new key/value pairs for intermediate storage processing to minimize.

Reduce: Parallel to all values of each independent map output key, a reduction function is implemented to emit output key/value pairs. The following equation shows the MapReduce equation:

$$f(wi|wj) = \frac{N(wi, wj)}{\Sigma w' N\left(wi, w'\right)} \tag{9}$$

3.4 Dataset

The Iris flowers

The Iris flower data set is the use of multi-measured measurements in taxonomic problems is a multivariate data set introduced in 1936 by British statistics and biologist Ronald Fisher. The fact that Edgar Anderson has gathered the data to quantified the morphological variation of Brisson flowers from three different species also is called Anderson's Iris data collection. The compilation of data contains 50 samples of three Iris species (Iris Setosa, Iris virginica, and Iris versicolor). The length and width of sepals and petals in centimeters were four characteristics measured for each sample. That dataset became a standard test case for many techniques of machine learning statistical classification such as vector support. The Dataset ben used under the name of Augmented_Iris contains more than the million records with five columns, this amount of data helps to extract and shows better results from the optimizer and proposed method.

MNIST Dataset

This is a database of handwritten digits. It contains 60,000 training images and 10,000 testing images. This is a perfect dataset to start implementing image classification where you can classify a digit from 0 to 9.

The MNIST training set is composed of 30,000 patterns from SD-3 and 30,000 patterns from SD-1. Our test set was composed of 5,000 patterns from SD-3 and 5,000 patterns from SD-1. The 60,000-pattern training set contained examples from approximately 250 writers. We made sure that the sets of writers of the training set and test set were disjoint.

The Boston Housing Dataset

This is a popular dataset used in pattern recognition. It contains information about the different houses in Boston based on crime rate, tax, number of rooms, etc. It has 506 rows and 14 different variables in columns. You can use this dataset to predict house prices. Data Science Project Idea to Predict the housing prices of a new house using linear regression. Linear regression is used to predict values of unknown input when the data has some linear relationship between input and output variables.

Wine quality dataset

The dataset contains different chemical information about the wine. It has 4898 instances with 14 variables each. The dataset is good for classification and regression tasks. The model can be used to predict wine quality.

The two datasets are related to red and white variants of the Portuguese "Vinho Verde" wine.d Due to privacy and logistic issues, only physicochemical (inputs) and sensory (the output) variables are available (e.g. there is no data about grape types, wine brand, wine selling price, etc.).

These datasets can be viewed as classification or regression tasks. The classes are ordered and not balanced (e.g., there are many more normal wines than excellent or poor ones). Outlier detection algorithms could be used to detect the few excellent or poor wines. Also, we are not sure if all input variables are relevant, it could be interesting to test feature selection methods.

Parkinson Dataset

Parkinson's is a nervous technique infection that influences people's movement. The dataset used in this study includes 195 records of individuals with 23 distinct attributes, including biomedical measurements. The data separates healthy people from people with Parkinson's disease. The model can differentiate healthy people from people with Parkinson's disease. The practical algorithm for this objective is XGboost which stands for robust gradient increasing based on decision trees.

Titanic Dataset

On 15 April 1912, the unsinkable Titanic ship sank and killed 1502 passengers out of 2224. The dataset contains information like name, age, sex, number of siblings aboard, etc. of about 891 passengers in the training set and 418 passengers in the testing set. Build a fun model to predict whether a person would have survived the Titanic or not. You can use linear regression for this purpose.

Credit Card Fraud Detection Dataset

The dataset includes transactions made by credit cards, marked as fraudulent or genuine, which is essential for enterprises with transaction systems to construct a model for catching illegal activities. Implement algorithms like logistic regression, artificial neural networks, and decision trees to acquire better accuracy. Compare the results of each algorithm and comprehend the behavior of samples.

Table 2 Dataset description and information summary

Dataset	Object	Cluster	Instance
IRIS	150	5	3
MNIST dataset	500	3	4
The boston housing dataset	250	14	2
Wine quality dataset	5847	13	2
Parkinson dataset	8400	42	2
Titanic dataset	890	11	7
Credit card fraud detection dataset	284,807	492	28
Chars74k dataset	255	32	8

Chars74k Dataset

The dataset includes characters utilized in the English and Kannada languages. It has 7.7 k characters from natural pictures, 64 categories (0–9, A-Z, a-z), 62 k computer-synthesized fonts, and 3.4 k hand-drawn characters. Implement character recognition in natural languages. Character recognition is the procedure of automatically recognizing characters from written printed texts or papers. Dataset's definitions and information are given in Table 2.

From Table 2. the dataset that has been used in the proposed method are various, well-known dataset and serve different type of classification and pre-processing action that make using the datasets are significant for the method.

3.5 Experimental Setup

The system is designed to evaluate the results and performance of the kNN with and without optimization, the experiment had two main parts, software, and hardware, as discussed below, the idea is to use the optimizer with MapReduce algorithms to obtain better results from the kNN.

3.6 Training and Testing

A training set is put together to construct a model in a data set, while the built-in model is tested by the evaluation (or validation set. Data points are omitted from the evaluation (validation) set in the training set. In most instances, a dataset is split into a training set, a validation set for every iteration, or a validation set and a test set for each iteration (some people use "test set.". We try to build a model for test data in machine learning. We then use the training data to match the model and validate it.

The created models are used to predict the uncertain results that are called the test set. The datasets are split into trains and experiments to validate the accuracy, precision, and measuring through preparation.

Training Set: The training to train the model of the data with more data to train a better result can be obtained.

Validation Set: This is important for your estimator to choose the right parameters. We may split the collection into a set of trains and validation systems. The model can be conditioned based on the validation data (for instance, changing parameters, classifiers). That helps us to get the best model possible.

Testing Set: Here, once the model is obtained, you can predict using the model obtained on the training set.

In this work the training, validate the data, and testing of the data will be performed, the training data size will be near 70% of the data total, 20% for testing, and 10% for validation, also, in another test the training data size will be near 50% from the data total, 50% for testing, and 10% for validation.

3.7 Evaluation Methods

Such these include many machine learning algorithms in the work using Python at each point, beginning with loading the data, filtering, extraction, and finishing with classification and precision, and using machine learning algorithms of different types to render the best possible testing of the data set and to compare the results to produce the best performance, in the following the metrics that been used to measure the accuracy for the classifiers with each feature extraction algorithm. We can define the metrics like the following:

Several terms are commonly used along with the description of recall, precision, and accuracy. They are true positive (TP), true negative (TN), false negative (FN), and false-positive (FP). If the existence of the disorder is shown in a patient, the test also demonstrates that the disease occurs, the diagnosis test findings are known to be a true positive. Similarly, if a patient has a demonstrated absence of a disorder, the diagnosis examination indicates that the disease is lacking as well (TN). A satisfactory finding between the diagnostic test and the proven condition will mean both true positive and negative outcomes (also called the standard of truth). But there is no flawless diagnostic exam. When a patient who has no illness is revealed by the screening test, the findings are false-positive the existence of disease (FP). Likewise, if the diagnostic test outcome indicates that a patient with the disorder is not present for sure, the test result is incorrect (FN). The test outcomes are contrary to the real conditions: false positive and false negative.

Confusion Matrix: Confusion Matrix As the title suggests, it gives us an output matrix describing the full model execution.

Precision = TP/ (TP + FN).
Recall = TN/ (TN + FP).
Accuracy = (TN + TP)/(TN + TP + FN + FP).

The F1 Score is the 2*((precision*recall)/(precision + recall)). The F score or F calculation is often referred to it. F1 indicates the balance between reminder and accuracy.

Mean Absolute Error: The cumulative deviation between the fundamental and expected values is the mean absolute error. It informs us of the degree to which the projections were based on the current outcome. However, they do not indicate the error path, i.e., whether we forecast the data or predict the data.

Mean Squared Error: The mean squared error (MSE) was similar to the mean absolute error. The only distinction was that MSE took the average difference square of the initial value to the estimated values. MSE has the advantage of estimating the gradient simpler, whereas Mean Absolute Error requires sophisticated linear programming techniques to measure the gradient. As the error is quadrated, more big mistakes than lesser mistakes are caused so that the algorithm can concentrate more on the more significant errors.

4 Experimental Results

4.1 Results Metric and Dataset Training

In this section, the metrics that will be used to test the proposed model with the classifiers will be discussed with the main formulas that need to be taken into consideration during the evaluation of the model. The first section talking more about the metrics and the second section will talking more about the training sizes and the dataset.

4.2 Results

4.2.1 KNN Results

The kNN technique starts by loading math libraries and data science as Numpy, Pandas, and matplotlibrary to organize that data and plot those results. The Sklearn's machine learning has been loaded by importing the KNeighborsClassifier for loading the class labels of the training data for classification of the test Binary_Hash_Training vector is classified by Creates 3 matrices to store STFTs of each trial, the PCA algorithm Perform to Calculate and subtract the mean and Eigenvalue decomposition on Training Data and Finlay to Find the covariance, the parameters for the kNN are 100 times of iteration, 3, 5,7 and 11 distance in the cluster. The Test results of

Table 3 kNN classifier confusion matrix IRIS dataset

K	Accuracy %	Precision %	Recall %	f1-score %
3	84.66	87	83	84
5	84.73	88	83	85
7	84.59	88	88	88
11	84.84	88	82	86

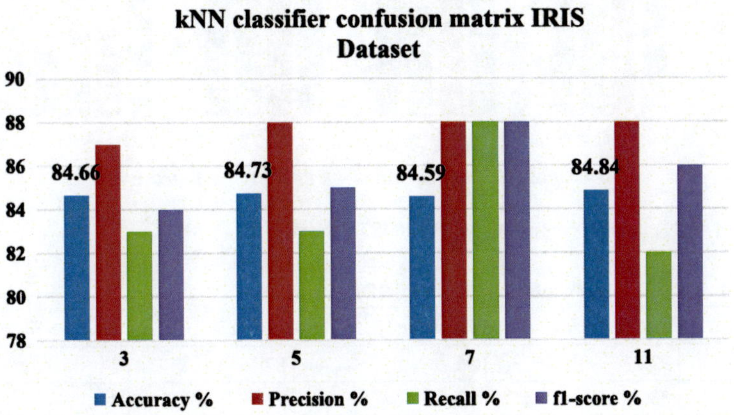

Fig. 10 kNN classifier confusion matrix IRIS dataset

classification compared to ground truth data calculates the correction classifications as a decimal resulting plot of the different values K.

The testing size for the results was 30% of the amount of the data, the results and tables are all based on 70% for training, and 30% for testing. The confusion matrix that been extracted and calculated from performing the kNN classifier comparing with the four-K values as the following Table 3 and Fig. 10.

As the results show, the kNN had poor accuracy results for the dataset, the dataset with k = 11 had 84% accuracy which considered the best accuracy, the accuracy chart shows un-patterned and increase of the accuracy with increasing of K. the kNN has been proofed that need to be supported from other optimizers to show better results (Fig. 11 and Table 4).

As the results show, for the dataset kNN had low accuracy, the data set was 84.68% accurate in the best with using K = 7, the accuracy graph was unpattern and the accuracy increased as K increased. The kNN data set indicates poor accuracy. The kNN has been shown that other optimizers require help to produce better results (Fig. 12 and Table 5).

The kNN had poor accuracy results for the dataset, as seen in the results; but, the dataset with k = 7 had 85.74% accuracy, which was considered the maximum accuracy; the accuracy chart showed un-patterned and increasing accuracy as K

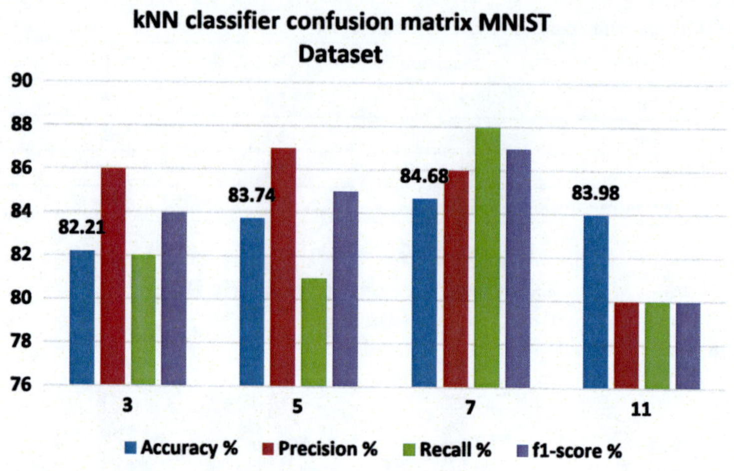

Fig. 11 kNN classifier confusion matrix MNIST dataset

Table 4 kNN classifier confusion matrix MNIST dataset

K	Accuracy %	Precision %	Recall %	f1-score %
3	82.21	86	82	84
5	83.74	87	81	85
7	84.68	86	88	87
11	83.98	80	80	80

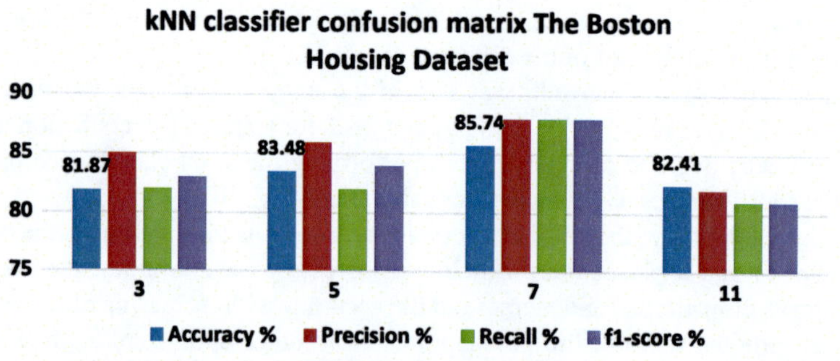

Fig. 12 kNN classifier confusion matrix the boston housing dataset

Table 5 kNN classifier confusion MNIST dataset

K	Accuracy %	Precision %	Recall %	f1-score %
3	81.87	85	82	83
5	83.48	86	82	84
7	85.74	88	88	88
11	82.41	82	81	81

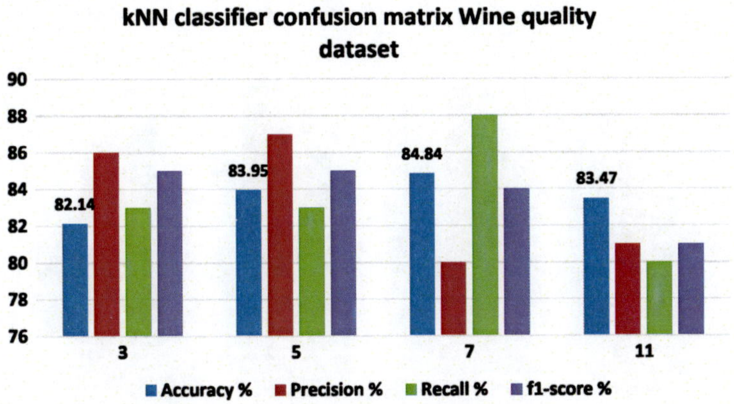

Fig. 13 kNN classifier confusion matrix Wine quality dataset

Table 6 kNN classifier confusion the Boston housing dataset

K	Accuracy %	Precision %	Recall %	f1-score %
3	82.14	86	83	85
5	83.95	87	83	85
7	84.84	80	88	84
11	83.47	81	80	81

increases. The kNN has been shown to benefit from the feedback of other optimizers in order to achieve better results (Fig. 13 and Table 6).

As the results show, the kNN had low accuracy results for the dataset; the dataset with k = 7 had 84.48% accuracy, which was deemed the highest accuracy; the accuracy curve shows un-patterned and increasing accuracy with increasing K. The kNN has been shown to require the assistance of other optimizers to produce better performance (Fig. 14 and Table 7).

As seen in the results, the kNN dataset had poor accuracy results, while the dataset with k = 11 had 86.74% accuracy, which was considered the best accuracy. With increasing K, the precision curve shows unpatterned and increasing accuracy. The kNN has been shown to need the feedback of other optimizers in order to achieve better efficiency (Fig. 15 and Table 8).

kNN's accuracy results for the dataset were poor, while the dataset with k = 11 had 86.41 percent accuracy, which was called the highest accuracy. With the K, the accuracy curve shows an unpatroned increase in accuracy. To attain better efficiency, the kNN has been shown to require assistance from other optimizers (Fig. 16 and Table 9).

As the results show, the kNN had low accuracy results for the dataset; the dataset with k = 11 had 87.74 percent accuracy, which was deemed the highest accuracy; the accuracy map indicates an un-patterned improvement in accuracy as K increases.

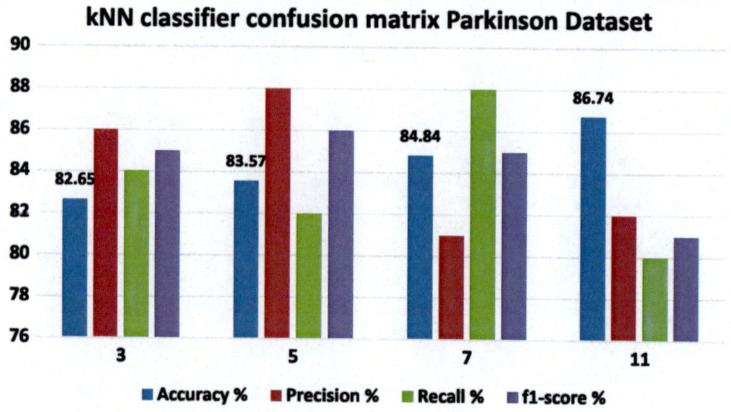

Fig. 14 kNN classifier confusion matrix parkinson dataset

Table 7 kNN classifier confusion matrix Wine quality dataset

K	Accuracy %	Precision %	Recall %	f1-score %
3	82.65	86	84	85
5	83.57	88	82	86
7	84.84	81	88	85
11	86.74	82	80	81

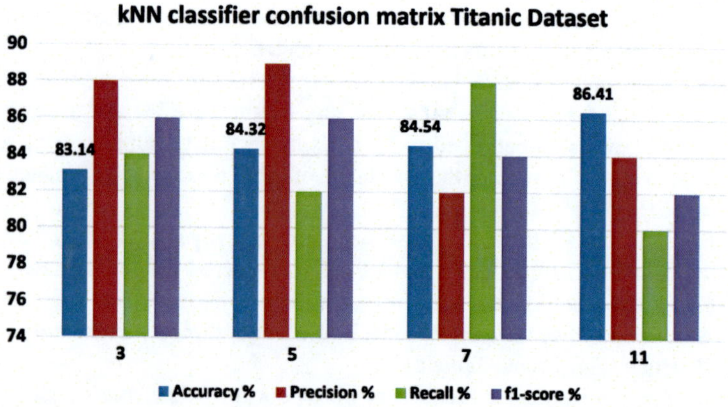

Fig. 15 kNN classifier confusion matrix titanic dataset

The kNN has been shown to need further optimization by other optimizers to achieve better performance (Fig. 17 and Table 10).

As the results show, The kNN had low accuracy results for the dataset, while the dataset with k = 11 had 88.21 percent accuracy, which was deemed the highest

Table 8 kNN classifier confusion matrix Titanic dataset

K	Accuracy %	Precision %	Recall %	f1-score %
3	83.14	88	84	86
5	84.32	89	82	86
7	84.54	82	88	84
11	86.41	84	80	82

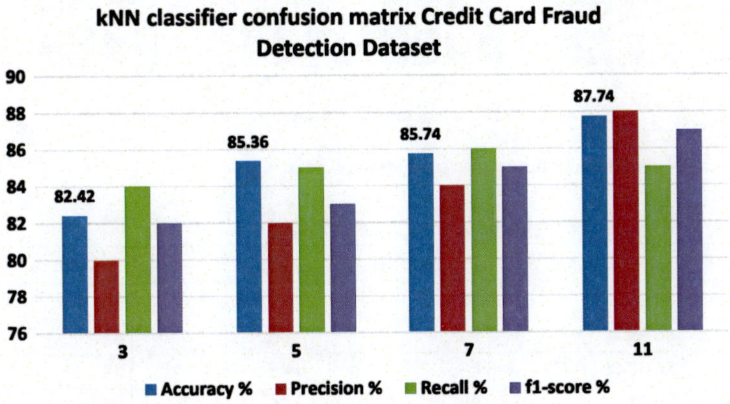

Fig. 16 kNN classifier confusion matrix credit card fraud detection

Table 9 kNN classifier confusion matrix credit card fraud detection

K	Accuracy %	Precision %	Recall %	f1-score %
3	82.42	80	84	82
5	85.36	82	85	83
7	85.74	84	86	85
11	87.74	88	85	87

accuracy. The accuracy curve indicates an un-patterned improvement in accuracy with increasing K. The kNN has been shown to need help from other optimizers to achieve better performance.

In this section the overall results will be explained and presented for all datasets and accuracy rates (Fig. 18).

4.2.2 KNN & HHO—MapReduce Results

Centered on the parallel machine architecture for MapReduce. For any partition, the nearest K neighbor inquiry is built and the outcome of the individual partition

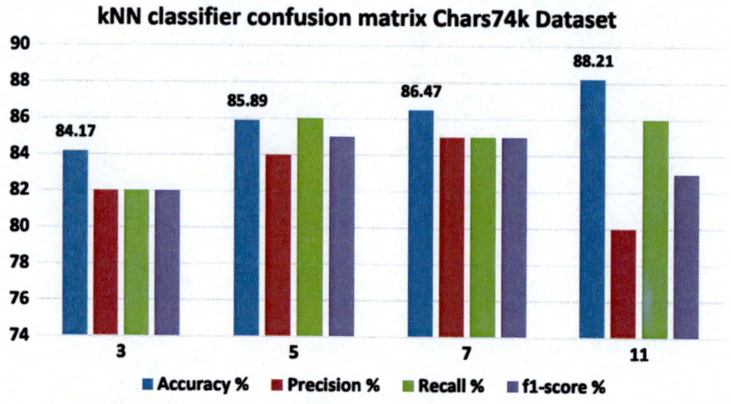

Fig. 17 kNN classifier confusion matrix Chars74k dataset

Table 10 kNN classifier confusion matrix Chars74k dataset

K	Accuracy %	Precision %	Recall %	f1-score %
3	84.17	82	82	82
5	85.89	84	86	85
7	86.47	85	85	85
11	88.21	80	86	83

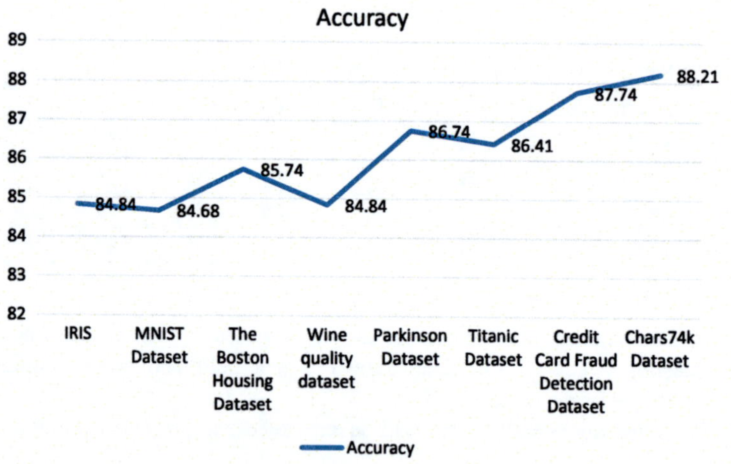

Fig. 18 kNN classifier with all dataset accuracy

queries is applied to be the final results. As previously mentioned, the MapReduce sorting process is independent of the previous query phase. KNN algorithm utilizes the functionality, sorting method, and calculation process at various cluster nodes that increase device efficiency significantly. If the data size and the value K are greater, this function is more popular. The fundamental principle of MapReduce-based parallel KNN query is: First, the road network is segregated by the number of Hadoop experimental cluster's measurement nodes. The partitions number is the same as the machine nodes number. The respective machine node copies each partitioned data. Second, by querying the index structure of R-Tree for determining the query point q position and to determine the candidate K sets as the output of the map function in each partition, using the list of adjacencies saved at each computes node. Finally, export to the challenge Minimize any map and get the results.

The testing size for the results was 30% of the amount of the data, the results and tables are all based on 70% for training, and 30% for testing.

The confusion matrix that been extracted and calculated from performing the kNN classifier comparing with the four-K values as the following Fig. 19 and Table 11.

As the results show, by using the IRIS dataset with the optimizer, the accuracy improved with using K = 11, the optimizer helped in increase in the accuracy by more than 10%. The iris with the three major objects optimized using the proposed method (Fig. 20 and Table 12).

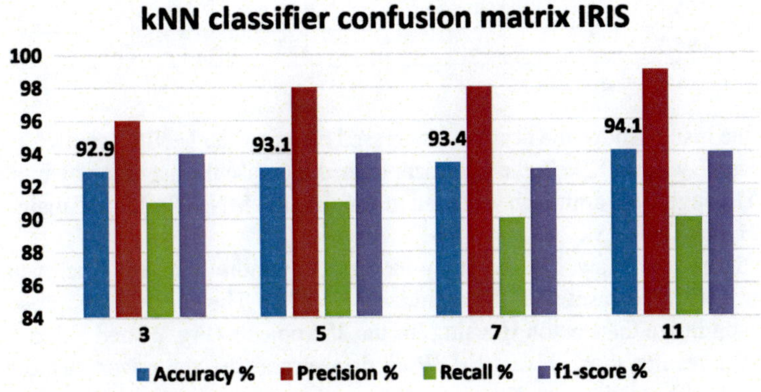

Fig. 19 kNN classifier confusion matrix

Table 11 kNN classifier confusion matrix IRIS

K	Accuracy %	Precision %	Recall %	f1-score %
3	92.9	96	91	94
5	93.1	98	91	94
7	93.4	98	90	93
11	94.1	99	90	94

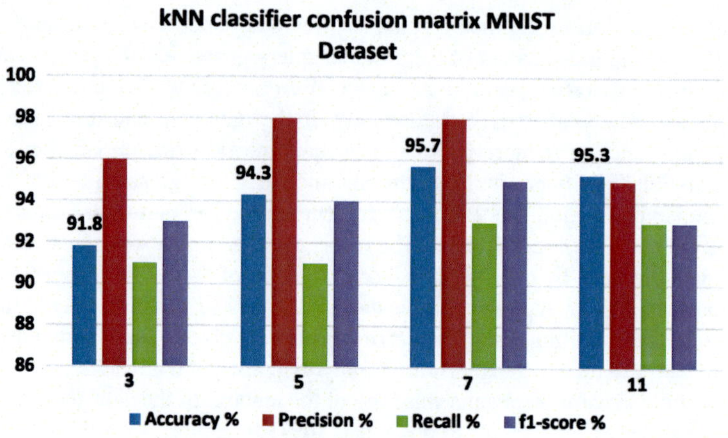

Fig. 20 kNN classifier confusion matrix MNIST dataset

Table 12 kNN classifier confusion matrix MNIST dataset

K	Accuracy %	Precision %	Recall %	f1-score %
3	91.8	96	91	93
5	94.3	98	91	94
7	95.7	98	93	95
11	95.3	95	93	93

As the results show, the accuracy increased by using the MNIST dataset with the optimizer. By using K = 7, the optimizer helped increase the accuracy by more than 11%. The suggested approach was used to simplify the MNIST with 500 main pieces (Fig. 21 and Table 13).

As the results show, The accuracy increased by more than 8% when using the Boston Housing dataset with the optimizer (K = 11). The suggested approach was used to optimize the Boston Housing for the 250 objects (Fig. 22 and Table 14).

As the results show, Using the Wine dataset and the optimizer, the accuracy increased with K = 11, and the optimizer assisted in increasing the accuracy by more than 10%. The Wine with 5000 main artifacts that were optimized using the proposed process (Fig. 23 and Table 15).

As the results show, the accuracy improved with K = 11 by using the Wine dataset and the optimizer, and the optimizer helped in increasing the accuracy by more than 10%. The wine has 198 key objects that have been optimized using the proposed technique (Fig. 24 and Table 16).

As the results show, Using the Titanic dataset and the optimizer, the accuracy increased with K = 11, and the optimizer assisted in increasing the accuracy by more than 11%. The proposed methodology has been used to refine 890 primary objects on the Titanic (Fig. 25 and Table 17).

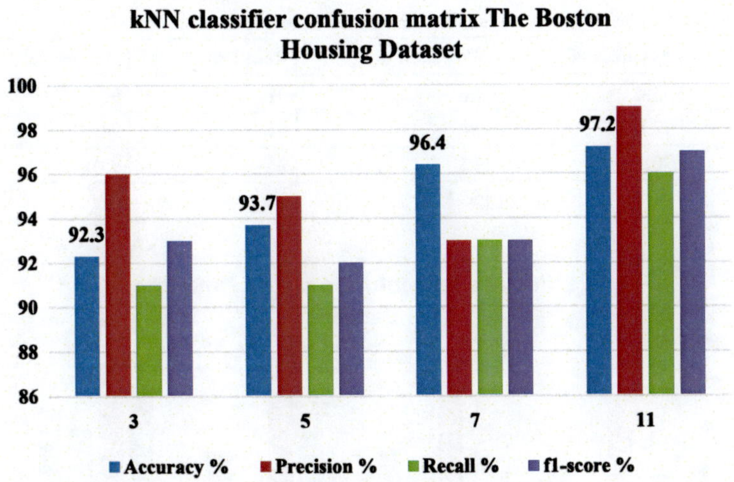

Fig. 21 kNN classifier confusion matrix the boston housing dataset

Table 13 kNN classifier confusion matrix the boston housing dataset

K	Accuracy %	Precision %	Recall %	f1-score %
3	92.3	96	91	93
5	93.7	95	91	92
7	96.4	93	93	93
11	97.2	99	96	97

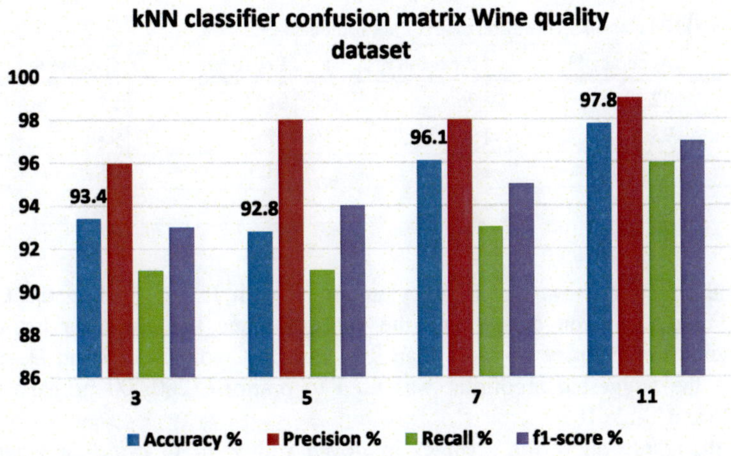

Fig. 22 kNN classifier confusion matrix Wine quality dataset

Table 14 kNN classifier confusion matrix the Boston housing dataset

K	Accuracy %	Precision %	Recall %	f1-score %
3	93.4	96	91	93
5	92.8	98	91	94
7	96.1	98	93	95
11	97.8	99	96	97

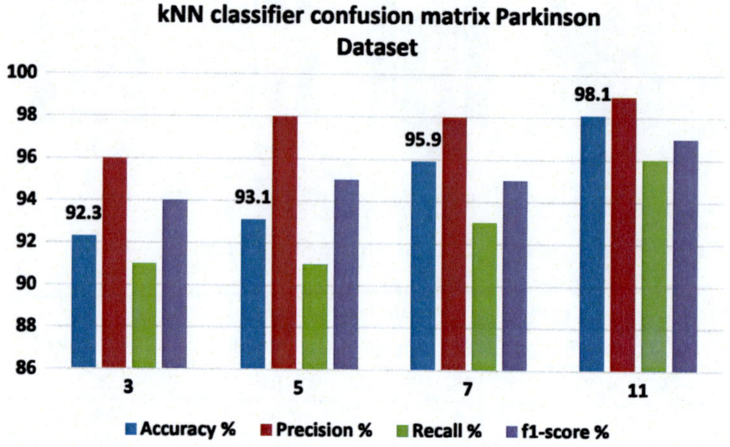

Fig. 23 kNN classifier confusion matrix Parkinson dataset

Table 15 kNN classifier confusion matrix Parkinson dataset

K	Accuracy %	Precision %	Recall %	f1-score %
3	92.3	96	91	94
5	93.1	98	91	95
7	95.9	98	93	95
11	98.1	99	96	97

As the results show, the accuracy improved with K = 11 using the Credit Card Fraud Detection dataset and the optimizer, and the optimizer helped in improving the accuracy by more than 9%. On the Credit Card Fraud Detection dataset, the suggested algorithm was used to optimize 200.000 primary items (Fig. 26 and Table 18).

As the results show, the accuracy improved with K = 11 using the Chars74k dataset and the optimizer, and the optimizer helped in improving the accuracy by more than 20%. On the Chars74k Dataset, the suggested algorithm was used to optimize 255 primary items.

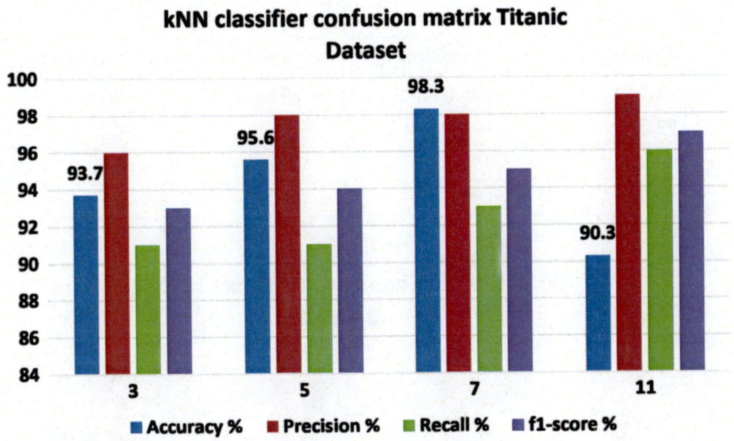

Fig. 24 kNN classifier confusion matrix Titanic dataset

Table 16 kNN classifier confusion matrix Titanic dataset

K	Accuracy %	Precision %	Recall %	f1-score %
3	93.7	96	91	93
5	95.6	98	91	94
7	98.3	98	93	95
11	90.3	99	96	97

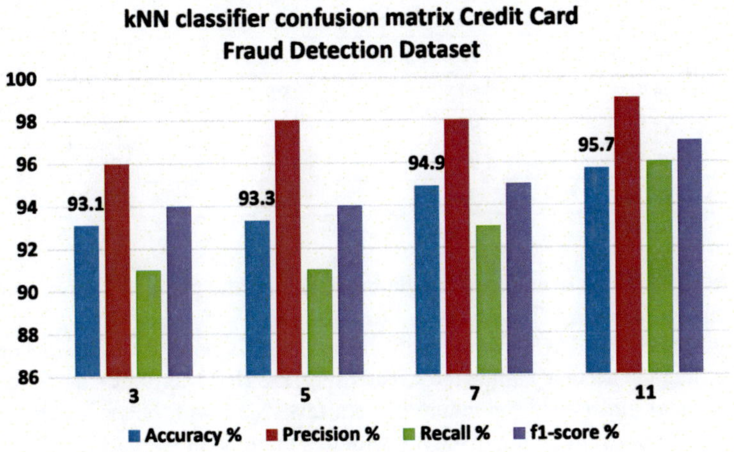

Fig. 25 kNN classifier confusion matrix credit card fraud detection

Table 17 kNN classifier confusion matrix Credit card fraud detection

K	Accuracy %	Precision %	Recall %	f1-score %
3	93.1	96	91	94
5	93.3	98	91	94
7	94.9	98	93	95
11	95.7	99	96	97

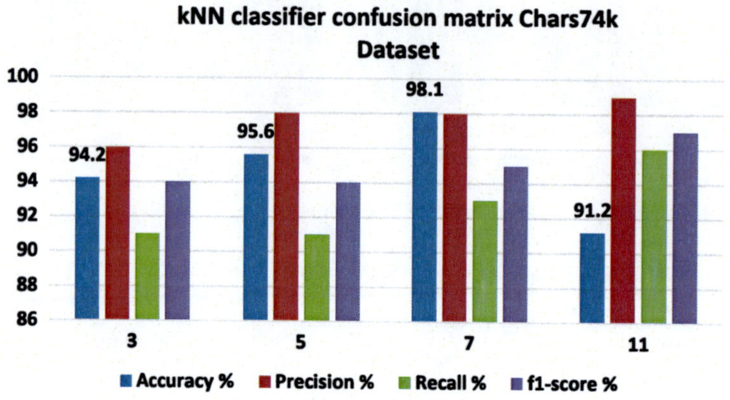

Fig. 26 kNN classifier confusion matrix Chars74k dataset

Table 18 kNN classifier confusion matrix Chars74k dataset

K	Accuracy %	Precision %	Recall %	f1-score %
3	94.2	96	91	94
5	95.6	98	91	94
7	98.1	98	93	95
11	91.2	99	96	97

In this section the overall results will be explained and presented for all datasets and accuracy rates (Fig. 27).

By using HHO and MR with the kNN the results show better accuracy and pattern of increasing the accuracy while the k increasing, the HHO and MR supported kNN to cluster and understand the nature of data faster and with better accuracy. The accuracy still needs to be improved but with this increase of the data, the HHO and MR show and proved that the experiments had a better result.

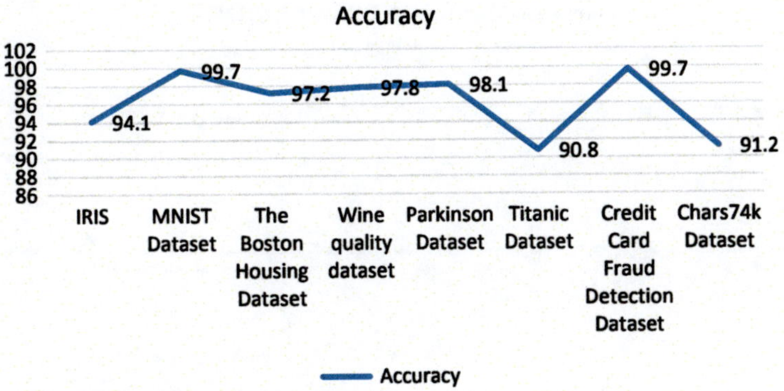

Fig. 27 kNN classifier with all dataset accuracy

4.2.3 Performance Results

Using KNN as classifier and integrated it with KNN as Kmeans algorithms to break what is the data and more understand for the labeled dataset that been used in the proposed method, map reduces and HH optimized that data and facilated the KNN and Kmeans to understand the data and how to break it and better results in classification and results as showed in the classification reports. In this section, the performance of both kNN and kNN with MapReduce algorithms have been compared and resulted. As it can be seen in the performance comparisons, kNN - MapReduce performs better almost in all of the K values compared to the basic kNN based proposals.

The dataset accuracy improved significantly with using the Harris hawk optimizer, the accuracy rate increased by more than 11% from the normal kNN machine learning big data classifier. The results from the previous section show the change in parameters and accuracy of the proposed model (Fig. 28).

By comparing the taken time kNN with and without using HHO -MR. when k = 3 the kNN had better results due the k is small and the processes with HHO takes more time, but when k = 5, 7 and 11 HHO and MR had better result due to sampling the data and support the kNN to understand the data with the support of performance, accuracy and the run time.

4.2.4 Comparison Results

In this section the comparison between the proposed method and other authors' results regards the same researching field, the results are presented in the following table and figure (Table 19).

As the results in Table 19. The proposed method has been compared with the other researches and authors in this field the proposed method shows better results in increasing the accuracy of the kNN with using the optimizer with increasing 11% which is significant change in the results the proposed method has been proved a

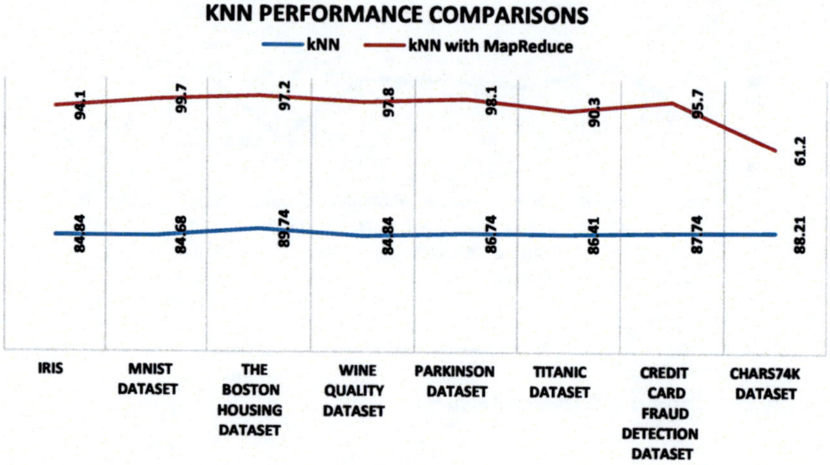

Fig. 28 kNN performance comparisons

Table 19 Comparison results

Author	Improvement percentage (%)	Algorithm
El-Hasnony et al. (2020)	91	kNN
Chatzigeorgakidis et al. (2019)	93	kNN
Ding et al. (2019)	94	kNN
Seyfollahi et al. (2019)	94	kNN
Sihwail et al. (2020)	94.7	kNN
Ye et al. (2020)	95	kNN
Our proposed method	97	kNN

successful implementation and results. From the table and figure, we can notice that the proposed model had been improved the results and kNN classification for more than 11%. Which is a significant change regarding the other researchers in the optimization field.

5 Conclusions and Future Work

Big data analysis is a "technology-enabled strategy" which enables users and company to gain a fertile, broad, and more precise vision and eventually gain competitive advancement. A new hybrid optimization algorithm is being developed in the analysis to cluster the big data and to concentrate on the best clusters. This proposes a modern Hybrid Harris Hawks Optimizer that leverages the search space to deliver maximal performance. The idea of a MapReduce parallel system was then implemented since it provides fault tolerance, load balance, and data position. In addition to the clustering algorithm, the incorporation of the MapReduce method improves the power of the clustering mechanism and makes it optimal for automated parallelism and delivery. The MapReduce principle also tolerates fault with the clustering algorithm. In comparison to other well-known optimizers, the findings suggest that HHO is capable of discovering outstanding solutions. HHO can be developed in the future and used to solve a variety of challenges in engineering and other industries. In future work, the research focuses more on different types of Clustering and unsupervised fields and studies with expanding the scope of the study by adding Deep learning classifications to classify the data based on the powerful deep learning LSTM and RNN algorithms [46].

- This research was constrained to use the Harris Hawks Optimizer (HHO) is hybrid with the K-means clustering technique.
- The K-means are used to generate the solutions of the initial of the HHO. Then, the HHO will use the general solution and seeks for finding the optimal solution through the course of iterations.
- The solution can be presented as a vector of dimension, from which the suggested HHO detects the ideal centroids.
- Properties, such as continuous streaming and the massive volume of the data, can be fixed with the use of the MapReduce framework with the clustering algorithm.
- Among the limitations that faced the researcher learning programming using Python and the difficulty of learning the programming accompanying the program.
- The limited time available to work on the research, as the research and programming process requires a long time and careful study.
- The study relied on the use of scale and not on real data.

References

1. T.K. Das, P.M. Kumar, Big data analytics: a framework for unstructured data analysis. Int. J. Eng. Sci. Technol. **5**(1), 153 (2013)
2. M.A. Shinwan, K. Chul-Soo, Enhanced mobile packet core network scheme for next-generation mobile communication systems. Int. J. Electron. Commun. Comput. Eng. **8**, 56–61 (2017)
3. M. Al Shinwan, T.-D. Huy, K. Chul-Soo, A flat mobile core network for evolved packet core based SAE mobile networks. J. Comput. Commun. **5**(5), 62–73 (2017)
4. M. Al Shinwan, K. Chul-Soo, A future mobile packet core network based on ip-in-ip protocol. Int. J. Comput. Networks Commun. **10** (2018)
5. X. Cui, P. Zhu, X. Yang, K. Li, C. Ji, Optimized big data K-means clustering using MapReduce. J. Supercomput. **70**(3), 1249–1259 (2014)
6. S. De, S. Dey, S. Bhattacharyya, Recent advances in hybrid Metaheuristics for data clustering (2020)
7. D. Singh, C.K. Reddy, A survey on platforms for big data analytics. J. big data **2**(1), 1–20 (2015)
8. A.S. Shirkhorshidi, S. Aghabozorgi, T.Y. Wah, T. Herawan, Big data clustering: a review, in *International Conference on Computational Science and its Applications* (2014), pp. 707–720

9. H.-G. Li, G.-Q. Wu, X.-G. Hu, J. Zhang, L. Li, X. Wu, K-means clustering with bagging and mapreduce, in *2011 44th Hawaii International Conference on System Sciences* (2011), pp. 1–8
10. T. Condie, N. Conway, P. Alvaro, J.M. Hellerstein, K. Elmeleegy, R. Sears, MapReduce online, in *Nsdi* (2010), vol. 10, no. 4, p. 20
11. M. Al Shinwan et al., An efficient 5G data plan approach based on partially distributed mobility architecture. Sensors **22**(1), 349 (2022)
12. L.M. Abualigah, A.T. Khader, M.A. Al-Betar, O.A. Alomari, Text feature selection with a robust weight scheme and dynamic dimension reduction to text document clustering. Expert Syst. Appl. **84**, 24–36 (2017)
13. J. Fan, F. Han, H. Liu, Challenges of big data analysis. Natl. Sci. Rev. **1**(2), 293–314 (2014)
14. L. Abualigah et al., Hybrid Harris Hawks optimization with differential evolution for data clustering, in *Metaheuristics in Machine Learning: Theory and Applications* (Springer, 2021), pp. 267–299
15. A. Gupta, H.K. Thakur, R. Shrivastava, P. Kumar, S. Nag, A big data analysis framework using apache spark and deep learning, in *2017 IEEE international conference on data mining workshops (ICDMW)* (2017), pp. 9–16
16. J. Qiu, Q. Wu, G. Ding, Y. Xu, S. Feng, A survey of machine learning for big data processing. EURASIP J. Adv. Signal Process. **2016**(1), 1–16 (2016)
17. S. Sagiroglu, D. Sinanc, Big data: a review, in *2013 International Conference On Collaboration Technologies And Systems (CTS)* (2013), pp. 42–47
18. M. Alshinwan, L. Abualigah, C.-S. Kim, H. Alabool, Development of a real-time dynamic weighting method in routing for congestion control: application and analysis. Wirel. Pers. Commun. **118**(1), 755–772 (2021)
19. M. Al Shinwan, L. Abualigah, N.D. Le, C. Kim, A.M. Khasawneh, An intelligent long-lived TCP based on real-time traffic regulation, Multimed. Tools Appl. **80**(11), 16763–16780 (2021)
20. L. Abualigah, M. Shehab, M. Alshinwan, H. Alabool, Salp swarm algorithm: a comprehensive survey. Neural Comput. Appl. **32**(15), 11195–11215 (2020)
21. L. Abualigah, M. Shehab, M. Alshinwan, S. Mirjalili, M. Abd Elaziz, Ant lion optimizer: a comprehensive survey of its variants and applications. Arch. Comput. Methods Eng. **28**(3), 1397–1416 (2021)
22. M. Shehab, L. Abualigah, H. Al Hamad, H. Alabool, M. Alshinwan, A.M. Khasawneh, Moth--flame optimization algorithm: variants and applications. Neural Comput. Appl. 32(14), 9859–9884 (2020)
23. L. Abualigah et al., Advances in meta-heuristic optimization algorithms in big data text clustering. Electronics 2021, 10, 101." s Note: MDPI stays neu-tral with regard to jurisdictional clai-ms in~..., (2021)
24. L. Abualigah et al., Nature-inspired optimization algorithms for text document clustering—a comprehensive analysis. Algorithms **13**(12), 345 (2020)
25. S. Lohr, The age of big data. N.Y. Times **11**, 2012 (2012)
26. E. Slack, Storage infrastructures for big data workflows. Storage Switch. LLC, Tech. Rep (2012)
27. Z. Zheng, J. Zhu, M.R. Lyu, Service-generated big data and big data-as-a-service: an overview, in *2013 IEEE International Congress on Big Data* (2013), pp. 403–410
28. H.N. Alshaer, M.A. Otair, L. Abualigah, M. Alshinwan, A.M. Khasawneh, Feature selection method using improved CHI Square on Arabic text classifiers: analysis and application. Multimed. Tools Appl. **80**(7), 10373–10390 (2021)
29. S. Tiwari, H.-M. Wee, Y. Daryanto, Big data analytics in supply chain management between 2010 and 2016: insights to industries. Comput. Ind. Eng. **115**, 319–330 (2018)
30. L. Mohammad Abualigah et al., Hybrid harmony search algorithm to solve the feature selection for data mining applications. *Recent Adv. Hybrid Metaheuristics Data Clust.* 19–37 (2020)
31. L. Abualigah, B. Alsalibi, M. Shehab, M. Alshinwan, A.M. Khasawneh, H. Alabool, A parallel hybrid krill herd algorithm for feature selection. Int. J. Mach. Learn. Cybern. **12**(3), 783–806 (2021)

32. L.M. Abualigah, E.S. Hanandeh, A.T. Khader, M.A. Otair, S.K. Shandilya, An improved b-hill climbing optimization technique for solving the text documents clustering problem. Curr. Med. imaging **16**(4), 296–306 (2020)
33. M.R. Naqvi, M.A. Jaffar, M. Aslam, S.K. Shahzad, M.W. Iqbal, A. Farooq, Importance of big data in precision and personalized medicine, in *2020 International Congress on Human-Computer Interaction, Optimization and Robotic Applications (HORA)* (2020), pp. 1–6
34. B.M. Balachandran, S. Prasad, Challenges and benefits of deploying big data analytics in the cloud for business intelligence. Procedia Comput. Sci. **112**, 1112–1122 (2017)
35. L. Abualigah et al., Ts-gwo: Iot tasks scheduling in cloud computing using grey wolf optimizer, in *Swarm Intelligence for Cloud Computing*, Chapman and Hall/CRC (2020), pp. 127–152
36. L. Barthelus, Adopting cloud computing within the healthcare industry: opportunity or risk? Online J. Appl. Knowl. Manag. **4**(1), 1–16 (2016)
37. N. Ilyasova, A. Kupriyanov, R. Paringer, D. Kirsh, Particular use of BIG DATA in medical diagnostic tasks. Pattern Recognit. Image Anal. **28**(1), 114–121 (2018)
38. M.M. Najafabadi, F. Villanustre, T.M. Khoshgoftaar, N. Seliya, R. Wald, E. Muharemagic, Deep learning applications and challenges in big data analytics. J. Big Data **2**(1), 1–21 (2015)
39. E. Dumbill, What is big data?—An introduction to the big data landscape. (2012), [Online]. Available: http://radar.oreilly.com/2012/01/what-is-big-data.html
40. H. Rashaideh et al., A grey wolf optimizer for text document clustering. J. Intell. Syst. **29**(1), 814–830 (2020)
41. W. Zhao, H. Ma, Q. He, Parallel k-means clustering based on mapreduce, in *IEEE International Conference on Cloud Computing* (2009), pp. 674–679
42. S.B. Elagib, A.R. Najeeb, A.H. Hashim, R.F. Olanrewaju, Big data analysis solutions using MapReduce framework, in *2014 International Conference on Computer and Communication Engineering* (2014), pp. 127–130
43. L. Chen, X. Huo, G. Agrawal, Accelerating mapreduce on a coupled cpu-gpu architecture, in *SC'12: Proceedings of the International Conference on High Performance Computing, Networking, Storage and Analysis* (2012), pp. 1–11
44. V. López, S. Del R\'\io, J. M. Ben\'\itez, F. Herrera, Cost-sensitive linguistic fuzzy rule based classification systems under the MapReduce framework for imbalanced big data. Fuzzy Sets Syst. **258**, 5–38 (2015)
45. J. Dean, S. Ghemawat, MapReduce: simplified data processing on large clusters. Commun. ACM **51**(1), 107–113 (2008)
46. M. Abd Elaziz et al., Advanced metaheuristic optimization techniques in applications of deep neural networks: a review. Neural Comput. Appl., 1–21 (2021)

Deep Neural Network for Virus Mutation Prediction: A Comprehensive Review

Takwa Mohamed, Sabah Sayed, Akram Salah, and Essam Halim Houssein

Abstract Artificial intelligence (AI) and Deep Learning Algorithms are potential methods for preventing the alarmingly widespread RNA viruses and ensuring pandemic safety, they have become an integrative part of the modern scientific methodology, offering automated procedures for the prediction of a phenomenon based on past observations, unraveling underlying patterns in data and providing insights about the problem. With the continuous growth in the number of RNA Virus COVID-19 patients, likely, doctors and healthcare personnel won't be helpful in treating every case. Thus, data scientists can help in the battle against RNA Viruses Mutations by implementing more innovative solutions in order to accomplish controlling severe acute respiratory syndrome quickly RNA Viruses are viruses that are made up of strands of RNA. This work studies the induction of machine learning models and motivating their design and purpose whenever possible. In the second part of this work, we analyze and discuss the biological data in the eyes of deep learning models. The core of our contributions rests in the role of machine learning in viruses pandemics.

Keywords Machine learning algorithms · Deep learning algorithms · RNA viruses mutation problems · Corona virus · Spike protein · RNA genome

1 Introduction

Annual epidemics and pandemics do occur from time to time are caused by RNA viruses, which pose a constant threat to public health. Due to fast mutations, the RNA viruses evolution remains the principal impediment to the efficiency of antiviral ther-

T. Mohamed · E. H. Houssein (✉)
Faculty of Computers and Information, Minia University, Minia, Egypt
e-mail: essam.halim@mu.edu.eg

S. Sayed · A. Salah
Faculty of Computers and Artificial Intelligence, Cairo University, Cairo, Egypt

E. H. Houssein et al. (eds.), *Integrating Meta-Heuristics and Machine Learning for Real-World Optimization Problems*, Studies in Computational Intelligence 1038,
https://doi.org/10.1007/978-3-030-99079-4_9

apy. Modeling the sequential RNA viruses temporality and dimensionality strains, as well as interpreting the prediction results, is one of the primary issues [1].

1.1 Virus Structure and Classification

Viruses can be classified in several ways, such as particle morphology, genome type, replication strategy, host organism, mode of transmission, or by the type of disease they can cause. The two most commonly used classification systems are the Baltimore classification and the International Committee on Taxonomy of Viruses (ICTV) system. The Baltimore system is based on replication strategy, i.e. the nature of the virus genome and its mode of expression and divides the viruses into seven categories; Single-stranded DNA (+), Double-stranded DNA, Single-stranded RNA (+), Double-stranded RNA, Single-stranded RNA (−), Double-stranded DNA with a RNA intermediate, and Single-stranded RNA (+) with a DNA intermediate [2].

The ICTV system uses the hierarchical taxonomic levels of order (-virales), genus (-virus), family (-viridae), sub-family (-virinae) and species (There is no set ending for species names, and they can contain more than one word) based on the Linnaean System of Taxonomy [3] as shown in Table 1.

There are three orders, 233 genera, 9 subfamilies, and 56 families in this system. Although the ICTV recognizes approximately 1,550 virus species, virologists are monitoring approximately 30 thousands of virus strains [2].

1.2 RNA Viruses

RNA is a polymer made from repeating units called ribonucleotides covalently linked together. The ribose in RNA is hydroxylated, and the pyrimidine nucleotide thymine is in place of uracil, which is a chemical difference between DNA and RNA. RNA viruses and retroviruses employ RNA as their genetic material, which means that the RNA acts as a genome template rather than an intermediary template for gene

Table 1 The seven taxonomic levels of the ICTV system

Taxonomic level		
Taxonomic level	Suffix	Example
Order	-Virals	Mononegavirales
Family	-Viridac	Paramyxoviridae
Subfamily	-Virinac	Paramyxovirinae
Genus	-Virus	Rubulavirus
Species	A single kind of virus	Mumpsvirus

translation into proteins. Due to the low efficiency of RNA-dependent RNA/DNA polymerases, many RNA viruses have a lot of sequence variability, which is partly caused by point mutations incorporated into the viral genome during replication. Because of polymerase proofreading activity, DNA viruses have a low mutation rate, but RNA/DNA polymerases do not have this error correction [4].

There are 350 distinct human infections are caused by RNA viruses. They're to blame for a slew of new ailments. Researchers are struggling to find an effective medical treatment for RNA virus infection since many RNA viruses only take a few weeks to escape the immune system or evolve drug resistance. A short generation time and a tiny genome contribute to variation in the viral population, meaning that any error or mutation can result in phenotypic change. Alteration of nucleic acid sequence can also occur through recombination, which is an exchange of genetic information between existing variants. New variants are generated through recombination when two different viruses coincidentally infect the same cell and their genes are mixed, and a third variant is produced with features of the two. The obvious consequence of genetic diversity resulting from point mutation is that RNA viruses possess a great potential for rapid evolution and adaptation [5].

RNA Viruses Replication Strategies Because the host cell lacks an RNA polymerase that relies on RNA, viral RNA genome replication is unique in that most RNA viruses encode their RNA polymerase to overcome this barrier. Some viruses produce polymerase after infection, whereas others pack it onto the virus particle. RNA viruses have a variety of genome structures, including unimolecular or segmented, positive-stranded, negative-stranded, double-stranded, or circular genomes [6]. The genome of RNA of positive-stranded RNA viruses is the same as polycistronic mRNA. The viral mRNA is translated as soon as the host cell has been infected into a polyprotein product. The mature proteins are formed from the cleavage of the polyprotein. Coronaviridae, flaviviridae and picornaviridae are examples of this class. The negative-stranded RNA viruses can have either segmented or non-segmented genomes [7].

Conserved Secondary Structures in Viral RNA Genomes RNA viruses' genomes contain not just encodes only proteins, but also active RNA structures. Functional secondary structures evolve at a considerably slower rate than the underlying sequences and will almost certainly play a crucial role in the life cycle of the virus [8]. In general, functionally active secondary RNA structures are susceptible to point mutations, and do not differ from the ones formed by random sequences: computer simulations reveals that large changes in the secondary structures come up from a tiny group of point mutations [9].

Viruses use different approaches to ensure their genome amplification. This can be achieved in an efficient and in some cases, a cell-type-specific manner, The plurality of RNA viruses have their polymerases that discriminately amplify their genomes; however, other viruses have evolved a range of strategies to compete directly with the host cell for components required for viral gene replication and packaging. Viruses must contend with the host for both the ribosome and the small number of available resources eukaryotic initiation factors (eIFs) essential for ribosome recruitment to viral and cellular mRNAs during the phase of the illness. The internal ribosome entrance site has developed in some viruses to recruit ribosomes via cis-acting elements (IRES). The IRES elements were first described within the RNAs from various picornaviruses including poliovirus, human rhinoviruses, encephalomyocarditis virus hand foot-and-mouth disease virus [10].

The IRES is an extremely specialized RNA element that is often located at the 5' end of the viral genomes. It enables the translational machinery to be assembled near or immediately at the start codon without a 5' cap structure. IRES is conserved and is, therefore, a useful target for broadly targeted PCRs [11].

Pathogenicity and Virulence of RNA Viruses When the terms pathogenicity and virulence are applied to viruses, it is important not to use them without reference to the host. These terms always refer to the virus's relationship with the infected host. Pathogenicity of a virus is its ability to infect and induce a host's infection [12]. On the other hand, virulence refers to the disease's seriousness once it has been infected. The case fatality rate, or the proportion of clinical cases that develop serious disease, is a measure of virulence [13].

Pathogenic steps of the viral disease include Virus implantation at the entry point, local replication, dissemination to address disease sites, and spread to viral shedding sites in the atmosphere. The pathogenic mechanisms can be affected by factors such as The virus's accessibility to tissue, susceptibility of cells to virus replication, and Virus resistance to host defenses are all factors to consider. RNA viruses in a variety of shapes and sizes are very infectious and pathogenic in the short term, and every year, they cause a large amount of morbidity and mortality in humans. Newly emerging RNA viruses may endanger public health [14].

Transmission of Enveloped RNA Viruses Enveloped RNA viruses infect their hosts via special interactions of binding between RNA Viruses glycoproteins of the envelope and cell receptors. During the viral replication cycle, they must cross through the host cell's plasma membrane twice: when the virus enters the body and when the particles are released [15]. Entry is usually accomplished through a membrane fusion event that occurs either directly on the cell surface after particle binding or in reduced endosomes after endocytosis of bound virions [16].

Enveloped viruses have developed a range of methodologies for escaping the cell [15]. Common pathways of virus entry include skin, genitourinary tract, gastrointestinal tract, respiratory tract, and conjunctiva. Most RNA viruses transmit via the fecal-oral route and/or the respiratory tract, they spread disease by multiplying in epithelial cells [17].

A respiratory infection can be caused by viruses found in aerosols inhaled by the recipient host or viruses found in nasopharyngeal fluids that are spread by touch contact. Viral infections can be localized to the site of inoculation. However, virus propagation in the blood is the most important route since viruses can potentially be carried to any site in the body. This is referred to as viremia [18]. The approach of transmission varies for each virus. For instance, retroviruses integrate their genome into host DNA and when integrated into germline cells it becomes endogenized, e.g. endogenous retroviral sequences (ERVs). Understanding the routes of entry and modes of transmission is important for the selection of the right sample material [19].

1.3 Human RNA Viruses

Influenza Virus Influenza viruses, the Orthomyxoviridae Influenza viruses are a single-stranded RNA (-), belonging to the family Orthomyxoviridae. They are characterized by a high mutation rate resulting in antigenic drift and a high frequency of genetic rearrangement resulting in the antigenic shift. These properties lead to the antigenicity of viral surface glycoproteins changes. Influenza viruses classified into 3 classes, antigenic variations in viral hemagglutinin and composite proteins classify viruses as A, B, or C [20].

Influenza RNA is categorized into eight different types (7 for types C), coding for ten different proteins. Influenza A is split into subtypes based on the surface proteins antigenic differences neuraminidase (NA) and hemagglutinin (HA) [21]. There are currently 16 special HA ranges between H1 to H16 and NA ranges between N1 to N9 sub-types [22]. Influenza A displays the highest genetic variability of the three influenza types. It largely results from antigenic drift, which enables the virus to avoid neutralization of antibodies. In antigenic shift, two influenza A viruses combine to form a novel sub-type by exchange of genomic segments. The new subtype has a mixture of segments from two viruses [23].

Influenza is a virus that causes illness. In birds and some animals, a virus causes influenza, and these influenza viruses are found naturally among birds. The viruses are carried in the intestines of wild anseriform birds (such as ducks) all over the world, but they rarely cause illness. Such lowly pathogenic (LPAI) viruses can become highly pathogenic (HPAI) avian influenza viruses either by insertion, mutation, or recombination [24]. If highly pathogenic strains are received from wild aquatic birds to domestic poultry, including chickens, ducks, and turkeys, it could make them very sick and kill them. During a crisis of avian influenza among poultry, persons who come into touch with diseased birds or surfaces contaminated with secretions or diseased birds' excretions are in danger. This could potentially cause a major outbreak in humans, in the worst case a new human influenza pandemic. However, aside from relatively few incidents, this has not occurred. The term "human influenza virus" usually refers to strains of influenza that have extended among people. There are only three subtypes of influenza A viruses Currently deployed in humans (H3N2, H1N2, and H1N1) [25].

Three pandemics have occurred between 1900 and the present, caused by three different influenza A subtypes, Hong Kong flu, Asian flu and Spanish flu [26] as shown in Table 2.

Influenza B and C show lower genetic variability than influenza A. Up to now, the mechanisms responsible for changes in these viruses are not well characterized (although it is likely to exist). Although genetic variability of influenza A and B is

Table 2 Influenza pandemics occurred in 1900s

Name	Time period	Deathes
Hong Kong Flu (H3N2)	1968	1 million
Asian Flu (H2N2)	1957	2 million
Spanish Flu (H1N1)	1918	40–50 million

higher there have been reports showing that influenza C causes genetic mutations such as influenza A or B [27].

Influenza A and B are contagious and can cause critical illness, hospitalizations, and death in the elderly [28]. Although influenza C was isolated more than 60 years ago, there are few reports describing its clinical features. Influenza C is linked to a common cold-like illness that primarily affects youngsters [29].

Newcastle Disease Virus The Newcastle disease virus (NDV) is a member of the avulavirus genus within the Mononegavirales order, the family of Paramyxoviridae [30]. The virus has an RNA genome that is single-stranded (-) wrapped around it. The genome is about 15 kb long and follows the "rule of six," which is required for effective viral replication [31]. The hemagglutinin-neuraminidase (HN), fusion protein (F), phosphoprotein (P), matrix protein (M), nucleoprotein (NP), and RNA large polymerase (L) proteins are all genetically encoded (3' to 5'). The NP, P, and M proteins make up the viral inner surface, while the L protein, along with the NP and P proteins, makes up the viral nucleocapsid. HN and F are two surface glycoproteins that attach to sialic acid receptors on host cells and connect the envelope to the host cell membrane, respectively [32]. The NDV has a high protein-coding capacity as a family trait, which is further boosted by a technique known as "RNA editing". During the transcription of P gene mRNA, as a result of this method the creation of V and W proteins with one insertion or guanine (G) respectively [33].

NDV strains can be categorized as pathotypes founded on Vivo pathogenicity indices for chickens. The viscerotropic velogenic is very pathogenic and produces intestinal infection with a high fatality rate, whereas the neurotropic velogenic NDV causes respiratory and neurological system symptoms with a high mortality rate. The pathogenicity of mesogenic strains is lower, causing acute respiratory and neurological symptoms in many people but with a low mortality rate. NDV strains that are lentogenic produce minor respiratory infections. It is an asymptomatic intestinal variant in which the host lives longer and the virus has exclusive replication and shedding rights [32]. The discrepancies in pathogenicity are mostly due to differences in the F protein's cleavage location. This protein is made as a non-functional precursor (F0), which is then cleaved into two polypeptides with functional activity by host proteases (F1 and F2). Within the F protein, all mesogenic and velogenic NDV strains have the sequence of amino acids of 112R/K-R-Q-R/K-R-F117 whereas lentogenic viruses have 112 G/E-K/R-Q-G/E-R-L117 [34].

Based on the analysis of phylogenetic nucleotide sequences of partial hypervariable from the gene (F), strains of NDV have ten types of genotypes (I-X). The first five genotypes are old (1930–1960), while the last five genotypes are new (after 1960). In their hosts, however, they are all harmful in the same way. The seventh genotype is split into seven subgenotypes and the eighth genotype is split into five subgenotypes [35]. In the meantime, Aldous et al. [36] offered an alternate pattern for NDV classification During studies on a huge number of NDV isolates obtained from various nations. The criterion is that the NDV has six types of genetic lineages, each of which has many sublineages. There are around 55 full genomes of various NDV strains [37].

Newcastle disease (ND) is an OIE-recognized disease, and any outbreak must be reported to the OIE [38]. ND is found all over the world and is reported regularly from all continents. During the year, a sporadic form of the disease appears in Pakistan, and only a few outbreaks are officially reported or unofficially reported each year. Despite the widespread and imported vaccinations that can be used without restriction, NDV continues to be the most common poultry illness in Pakistan's rural and commercial chickens [39]. The failure of the vaccination is due to incompatibility between field and vaccine strains, as well as the emergence of novel NDV strains. Furthermore, the involvement of poultry in remote areas in the country's NDV epizootiology has always been a mystery. The full genome of NDV was characterized genetically, phylogenetically, and biologically to assess the genetic diversity degree of Strains of NDV found in poultry in the backyard and to calculate relationships to NDV currently being circulated in the area [38].

HIV Virus The human immunodeficiency virus is defined as HIV is classified as a member of the Lentivirus genus in the Retroviridae family, subfamily Orthoretrovirinae [40]. HIV is split into categories 1 and 2 based on genetic features and changes in viral antigens (HIV-1, HIV-2). Nonhuman primate immunodeficiency viruses (simian immunodeficiency virus, SIV) are also classified as Lentiviruses. HIV was first detected in the human population between 1920 and 1940, according to epidemiologic and phylogenetic evidence. HIV1 and HIV2 are viruses that attacks humans and animals from Central African chimpanzees (SIVcpz) and West African cats (SIVsm) respectively [41, 42].

When you look inside of the virus particle, you will find, the HIV genome is made of 2 single-stranded RNA molecules that are identical. The HIV provirus genome, also known as proviral DNA, occurs when a virus converts the genome into DNA, the RNA is destroyed, and HIV double-stranded DNA is incorporated into the human genome. On both ends of the DNA genome, long terminal repeat (LTR) sequences are found. In the 5' LTR region, the promotor for viral gene transcription is encoded. The gag gene contains nucleocapsid proteins (NC, p 7), peripheral proteins (MA, p 17), capsid proteins (CA, p 24), and lowered proteins that impair the stability of nucleic acids. The reading stop bezel is located behind the pol reading bezel. It combines and combines proteins, transcriptome, RNase or RT, and RNase H. Glycoprotein (surface protein, SU) and (membrane protein, TM) to read both packages. Border environment near many the pol gene [43].

The HIV genome codes for a variety of regulatory proteins in addition to protein structure: The regulatory proteins Vif (viral infectivity factor), Nef (negative regulating factor), Vpu (virus protein unique), and Vpr (virus protein r) influence viral replication. Tat (transactivator protein) and Rev (RNA splicing-regulator) are required for HIV replication to begin, while the other regulatory proteins Vif (viral infectivity factor), Nef (negative regulating factor), Vpu (In HIV-2, Vpx (viral protein x), and Vpr (virus protein r) is coded for instead of Vpu, contributing to the virus's lower pathogenicity [44]. Chimp immunodeficiency viruses (SIVcpz) and gorilla immunodeficiency viruses (SIVgor) have the same genetic structure as HIV-1 [45].

SARS-CoV-2 Virus SARS-CoV-2 is a zoonotic beta coronavirus with positive-sense single-stranded RNA genomes that are enveloped. Their lengths range from 80 to 160 nm and their forms range from spherical to pleomorphic [46]. Envelope (E), Nucleocapsid (N), Spike (S), and Membrane (M) are the four structural proteins found in SARS-CoV-2 (N). The envelope of this virus is made up of the S, M, and E proteins. SARS-CoV-2 is produced and matured in part by the E protein, which is the smallest protein in the structure. During virus replication, the S and M proteins are also important in virus attachment. Inside the envelope, the N proteins continue to be dominated by the RNA, forming a nucleocapsid. Other phases of the cycle of virus replication and the host-cell reply to viral infection are also affected by N. The S protein of this virus has a crown-like shape when viewed under a microscope, earning it the name coronavirus [47].

SARS-CoV-2 is a virus that can be transmitted from bats and humans. The ACE2 receptors found in many organs such as the lungs, heart, kidneys, and gastrointestinal system allow this virus to enter the human body. As a result, ACE2 aids the virus's entry into target cells [48]. The CoV infection process starts when the S protein, which is made up of S1 and S2 subunits, attaches to the ACE2 receptor in the host cell [49]. As a result, infected individuals not only suffer from respiratory difficulties like pneumonia, which can progress to Acute Respiratory Distress Syndrome (ARDS), but also from heart, kidney, and digestive system problems. The virus attaches to host cells more strongly than other the same origin viruses because of the S protein's tight ridges [50].

The rest of the chapter is structured as follows: Sect. 2 introduces an overview of the RNA viruses mutations. Section 3 presents the The machine learning techniques used in the virus mutations prediction. Related work for virus mutations prediction using machine learning is discussed in Sect. 4. There are several challenges in this research area are included in Sect. 5. Section 6 concludes the chapter and future work.

2 RNA Virus Mutations

Virus mutations produce biological diversity, which is subject to the competing forces of variety and irregular genetic drift, which are both inspired by the size of the virus population [51].

Changes in genetic information are more likely to occur down to the following generation is defined as an organism's mutation rate. A generation of viruses is commonly described as a cycle of infection of the cell, that comprises attachment to the surface of the cell, gene expression, entrance, encapsidation, replication, and infectious particle discharge. Mutations aren't limited to replication; they also can happen as a result of genetic material editing or spontaneous nucleotide damage [52].

The frequency in a particular viral population, which mutations are present should not be associated with the mutation rate. The latter is a measure of genetic variation that is influenced by recombination, random genetic drift, natural selection, and other factors. Increased rates of mutations contribute to greater genetic variety, but mutation rates cannot be inferred directly from observable population mutation frequencies, unless in exceptional circumstances. Although a variety of factors influence genetic diversity, the mutation rate is of particular relevance because of the most important source of genetic diversity. Similarly, mutation rates and molecular evolutionary rates should not be conflated [53].

According to evolutionary theories that do not take into account the neutral theory, these two effects have a linear relationship, even though molecular evolution defines the establishment of novel genes in populations, whereas mutation is a biochemical/genetic process [53]. Knowing and managing medication resistance, vaccination, immunological evasion, pathogenesis, and the creation of novel diseases depends on understanding and managing the processes behind viral mutation rates. The significance of viral mutation rates in clinics can be demonstrated by the historical past of anti-HIV therapies. The first anti-HIV treatment to be approved was the nucleoside analog azidothymidine (AZT), but the emergence of drug-resistant variations quickly rendered this monotherapy ineffective. HIV-1 is a rapidly-mutating virus that generates all potential single-base substitutions (AZT resistance mutations, for example) within a patient daily [54].

The following highly active antiretroviral therapy's success was based not just on increasing drug potency, but also on mixing multiple medications (including AZT) to reduce the likelihood of resistance mutations arising. The same logic applies qualitatively to other fast-evolving viruses, such as the hepatitis C virus (HCV). Multiple resistances to new HCV treatments have already been described [55], and population sequence analysis has discovered that protease inhibitor and non-nucleoside polymerase inhibitor resistance occurs naturally. In treatment-naive patients, i.e., in the absence of selection favoring these mutations [56].

Combination therapy is currently the only viable therapeutic option for fast-mutating virus-related chronic illnesses. Antiviral immunity can be depicted similarly. Viruses with a high mutation rate are more likely to evade immunity. There are multiple reports of antibody evasion in cytotoxic T lymphocyte (CTL), hepatitis B virus (HBV), HCV, and HIV-1, three RNA Viruses that are fast-mutating viruses that cause chronic infection. A sequence of genetic variations has been linked to immunological avoidant and failure of vaccination in HBV, the much more popular reason of hepatitis massive global, with about more than 300 million chronically people with the disease [57]. Immune escape occurs at the level of the host population rather than on an intra-host basis in acute viruses. The virus's ability to re-infect previously infected hosts or that viruses hosts that recognize identical antigens are the benefit of escape in this scenario. The most well-known sample is the virus of influenza, which changes antigenicity regularly and so necessitates yearly vaccine updates. Currently, efforts are focused on producing vaccines that target more evolutionarily conserved influenza strains but immunogenic protein domains [58].

The genetic diversity of viruses, which is ultimately defined by mutation rates, has a significant impact on antiviral strategy development. Virus mutation rates are influenced not just by polymerase errors, but also by a virus's ability to rectify DNA mismatches through post-replicative repair and/or proofreading. Other mutation sources include spontaneous nucleotide damage, host enzymes, and even particular genetic components found inside the genomes of certain viruses whose sole purpose is to generate novel mutations [59].

2.1 RNA Viruses Versus DNA Viruses

The Baltimore classification based on genetic information, viruses classify themselves into the categories contained in the virion: positive-strand RNA viruses (e.g. tobacco mosaic virus, rhinoviruses), negative-strand RNA viruses (rabies virus, influenza viruses), double-strand RNA viruses (rotaviruses), retroviruses (human T cell leukemia virus, HIV), single-stranded DNA viruses (parvoviruses), and double-stranded DNA viruses (herpesviruses, poxviruses) [59].

Viruses have the most diverse mutation rates of any biological system, with the greatest variances between RNA and DNA viruses. Table 1 provides a summary of mutation rates for various viruses. As previously noted, numerous causes of estimating error and bias affect the dependability of certain of these rates [53]. Despite these uncertainties, it may be estimated that The number of substitutions per nucleotide per cell infection varies between 10-8 and 10-4, with DNA viruses falling between 10-8 and 10-6 and RNA viruses falling between 10-6 and 10-4. There are various mechanisms for these variances. For starters, RNA viral polymerases lack 30 exonuclease proofreading functions, making them more error-prone than DNA virus polymerases [60].

Coronaviruses, a class of positive-strand RNA viruses that encode a complex RNA-dependent RNA polymerase with a 30 exonuclease domain, are an exception to this norm [61]. 30 exonuclease activity is similarly absent in reverse transcriptases (RTs) [62]. As a result, retroviruses (those with RNA-containing cellular DNA stagevirions) and paraviruses (those with DNA-containing a cellular RNA stage and virions) change and evolve at the same pace as non-reverse transcribing RNA viruses (riboviruses). While the distinction between RNA/RT and DNA viruses is well known from a genetic and mechanistic standpoint, distinctions in molecular evolution are less obvious [63].

This highlights the reality that evolution is influenced by a variety of factors other than mutation rate, as well as the fact that mutation rates for many DNA viruses are unknown and it may be higher than previously assumed. A genome-wide average of 2 9 10-7 s/n/c was discovered in recent work with human CMV, even though this estimate was indirect, which is a little higher than before anticipated for a big double-strand DNA virus. Given that many RNA viruses and DNA viruses have comparable lifestyles, it's unclear why mutation rates in these two groups have developed so differently [64].

2.2 *Viruses that Are Single-Stranded Have a Higher Mutation Rate Than Viruses that Are Double-Stranded*

Although this distinction is based on studies with bacteriophages, for eukaryotic single-strand DNA viruses, no mutation rate estimates have been produced, DNA viruses that single-stranded seem to mutate quicker than DNA viruses that double-stranded [53]. There are no clear differences in mutation rate within Baltimore classes within RNA viruses. The processes that underpin these disparities are unknown. Nucleic acids that are single-stranded are more susceptible to deoxidizing amine and other kinds of chemical degradation, which could explain the variations between single and double-strand viruses [59]. During viral infections, high amounts of Molecular mutations can be caused by reactive oxygen species (ROS) and other biological metabolites in both the virus and the host cell. For example, ethanol is expected to work in concert with virus-contained oxidative stress to boost HCV mutation rates [65].

Differences in access to post-replicative repair between single-strand and double-stranded DNA viruses can also be explained. Work with the bacteriophage /X174 has shown some intriguing results. Dam methylase and MutHLS proteins execute (MMR) methyl-directed mismatch repair in enterobacteria. Dam methylation of motifs of GATC sequence is essential to execute mismatch repair since it is used to distinguish the daughter DNA strands and the template DNA strands [66]. MutS recognizes mismatches and interacts with MutL, causing the MutH endonuclease to be activated, excising the daughter strand. Even though around by chance, 20 such locations are expected, the bacteriophage /X174 genome has no motifs of GATC sequence. As a result, MMR cannot be performed on the /X174 DNA. This helps to explain the virus's genetic variation will be in the range of 10-6 s/n/c, which is three magnitudes greater than E. coli's and the greatest within DNA viruses [67].

Avoidance of GATC motifs could be due to mutation rate selection, but it could also be due to other selective causes. Methylation of phage DNA is inefficient, for example, may make it vulnerable to MutH cleavage, putting a selection pressure on GATC sequence motifs [68]. In contrast to bacteriophage /X174, the relationship between mutation rate and post-replicative repair in eukaryotic viruses is still unknown. Viruses connect with DNA damage response (DDR) paths by modifying the localization or encouraging the degeneracy of DDR components, according to numerous studies [69]. For example, the adenoviral E4orf6 protein causes TOPBP1, a DDR component, to be degraded by proteasomes. DDR activity can happen as a result of infection-induced cellular stress or as part of an antiviral response that is then counteracted by viruses. However DNA viruses are known to cause instability in the genome in their hosts, it is unknown if DDR dysregulation influences DNA virus mutation rates [70].

3 Machine Learning Techniques

Machine learning approaches can be divided into three broad categories depending on the type of data they need to process: reinforcement learning, unsupervised and supervised [71–73].

The training data is used in supervised learning consists of a collection of inputs and intended outputs. Techniques of supervised learning look at the training dataset and create an assumed function that may be used to map novel samples [74]. Unsupervised learning is easier to understand based on the name. This class does not have a learning label, so the algorithm has to find the structure itself. To learn by reinforcement, the computer program must interact with their dynamic environment [75].

3.1 *Logistic Regression*

Predictors should, in particular, assign a weighting to each test based on the likelihood that it will fail, so that we can decide whether to run it or not. Essentially, this means we favor a value between 0 and 1 [76].

The logistic or sigmoid function is selected as the predictive function in the context of logistic regression, as it is described as

$$h_\theta(k) = \frac{1}{1 + e^{-\theta^T k}} \tag{1}$$

where $h_\theta(k)$ is the estimated probability that output equal 1 on input k.

This function will return a number between 0 and 1 as a result. The next step is to calculate the logistic function's parameters so that the algorithm, provided the features for each set of data, outputs a value that is comparable to the given predicted value. After that, the ideal parameter value is then calculated by several optimization approaches to fixing this problem.

Scikit-learn is a Python library for machine learning that's used to develop this algorithm. The simplicity and flexibility of this software, and its open-source nature, leading to its selection [77]). It is noteworthy that the generated predictor returns values ranging from 0 to 1.

If this number is more than 0.5, the expected result is 1 by default (the test will fail). Otherwise, the result of the prediction will be 0, which indicates that the test will succeed. A simple threshold produces poor rating results only in rare cases.

3.2 Random Forest

Before defining random forest, we must first explain decision trees. The model which is based on splitting the feature set and storing the distribution on the class label for each region is called a decision tree [78].

This can be reproduced by a tree, implying that the hypothesis or predictor function of this learning algorithm is a tree. The probability distribution for every class is then assigned to each leaf in that tree that matches to each area. We may be using decision trees to create predictors for each experiment depending on this definition. Even while decision trees are simple to understand and use, their results are not as reliable as those based on logistic regression or other predictors. Additionally, the tree structure is very dependent on the data presented. This indicates that even minor modifications to the data can have a significant impact on the tree structure that results [78].

Random forests are implemented to help solve this problem. These models can be thought of like a mash-up of many decision trees. To become more explicit, they are created by training various trees on various sets of data which are chosen at random and replaced. The outcome of a random forest prediction is then formed by mixing these decision trees [79].

3.3 Artificial Neural Networks

Artificial neural networks (ANNs) are modeled from the brain's organic neural networks. While mathematical algorithms are well-suited to linear programming, arithmetic, and logic computations, ANNs are more effective in pattern recognition and matching, clustering, and classification tasks [80].

An artificial neural network is a classifier that is modeled after the human brain, which is considerably different from how computer code is generally written. The human brain is densely packed with nerve cells and neurons. Each of these cells is linked to a large number of other similar cells, forming a complicated signal transmission network. Each cell collects input from all of the other brain cells to which it is linked, and when it reaches a specific threshold, it sends a signal to all of the cells to which it is attached [81].

The output signal is y, the activation function is, the number of perceptron connections is n, the weight associated with the connection (i) is w(i), and the value of the connection (i) is x(i). In the diagram, the letter b stands for the threshold [82].

The perceptron may aggregate many input values that are weighted and activate and transmit an output when the sum of the inputs reaches a threshold. The activation function determines which output it transmits, which is usually within 0 and 1 or -1 and 1. Because the activation function's derivative is frequently employed in network training, it's more convenient if the derivative can be represented in terms of the actual function value, as this requires fewer unnecessary calculations [81].

A perceptron's equation is as follows:

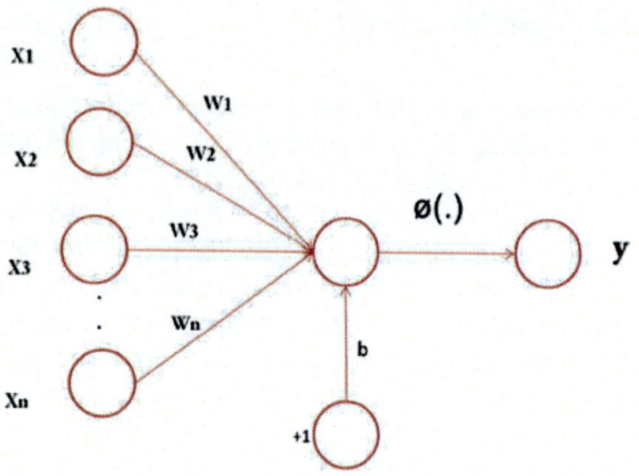

Fig. 1 A simple perceptron is depicted graphically

$$y_{out} = \phi(\sum_{i=1}^{n}(w_i x_i) + b) \tag{2}$$

There's a graph for perceptron in Fig. 1. A neuron with −1 is a constant value is used to represent the threshold b. A dynamic threshold once the perceptron activation is produced by enabling the network to adjust the weight affiliated with b. This is a fairly straightforward design, but when numerous perceptrons are combined and operate together, its strength is revealed.

Layers of perceptrons are common, the input of each layer is from the previous, applying weights, and then signals to the next layer if necessary see in Fig. 2. As previously stated, a classifier had to be capable of learning from and adapt to examples. This is accomplished in an ANN by adjusting the weights connected with the layers' relationships. There are many techniques to do this, but most of them encompass initializing the weights and feeding an example to the network.

The network's output error is then determined and fed backward through a technique known as back-propagation. This technique is then utilized to adjust the weights, and the network will try to differentiate between multiple different classes through the frequent application of this process. Techniques such as momentum can be employed to improve the efficiency of the training. The proper update phase for the weights is determined by using momentum. The network will require a lengthy time to converge if the step is too tiny. The step may never converge and begin vibrating if it is too large. When employing swing, stride length is determined dynamically while running, and weight, which frequently changes in the same direction, should be large and change quickly [83].

Over-fitting of the data, which occurs when the classifier becomes too strong at classifying the training data but unable to classify a generalized input, is one issue with the ANN technique. This may be prevented by using cross-validation, which involves training the network on samples of data and then evaluating it on a different set. The network may be over-fitted if the error in the validation set starts to rise.

Fig. 2 A graphical illustration of a single hidden layer artificial neural network

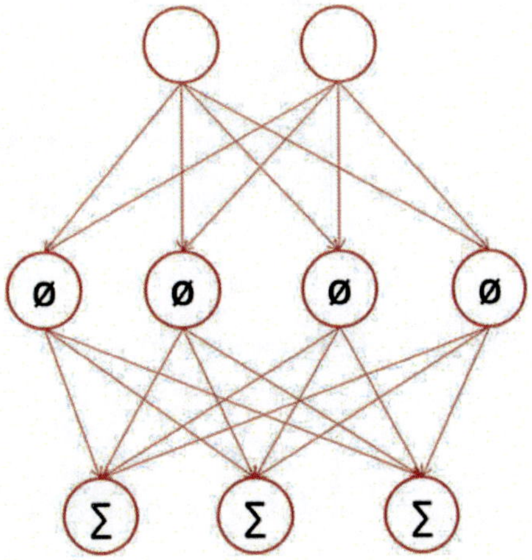

If prior networks have been saved, the network can be reverted to the one with the smallest error [84].

3.4 Deep Learning

Within the machine learning area, whenever neural networks are referred to, we refer to artificial neural networks. These neural networks consist of neurons that are ordered in layers. Because the network consists of multiple layers stacked upon each other, the network is called deep. We often refer to this as deep learning [85].

The layers of a deep learning network are connected with other layers and send signals to each other when they fire. There are numerous ways to connect layers. A neural network is composed of three layers which are an input layer, hidden layers, and an output layer. A layer is called hidden when the output of that layer cannot directly be observed [86].

An example of a neural network can be found in Fig. 3. Each neuron consists of a variable value and a bias parameter. The connections between the layers have weight parameters. An activation function determines the values of the neurons, the weights, and the biases. This activation function is a non-linear function, such as Rectified Linear Unit (ReLu) [87].

The goal of training a neural network is to optimize the weights corresponding with the connections between the neurons and their biases, such that the inputs will transform into their corresponding outputs. Thus, a neural network can be seen as a learned non-linear function that is able to map an input to a given output. During training, there will be a discrepancy between the target output and the predicted output [88].

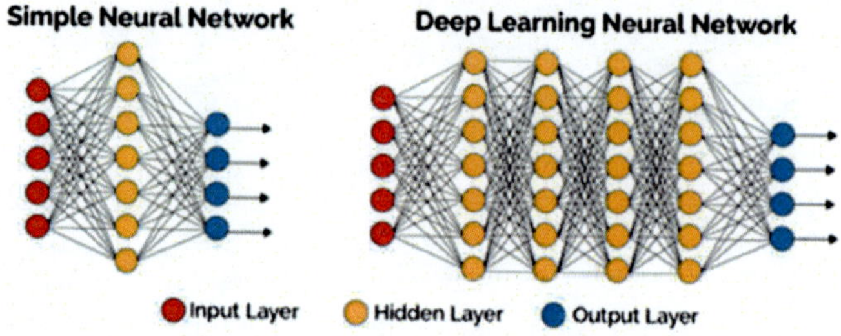

Fig. 3 Example of a neural network

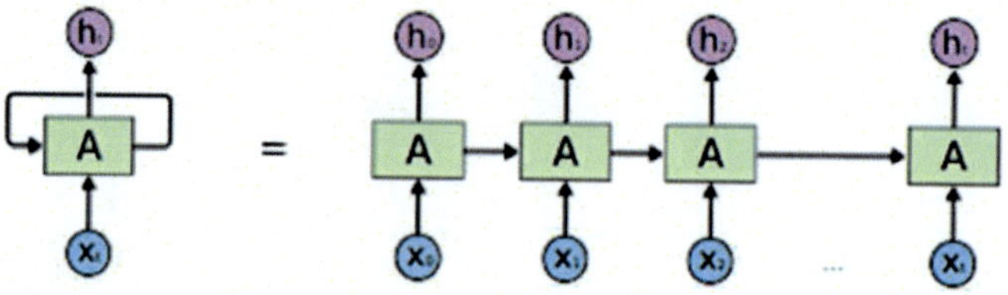

Fig. 4 Example of a RNN

A loss function aims to quantify this discrepancy. By calculating the gradient descent of the loss function, the algorithm knows how to adjust the weights of the network. There are many different types of loss functions. Choosing the correct loss function will depend on the task. This is done at the output of the network. Because it distributes the values back through the network, this phenomenon is called backpropagation. It should be noted that we need the correct outputs in order to train the model in this way. This is called supervised learning [89].

Recurrent Neural Networks Recurrent Neural Networks (RNNs) are a special instance of neural networks. RNNs have an input and output connection, together with recurrent connections to themselves (Fig. 4). In sentences, a phrase at the beginning of the sentence can be related to a phrase later on in the sentence. To capture these dependencies, we can unfold the RNN, which creates a sequence of RNNs that are connected (Fig. 4). This connected chain makes them especially appropriate for modeling sequences such as sentences. Each of these RNNs retrieves an object as input at each unfolded time step. The output of an RNN can be used for classification (e.g., negative or positive tweets) or even for mapping a sequence to a new sequence (e.g., translation) [90].

In an RNN, each unfolded RNN represents a layer of the neural network. Because there are multiple layers stacked upon each other, an RNN is a deep neural network. They differ from other neural networks in the sense that the hidden layers are linked to time. In theory, there is no limit to the length of a sentence. Because sentences can be long and the beginning and end of a sentence can have a dependency, information has to be kept for a fairly long time. The longer the causal dependencies between two points that influence a process, the more time steps there are in between the relevant

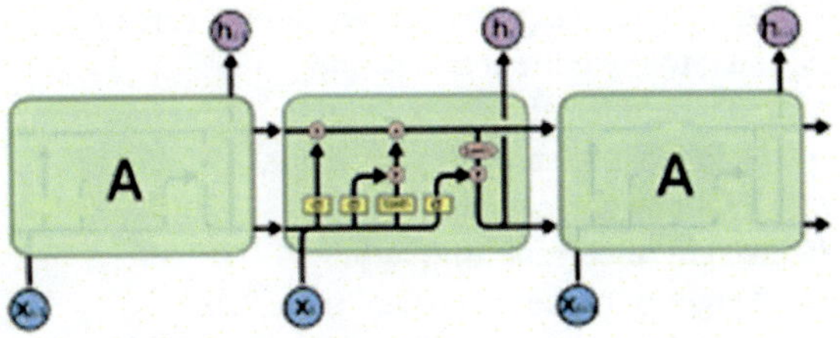

Fig. 5 The structure of an LSTM block

data. Because sentences have arbitrary length and there are no clear boundaries that limit phrases to have long-term dependencies, we cannot capture meaning properly if the model is too short.

A model that captures sentence meaning should therefore be sufficiently long. However, making an RNN longer causes different problems. Each step through a layer in the backpropagation adds another fraction to the chain rule in the gradient calculation. Thus, the longer the network, the more the gradient can be adjusted. This leads to exploding or vanishing gradients (Hochreiter (1998) [91], Bengio et al. (1994) [92]). Exploding gradients are easy to handle: you just limit the maximum value of the gradient, preventing it from exploding. Vanishing gradients are more problematic; if the gradient becomes too small, it will no longer propagate meaningfully through further layers [93].

Long Short-Term Memory Long Short-Term Memory networks (LSTMs) are an extension of RNNs (Hochreiter and Schidhuber (1997). As with RNNs, LSTMs form a chain of layers that are linked in time. In addition, LSTM blocks have feature gates that determine which information from the previous step flows through to the next step in time. This allows the network to learn when to truncate the gradient. This significantly reduces the training time needed for long-term dependencies [94].

Each LSTM network consists of three basic gates which are two gates for input gate ($input_t$) and output gate ($output_t$), and a forget gate ($forget_t$). Furthermore, an LSTM consists of a cell state and a hidden state (h_t). The input gate make a decision of which values should be updated, while the forget gate controls which part of the previous cell state (c_{t-1}) and the input should be forgotten. Based on the input and forget gate, the new cell state (c_t) is calculated. In the output gate, it is calculated which information should be handed over to the next cell, that is, it calculates the cell hidden states. Figure 5 shows the construction of an LSTM block [90].

The generalized formulas of the LSTM can be found in Eq. 3.

$$\begin{aligned}
input_t &= \sigma(W_i.[h_{t-1}, x_t] + b_i) \\
forget_t &= \sigma(W_f.[h_{t-1}, x_t] + b_f) \\
output_t &= \sigma(W_o.[h_{t-1}, x_t] + b_o) \\
c_t = forget_t * C_{t-1} &+ input_t.\tanh(W_c.[h_{t-1}, x_t] + b_c) \\
h_t &= output_t.\tanh(c_t)
\end{aligned} \tag{3}$$

Let's try to understand the equations, here
x_t: input to the current timestamp.
h_{t-1}: The hidden state of the previous timestamp.
b_f: weight associated with the input.
b_i: weight matrix of input.
b_c: Weight of current time step.
b_o: Weight of output in current time step.
W_f: It is the weight matrix associated with hidden state.
W_i: Weight matrix of input associated with hidden state.
W_c: Weight matrix of current time step.
W_o: Weight matrix of output in current time step.

4 Machine Learning for Virus Mutations Prediction

4.1 RNA Genome Mutations

Predicting methods have long been used in the study of genetics, and one of these aspects is the description of influenza virus genome sequence changes after it has infected humans from other hosts of animals. Based on the structure of a little number of oligonucleotide-based hosts, a method for anticipating changes to the virus's directive sequence and it is recommended to use non-human sources to regulate strains that could be hazardous when introduced to human populations [95].

Another direction is based on the association of mutations between 241 H5N1 hemagglutinins of the influenza A virus, according to the authors of [96] prediction, their amino acid, and RNA codon sequences were used to make the decision. Using logistic regression to combine these six independent features with the likelihood or absence of mutation as a dependent feature.

Generally, logistic regression may identify this association; but, only when just a few mutations are included in the regression, this link cannot be caught. This appears to mean that with regression analysis, only those mutations in the hemagglutinin sequence should be grouped into the first hemagglutinin sequence, preceded by the mean overall independents in a single hemagglutinin sequence, rather than all mutations in the hemagglutinin sequence with their mutations backward.

The authors of [97, 98] travel in a different route depending on the randomness. Attempting to utilize a standard neural network to determine the correlation between reason and mutation in order to predict probable mutation sites, the chance of amino acid mutation is a tool for predicting amino acids that may form in the expected places. The findings limited the ability to utilize the relationship and reason of mutations with a neural network model to predict mutation positions and the possibility of using mutant amino acids to predict amino acids that could be produced.

In [99] the 2324 genomic modules were obtained. A genomic component was considered dominant if it was found in at more than 50% of the strains in a minimum one season between 1968 and 2006, and 417 out of 2324 modules were identified as prevalent modules, it was considered a dominating module Of the predominant modules, From 1968 to 2006, 153 strain were designated as preserved because they were found primarily in all seasons. If they shared at least one or two of the same locations occupied by diverse nucleotides, the remaining 264 nonconserved prominent modules were designated as transition modules. They were able to obtain 114 transition modules. Transient modules were assigned to the remaining 39 modules. They also found the amino acid replacements depending on modules that corresponded to the nucleotide modifications in transition and transient modules for each module.

This study [100] relies on the random forest that is a machine learning approach to predict host tropism, and 11 influenza protein computational models were constructed to do so. On influenza protein sequences extracted from both avian and human samples, prediction models were trained. Using the properties of 11 proteins, the model was able to calculate the level of host distension for each influenza protein, allowing it to develop a virus-host influenza prediction model.

In [101] the other direction is based on the foundations of its ability to find hotspot combination and its association strategy in influenza evolution by developing a model capable of classification of nucleotide and protein levels pandemic and nucleotide and protein levels non-pandemic.

In [102] the alignments of primary RNA sequence-structure are projected to have potential point mutations. To predict novel strains, they used a neural network technique to predict each nucleotide's genotype in the RNA sequence and established that nucleotides in the RNA sequence had an impact on nearby nucleotide mutations in the sequence, and then used rough set theory to predict mutation patterns. Several time-series species of the Newcastle virus have been identified using aligned RNA isolates were used in this model. For model validation, two data sets from two separate sources were employed. The prediction accuracy of this study to predict nucleotides in the next generation is greater than 75%.

In [103] to predict the mutation of influenza A viruses, they suggest an efficient and prediction model of robust time-series mutation. They begin by using splittings and embeddings to create the sequential training samples. The model is capable of considering historical residue information by using recurrent neural networks (RNNs) with attention processes. By carefully focusing on sections of the residues, attention processes are increasingly being employed to enhance the performance of mutation prediction. Experiments on three influenza data sets reveal that the model outperforms previously used techniques in terms of predictive performance and provides fresh insights into viral mutation and evolution dynamics.

The combined convolution and recurrent neural networks and the combined convolution and bidirectional LSTM neural networks performed equally for different data of the genome in [104]. The performance of the prediction of mutations pipelines seems to be quite good, with the exception of a few spots (usually the first few of positions) in the data. In specifically, for positions 19610 through 19614, the two pipelines functioned admirably, maintaining accuracy of greater than 0.7. CNN-bidirectional LSTM worked somewhat better at positions 8447, 8448, and 8449, with an accuracy of 87.35%, 89.95%, and 92.69%, respectively. Both pipelines work equally for posi-

Table 3 RNA genome mutation problems

Author	Algorithm	Target
Wu and Yan [96]	Logistic regression	Possible mutations in influenza A virus H5N1 hemagglutitins prediction
Wu and Yan [97, 98]	Internal randomness	Mutations in influenza A virus H1 neuraminidases prediction
Du et al. [99]	Traditional approaches can be supplemented by a network model	Virus sequence evolution and influenza virus evolution are studied
Eng et al. [100]	Random forest	Prediction host tropism in influenza virus
Kargarfard et al. [101]	CBA algorithm	Identifying hotspot combination in influenza
Salama et al. [102]	RNN	The prediction of primary RNA sequence structure possible point mutations
Yin et al. [103]	RNN	The mutation prediction of influenza A viruses
Sawmya et al. [104]	CNN-LSTM	The similarly for different SoIs of the genome

tions 24067, 24068, and 24069, obtaining 92.33%, 87.43%, and 92.48% accuracy, respectively. CNN-bidirectional LSTM outperformed at places 23827 and 23828, reaching 84.87% and 95% accuracy, respectively, while both worked similarly at position 23829, with an accuracy of 89.75% (Table 3).

4.2 *Spike Protein Mutations*

In this study [105], they looked at how two areas in the S protein affect fusion. During transit to the plasma membrane, the 180-kDa mature S protein is partially split into two 90-kDa subunits. They've discovered a few amino acids that are crucial for S cleavage, and we show that cleavage isn't required for fusion. However, the fusion kinetics appear to be influenced by the degree of cleavage. The spike protein was fully cleaved after an arginine was introduced at position P2 to imitate the MHV-JHM cleavage site. They also looked at the impact of mutations in the S protein's transmembrane (TM) region. The mutant proteins' maturation and cell surface expression were unaffected, and they all got acylated. The mutant with the shorter anticipated transmembrane domain did not cause syncytia. One mutant failed to cause syncytia, another had delayed syncytia development, and the third mutant produced syncytia in about the same manner as the wild-type protein. The transmembrane domain's potential significance infusion is considered.

In this study [106], they used random analysis to look at mutations in two types of spike proteins from the human coronavirus strains 229E and OC43 impact amino acid pairings to get insight into probable alterations in the SARS-CoV spike protein. The findings show that the mutations are more vulnerable to amino acid pairings that are randomly unpredictable. The greater the discrepancy between actual and projected rates, the more likely it is that a mutation will occur. Mutations have the effect of narrowing the gap between actual and expected frequencies. Amino acid pairings are probable to be targeted by mutations whose targeted frequencies exceed their output frequencies, whereas amino acid pairings whose targeted frequencies are lower than expected are more probably to arise following mutations.

In this study [107], several sequences Alignment of COVID-19 spike protein sequences from the United States of America previously revealed numerous mutations at a few common places, despite the fact that some of the protein's components were stable. List of mutations observed in isolates of USA. Although Few mutations occurred more often in the spike protein sequence, others were scattered at diverse locations. The mutation D to G at position 614 was found in 99 of the isolates, indicating that it is a relatively common mutation.

In this [108] multiple sequence alignment was used to compare human spike protein sequences from Europe, Africa, Asia, South America, North America, and Oceania to the reference SARS-CoV-2 protein sequence from Wuhan-Hu-1, China. 8155 proteins out of 10,333 sequences of spike protein tested had one or more mutations. A total of 9654 mutations were discovered, with 400 different mutation sites. The receptor-binding domain (RBD), which interacts with the human enzyme-2 (ACE-2) receptor and induces infection that leads to COVID-19 disease, had 44 mutations, with residues within 3.2 interacting distance from the ACE-2 receptor. The mutations found in spike proteins are examined in terms of their geographic distribution, mutation types, number of mutations at glycosylation sites, and mutation sites.

In this study [109], Protein-protein docking and binding free energy calculations were used to assess the binding effect of these RBD variants to ACE2. A pan-proteomic study revealed 113 mutations, 33 of which are sparse. Positive selection bias is found in five RBD variants: V483A, V367F, G476S, A348T, and S494P, according to evolutionary research. The ACE2 binding affinity changes when these locations change. When compared to the Wuhan SARS-CoV2 spike protein, the A348T, V483A, and G476S variants had a lower affinity for ACE2. While the S494P and V367F population variants have a higher binding affinity for human ACE2, the S494P and V367F population variants have a lower binding affinity for human ACE2. Reorientation of many critical residues at the RBD-ACE2 interface allows the V367F variation to generate more hydrogen bonds, increasing the binding energy during ACE2 recognition. The increased binding affinity of S494P, on the other hand, is attributed to significant interfacial complementarity between the RBD and ACE2.

In this study [110], using bioinformatics approaches results that Spike protein's S1 sub-unit was found with greater sequence specificity. Bioinformatics study led to the selection of three immunodominant segments at the S1 sub-unit: Spike56-94, Spike199-264, and Spike577-612. On spike protein, glycosylation sites and mutation

Table 4 Spike protein mutation problems

Author	Algorithm	Target
Bos et al. [105]	Site-directed mutagenesis	They looked at how two areas in the S protein affect fusion
Wu and Yan [106]	Mathematical calculations	Random analysis was used to determine whether amino acid pairs in two spike proteins from 229E and OC43 human coronaviruse were altered by mutations
Banerjee et al. [107]	Multiple sequences alignment	Revealing many mutations in a few common places, despite the fact that some of the protein's portions were found to be conserved
Guruprasad [108]	Multiple sequence alignment	The human spike protein sequences Mutations analysis by comparing with the reference from Wuhan-Hu-1, China
Chakraborty [109]	Binding free energy calculations	The binding efficacy of these RBD variants to the ACE2 analyzing
Zhuang et al. [110]	Bioinformatics approaches	Spike protein's S1 subunit was found with greater sequence specificity

sites with high-frequency were avoided in the antigen design. Surface accessibility, hydrophilicity, and antigenicity are all present in all of the fragments found. Finally, a length of recombinant antigen from 194 amino acids was created, containing the immunodominant regions as well as a universal Th epitope (Table 4).

4.3 *The Machine Learning Role with the Novel Corona Virus*

During recent events generally in the world. Scientists around the world continue in search of new technologies to assist in processing the Covid-19 pandemic [111–113]. Prove for Machine Learning (ML) and Artificial Intelligence Application (AI) [114–116] on the past epidemic empowers analysts to donate a modern point to battling the novel Coronavirus flare-up [117].

Support Vector Machine (SVM), logistic regression, and clustering are examples of machine learning algorithms. Machine learning is a field of artificial intelligence that relies on how the system learns from prior data, sees patterns, and makes judgments with minimum human intervention. Deep learning (DL) is a subset of machine learning that may be described as a computational model [118] that consists of numerous processing layers for learning to represent data and extract characteristics with multi-level abstraction, as well as in terms of medical images [119].

The [120] has suggested a deep learning-based technique called deep transfer learning that can automatically predict individuals with Coronavirus disease. It makes use of chest X-ray images from both coronavirus sufferers and healthy people. They also mentioned that we are sure in our conclusions, this can help workers in the field of public health care decision-making in clinical practice because of its performance and high accuracy. Also, for coronavirus patients, early prediction of infection can avoid the rapid spread of disease. Deep learning neural network plays an important role in this epidemic.

In [121] have utilized deep neural networks such as long short term memory (LSTM) for the possible time prediction for ending the pandemic of coronavirus in CANADA. They estimate On the basis of their LSTM model that the time it will take to end this epidemic is around three months.

In [122] genetic programming-based models have been developed to predict a total number of known diseases and deaths in hard-hit India's countries as well as the entire country. They have detailed that the model is less delicate to the variables but it is exceedingly dependable within the confirmed cases and deaths prediction.

In [123] the generalized adaptive model was utilised for the associations of the average relative humidity and daily temperature with COVID-19 cases daily in China's various provinces. Their research discovered that as the average daily temperature rises and the relative humidity average rises, the cases of COVID-19 decrease. However, in the main region of China, there are almost no apparent statistics for cases of COVID-19.

In conducting a study [124] compared on the basis of models ARIMA and LSTM neural networks nonlinear (NARNN) for cases COVID-19 in Belgium, Denmark, France, Germany, Turkey, Switzerland, Finland, and the United Kingdom prediction. They Conclude that the LSTM results are the lowest Root Mean Square Error (RMSE) from other models.

In [125] they sought to predict COVID-19 pandemic mortality cases that have been confirmed, publicized, and found to be negative using LSTM, Recurrent Neural Networks (RNN), Gated Recurrent Unit (GRU). Their study proved that the combination of LSTM and the RNN model excelled in some other independent models in terms of prediction.

In [126] they have used curve fitting and LSTM to predict the COVID-19 cases number in India for 30 days ahead, On the spread of COVID-19, the effect of the preventive measure like lockdowns and social alienation. Their results clear the social distancing and implementation of lockdowns importance. Actually, the daily cases number reduced as the Fitting Methods Curve and LSTM expectation.

In [127] they utilized Recurrent neural networks especially LSTM neural network, Seasonal Autoregressive Integrated Moving Average (SARIMA), and Holt's exponential smoothing in the winter and moving average strategies to predict Iran COVID-19 cases Their study proved that the LSTM model results in less error value for Iran infection development from other models.

Another study [128] employed deep learning for predicting age-related macular degeneration (AMD), which is the major cause of blindness in the elderly. The average area under the curve (AUC) value is 0.85 in the results. On the other hand, [129] employed a prediction approach based on deep learning to predict disease-associated mutations of metal-binding regions in proteins. The accuracy of the prediction is 0.82 and the AUC is 0.90 (Table 5).

Table 5 Machine learning in novel Corona virus problems

Author	Machine learning algorithm	Problem
Apostolopoulos and Mpesiana [120]	Deep transfer learning	Prediction of patients with Coronavirus disease automatically
Chimmula and Zhang [121]	Long short term memory neural networks	The possible time prediction for ending the pandemic in CANADA
Salgotra et al. [122]	Genetic programming algorithms	Prediction of known diseases and deaths numbers in hard-hit countries of India and the entire country
Qi et al. [123]	Introduced new adaptive generalized model	The associations of the average of relative humidity and daily temperature with COVID-19 cases on a daily basis in China's various provinces
Kırbaş et al. [124]	The basis of models ARIMA and LSTM neural networks nonlinear (NARNN)	COVID-19 cases in Finland, Belgium, Germany, Turkey, the United Kingdom, Denmark, Switzerland and France prediction
Bandyopadhyay and Dutta [125]	LSTM, grated recurrent unit (GRU), recurrent neural networks (RNN)	They sought to predict the confirmed, released, deaths and found to be negative cases of COVID-19 pandemic
Tomar and Gupta [126]	They have used curve fitting and LSTM	Prediction of COVID-19 cases number in India for the next 30 days and the precautionary measures effects like lockdowns and social alienation on COVID-19 spread
Azarafza et al. [127]	RNN, LSTM, average of seasonal autoregressive integrated moving, Holt's exponential smoothing in the winter and the average of moving strategies	Prediction of Iran COVID-19 cases
Koohi-Moghadam et al. [129]	Introduced a deep learning approach	Prediction of disease associated mutations of metal-binding regions in proteins

5 Open Issues and Challenges

There are several challenges facing the work in this research area including:

1. Using SHAPE and sequencing, structural investigation of viral RNAs at greater sizes has just been possible. The interpretation of this data necessitates both detailed processing of the raw SHAPE data and the insertion of these data as constraints into RNA structure prediction algorithms.
2. The machine learning models treat the RNA sequence as a string of four different stages, and so watch nucleotide changes as the sequence progresses. The assumption in these studies was that the sequence's different nucleotides autonomously and indistinguishably, which was deserved using the neutral nucleotide evolution scenario.
3. These are black-box techniques, meaning they can't be utilized to deduce any information about the rules that were used to make the prediction.
4. As representation for twenty separate amino acid condons, these techniques use 20-bit binary integers rather than characters. This raises the input size to the utilized algorithm, which increases the prediction process's complexity.
5. The prediction of single nucleotide variations (SNV) is the third study direction; while this strategy is correct for a huge number of sequences data, due to a low number of sequences obtained, it is not always the case. Furthermore, it just forecasts the occurrence of the SNV in the future, not the sequence.
6. The size of every nucleotide will be scaled to four binary bits. However, converting letters to numbers, such as The nucleotide codes of [A, C, G, T] to [0, 1, 2, 3]. Since the distance of every two nucleotides is the same, but not between [0–3] and [0–1].

6 Conclusion and Future Works

A review of RNA Viruses Mutation problems is provided in this chapter, an introduction about RNA Viruses and Machine Learning algorithms, and their features. This will give researchers a summary of the RNA viruses that are pointing them in the right direction. Further, the chapter reviews the usage of machine learning in Human RNA Viruses, and machine learning algorithms in solving mutation problems. Besides previously discussed the RNA Viruses evolution Mutations are still being researched as a discipline, and new techniques for handling the problems of mutations are being created.

There is a lot of scope for Machine Learning in Healthcare. For Future work, it is recommended to work on calibrated and ensemble methods that could resolve quirky problems faster with better outcomes than the existing algorithms. Also an AI-based application can be developed using various sensors and features to identify and help diagnose diseases. As healthcare prediction is an essential field for future, A prediction system that could find the possibility of outbreak of novel diseases that could harm mankind through socio-economic and cultural factor consideration can be developed.

Conflict of Interest The authors declare that there is no conflict of interest.

References

1. T. Mohamed, S. Sayed, A. Salah, E.H. Houssein, Long short-term memory neural networks for RNA viruses mutations prediction. Math. Probl. Eng. **2021** (2021)
2. S. Muradrasoli, Detection and quantification of variable viral RNA by real-time PCR assays. Ph.D. Dissertation, Acta Universitatis Upsaliensis (2008)
3. M. Ereshefsky, Names, numbers and indentations: a guide to post-Linnaean taxonomy. Stud. Hist. Philos. Sci. Part C: Stud. Hist. Philos. Biol. Biomed. Sci. **32**(2), 361–383 (2001)
4. V.K. Pathak, H.M. Temin, Broad spectrum of in vivo forward mutations, hypermutations, and mutational hotspots in a retroviral shuttle vector after a single replication cycle: substitutions, frameshifts, and hypermutations. Proc. Natl. Acad. Sci. **87**(16), 6019–6023 (1990)
5. G. Dahourou, S. Guillot, O. Le Gall, R. Crainic, Genetic recombination in wild-type poliovirus. J. Gen. Virol. **83**(12), 3103–3110 (2002)
6. G.C. Sen, S.N. Sarkar, Transcriptional signaling by double-stranded RNA: role of TLR3. Cytokine Growth Factor Rev. **16**(1), 1–14 (2005)
7. L. Contu, G. Balistreri, M. Domanski, A.-C. Uldry, O. Mühlemann, Characterisation of the Semliki forest virus-host cell interactome reveals the viral capsid protein as an inhibitor of nonsense-mediated mRNA decay. PLoS Pathog. **17**(5), e1009603 (2021)
8. W. Fontana, D.A. Konings, P.F. Stadler, P. Schuster, Statistics of RNA secondary structures. Biopolym. Orig. Res. Biomol. **33**(9), 1389–1404 (1993)
9. I.L. Hofacker, W. Fontana, P.F. Stadler, L.S. Bonhoeffer, M. Tacker, P. Schuster, Fast folding and comparison of RNA secondary structures. Monatshefte Chem. - Chem. Mon. **125**(2), 167–188 (1994)
10. J. Pelletier, N. Sonenberg, Internal initiation of translation of eukaryotic mRNA directed by a sequence derived from poliovirus RNA. Nature **334**(6180), 320–325 (1988)
11. M. Vallejos, P. Ramdohr, F. Valiente-Echeverría, K. Tapia, F.E. Rodriguez, F. Lowy, J.P. Huidobro-Toro, J.A. Dangerfield, M. Lopez Lastra, The 5-untranslated region of the mouse mammary tumor virus mRNA exhibits cap-independent translation initiation. Nucleic Acids Res. **38**(2), 618–632 (2010)
12. A. Casadevall, L.-A. Pirofski, Host-pathogen interactions: redefining the basic concepts of virulence and pathogenicity. Infect. Immun. **67**(8), 3703–3713 (1999)
13. T.E. Love, B. Jones, Introduction to pathogenic bacteria, in *Principles of Bacterial Detection: Biosensors, Recognition Receptors and Microsystems* (Springer, 2008), pp. 3–13
14. S. Baron, M. Fons, T. Albrecht, Viral pathogenesis, in *Medical Microbiology*, 4th edn. (1996)
15. E.O. Freed, Mechanisms of enveloped virus release. Virus Res. **106**(2), 85–86 (2004)
16. M.A. Barocchi, V. Masignani, R. Rappuoli, Cell entry machines: a common theme in nature? Nat. Rev. Microbiol. **3**(4), 349–358 (2005)
17. M. Bomsel, A. Alfsen, Entry of viruses through the epithelial barrier: pathogenic trickery. Nat. Rev. Mol. Cell Biol. **4**(1), 57–68 (2003)
18. D.M. Knipe, P.M. Howley et al., *Fundamental Virology*, 4th edn. (Lippincott Williams & Wilkins, 2001)
19. P. Jern, G.O. Sperber, J. Blomberg, Use of endogenous retroviral sequences (ERVs) and structural markers for retroviral phylogenetic inference and taxonomy. Retrovirology **2**(1), 1–12 (2005)
20. M.R. Hilleman, Realities and enigmas of human viral influenza: pathogenesis, epidemiology and control. Vaccine **20**(25–26), 3068–3087 (2002)

21. R.G. Webster, W.J. Bean, O.T. Gorman, T.M. Chambers, Y. Kawaoka, Evolution and ecology of influenza a viruses. Microbiol. Rev. **56**(1), 152–179 (1992)
22. R.A. Fouchier, V. Munster, A. Wallensten, T.M. Bestebroer, S. Herfst, D. Smith, G.F. Rimmelzwaan, B. Olsen, A.D. Osterhaus, Characterization of a novel influenza a virus hemagglutinin subtype (H16) obtained from black-headed gulls. J. Virol. **79**(5), 2814–2822 (2005)
23. M. Yuan, D. Huang, C.-C.D. Lee, N.C. Wu, A.M. Jackson, X. Zhu, H. Liu, L. Peng, M.J. van Gils, R.W. Sanders et al., Structural and functional ramifications of antigenic drift in recent SARS-CoV-2 variants. Science (2021)
24. M.L. Perdue, M. Garcıa, D. Senne, M. Fraire, Virulence-associated sequence duplication at the hemagglutinin cleavage site of avian influenza viruses. Virus Res. **49**(2), 173–186 (1997)
25. R.A. Weinstein, C.B. Bridges, M.J. Kuehnert, C.B. Hall, Transmission of influenza: implications for control in health care settings. Clin. Infect. Dis. **37**(8), 1094–1101 (2003)
26. J.K. Taubenberger, A.H. Reid, T.A. Janczewski, T.G. Fanning, Integrating historical, clinical and molecular genetic data in order to explain the origin and virulence of the 1918 Spanish influenza virus. Philos. Trans. R. Soc. Lond. Ser. B: Biol. Sci. **356**(1416), 1829–1839 (2001)
27. H. Goto, T. Tanaka, K. Tobita, Comparison of nine strains of influenza C virus in growth characteristics and viral polypeptides. Adv. Virol. **82**(1–2), 111–117 (1984)
28. D.J. Weber, W.A. Rutala, W.A. Fischer, H. Kanamori, E.E. Sickbert-Bennett, Emerging infectious diseases: focus on infection control issues for novel coronaviruses (severe acute respiratory syndrome-CoV and middle east respiratory syndrome-CoV), hemorrhagic fever viruses (Lassa and Ebola), and highly pathogenic avian influenza viruses, A(H5N1) and A(H7N9). Am. J. Infect. Control **44**(5), e91–e100 (2016)
29. S. Katagiri, A. Ohizumi, S. Ohyama, M. Homma, Follow-up study of type C influenza outbreak in a children's home. Microbiol. Immunol. **31**(4), 337–343 (1987)
30. O. de Leeuw, B. Peeters, Complete nucleotide sequence of newcastle disease virus: evidence for the existence of a new genus within the subfamily paramyxovirinae. J. Gen. Virol. **80**(1), 131–136 (1999)
31. D. Kolakofsky, L. Roux, D. Garcin, R.W. Ruigrok, Paramyxovirus mRNA editing, the 'rule of six' and error catastrophe: a hypothesis. J. Gen. Virol. **86**(7), 1869–1877 (2005)
32. R.A. Lamb, Paramyxoviridae: the viruses and their replication. Fields Virol. (2001)
33. M. Steward, I.B. Vipond, N.S. Millar, P.T. Emmerson, RNA editing in newcastle disease virus. J. Gen. Virol. **74**(12), 2539–2547 (1993)
34. M. Collins, J. Bashiruddin, D. Alexander, Deduced amino acid sequences at the fusion protein cleavage site of newcastle disease viruses showing variation in antigenicity and pathogenicity. Adv. Virol. **128**(3–4), 363–370 (1993)
35. B. Lomniczi, E. Wehmann, J. Herczeg, A. Ballagi-Pordany, E. Kaleta, O. Werner, G. Meulemans, P. Jorgensen, A. Mante, A. Gielkens et al., Newcastle disease outbreaks in recent years in western Europe were caused by an old (VI) and a novel genotype (VII). Adv. Virol. **143**(1), 49–64 (1998)
36. E. Aldous, J. Mynn, J. Banks, D. Alexander et al., A molecular epidemiological study of avian paramyxovirus type 1 (newcastle disease virus) isolates by phylogenetic analysis of a partial nucleotide sequence of the fusion protein gene. Avian Pathol. **32**(3), 237–255 (2003)
37. M. Munir, A.-M. Linde, S. Zohari, K. Ståhl, C. Baule, B. Engström, L.H. Renström, M. Berg, Whole genome sequencing and characterization of a virulent newcastle disease virus isolated from an outbreak in Sweden. Virus Genes. **43**(2), 261–271 (2011)
38. M. Munir, M. Abbas, M.T. Khan, S. Zohari, M. Berg, Genomic and biological characterization of a velogenic newcastle disease virus isolated from a healthy backyard poultry flock in 2010. Virol. J. **9**(1), 1–11 (2012)
39. T.A. Khan, C.A. Rue, S.F. Rehmani, A. Ahmed, J.L. Wasilenko, P.J. Miller, C.L. Afonso, Phylogenetic and biological characterization of newcastle disease virus isolates from Pakistan. J. Clin. Microbiol. **48**(5), 1892–1894 (2010)
40. P. Luciw, *Fundamental Virology* (1996)
41. F. Gao, E. Bailes, D.L. Robertson, Y. Chen, C.M. Rodenburg, S.F. Michael, L.B. Cummins, L.O. Arthur, M. Peeters, G.M. Shaw et al., Origin of HIV-1 in the chimpanzee pan troglodytes troglodytes. Nature **397**(6718), 436–441 (1999)

42. N.R. Faria, A. Rambaut, M.A. Suchard, G. Baele, T. Bedford, M.J. Ward, A.J. Tatem, J.D. Sousa, N. Arinaminpathy, J. Pépin et al., The early spread and epidemic ignition of HIV-1 in human populations. Science **346**(6205), 56–61 (2014)
43. P.K. Mozhi, D. Ganapathy, Awareness of structural biology of HIV among dental students. Eur. J. Mol. Clin. Med. **8**(1), 491–503 (2021)
44. E. Vicenzi, G. Poli, Novel factors interfering with human immunodeficiency virus-type 1 replication in vivo and in vitro. Tissue Antigens **81**(2), 61–71 (2013)
45. C. Kuiken, B. Foley, B. Hahn, P. Marx, F. McCutchan, J. Mellors, S. Wolinsky, B. Korber, *HIV Sequence Compendium 2001* (Theoretical Biology and Biophysics Group, Los Alamos National Laboratory, Los Alamos, NM, 2001)
46. A. Sapokta, Structure and genome of SARS-CoV-2 (Covid-19) with diagram. Microbe Notes, http://www.microbenotes.com/structure-and-genome-of-sars-cov-2 (2020)
47. D. Schoeman, B.C. Fielding, Coronavirus envelope protein: current knowledge. Virol. J. **16**(1), 1–22 (2019)
48. M. Cascella, M. Rajnik, A. Aleem, S. Dulebohn, R. Di Napoli, Features, evaluation, and treatment of coronavirus (Covid-19). StatPearls (2021)
49. I. Astuti et al., Severe acute respiratory syndrome coronavirus 2 (SARS-CoV-2): an overview of viral structure and host response. Diabetes Metab. Syndr. Clin. Res. Rev. **14**(4), 407–412 (2020)
50. H. Xu, L. Zhong, J. Deng, J. Peng, H. Dan, X. Zeng, T. Li, Q. Chen, High expression of ACE2 receptor of 2019-nCoV on the epithelial cells of oral mucosa. Int. J. Oral Sci. **12**(1), 1–5 (2020)
51. D. Shriner, R. Shankarappa, M.A. Jensen, D.C. Nickle, J.E. Mittler, J.B. Margolick, J.I. Mullins, Influence of random genetic drift on human immunodeficiency virus type 1 env evolution during chronic infection. Genetics **166**(3), 1155–1164 (2004)
52. S.L. Rutherford, From genotype to phenotype: buffering mechanisms and the storage of genetic information. BioEssays **22**(12), 1095–1105 (2000)
53. R. Sanjuán, M.R. Nebot, N. Chirico, L.M. Mansky, R. Belshaw, Viral mutation rates. J. Virol. **84**(19), 9733–9748 (2010)
54. A.S. Perelson, Modelling viral and immune system dynamics. Nat. Rev. Immunol. **2**(1), 28–36 (2002)
55. J.-M. Pawlotsky, Hepatitis C virus resistance to direct-acting antiviral drugs in interferon-free regimens. Gastroenterology **151**(1), 70–86 (2016)
56. N.J. Vickers, Animal communication: when i'm calling you, will you answer too? Curr. Biol. **27**(14), R713–R715 (2017)
57. N. Coppola, L. Onorato, C. Minichini, G. Di Caprio, M. Starace, C. Sagnelli, E. Sagnelli, Clinical significance of hepatitis B surface antigen mutants. World J. Hepatol. **7**(27), 2729–2739 (2015)
58. M. Schotsaert, A. García-Sastre, Influenza vaccines: a moving interdisciplinary field. Viruses **6**(10), 3809–3826 (2014)
59. R. Sanjuán, P. Domingo-Calap, Mechanisms of viral mutation. Cell. Mol. Life Sci. **73**(23), 4433–4448 (2016)
60. D.A. Steinhauer, E. Domingo, J.J. Holland, Lack of evidence for proofreading mechanisms associated with an RNA virus polymerase. Gene **122**(2), 281–288 (1992)
61. E.C. Smith, N.R. Sexton, M.R. Denison, Thinking outside the triangle: replication fidelity of the largest RNA viruses. Ann. Rev. Virol. **1**, 111–132 (2014)
62. L. Menéndez-Arias, Mutation rates and intrinsic fidelity of retroviral reverse transcriptases. Viruses **1**(3), 1137–1165 (2009)
63. R. Biek, O.G. Pybus, J.O. Lloyd-Smith, X. Didelot, Measurably evolving pathogens in the genomic era. Trends Ecol. Evol. **30**(6), 306–313 (2015)
64. N. Renzette, C. Pokalyuk, L. Gibson, B. Bhattacharjee, M.R. Schleiss, K. Hamprecht, A.Y. Yamamoto, M.M. Mussi-Pinhata, W.J. Britt, J.D. Jensen et al., Limits and patterns of cytomegalovirus genomic diversity in humans. Proc. Natl. Acad. Sci. **112**(30), E4120–E4128 (2015)

65. S. Seronello, J. Montanez, K. Presleigh, M. Barlow, S.B. Park, J. Choi, Ethanol and reactive species increase basal sequence heterogeneity of hepatitis C virus and produce variants with reduced susceptibility to antivirals. PLoS ONE **6**(11), e27436 (2011)
66. J. Jiricny, Postreplicative mismatch repair. Cold Spring Harb. Perspect. Biol. **5**(4), a012633 (2013)
67. J.M. Cuevas, S. Duffy, R. Sanjuán, Point mutation rate of bacteriophage ΦX174. Genetics **183**(2), 747–749 (2009)
68. P. Deschavanne, M. Radman, Counterselection of GATC sequences in enterobacteriophages by the components of the methyl-directed mismatch repair system. J. Mol. Evol. **33**(2), 125–132 (1991)
69. M.A. Luftig, Viruses and the DNA damage response: activation and antagonism. Ann. Rev. Virol. **1**, 605–625 (2014)
70. A.N. Blackford, R.N. Patel, N.A. Forrester, K. Theil, P. Groitl, G.S. Stewart, A.M.R. Taylor, I.M. Morgan, T. Dobner, R.J. Grand et al., Adenovirus 12 E4orf6 inhibits ATR activation by promoting TOPBP1 degradation. Proc. Natl. Acad. Sci. **107**(27), 12,251–12,256 (2010)
71. D. Oliva, E.H. Houssein, S. Hinojosa, *Metaheuristics in Machine Learning: Theory and Applications*
72. E.H. Houssein, M. Dirar, K. Hussain, W.M. Mohamed, Assess deep learning models for Egyptian exchange prediction using nonlinear artificial neural networks. Neural Comput. Appl. **33**(11), 5965–5987 (2021)
73. E.H. Houssein, M.M. Emam, A.A. Ali, P.N. Suganthan, Deep and machine learning techniques for medical imaging-based breast cancer: a comprehensive review. Expert Syst. Appl. 114161 (2020)
74. M. Mohri, A. Rostamizadeh, A. Talwalkar, *Foundations of Machine Learning* (2012)
75. T.-M. Huang, V. Kecman, I. Kopriva, *Kernel Based Algorithms for Mining Huge Data Sets*, vol. 1 (Springer, 2006)
76. S. Dreiseitl, L. Ohno-Machado, Logistic regression and artificial neural network classification models: a methodology review. J. Biomed. Inform. **35**(5–6), 352–359 (2002)
77. F. Pedregosa, G. Varoquaux, A. Gramfort, V. Michel, B. Thirion, O. Grisel, M. Blondel, P. Prettenhofer, R. Weiss, V. Dubourg et al., Scikit-learn: machine learning in Python. J. Mach. Learn. Res. **12**, 2825–2830 (2011)
78. L. Rokach, O. Maimon, Decision trees, in *Data Mining and Knowledge Discovery Handbook* (Springer, 2005), pp. 165–192
79. S. Boisvert, J.W. Sheppard, Quality diversity genetic programming for learning decision tree ensembles, in *EuroGP* (2021), pp. 3–18
80. P. Dell'Aversana, Artificial neural networks and deep learning. A simple overview (2019)
81. R.C. Eberhart, *Neural Network PC Tools: A Practical Guide* (Academic Press, 2014)
82. A. Honkela et al., Nonlinear switching state-space models. Master's Thesis, 2001
83. M. Puig-Arnavat, J.C. Bruno, Artificial neural networks for thermochemical conversion of biomass, in *Recent Advances in Thermo-Chemical Conversion of Biomass* (Elsevier, 2015), pp. 133–156
84. S.S. Haykin et al., Neural networks and learning machines/Simon Haykin (2009)
85. S. Al-Dabet, S. Tedmori, A.-S. Mohammad, Enhancing Arabic aspect-based sentiment analysis using deep learning models. Comput. Speech Lang. **69**, 101224 (2021)
86. D. Svozil, V. Kvasnicka, J. Pospichal, Introduction to multi-layer feed-forward neural networks. Chemom. Intell. Lab. Syst. **39**(1), 43–62 (1997)
87. G. Katz, C. Barrett, D.L. Dill, K. Julian, M.J. Kochenderfer, Reluplex: an efficient SMT solver for verifying deep neural networks, in *International Conference on Computer Aided Verification* (Springer, 2017), pp. 97–117
88. C.T. Shine, T.T.S. Nyunt, Feature selection and map reduce-based neural network classification for big data. Ph.D. Dissertation, University of Computer Studies, Yangon, 2018
89. P. Christoffersen, K. Jacobs, The importance of the loss function in option valuation. J. Financ. Econ. **72**(2), 291–318 (2004)

90. C. Olah, Understanding LSTM networks, August 2015, https://colah.github.io/posts/2015-08-Understanding-LSTMs (2018)
91. S. Hochreiter, The vanishing gradient problem during learning recurrent neural nets and problem solutions. Int. J. Uncertain. Fuzziness Knowl. Based Syst. **6**(02), 107–116 (1998)
92. Y. Bengio, P. Simard, P. Frasconi, Learning long-term dependencies with gradient descent is difficult. IEEE Trans. Neural Netw. **5**(2), 157–166 (1994)
93. A. Sherstinsky, Fundamentals of recurrent neural network (RNN) and long short-term memory (LSTM) network. Phys. D **404**, 132306 (2020)
94. S. Hochreiter, J. Schmidhuber, Long short-term memory. Neural Comput. **9**(8), 1735–1780 (1997)
95. Y. Iwasaki, T. Abe, Y. Wada, K. Wada, T. Ikemura, Novel bioinformatics strategies for prediction of directional sequence changes in influenza virus genomes and for surveillance of potentially hazardous strains. BMC Infect. Dis. **13**(1), 1–9 (2013)
96. G. Wu, S. Yan, Prediction of possible mutations in H5N1 hemagglutitins of influenza a virus by means of logistic regression. Comp. Clin. Pathol. **15**(4), 255–261 (2006)
97. G. Wu, S. Yan, Prediction of mutations engineered by randomness in H5N1 hemagglutinins of influenza a virus. Amino Acids **35**(2), 365–373 (2008)
98. G. Wu, S. Yan, Prediction of mutations in H1 neuraminidases from North America influenza a virus engineered by internal randomness. Mol. Divers. **11**(3), 131–140 (2007)
99. X. Du, Z. Wang, A. Wu, L. Song, Y. Cao, H. Hang, T. Jiang, Networks of genomic co-occurrence capture characteristics of human influenza a (H3N2) evolution. Genome Res. **18**(1), 178–187 (2008)
100. C.L. Eng, J.C. Tong, T.W. Tan, Predicting host tropism of influenza a virus proteins using random forest. BMC Med. Genomics **7**(3), 1–11 (2014)
101. F. Kargarfard, A. Sami, E. Ebrahimie, Knowledge discovery and sequence-based prediction of pandemic influenza using an integrated classification and association rule mining (cba) algorithm. J. Biomed. Inform. **57**, 181–188 (2015)
102. M.A. Salama, A.E. Hassanien, A. Mostafa, The prediction of virus mutation using neural networks and rough set techniques. EURASIP J. Bioinf. Syst. Biol. **2016**(1), 1–11 (2016)
103. R. Yin, E. Luusua, J. Dabrowski, Y. Zhang, C.K. Kwoh, Tempel: time-series mutation prediction of influenza a viruses via attention-based recurrent neural networks. Bioinformatics **36**(9), 2697–2704 (2020)
104. S. Sawmya, A. Saha, S. Tasnim, M. Toufikuzzaman, N. Anjum, A.H.M. Rafid, M.S. Rahman, M.S. Rahman, Analyzing hCov genome sequences: predicting virulence and mutation. *bioRxiv*, https://doi.org/10.1101/2020.06.03.131987 (2021)
105. E.C. Bos, L. Heijnen, W. Luytjes, W.J. Spaan, Mutational analysis of the murine coronavirus spike protein: effect on cell-to-cell fusion. Virology **214**(2), 453–463 (1995)
106. G. Wu, S. Yan, Prediction of amino acid pairs sensitive to mutations in the spike protein from SARS related coronavirus. Peptides **24**(12), 1837–1845 (2003)
107. A.K. Banerjee, F. Begum, U. Ray, Mutation hot spots in spike protein of Covid-19. Preprints **2020**, 2020040281 (2020)
108. L. Guruprasad, Human SARS CoV-2 spike protein mutations. Proteins Struct. Funct. Bioinform. **89**(5), 569–576 (2021)
109. S. Chakraborty, Evolutionary and structural analysis elucidates mutations on SARS-CoV2 spike protein with altered human ACE2 binding affinity. Biochem. Biophys. Res. Commun. **538**, 97–103 (2021)
110. S. Zhuang, L. Tang, Y. Dai, X. Feng, Y. Fang, H. Tang, P. Jiang, X. Wu, H. Fang, H. Chen, Bioinformatic prediction of immunodominant regions in spike protein for early diagnosis of the severe acute respiratory syndrome coronavirus 2 (SARS-CoV-2). PeerJ **9**, e11232 (2021)
111. E.H. Houssein, M.M. Emam, A.A. Ali, Improved manta ray foraging optimization for multi-level thresholding using Covid-19 *ct* images. Neural Comput. Appl. 1–21 (2021)
112. D.S. Abdelminaam, F.H. Ismail, M. Taha, A. Taha, E.H. Houssein, A. Nabil, CoAID-DEEP: an optimized intelligent framework for automated detecting Covid-19 misleading information on Twitter. IEEE Access **9**, 27,840–27,867 (2021)

113. E.H. Houssein, M. Ahmad, M.E. Hosney, M. Mazzara, Classification approach for Covid-19 gene based on Harris hawks optimization, in *Artificial Intelligence for COVID-19* (Springer, 2021), pp. 575–594
114. E.H. Houssein, D.S. Abdelminaam, H.N. Hassan, M.M. Al-Sayed, E. Nabil, A hybrid barnacles mating optimizer algorithm with support vector machines for gene selection of microarray cancer classification. IEEE Access **9**, 64,895–64,905 (2021)
115. Y.M. Wazery, E. Saber, E.H. Houssein, A.A. Ali, E. Amer, An efficient slime mould algorithm combined with k-nearest neighbor for medical classification tasks. IEEE Access (2021)
116. E.H. Houssein, D.S. AbdElminaam, I.E. Ibrahim, M. Hassaballah, Y.M. Wazery, A hybrid heartbeats classification approach based on marine predators algorithm and convolution neural networks. IEEE Access (2021)
117. S. Lalmuanawma, J. Hussain, L. Chhakchhuak, Applications of machine learning and artificial intelligence for Covid-19 (SARS-CoV-2) pandemic: a review. Chaos Solitons Fract. **139**, 110059 (2020)
118. A. Hamad, E.H. Houssein, A.E. Hassanien, A.A. Fahmy, Hybrid grasshopper optimization algorithm and support vector machines for automatic seizure detection in EEG signals, in *International Conference on Advanced Machine Learning Technologies and Applications* (Springer, 2018), pp. 82–91
119. F.A. Hashim, E.H. Houssein, K. Hussain, M.S. Mabrouk, W. Al-Atabany, A modified henry gas solubility optimization for solving motif discovery problem. Neural Comput. Appl. **32**(14), 10,759–10,771 (2020)
120. I.D. Apostolopoulos, T.A. Mpesiana, Covid-19: automatic detection from X-ray images utilizing transfer learning with convolutional neural networks. Phys. Eng. Sci. Med. **43**(2), 635–640 (2020)
121. V.K.R. Chimmula, L. Zhang, Time series forecasting of Covid-19 transmission in Canada using LSTM networks. Chaos Solitons Fract. **135**, 109864 (2020)
122. R. Salgotra, M. Gandomi, A.H. Gandomi, Time series analysis and forecast of the Covid-19 pandemic in India using genetic programming. Chaos Solitons Fract. **138**, 109945 (2020)
123. H. Qi, S. Xiao, R. Shi, M.P. Ward, Y. Chen, W. Tu, Q. Su, W. Wang, X. Wang, Z. Zhang, Covid-19 transmission in mainland china is associated with temperature and humidity: a time-series analysis. Sci. Total Environ. **728**, 138778 (2020)
124. İ. Kırbaş, A. Sözen, A.D. Tuncer, F.Ş Kazancıoğlu, Comparative analysis and forecasting of Covid-19 cases in various European countries with ARIMA, NARNN and LSTM approaches. Chaos Solitons Fract. **138**, 110015 (2020)
125. S.K. Bandyopadhyay, S. Dutta, Machine learning approach for confirmation of Covid-19 cases: positive, negative, death and release, in *MedRxiv* (2020)
126. A. Tomar, N. Gupta, Prediction for the spread of Covid-19 in India and effectiveness of preventive measures. Sci. Total Environ. **728**, 138762 (2020)
127. M. Azarafza, M. Azarafza, J. Tanha, Covid-19 infection forecasting based on deep learning in Iran, in *medRxiv* (2020)
128. Q. Yan, D.E. Weeks, H. Xin, A. Swaroop, E.Y. Chew, H. Huang, Y. Ding, W. Chen, Deep-learning-based prediction of late age-related macular degeneration progression. Nat. Mach. Intell. **2**(2), 141–150 (2020)
129. M. Koohi-Moghadam, H. Wang, Y. Wang, X. Yang, H. Li, J. Wang, H. Sun, Predicting disease-associated mutation of metal-binding sites in proteins using a deep learning approach. Nat. Mach. Intell. **1**(12), 561–567 (2019)

2D Target/Anomaly Detection in Time Series Drone Images Using Deep Few-Shot Learning in Small Training Dataset

Mehdi Khoshboresh-Masouleh and Reza Shah-Hosseini

Abstract Optimization problem is an important challenge in two-dimensional (2D) target/anomaly detection as a real-world application. As manual time series drone image interpretation is time-consuming and expensive, deep learning methods are of high interest for 2D target/anomaly detection. Despite 2D target and anomaly detection from time series drone images based on deep learning models is an active field in remote sensing engineering, but annotating remote sensing time series data is costly for training step. To build robust machine learning methods in remote sensing, deep few-shot learning approaches have been developed from real-world and real-time datasets based on drone images in small training data as an optimized solution. In this chapter, we focus on two real-world applications of 2D target/anomaly detection based on a new deep few-shot learning method, which can be widely used in urban management and precision farming. The experiments are based on two time series multispectral datasets, including traffic monitoring (as a target) and weed detection (as an anomaly). Compared with the few-shot learning with different backbones, the proposed method, called SA-Net, demonstrates better performance and good generalization ability for 2D target/anomaly detection.

Keywords 2D target/anomaly · Time series drone images · Deep few-shot learning · Traffic management · Weed detection

M. Khoshboresh-Masouleh (✉) · R. Shah-Hosseini
School of Surveying and Geospatial Engineering, College of Engineering, University of Tehran, 14399-57131 Tehran, Iran
e-mail: m.khoshboresh@ut.ac.ir

R. Shah-Hosseini
e-mail: rshahosseini@ut.ac.ir

E. H. Houssein et al. (eds.), *Integrating Meta-Heuristics and Machine Learning for Real-World Optimization Problems*, Studies in Computational Intelligence 1038,
https://doi.org/10.1007/978-3-030-99079-4_10

1 Introduction

1.1 Motivation

An optimized method for natural/artificial objects monitoring in urban and rural environments is a crucial challenge in decision-making for users [1]. In recent years, drone imaging is a good solution for real-time monitoring different objects on the local scale [2]. Two-dimensional target and anomaly detection from time series drone images in the local scale are important for traffic monitoring, and weed detection [3, 4]. The purpose of traffic monitoring and weed detection from time series drone images is to locate the object (target or anomaly) on the time series images by using a bounding box or semantic segmentation [5].

While many target and anomaly detection methods appear to understand single-time image processing based on big training dataset, real-time target/anomaly detection for the portable platforms potentially improves video understanding in real-world applications in small training dataset. Real-time target/anomaly detection is more difficult due to the variety of depth and scale objects in vertical/oblique view drone images [6]. Motivated by these facts, in this chapter, we consider the challenge of real-time target/anomaly detection for time series drone images based on small training dataset.

1.2 Vertical and Oblique Views in Time Series Drone Images

Lyu et al. 2020 [7] introduced a new oblique view drone dataset[1] (also known as UAVid) for real-time semantic segmentation, which brings good points of view, including temporal consistency preservation, large-scale object detection, and moving target segmentation. In this study, the mean Intersection over Union (mIoU) is about 35.6\%, 39.1\%, 40.9\%, and 42\% for fully convolutional network-8 s, Dilation-Net, U-Net and Multi-scale-dilation on the UAVid dataset, respectively. The exploration of previous works proves that there are still important problems such as generalization abilities for large-scale variation and moving object segmentation from oblique view drone videos which have not yet been considered well in the relevant studies.

In an oblique view drone image compared with a vertical view (Fig. 1), objects at different distances such as buildings, trees, humans, and cars are needed to be analyzed in different spectral-spatial features and scale-spaces. The most popular multi-scale networks such as DeepLabv2, DeepLabv3, PSPNet [8], and DUC [9] use the dilated convolution and some graphical models like Bayesian networks, and Conditional Random Fields (CRFs) as post-processing to preserve the spatial

[1] https://uavid.nl/.

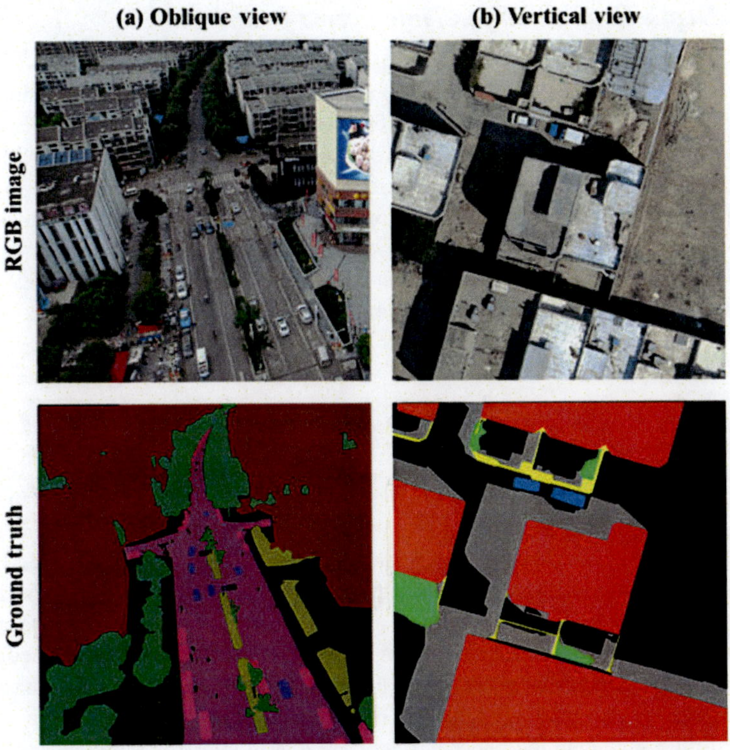

Fig. 1 An overview of the oblique and vertical views in drone images

size of the feature map. DeepLabv3 + designs a multi-scale convolutional architecture based on DeepLabv3 is proposed by a simple yet useful decoder to improve the semantic segmentation tasks [10]. Graphical convolution networks such as GAS allow the model to learn about local structures from neighbor nodes [11]. In Ghiasi and Fowlkes 2016 [12], a Laplacian pyramid module is proposed to combine multi-scale features in Convolutional Neural Network (CNN) to refine the semantic segmentation, especially along target/anomaly boundaries.

Recently, SANet designs a multi-scale network for indoor and outdoor datasets by merging outputs from different stages of backbone residual networks. In Khoshboresh-Masouleh, Alidoost, and Arefi 2020 [13], applies a "multi-kernel dilated convolution" module with the WBCE (Weighted Binary Cross-Entropy) which contains filters with various spatial sizes to learn effective relationships in multi-scale vertical/oblique view images for target detection. Moreover, a Multi-scale-dilation net designs a dilated convolutional network based on FCN-8 s to refine the multiple semantic segmentation especially along with oblique view drone images [7].

1.3 Depth Estimation for Time Series Drone Images

The various depths of objects in vertical and oblique views drone images will affect the target/anomaly detection, especially for object analysis in the complex scene [14]. It is important to learn an exchange task for unsupervised monocular depth estimation from time-series images that allows for the use of photometric warp losses. In Zhou et al. [15], an unsupervised method is proposed for monocular depth estimation from video sequences. In this regard, unsupervised depth, and ego-motion estimation are proposed based on differentiable three-dimension cost functions which can establish stability between the geometry of adjacent images [16].

1.4 Deep Domain Adaptation for Time Series Drone Images

Deep Domain Adaptation (DDA) has emerged to address the lack of massive amounts of annotated images and aims to learn more transferable features by embedding DDA in deep learning models for a better generalization at test time [17]. In Ouali, Hudelot, and Tami 2020 [18], a DDA method is proposed toward better generalization in semantic segmentation. In Bolte et al. 2019 [19], a DDA method is proposed to study in an unsupervised fashion for segmentation. In [20], a DDA method to translate aerial images from the target to the source domain is proposed based on two Generative Adversarial Network (GAN) networks.

Real-time segmentation plays an important role in applying the videos acquired from the aerial platforms to real-world applications requiring instant responses such as traffic management and weed monitoring. Real-time segmentation requires a fast algorithm to generate high-quality predictions [21–23]. In Wu, Shen, and Hengel [24], a two-column network and spatial sparsity are proposed to segmentation that reduces the processing costs by a factor of 25 with limited impact on the quality of segmentation results. The most popular real-time segmentation networks such as E-Net, ERFNet, ICNet, CGNet, ESPNet, BiSeNet, and LEDNet use a lightweight structure from scratch and deliver an extremely high speed.

1.5 Contributions

In this chapter, we aim to fill important gaps in the deep few-shot learning in the context of 2D target/anomaly detection of time series drone images. To estimate 2D target/anomaly and further provide assistance to object detection from time series images, we propose a new deep few-shot learning based on squeeze-and-attention mechanism. We summarize our contributions as follows:

1. The proposed squeeze-and-attention architecture, called SA-Net, is a novel deep few-shot learning method for time series drone images by reformulating the

convolutional networks and squeeze mechanism. The SA-Net is composed of two components, including the residual feature map, and the attention function that extracts the high-level representation for 2D target/anomaly detection based on small training dataset.

2. We examine the proposed model applicability for stand 2D target/anomaly detection when using time series images. The experiments show that the SA-Net achieves promising performance in time series drone images for traffic monitoring and weed detection.
3. To investigate the behaviour of the SA-Net, we conducted several ablation studies based on different backbones and challenges.

2 Experiments and Results

2.1 *Proposal Model*

A visual summary of the proposed model, called SA-Net, for 2D target/anomaly detection in time series drone images is presented in Fig. 2. The proposed model takes the MultiScaleNet (for target detection) [13] and DeepMultiFuse (for anomaly detection) [3] as the main backbone model.

MultiScaleNet and DeepMultiFuse are lightweight CNN models with a model size of 200 MB for pixel-wise binary segmentation in vehicle and weed detection.

MultiScaleNet is a new encoder–decoder ConvNet, consist of dilated filters with various spatial sizes and modified shortcut connections to improve the level of

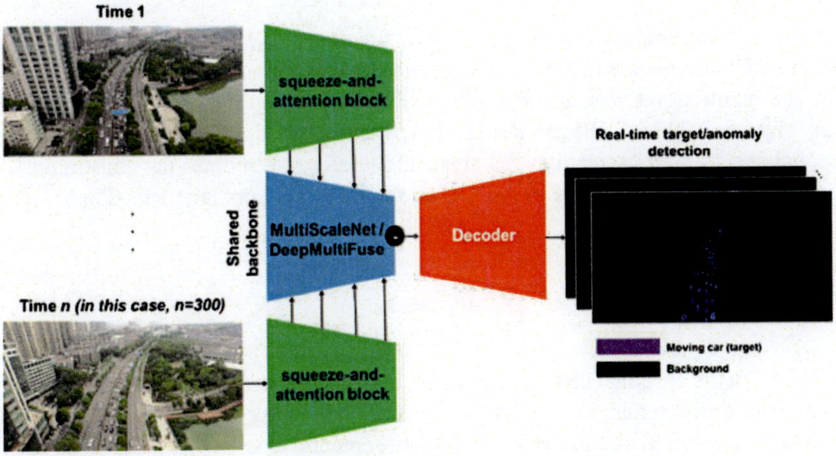

Fig. 2 The proposed squeeze-and-attention network for 2D target/anomaly detection from time series drone images

abstraction. In MultiScaleNet, weighted binary cross-entropy is used for the challenge of the target/anomaly imbalance in which all positive pixels get weighted by an amount close to one. The weighted binary cross-entropy is defined as follows [13]:

$$loss(cp, \widehat{cp}) = -(\gamma \log(\widehat{cp}) + (1 - cp) log(1 - \widehat{cp})) \quad (1)$$

where cp: conditional probability, $\widehat{cp}$: predicted categorical negative targets/anomalies based on the logistic function, and γ: number of negative pixels over total number of pixels.

DeepMultiFuse is a deep learning model based on the gated encoder-decoder convolutional networks and guided features. DeepMultiFuse is composed of the five modules, consist of guided feature, feature fusion block, dilation convolution, modified inception block, and gated encoder-decoder module that extracts the object-level feature. In DeepMultiFuse, the gated network is defined as follows [3]:

$$O_i = [B(L \otimes k_{3\times3})] \odot [UP(U \otimes k_{3\times3})] \quad (2)$$

where O_i: gated module output, B : batch normalization; L: lower encoder representations, U: upper encoder representations, and UP: upsampled function.

To aggregate multistage non-local features, we adopt squeeze-and-attention blocks on the multistage outputs of the MultiScaleNet/DeepMultiFuse model, resulting in better target/anomaly boundaries. The squeeze-and-attention block is defined as follows:

$$\Phi = U(RL(f_{att}(Pool(I)))) \times x_{res} + U(ReLU(f_{att}(Pool(I))) \quad (3)$$

where UP: upsampled function, RL : relu function, f_{att}: attention function, x_{res}: residual feature map, and I: input time-series images.

The input/output size, number of bands, kernel sizes, and activation functions are given in Table 1. Based on Table 1, the proposed model learns non-local spatial-spectral representations features and therefore overcomes the constraints of convolutional layers and masks generation for 2D target/anomaly detection.

2.2 Datasets

UAVid. For the 2D target detection scenario, we test our method on the UAVid. UAVid is a novel drone semantic segmentation dataset consisting of 30 time series drone images capturing in oblique views with eight semantic classes. UAVid is comprised of much larger dimensions (4096 × 2160 or 3840 × 2160) that acquired scenes in a larger range and with more scene complexity regarding the number of targets. In

Table 1 Structure of encoder and decoder in the proposed network. CNV: convolutional layer, DCNV: deconvolution layer, MP: max pooling, UP: upsampling, B: batch normalization, RL: ReLu, Φ: squeeze-and-attention, and SD: Stride

Encoder	Encoder blocks	Kernel size	SD	Result
Input data	–	–	–	$3@W_{Input} \times H_{Input}$
1	CNV1 + B + RL	64@3 × 3	1	$64@W_1 \times H_1$
	CNV2 + B + RL + Φ	64@3 × 3	1	$64@W_2 \times H_2$
	MP	2 × 2	2	–
2	CNV1 + B + RL	128@3 × 3	1	$128@W_3 \times H_3$
	CNV2 + B + RL + Φ	128@3 × 3	1	$128@W_4 \times H_4$
	MP	2 × 2	2	–
3	CNV1 + B + RL	256@3 × 3	1	$256@W_5 \times H_5$
	CNV2 + B + RL	256@3 × 3	1	$256@W_6 \times H_6$
	CNV3 + B + RL + Φ	128@3 × 3	1	$256@W_7 \times H_7$
	MP	2 × 2	2	–
4	CNV1 + B + RL	512@3 × 3	1	$512@W_8 \times H_8$
	CNV2 + B + RL	512@3 × 3	1	$512@W_9 \times H_9$
	CNV3 + B + RL + Φ	512@3 × 3	1	$512@W_{10} \times H_{10}$
	MP	2 × 2	2	–
5	CNV1 + B + RL	1024@3 × 3	1	$1024@W_{11} \times H_{11}$
	CNV2 + B + RL	1024@3 × 3	1	$1024@W_{12} \times H_{12}$
	CNV3 + B + RL + Φ	1024@3 × 3	1	$1024@W_{13} \times H_{13}$
	MP	2 × 2	2	–
Decoder	Decoder blocks	Kernel size	SD	Result
1	UP	2 × 2	2	–
	DCNV1 + B + RL	1024@3 × 3	1	$1024@W_1 \times H_1$
	DCNV2 + B + RL	1024@3 × 3	1	$1024@W_2 \times H_2$
	DCNV3 + B + RL	512@3 × 3	1	$512@W_3 \times H_3$
2	UP	2 × 2	2	–
	DCNV1 + B + RL	512@3 × 3	1	$512@W_4 \times H_4$
	DCNV2 + B + RL	512@3 × 3	1	$512@W_5 \times H_5$
	DCNV3 + B + RL	256@3 × 3	1	$256@W_6 \times H_6$
3	UP	2 × 2	2	–
	DCNV1 + B + RL	256@3 × 3	1	$256@W_7 \times H_7$
	DCNV2 + B + RL	256@3 × 3	1	$256@W_8 \times H_8$
	DCNV3 + B + RL	128@3 × 3	1	$128@W_9 \times H_9$
4	UP	2 × 2	2	–
	DCNV1 + B + RL	128@3 × 3	1	$128@W_{10} \times H_{10}$
	DCNV2 + B + RL	64@3 × 3	1	$64@W_{11} \times H_{11}$

(continued)

Table 1 (continued)

Encoder	Encoder blocks	Kernel size	SD	Result
5	UP	2 × 2	2	–
	DCNV1 + B + RL	64@3 × 3	1	$64@W_{12} \times H_{12}$
	DCNV2 + B + RL	2@3 × 3	1	$2@W_{13} \times H_{13}$
	WBCE	–	–	–

Table 2 List of datasets for deep few-shot learning in drone imaging sensors

Reference	Data source	Type	View	Texture distortion	Semantic annotation	Target/anomaly class
UAVid	Drone	Video	Oblique	No	Yes	Vehicle
WeedMap	Drone	Orthophoto	Vertical	Low	Yes	Weed

conclusion, UAVid is a remote sensing dataset with important challenges for real-world applications (e.g. tiny and moving objects, temporal consistency preservation, large-scale variation) [7].

WeedMap. For the 2D anomaly detection scenario, we test our approach on the WeedMap set. WeedMap dataset includes multispectral images using the Sequoia sensor based on the Mavic pro platform in Eschikon and the RedEdge-M sensor based on the Inspire-2 platform in Rheinbach. The Sequoia is a multispectral sensor with four spectral channels (e.g., red, green, red edge, and NIR) and 1280 × 960 resolution. The RedEdge-M is a multispectral sensor that captures five discrete spectral bands with a 9.4 cm × 6.3 cm sensor size [25].

An overview of existing datasets for deep few-shot learning can be found in Table 2.

2.3 *Implementation Details*

Our method was implemented with Keras and used the Adam algorithm for the optimization stage with the *initial learning rate*, $\beta 1$, $\beta 2$, and ε fixed to 0.01, 0.9, 0.999, and 10^{-8}, respectively. Moreover, the proposed network is fed with a mini-batch of 64 pair images, and the training images are augmented using horizontal and vertical flipping. The detail of training and testing sets are listed in Table 3.

Table 3 Detail of training and testing sets for 2D target and anomaly detection

Dataset	Patch size (pixels)	Spectral bands	Samples		
			Training	Validation	Testing
UAVid	3840 × 2160	Red, Green, Blue	1000	420	40
WeedMap	480 × 360	Red, Green, Blue, Red edge	1530	550	120

2.4 Experimental Results

2.4.1 Assessment Metrics

In this study, the mean Jaccard index or the Intersection-over-Union (mIoU) is the standard metric used in evaluating 2D target/anomaly detection based on ground truth and predicted segments, which are formulated as:

$$IoU = \frac{|p \cap g|}{|p \cup g|} \tag{4}$$

$$F1 = \frac{2 \times TP}{2 \times TP + FN + FP} \tag{5}$$

$$Completeness = \frac{TP}{TP + FN} \tag{6}$$

$$Correctness = \frac{TP}{TP + FP} \tag{7}$$

where g: ground truth, p: predicted map, TP: correctly extracted target/anomaly, FP: targets mislabeled as the target/anomaly, and FN: targets mislabeled as the non-targets/non-anomalies.

2.4.2 Target/Anomaly Detection Results

In this study, the performance of the SA-Net analyzed for important challenges in 2D target/anomaly detection in time series drone images. In remote sensing and computer vision, developing an assessment of 2D target/anomaly detection is classified into three separate challenges, including shadows areas, vegetation covers, and dense objects. To investigate the behaviour of the SA-Net, we conducted several ablation studies are as follows.

Shadows areas. Due to building shadow in urban regions, the pixels of target/anomaly can be changed in the time-series data, and it becomes difficult to the target/anomaly detection. The SA-Net is trained using shadow-included time-series data from various drone images to learn how the model tackles that issue.

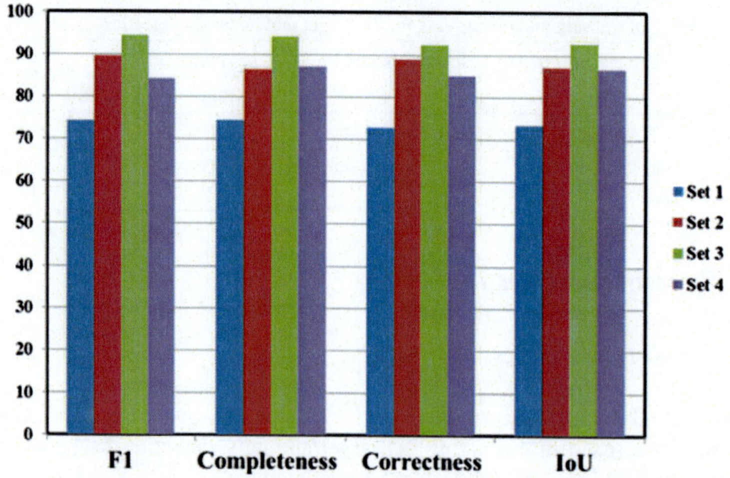

Fig. 3 Accuracy assessment of vehicle monitoring for different sample images with the SA-Net based on MultiScaleNet

Vegetation covers. Vegetated areas are the important issues in target/anomaly detection, particularly where some parts of targets/anomalies are covered by grass and tree covers. In this regard, to improve the training dataset, we are adding time-series data that include various grass and tree covers to make the model robust to target/anomaly detection in the presence of vegetation covers.

Dense areas. In dense areas, 2D target and anomaly segmentation are more complex due to the variety of spatial-spectral information of objects in urban and rural areas. The different objects (e.g., building, vehicle, crop and weed) can be detected as a target/anomaly if they are close together. The squeeze-and-attention blocks are used in the SA-Net to provide global views and larger neighborhoods for extracting the representation and to preserve the resolutions of representation maps.

Figures 3 and 4 show the numerical outputs of the test performed using the SA-Net, compared to the different accuracy metrics. When the DeepMultiFuse backbone is included, the average IoU and the average F1 scores could increase by approximately 4.5 and 5.0% compared to MultiScaleNet, respectively.

Table 4 shows the quantitative results for target and anomaly detection from time-series images. According to Table 3, significant accuracy improvements can be observed for the proposed method over set 3 for target detection (Fig. 5) and set 2 for anomaly detection (Fig. 6). In this regard, the mean IoU scores for target detection and anomaly detection scenarios are about 85 and 90%, respectively. Figures 3 and 4 show the target and anomaly detection based on the proposed method from time series drone images in vertical and oblique views.

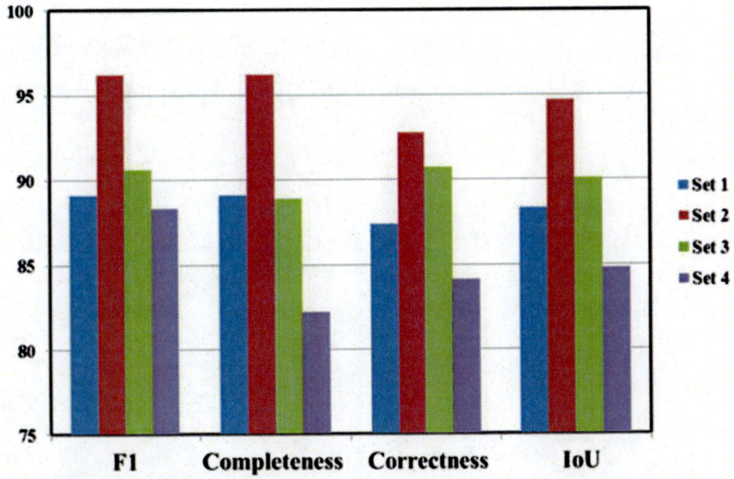

Fig. 4 Accuracy assessment of weed detection for different sample images with the SA-Net based on DeepMultiFuse

Table 4 2D target and anomaly detection comparisons on different types of test scenes based on the proposed method (SA-Net)

Dataset	Scenario	Time series samples	Model	Backbone	mIoU (%)
UAVid	Target detection	Set1, n = 10	SA-Net	MultiScaleNet	73.4
		Set 2, n = 10			87.1
		Set 3, n = 10			92.7
		Set 4, n = 10			86.7
WeedMap	Anomaly detection	Set1, n = 30	SA-Net	DeepMultiFuse	88.3
		Set 2, n = 30			94.7
		Set 3, n = 30			90.1
		Set 4, n = 30			84.8

3 Future Work

In time series drone images, multi-task learning allows deep learning methods that are composed of multiple blocks to learn features of the image with multiple levels of abstraction. Multi-task learning approaches have led to breakthroughs in different domains of remote sensing. In civil applications such as city monitoring based on time series drone images, multi-task learning is required for 2D/3D scene understanding due to improving generalization ability by using the domain-specific feature contained in the training set [6]. In the real world, when the number of tasks increases, duplicate data may exist across different tasks, and the enhancement becomes less

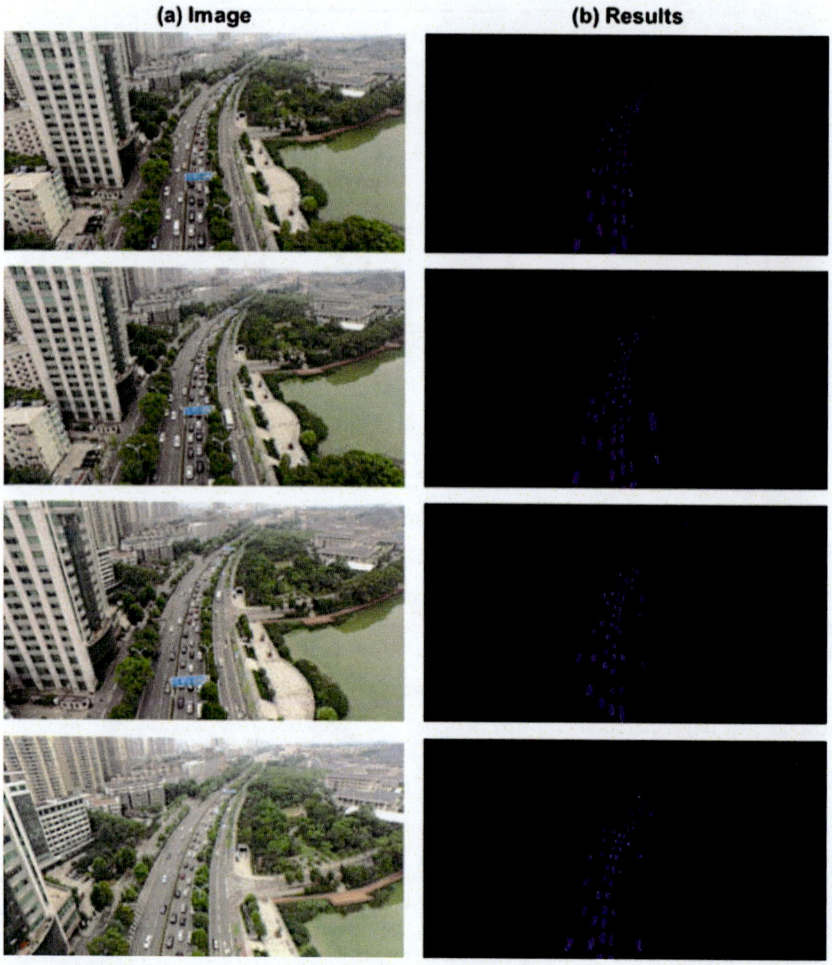

Fig. 5 Examples of target detection based on drone sensors for the test images by the proposed method

significant. Multi-tasking ability is an efficient strategy to knowledge-transfer problems which includes related tasks each with potentially limited training images. For the future, we would like to exploit our method based on multi-task scenario for 2D/3D target/anomaly segmentation.

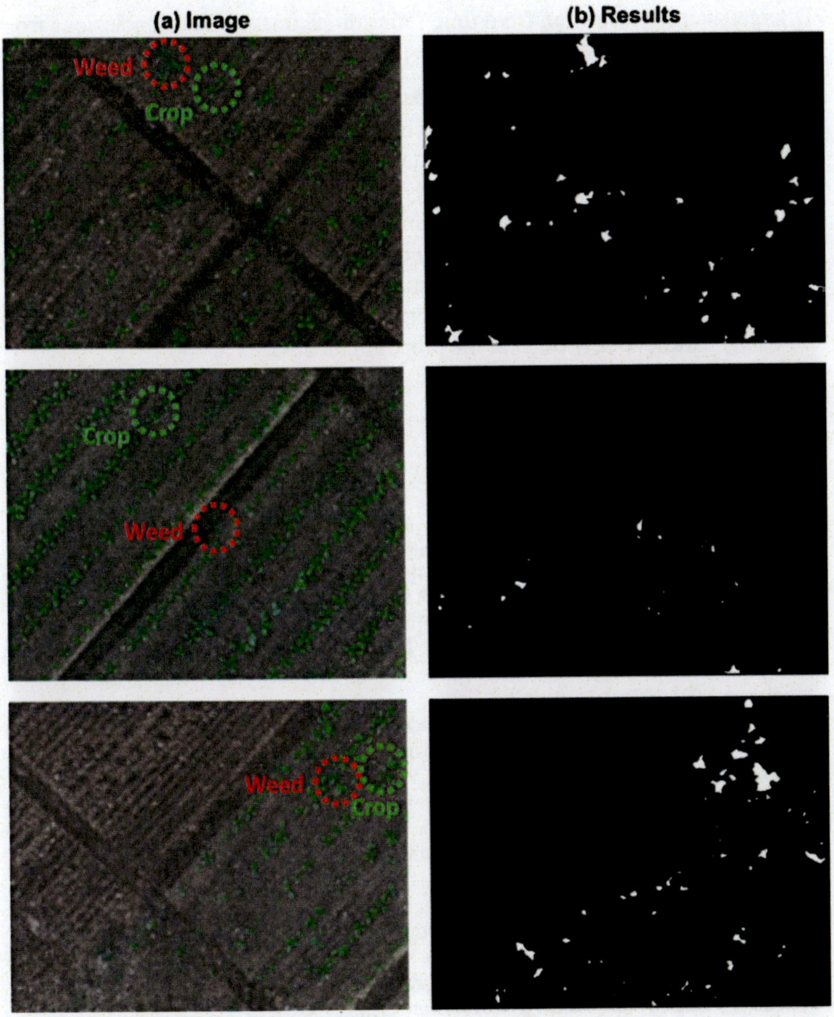

Fig. 6 Examples of anomaly detection based on drone sensors for the test images by the proposed method

4 Conclusions

In this chapter, we have performed a survey of target/anomaly segmentation research efforts applied in traffic monitoring and weed detection. We have identified different challenges in this field including shadows areas, vegetation covers, dense areas, vertical and oblique views based on time series drone images. Therefore, this chapter presents a new deep learning method, called SA-Net (subsection 2.1), for

2D target/anomaly detection from time series drone images. The conclusions from this chapter are as follows:

1. This study demonstrated that the impact of the proposed network can be investigated into the squeeze-and-attention block is important to understand the effect of high-level representations on 2D target/anomaly detection.
2. Compared with the target and anomaly detection scenarios, the proposed method achieved the highest mIoU for anomaly detection due to using multispectral images for training. Our findings indicate that the proposed approach offers better performance in the generalization ability for anomaly detection.
3. We also showed the effect of sharing the MultiScale-Net and DeepMultiFuse as the backbone networks for the proposed method.

References

1. S. Famiyeh, E. Adaku, K. Amoako-Gyampah et al., Environmental management practices, operational competitiveness and environmental performance: empirical evidence from a developing country. J. Manuf. Technol. Manag. **29**, 588–607 (2018). https://doi.org/10.1108/JMTM-06-2017-0124
2. L. Tan, X. Lv, X. Lian, G. Wang, YOLOv4_Drone: UAV image target detection based on an improved YOLOv4 algorithm. Comput. Electr. Eng. **93**(2021). https://doi.org/10.1016/j.compeleceng.2021.107261
3. M. Khoshboresh-Masouleh, M. Akhoondzadeh, Improving weed segmentation in sugar beet fields using potentials of multispectral unmanned aerial vehicle images and lightweight deep learning. JARS **15**(2021). https://doi.org/10.1117/1.JRS.15.034510
4. M. Khoshboresh Masouleh, R. Shah-Hosseini, Development and evaluation of a deep learning model for real-time ground vehicle semantic segmentation from UAV-based thermal infrared imagery. ISPRS J. Photogramm. Remote. Sens. **155**, 172–186 (2019). https://doi.org/10.1016/j.isprsjprs.2019.07.009
5. J, Fu, H. Zhang, H. Wei, X. Gao, Small bounding-box filter for small target detection. OE **60**, 033107 (2021). https://doi.org/10.1117/1.OE.60.3.033107
6. M.R. Bayanlou, M. Khoshboresh-Masouleh, Multi-task learning from fixed-wing UAV images for 2D/3D city modeling. Int Arch Photogramm Remote Sens Spatial Inf Sci XLIV-M-3–2021:1–5 (2021). https://doi.org/10.5194/isprs-archives-XLIV-M-3-2021-1-2021
7. Y. Lyu, G. Vosselman, G.-S. Xia et al., UAVid: a semantic segmentation dataset for UAV imagery. ISPRS J. Photogramm. Remote. Sens. **165**, 108–119 (2020). https://doi.org/10.1016/j.isprsjprs.2020.05.009
8. H. Zhao, J. Shi, X. Qi et al., Pyramid scene parsing network (2017). arXiv:161201105 [cs]
9. P. Wang, P. Chen, Y. Yuan et al., Understanding convolution for semantic segmentation (2018). arXiv:170208502 [cs]
10. L.-C. Chen, Y. Zhu, G. Papandreou et al., Encoder-decoder with atrous separable convolution for semantic image segmentation (2018). arXiv:180202611 [cs]
11. J. Tompson, A. Jain, Y. LeCun, C. Bregler, Joint training of a convolutional network and a graphical model for human pose estimation (2014). arXiv:14062984 [cs]
12. G. Ghiasi, C.C. Fowlkes, Laplacian pyramid reconstruction and refinement for semantic segmentation (2016). arXiv:160502264 [cs]
13. M. Khoshboresh-Masouleh, F. Alidoost, H. Arefi, Multiscale building segmentation based on deep learning for remote sensing RGB images from different sensors. JARS **14**(2020). https://doi.org/10.1117/1.JRS.14.034503

14. M.R.R. Bayanlou, M. Khoshboresh-Masouleh, SAMA-VTOL: a new unmanned aircraft system for remotely sensed data collection, in *SPIE Future Sensing Technologies. SPIE* (2020), pp. 169–175
15. T. Zhou, M. Brown, N. Snavely, D.G. Lowe, Unsupervised learning of depth and ego-motion from video (2017). arXiv:170407813 [cs]
16. R. Mahjourian, M. Wicke, A. Angelova, Unsupervised learning of depth and ego-motion from monocular video using 3D geometric constraints (2018). arXiv:180205522 [cs]
17. A. Rozantsev, M. Salzmann, P. Fua, Residual parameter transfer for deep domain adaptation (2018), pp. 4339–4348
18. Y. Ouali, C. Hudelot, M. Tami, Semi-supervised semantic segmentation with cross-consistency training (2020). arXiv:200309005 [cs]
19. J.-A. Bolte, M. Kamp, A. Breuer et al., Unsupervised domain adaptation to improve image segmentation quality both in the source and target domain, in *2019 IEEE/CVF Conference on Computer Vision and Pattern Recognition Workshops (CVPRW)* (2019), pp. 1404–1413
20. B. Benjdira, A. Ammar, A. Koubaa, K. Ouni, Data-efficient domain adaptation for semantic segmentation of aerial imagery using generative adversarial networks. Appl. Sci. **10**, 1092 (2020). https://doi.org/10.3390/app10031092
21. Y. Tarabalka, T.V. Haavardsholm, I. Kåsen, T. Skauli, Real-time anomaly detection in hyperspectral images using multivariate normal mixture models and GPU processing. J. Real-Time Image Proc. **4**, 287–300 (2009). https://doi.org/10.1007/s11554-008-0105-x
22. M. Khoshboresh-Masouleh, M. Hasanlou, Improving hyperspectral sub-pixel target detection in multiple target signatures using a revised replacement signal model. Eur. J. Remote Sens. **53**, 316–330 (2020). https://doi.org/10.1080/22797254.2020.1850179
23. B. Yang, M. Yang, A. Plaza et al., Dual-mode FPGA implementation of target and anomaly detection algorithms for real-time hyperspectral imaging. IEEE J. Selec. Topics Appl. Earth Observ. Remote Sens. **8**, 2950–2961 (2015). https://doi.org/10.1109/JSTARS.2015.2388797
24. Z. Wu, C. Shen, A. Hengel van den, Real-time Semantic Image segmentation via spatial sparsity (2017). arXiv:171200213 [cs]
25. I. Sa, M. Popović, R. Khanna et al., WeedMap: a large-scale semantic weed mapping framework using aerial multispectral imaging and deep neural network for precision farming. Remote Sens. **10**, 1423 (2018). https://doi.org/10.3390/rs10091423

Hybrid Adaptive Moth-Flame Optimizer and Opposition-Based Learning for Training Multilayer Perceptrons

Benedict Jun Ma

Abstract This chapter is dedicated to improving the optimization capability of moth-flame optimizer (MFO), which works based on a spiral operation mimicking the moths' navigation behavior and dynamic coefficients. However, such a basic version can be easily trapped in the local optima and is associated with an unstable balance between exploratory and exploitative cores. To mitigate the shortcomings of slow convergence and local stagnation, an adaptive moth flame optimization with opposition-based learning (AMFOOBL) is presented, employing a new adaptive structure and the opposition-based learning (OBL) strategy. This original adaptive tool is devised to reduce the number of flames around which agents update their positions for balancing the exploration and exploitation stages more effectively. The performance of AMFOOBL is evaluated in two experiments: First, the quantitative results of 23 benchmark function tests show that AMFOOBL outperforms AMFO, followed by MFO, validating the effectiveness of the proposed approach in terms of accuracy and convergence rate. Second, AMFOOBL is demonstrated on multi-layer perceptron's (MLP) structural realization and training compared with nine advanced algorithms. The simulation on eight datasets for pattern classification and function approximation reveals outstanding performance in the AMFOOBL-based trainer concerning classification accuracy and test error. Our findings suggest that AMFOOBL is a superior algorithm, and the developed evolutionary-enhanced MLP can be considered a helpful tool.

Keywords Swarm-intelligence · Moth-flame optimization · Multilayer perceptron · Opposite-based learning · Soft computing

---

The original version of this chapter was revised: Author provided new figures for Figs. 1 to 6 are updated. The correction to this chapter can be available at
https://doi.org/10.1007/978-3-030-99079-4_20

B. J. Ma (✉)
Department of Industrial and Manufacturing Systems Engineering, The University of Hong Kong, Hong Kong SAR, China
e-mail: bjma@connect.hku.hk

E. H. Houssein et al. (eds.), *Integrating Meta-Heuristics and Machine Learning for Real-World Optimization Problems*, Studies in Computational Intelligence 1038,
https://doi.org/10.1007/978-3-030-99079-4_11

1 Introduction

Artificial neural networks (ANNs), mathematically developed by McCulloch and Pitts with an inspiration of biological neural systems in human brains [1], are one of the most significant inventions in the sphere of computational intelligence and soft computing [2, 3]. As a powerful and popular tool, some different variants have been devised subsequently in the literature, such as feedforward neural network (FNN) [4], recurrent neural network (RNN) [5], and convolutional neural network (CNN) [6]. Regardless of further progress, learning is an indispensable part of any neural network, and developing the learning process is considered a challenging task that has attracted much attention in recent years [7, 8].

An ANN is generally constructed by an input layer, one or more hidden layers, with an output layer. Among various types of ANNs, FNN is most widely employed in which information transfers in a unidirectional manner from input to output. Actually, single-layer perceptron (SLP) and multilayer perceptron (MLP) are two typical types of FNN [9]. In the SLP, nodes (neurons) in the input layer are connected directly to the output layer without any hidden layers, which prefers to solve binary linear problems. On the other hand, MLP contains at least one hidden layer that makes it capable of classifying separable patterns as well as approximating both continuous and discrete functions within nonlinear systems.

The training algorithm in MLPs can be classified into two groups: deterministic versus stochastic [10]. Most deterministic trainers are based on mathematical optimization techniques, whereas stochastic trainers use random optimization techniques [11, 12]. Gradient-based methods typically belong to the deterministic group, in which the standard backpropagation (BP) algorithm [13] alongside its variants [14] are the most commonly applied. The advantages of deterministic tools are simplicity and fast computation rate, especially in linear and straightforward problems. However, deterministic methods have drawbacks in the high possibility of local optima stagnation, which means the output (i.e., error) of MLP frequently stays large without further changes for an extended period [15]. In addition, the quality of obtained solutions in gradient descent algorithms is highly dependent on the initial values of critical parameters like weight and learning rate [16]. As the learning process enhancement targets optimizing the structural values of networks to realize a minimum error [17], the above limitations make deterministic techniques unreliable for complex applications practically. Therefore, stochastic-based methods, such as metaheuristic algorithms (MAs), have been motivated as alternatives to deterministic-based for training MLPs. And this chapter is dedicated to proposing an effective MAs-based approach to train MLPs.

2 Related Works

Real-world phenomena always inspire MAs among biology, physics, and ethology, providing general optimization frameworks with stochastic features that can iteratively improve current solutions [18–20]. According to the optimization structure, MAs are mainly categorized as evolutionary algorithms (EAs) and swarm intelligence (SI) [21, 22]. SI can also be stratified into biology-based, physics-based, and human-based, considering the source of inspiration. MAs initiate the optimization process with generated solutions stochastically without gradient information, making algorithms highly practicable to solve nonlinear problems under limited circumstances when the derivative information is unknown [23, 24].

Since MLP training steps can be modeled as an optimization problem, many related works based on MAs have been studied in the literature. Mirjalili et al. [17] employed the magnetic optimization algorithm (MOA) that is originated from the magnetic field theory for training MLP. Experimental results suggested that MOA had high performance compared to PSO and GA-based learning algorithms. Moallem et al. [25] applied invasive weed optimization (IWO) to MLP training for potato color image segmentation. Experiments on over 500 potato images showed their method could significantly improve MLP training performance than traditional BP algorithms. Mirjalili et al. [2] then introduced biogeography-based optimization (BBO), inspired by biological organisms study regarding geographical distribution, for training MLPs to reduce local entrapment minima. Compared to other well-known MAs in terms of accuracy and convergence rate, the results indicated that training MLPs by BBO was substantially better than BP and five other MAs. Alboaneen et al. [26] investigated the glowworm swarm optimization (GSO) algorithm in training MLPs. The proposed method was evaluated on five classification datasets and compared with four other MAs; conclusions validated that GSO achieved a better classification accuracy rate in most datasets. Zhao et al. [27] proposed OISHO using selfish herd optimization (SHO) and orthogonal design and information update, which was then used to train MLPs for estimating its effectiveness. Twenty standard datasets from the UCI machine learning repository were adopted, and experimental results proved that OISHO had accurate, fast, and stable characteristics than other competitors.

Although various MAs have been proposed, none can make sure to find the global optimum for all kinds of optimization problems according to the No-Free-Lunch theory [12, 28], which motivates scholars to continuously create and modify new algorithms and employ them in different ways applications. Based on this view, this chapter introduces moth-flame optimization (MFO) designed in 2015 [28] to train MLPs. Prior to this study, MFO has been applied to solve many other problems where its performance has been demonstrated widely to some extent [24, 29]. For example, Yıldız et al. [30] validated the effectiveness of MFO in maximizing the profit rate for multi-tool milling operators under challenging constraints in the manufacturing industry. Li et al. [31] hybridized the annual power load forecasting model based on the least squares support vector machine (LSSVM) with MFO, where MFO optimally determines parameters of LSSVM. The experimental results indicated that the MFO-LSSVM model could significantly improve forecasting accuracy compared to PSO and fruit-fly optimization algorithm (FOA)-based. Wang et al. [32] introduced a novel learning scheme for the kernel extreme learning machine (KELM) based on practically changing the chaotic MFO strategy for medical diagnosis. The new method was compared with the original MFO, PSO, and GA, and conclusions showed that it could be a valuable and efficient medical decision-making tool. Jia et al. [11] modified the standard MFO algorithm based on an original self-adaptive weight and thresholding heuristic (TH) to propose SAMFO-TH for multilevel thresholding in color satellite images. It suggested that SAMFO-TH outperformed another five competitive MAs in stability, accuracy, and convergence rate. Zawbaa et al. [33] utilized MFO in feature selection, and a set of UCI datasets was used to compare different assessment indicators. The related results revealed that MFO was more effective than PSO and GA.

Based on the above illustration, MFO has been modified and employed to enhance its performance. Nevertheless, due to the non-optimal trade-off between the exploration and exploitation of searching, it is still easy for MFO to suffer from local optima and premature convergence troubles. Therefore, AMFOOBL is enhanced based on the raw MFO to deal with MLP training in this work.

The rest of the chapter is organized as follows: Sect. 3 briefly introduces multilayer perceptron training. Section 4 illustrates the mathematical model of MFO alongside its modification. Section 5 introduces the experimental setup on benchmark function tests and MLP training. Section 6 provides statistical analysis and discussion. Section 7 concludes the work and suggests directions for future study.

3 Multilayer Perceptron Training

MLP with three layers is the simplest structure: an input layer followed by a hidden layer and ends with an output layer in one direction, as is shown in Fig. 1. Furthermore, Fig. 2 presents an example of MLP with two hidden layers.

In the MLP, each node in tth layer receives data from nodes in $t-1^{th}$ layer and delivers data to nodes in $t+1^{th}$ layer. More specifically, inputs given in the input layer are multiplied by connection weights (outputs of the input layer) and sent as inputs to the hidden layer. Hidden layers provide computational processing in the network to produce outputs calculated with connection weights and then transited as inputs to the output layer. The training process of MLP with three layers can be expressed below.

(1) The weighted sums of inputs are calculated by Eq. (3.1).

$$s_j = \sum\nolimits_{i=1}^{n} \left(\omega_{ij} \bullet X_i\right) - \theta_j,\ j = 1, 2, \cdots, h \tag{3.1}$$

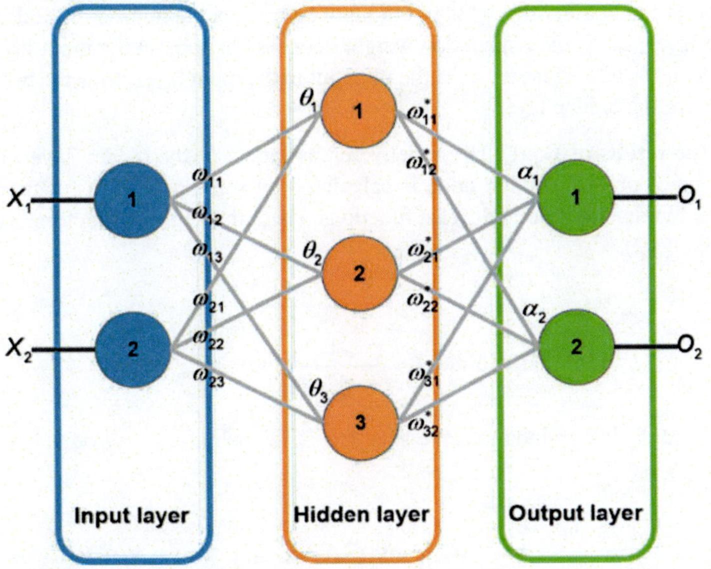

Fig. 1 MLP with the structure of 2-3-2

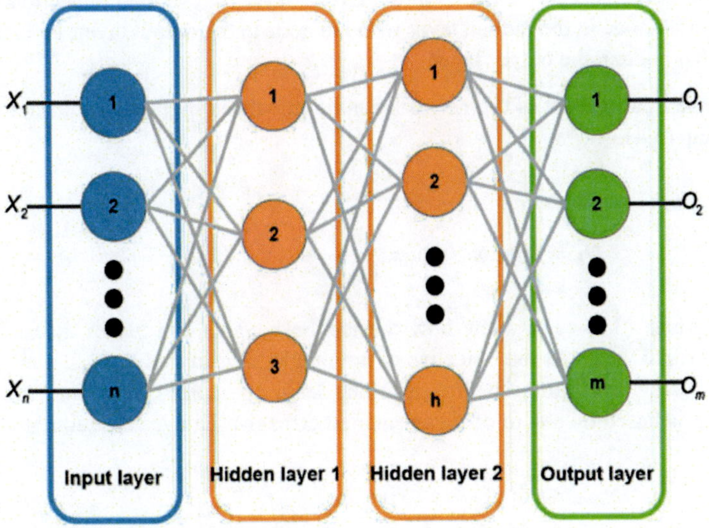

Fig. 2 MLP with the structure of n-3-h-m

where n is the number of nodes in the input layer; h is the number of nodes in the hidden layer; ω_{ij} is the connection weight between ith node in the input layer with jth node in the hidden layer; X_i is the ith input in the input layer; θ_j is the bias of the jth node in the hidden layer.

(2) The results of Eq. (3.1) are delivered as inputs of the hidden layer, and the output of each hidden node is calculated by some activation function, such as bipolar sigmoid and ReLu functions. Here, the sigmoid function is used in Eq. (3.2).

$$S_j = \text{sigmoid}(s_j) = \frac{1}{(1 + e^{s_j})}, j = 1, 2, \cdots, h \tag{3.2}$$

(3) Then, S_j is weighted to serve as inputs of the output layer through Eq. (3.3).

$$o_k = \sum\nolimits_{j=1}^{h} (\omega_{jk}^* \bullet S_j) - \alpha_k, k = 1, 2, \cdots, m \tag{3.3}$$

where m is the number of nodes in the output layer; ω_{jk}^* is the connection weight between jth node in the hidden layer with kth node in the output layer; α_k is the bias of the kth node in the output layer.

(4) The final outputs of the network can be calculated by the sigmoid function in Eq. (3.4).

$$O_k = \text{sigmoid}(o_k) = \frac{1}{(1 + e^{o_k})}, k = 1, 2, \cdots, m \tag{3.4}$$

Equations (3.1)–(3.4) show how connection weights and biases determine the outputs' final value. As the objective of training MLP is to find optimal weights and biases, achieving a minimum error regarding desired outcomes. Therefore, the fitness function is based on a sum of mean squared error (MSE) over all training samples as follows:

$$MSE = \sum\nolimits_{i=1}^{Q} \frac{\sum_{k=1}^{m} \left(O_i^k - D_i^k\right)^2}{Q} \tag{3.5}$$

where Q is the number of training inputs; O_i^k is the actual output of ith training sample in kth output node; D_i^k is the desired output of the ith training sample in kth output node.

When training MLPs with metaheuristic algorithms, Eq. (3.5) is regarded as the fitness function in both proposed and other competitive optimization algorithms to find a proper combination of weights and biases, providing the minimum MSE value.

4 Proposed Method

The MFO algorithm has attracted extensive attention due to its few parameters and robust searching ability. Nonetheless, MFO still faces some trouble avoiding falling into local sites, especially in dealing with complex and hyper-dimensional problems. To further overcome this problem, in this chapter, MFO is developed based on a new mechanism of flame adaption and the OBL strategy to improve its performance in both exploration and exploitation stages.

4.1 Standard MFO

People can always acquire inspiration from nature; MFO is a strong metaheuristic algorithm simulating the navigation behavior of moths. In fact, moths have their unique navigation methods at night called transverse orientation to maintain a fixed angle with respect to the moon. Since the moon is far away from moths, taking transverse orientation can guarantee to fly in a straight line for them. However, moths are often tricked and distracted by artificial lights such as flames; if they still rely on the previous navigation to keep a fixed angle with flames that are close to themselves, they will fly in a spiral path towards the flame eventually. In particular, the general behavior of MAs can be systematically shown in Fig. 3 [34].

The mathematical models of MFO are established in the following. First, the position of moths in the search space is presented as:

$$M = \begin{bmatrix} M_1 \\ M_2 \\ \vdots \\ M_n \end{bmatrix} = \begin{bmatrix} m_{11} & m_{12} & \cdots & m_{1d} \\ m_{21} & m_{22} & \cdots & m_{2d} \\ \vdots & \vdots & \ddots & \vdots \\ m_{n1} & m_{n2} & \cdots & m_{nd} \end{bmatrix} \tag{4.1}$$

where the number of moths is n; the space dimension of each moth is d.

The matrix in Eq. (4.1) is calculated by the fitness function in which each row will obtain a fitness value, considering in another matrix as:

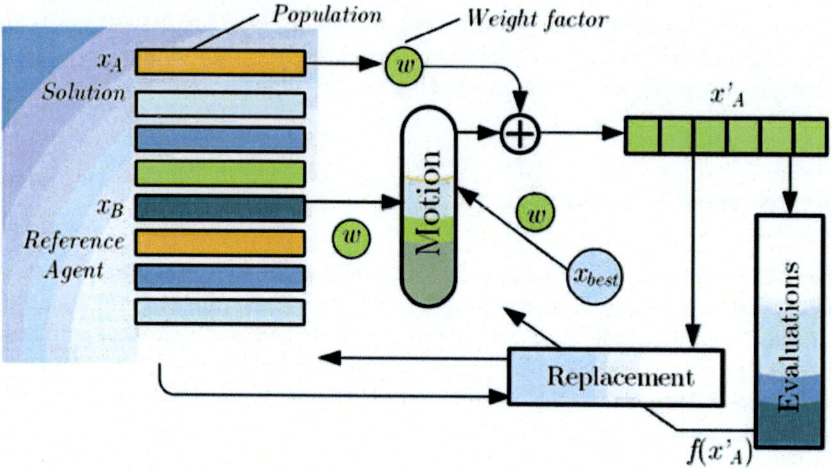

Fig. 3 The general structure of MAs

$$OM = \begin{bmatrix} OM_1 \\ OM_2 \\ \vdots \\ OM_n \end{bmatrix} \tag{4.2}$$

Another essential component in MFO is flame, and the position of flames with their values are expressed as:

$$F = \begin{bmatrix} F_1 \\ F_2 \\ \vdots \\ F_n \end{bmatrix} = \begin{bmatrix} f_{11} & f_{12} & \cdots & f_{1d} \\ f_{21} & f_{22} & \cdots & f_{2d} \\ \vdots & \vdots & \ddots & \vdots \\ f_{n1} & f_{n2} & \cdots & f_{nd} \end{bmatrix} \tag{4.3}$$

$$OF = \begin{bmatrix} OF_1 \\ OF_2 \\ \vdots \\ OF_n \end{bmatrix} \tag{4.4}$$

In MFO, moths update their positions in the hyper-dimensional space. It should be noted that moths and flames in the space are both solutions to the problem; however, the distinction among them is different ways to update the positions. More specifically, moths are actual search agents in the space, while flames are the best location that all moths obtained so far.

A logarithmic spiral is defined as the major update of moths' positions as follows:

$$M_i = D_i \bullet e^{b\theta} \bullet \cos(2\pi\theta) + F_j \tag{4.5}$$

where M_i is the ith moth; F_j is the jth flame; $D_i = \left|F_j - M_i\right|$ is the distance between ith moth and jth flame; b is a constant to define the shape of the logarithmic spiral; θ is a random value in the range of [-1,1].

It is shown in Eq. (4.5), the spiral flying path of moths is simulated where a moth's position in each iteration is related to a flame. The parameter θ defines how close the moth's next position is to the flame: $\theta = -1$ indicates the closest position to the flame and $\theta = 1$ indicates the farthest. Nevertheless, this formulation only shows the process of moths flying towards the flame; hence it is easy to make MFO converge fast and fall into local optimum. To avoid the above situation, per moth is required to update its position concerning only one flame. In each iteration, flames are sorted based on the fitness values of moths. In this view, the first moth always updates its position concerning the best flame, whereas the last moth updates relatively to the worst one. According to this mechanism, MFO can effectively avoid falling into the local optimum to a certain extent.

Moreover, to improve the search efficiency and accuracy of the optimum, an adaptive mechanism is proposed for the number of flames in Eq. (4.6).

$$flame_{num} = \text{round}\left(N - t \bullet \frac{N-1}{T}\right) \tag{4.6}$$

where N is the maximum number of flames (equal to the number of moths); t is the current number of iterations; T is the maximum number of iterations.

As the number of iterations increases until achieving the maximum iteration, the final flame left is the optimal solution obtained globally.

4.2 Improved MFO

In the standard MFO algorithm, flames are generated by sorting the best moths, and moths update their positions concerning flames as well. In this way, one salient disadvantage in MFO is the low population diversity that causes the loss of some promising individuals. In this chapter, the OBL strategy is employed for increasing the diversity of moths.

On the other hand, although a mechanism in the decreasing number of flames is utilized to improve the exploitation ability in Eq. (4.6), it leads imbalance of the exploration stage. Consequently, in this chapter, a new adaptive mechanism in decreasing the number of flames is created to enhance the balance between exploration and exploitation cores.

4.2.1 Opposition-Based Learning

OBL [35] is a popular and effective technique employed in many optimization algorithms to boost performance. For example, Ewees et al. [36] modified the grasshopper optimization algorithm (GOA) based on the OBL strategy and generated OBLGOA. Their experimental results based on 23 standard functions and four engineering problems proved that OBLGOA was superior to ten well-developed algorithms, including standard GOA. Gupta et al. [37] improved the sine cosine algorithm (SCA) with OBL called m-SCA, which was tested on two sets of benchmarks (classical and CEC2014) and five engineering problems to demonstrate the algorithm's efficiency. Yu et al. [38] enhanced the firefly algorithm (FA) using generalized OBL; the experiments on 16 benchmark functions validated that OBL can improve the accuracy compared to the primary FA.

The OBL approach works to diversify the population by producing opposite positions according to current positions in the search space. Assume x is defined as a real number over the interval $x \in [lb, ub]$ and the opposite number denoted by $\tilde{x}$ is calculated by Eq. (4.7).

$$\tilde{x} = lb + ub - x \tag{4.7}$$

The above definition can be generalized to a d-dimension space with N-moth (agent) in a matrix Eq. (4.8), and its inverse position is presented in Eq. (4.9) calculated by Eq. (4.10).

$$X = \begin{bmatrix} x_1^1 & x_1^2 & \cdots & x_1^d \\ x_2^1 & x_2^2 & \cdots & x_2^d \\ \vdots & \vdots & \ddots & \vdots \\ x_N^1 & x_N^2 & \cdots & x_N^d \end{bmatrix} \tag{4.8}$$

$$\tilde{X} = \begin{bmatrix} \tilde{x}_1^1 & \tilde{x}_1^2 & \cdots & \tilde{x}_1^d \\ \tilde{x}_2^1 & \tilde{x}_2^2 & \cdots & \tilde{x}_2^d \\ \vdots & \vdots & \ddots & \vdots \\ \tilde{x}_N^1 & \tilde{x}_N^2 & \cdots & \tilde{x}_N^d \end{bmatrix} \tag{4.9}$$

$$\tilde{x}_i^j = lb + ub - x_i^j, i = 1, 2, \cdots, N; j = 1, 2, \cdots, d \tag{4.10}$$

Regarding $f(\bullet)$ as the fitness function, and if $f(\tilde{x}_i)$ is superior to $f(x_i)$ in a particular problem, then $x_i \leftarrow \tilde{x}_i$; otherwise, $x_i \leftarrow x_i$. This is a common method that has been used in most optimization algorithms. However, owing to there is only one place to calculate fitness values in the coding of MFO. Thereby in this chapter, the original positions generated randomly are combined with the inverse positions to increase doubly the moths' population. Then the new position of all agents is expressed as Eq. (4.11).

$$X = \left[X; \tilde{X}\right] = \begin{bmatrix} x_1^1 & x_1^2 & \cdots & x_1^d \\ x_2^1 & x_2^2 & \cdots & x_2^d \\ \vdots & \vdots & \ddots & \vdots \\ x_N^1 & x_N^2 & \cdots & x_N^d \\ \tilde{x}_1^1 & \tilde{x}_1^2 & \cdots & \tilde{x}_1^d \\ \tilde{x}_2^1 & \tilde{x}_2^2 & \cdots & \tilde{x}_2^d \\ \vdots & \vdots & \ddots & \vdots \\ \tilde{x}_N^1 & \tilde{x}_N^2 & \cdots & \tilde{x}_N^d \end{bmatrix} \quad (4.11)$$

To clearly display the diversity of OBL performed, suppose that 20 search agents (N) within the limited interval [0,50] and each agent include ten elements (d). Figures 4, 5, and 6 display the current position distribution, opposite position distribution, and both, respectively. In Fig. 4, each star symbol in a straight line represents an agent's element, and there are a total of ten elements in an agent serving as its position in the 10-dimensional search space. As calculated by Eq. (4.10), opposite positions of all agents in Fig. 4 are shown in Fig. 5; in other words, they can also be regarded as a group of new agents. Figure 6 combines the positions of both current agents and opposite agents to generate a more diversified population.

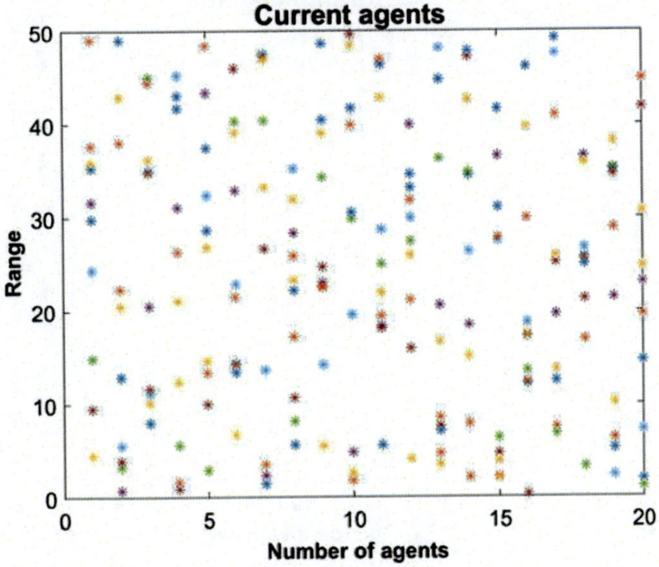

Fig. 4 Current positions of agents

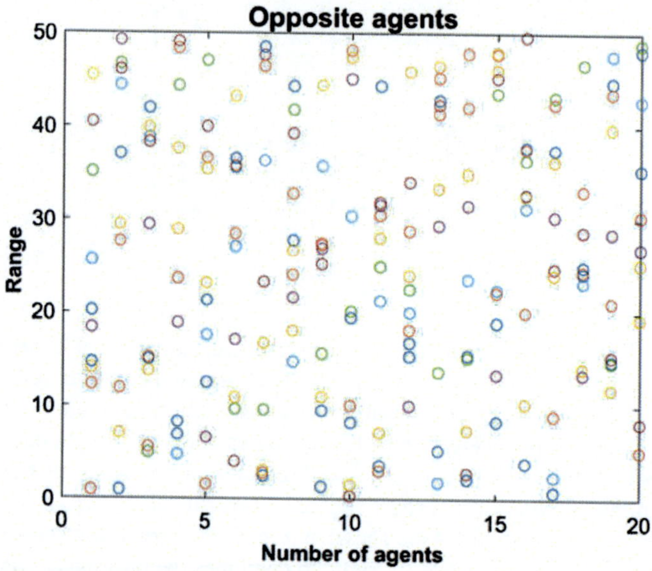

Fig. 5 Opposite positions of agents

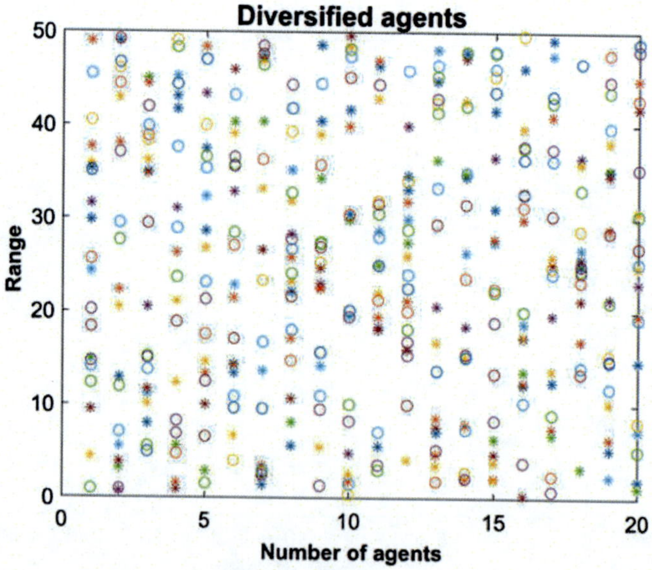

Fig. 6 Both current and opposite positions of agents

4.2.2 Flame Number Adaptive Mechanism

Since moths update their position with respect to flames, simultaneously, the only flame left when reaching the maximum iteration is the obtained optimal solution, which implies the quality of flames plays an essential role during the iterative process. Accordingly, a novel enhanced mechanism in decreasing the number of flames is proposed to provide better matching between exploration and exploitation, which is expressed as follows:

$$flame_{num} = \text{round}(N - \mu \bullet (N - 1))$$

$$\mu(t) = \frac{\arctan(\delta \bullet (t/T)^{\tau})}{\arctan(\delta)} \tag{4.12}$$

where t is the current number of iterations; T is the maximum number of iterations; δ and τ are two parameters determining the characteristics of convexity and concavity.

To determine the exact value of δ and τ, a set of experiments are required to find the optimal combination of these two variables. In addition, the difference regarding function trends between μ and $\frac{t}{T}$ should be investigated and analyzed deeply. Table 1 presents twenty different combinations between δ and τ in which the blue line

Table 1 Different combinations between δ and τ

	$\delta = 10$	$\delta = 20$	$\delta = 30$	$\delta = 40$
$\tau = 1$	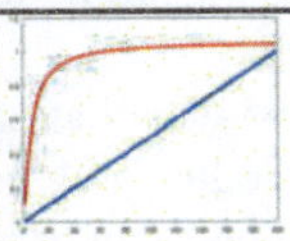	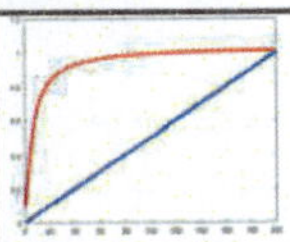	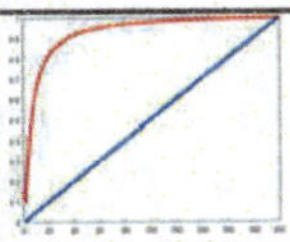	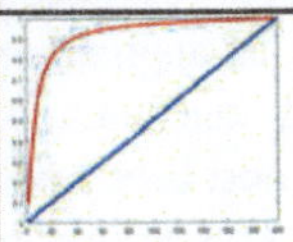
$\tau = 3$	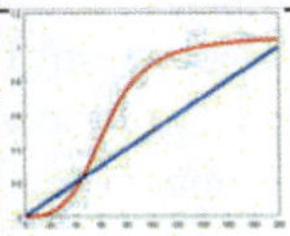	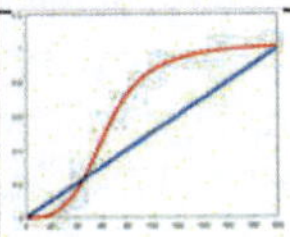	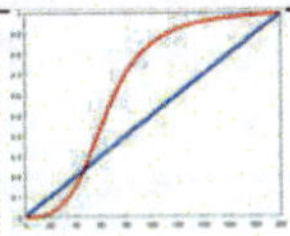	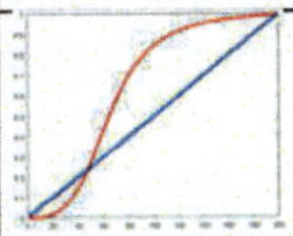
$\tau = 5$	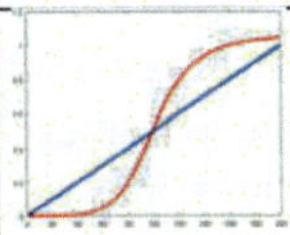	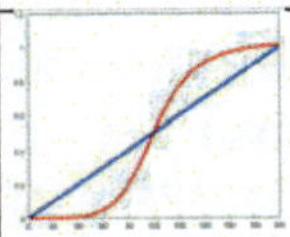	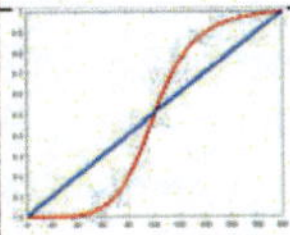	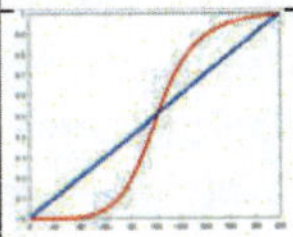
$\tau = 7$	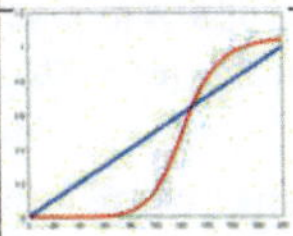	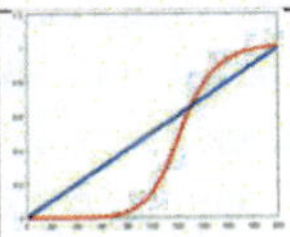	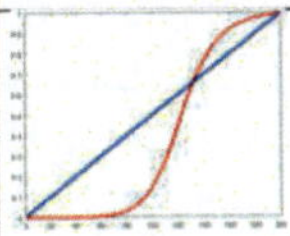	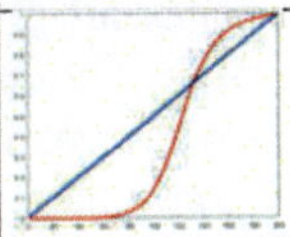
$\tau = 9$	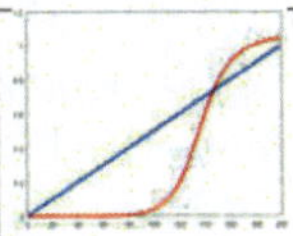	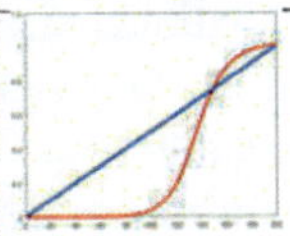	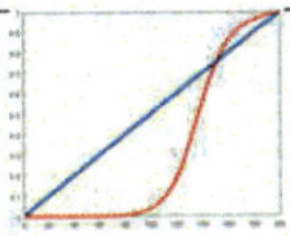	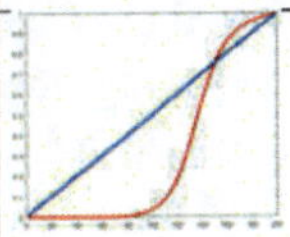

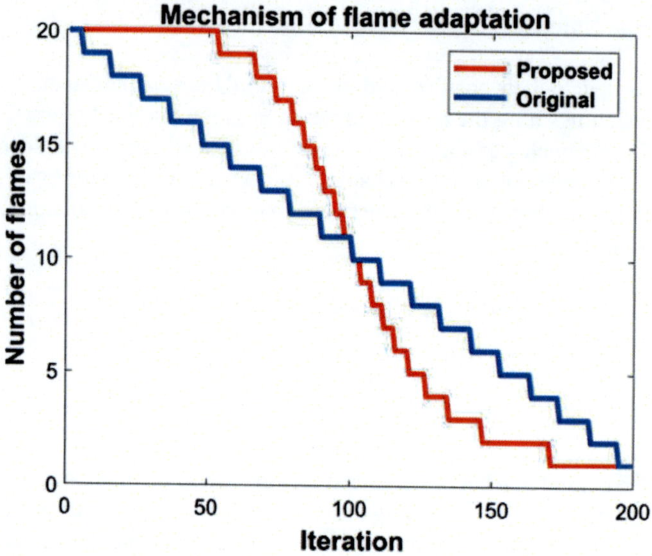

Fig. 7 Comparison of two decreasing mechanisms of flames

represents $\frac{t}{T}$ and the red line represents $\mu(t)$. From these pictures, we can find that the position of a turning point is dependent on τ. Besides, $\frac{t}{T} \to 1$ when $t \to T$. On the other hand, similarly, $\mu(t)$ should be close to 1 when t is close to T. Therefore, δ should be selected as 30 or 40 or above, and more analysis should be made on specific applications as well.

To keep an appropriate equilibrium between exploration and exploitation cores, the most balanced position of the turning point stands in the middle timeline during the iterative searching process. Furthermore, after some experiments on benchmark functions to determine actual values of δ and τ, it proved that a combination of $\delta = 30$ and $\tau = 5$ could yield good performance. Figure 7 compares the proposed and original mechanisms when the number of moths is 20, and the number of maximum iterations is set to 200.

Based on the exhibition in Fig. 7, the proposed mechanism can be regarded to split the whole iterative searching process into three parts: exploration, hybrid exploration and exploitation, and exploitation. In the exploration phase (iterations under 80), more flames are given for agents to update their positions in the space as completely as possible. On the contrary, in the exploitation stage (iterations over 120), fewer flames with good fitness values are left to limit the population's range to update. In the hybrid section (iterations around 100), the number of flames drops quickly to improve the convergence rate for less computational time.

Employing the above two innovations in the standard MFO, a new algorithm named Adaptive Moth-Flame Optimization with Opposition-Based Learning (AMFOOBL) is proposed. The pseudo-code of AMFOOBL is shown in Algorithm 1.

Algorithm 1: Adaptive Moth-Flame Optimization with Opposition-Based Learning (AMFOOBL)

```
Input: N (moth population), d (dimension), T (max iteration), fitness (fitness function),
       ub (upper bound), lb (lower bound)
            %%%Randomly initialize each individual in moths%%%
   for i = 1 : N
            M (i) = lb + rand (N, d)*(ub - lb);
   end
        %%%Generate opposite positions of moths using Eq. (3.10)%%%
   M*(i) = lb + ub – M (i);
                 %%%Reset the population of moths%%%
   M (i) = [M (i) ; M*(i)];
           %%%Update the position of each individual in moths%%%
   Initialize the iteration count t = 1;
   while t < T + 1
            Update the number of flames using Eq. (3.11) instead of Eq. (3.6);
            OM = fitness (M);
            if t == 1
                     F = sort (M); OF = sort (OM);
            else
                     F = sort [M (t - 1), M (t)]; OF = sort [OM (t - 1), OM (t)];
            end
            Best_flame = OF (1);                  %%Supposed to find the minimum
            Best_pos = F (1);
            for i = 1 : N
                     for j = 1 : d
                              Update the position of each moth using Eq. (3.5);
                     end
            end
            t = t + 1;
   end
Output: Best_flame, Best_pos
```

4.3 AMFOOBL for Training MLPs

There are three standard methods for training MLPs with metaheuristic techniques:

1. Finding a combination of connection weights and biases to achieve the minimum error.
2. Building a proper structure of the MLP in a given problem.
3. Determining the parameters of other learning algorithms, such as learning rate and momentum in gradient-based methods.

AMFOOBL is a new algorithm developed and utilized to train MLPs based on the first-mentioned method to optimize all weights and biases of the network. Figure 8 shows the basic structure of the AMFOOBL-based trainer.

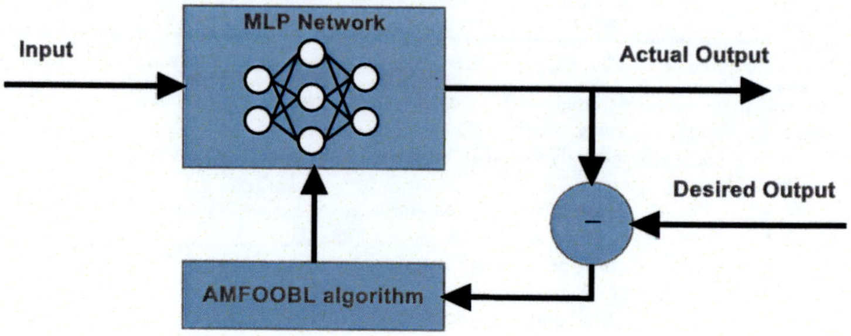

Fig. 8 AMFOOBL-based MLP trainer

4.3.1 Encoding Strategy

To represent the weights and biases in the encoding work, there are three strategies in the algorithms: vector, matrix, and binary [2, 17]. In this chapter, the vector encoding strategy is chosen because of its simplicity and high efficiency. Take the MLP network in Fig. 1 for example; the encoding can be shown as follows:

$$\text{moth} = \left[\omega_{input\_hidden}, \omega^*_{hidden\_output}, \theta_{hidden}, \alpha_{output}\right] \tag{4.13}$$

$$\omega_{input\_hidden} = \{\omega_{11}, \omega_{12}, \omega_{13}, \omega_{21}, \omega_{22}, \omega_{23}\} \tag{4.14}$$

$$\omega^*_{hidden\_output} = \left\{\omega^*_{11}, \omega^*_{12}, \omega^*_{13}, \omega^*_{21}, \omega^*_{22}, \omega^*_{23}\right\} \tag{4.15}$$

$$\theta_{hidden} = \{\theta_1, \theta_2, \theta_3\} \tag{4.16}$$

$$\alpha_{output} = \{\alpha_1, \alpha_2\} \tag{4.17}$$

where ω_{input_hidden} indicates connection weights between the input layer and hidden layer; $\omega^*_{hidden\_output}$ indicates connection weights between the hidden layer and output layer; θ_{hidden} indicates the biases in the hidden neurons; α_{output} indicates the biases in the output neurons.

When employing MAs in optimizing the weights and biases, it is necessary to determine the number of hidden layers and the number of neurons in each layer, which defines each search agent's dimension. The network structure is more complex with the increase of layers and neurons. In this part, the number of neurons in the input layer and the output layer is problem-decided, which will be illustrated in the experimental section. Besides, a fitness function is another essential component in MLP training with MAs, and according to Sect. 3, it is regarded as Eq. (3.5).

4.3.2 The Computation Complexity of AMFOOBL

In order to evaluate the running time of an algorithm, a rigorous analysis in assessing the computation complexity is necessary for AMFOOBL, which is defined based on the structure and implementation. For this task, the big-O notation is adopted [39], and the complexity of using AMFOOBL in MLP training is presented as:

$$O(MLP, AMFOOBL) = O(T(O(MLP) + O(AMFOOBL))) \quad (4.18)$$

$$O(AMFOOBL) = O(\text{quick sort}) + O(\text{position update}) \quad (4.19)$$

where T indicates the maximum iteration.

The complexity of the proposed method depends on the number of samples in datasets, the design of the MLP, the number of moths, the maximum number of iterations, and the sorting mechanism of flames in each iteration [2, 3, 9, 11, 17]. The computation complexity of the MLP with h hidden nodes, t training samples, and o outputs is equal to $O(t \times (h + o))$. Since quicksort is used to find the best flame, the sort's computational complexity is considered as $O(T \times N^2)$ in the worst case and $O(N\log N)$ the best case. Due to a combination of the OBL strategy in setting the moth population, the number of individuals is $2N$. Thus, the complexity of the sorting mechanism is $O(T \times 4N^2)$ in the worst case. Furthermore, the computation complexity of moths' positions updating is $O(T \times 2N \times d)$. Based on the above analysis, the final computation complexity of the proposed method is as follows:

$$O(MLP, AMFOOBL) = O\left(T\left(t \times (h + o) + \left(4N^2 + 2N \times d\right)\right)\right) \quad (4.20)$$

where N is the number of moths; d is the number of dimensions.

5 Experimental Simulations and Results

To validate the effectiveness of the proposed adaptive mechanism in decreasing the number of flames and to demonstrate the performance of AMFOOBL compared to standard MFO, the first experiment is conducted on a set of benchmark functions. Besides, the proposed algorithm is employed to train multilayer perceptron compared with some other advanced algorithms in the second experiment.

5.1 Experiment Setting

In the first experiment, 23 classical benchmark functions are utilized to test the usefulness of two involutions. As the OBL strategy has been largely verified in improving the performance of various MAs, the original adaptive approach in reducing the number of flames should be verified separately. Therefore, AMFOOBL is compared to both AMFO and MFO in finding the global optima of these functions. The number of populations is set to 20, and the number of maximum iterations is 200 for the mentioned three algorithms. Each algorithm is run 30 times to avoid stochastic characteristics and accidental errors. Moreover, six indexes: mean, standard deviation (Std), median, minimum, maximum, and computation time, are calculated for evaluation.

In the second experiment, AMFOOBL-MLP trainers are employed in five classification datasets from the UCI machine learning repository (XOR, Balloons, Iris, Breast cancer, and Heart disease) and three function approximation datasets (Sigmoid, Cosine, and Sines). In addition, another nine optimization algorithms are chosen to make a comparison, including ant lion optimizer (ALO) [40], salp swarm algorithm (SSA) [41], ant colony optimization (ACO) [42], whale optimization algorithm (WOA) [43], grey wolf optimizer (GWO) [44], Harris hawks optimization (HHO) [45], differential evolution (DE) [46], coefficient-based particle swarm optimization gravitational search algorithm (CPSOGSA) [47], and standard MFO [28]. Some of the specific parameters in these competitive algorithms are presented in Table 2. More descriptions of this experiment are illustrated in the subsection.

The aforementioned two experiments were implemented under MATLAB R2017b on the computer with a Windows 10 64bit, i7-8700 CPU, and 16 GB RAM.

5.2 Experiment 1: Benchmark Functions

5.2.1 Function Description

To evaluate the usefulness of the original mechanism in flame reduction and compare the overall performance of AMFOOBL with the standard MFO algorithm, 23 common test functions are adopted. All of these functions are minimization problems, which are different in size and complexity. Table 3 includes 7 unimodal functions; Table 4 presents 6 multimodal functions; Table 5 consists of 10 fixed-dimension multimodal functions. The above three tables show the functions' main characteristics in terms of the mathematical formulation, the range determining the search space's boundary, the search space's dimension, and the actual global optima. It should be noted that unimodal functions have a single optimum, while multimodal functions have more than one optimum. In addition, unimodal functions are mainly used to assess MAs' exploration ability, whereas multimodal functions are used to evaluate the exploitation ability more often.

Table 2 Specific parametric values in experimental algorithms

Algorithm	Symbol	Explanation	Value
AMFOOBL	δ	A constant in the flame adaptive mechanism	30
	τ	A constant in the flame adaptive mechanism	5
	b	A constant to define the shape of the spiral path	1
	θ	A random value to define the next position of moths	[-1,1]
ALO	ω	A constant defined the radius of ants' random walks	2,3,4,5,6
SSA	c_2, c_3	Random numbers regarding salps' position update	[0,1]
ACO	R_{ou}	Pheromone volatilization coefficient	0.3
	P_0	Evaporation rate	0.1
	T_{ou}	Initial pheromone	10
	Q	Pheromone update constant	1
WOA	$\vec{a}$	A linearly decreasing value	[0,2]
	b	A constant to define the logarithmic spiral shapes	1
	l	A random number	[-1,1]
GWO	$\vec{a}$	A linearly decreasing value	[0,2]
	r_1, r_2	Random vectors	[0,1]
HHO	β	A default constant in levy fight function	1.5
	E_0	Initial escaping energy of the prey	(-1,1)
DE	β_{max}	Maximum scaling factor	0.8
	β_{min}	Minimum scaling factor	0.2
	p_{CR}	Crossover probability	0.2
CPSOGSA	φ_1, φ_2	Control parameter	2.05
	φ	A parameter to define the constriction coefficient	4.1

Table 3 Unimodal benchmark functions

Unimodal functions	Dim	Range	f_{min}
$f_1(x) = \sum_{i=1}^{d} x_i^2$	10	[-100,100]	0
$f_2(x) = \sum_{i=1}^{d} \lvert x_i \rvert + \prod_{i=1}^{d} \lvert x_i \rvert$	10	[-10,10]	0
$f_3(x) = \sum_{i=1}^{d} \left(\sum_{j=1}^{i} x_j\right)^2$	10	[-100,100]	0
$f_4(x) = \max_i \lvert x_i \rvert, 1 \le i \le d$	10	[-100,100]	0
$f_5(x) = \sum_{i=1}^{d-1} [100(x_{i+1} - x_i^2)^2 + (x_i - 1)^2]$	10	[-30,30]	0
$f_6(x) = \Sigma_{i=1}^{d} ([x_i + 5])^2$	10	[-100,100]	0
$f_7(x) = \Sigma_{i=1}^{d} i \cdot x_i^4 + \text{rand}[0, 1]$	10	[-1.28,1.28]	0

Table 4 Multimodal benchmark functions

Multimodal functions	Dim	Range	f_{min}
$f_8(x) = \sum_{i=1}^{d} - x_i \sin(\sqrt{\lvert x_i \rvert})$	10	[-500,500]	-418.982
$f_9(x) = \Sigma_{i=1}^{d}\left[x_i^2 - 10\cos(2x_i\pi) + 10\right]$	10	[-5.12,5.12]	0
$f_{10}(x) = -20\exp\left(-0.2\sqrt{\frac{1}{d}\sum_{i=1}^{d} x_i^2}\right) - \exp\left[\frac{1}{d}\sum_{i=1}^{d}(cos2x_i\pi)\right] + 20 + e$	10	[-32,32]	0
$f_{11}(x) = \frac{1}{4000}\sum_{i=1}^{d} x_i^2 - \prod_{i=1}^{d}\cos\left(\frac{x_i}{\sqrt{i}}\right) + 1$	10	[-600,600]	0
$f_{12}(x) = \frac{\pi}{d}\left\{10\sin(\pi y_1) + \sum_{i=1}^{d-1}(y_i - 1)^2\left[1 + 10\sin^2(\pi y_{i+1})\right] + (y_d - 1)^2\right\} + \sum_{i=1}^{d} u(x_i, 10, 100, 4)$ $y_i = 1 + \frac{x_i + 1}{4}$ $u(x_i, a, k, m) = \begin{cases} k(x_i - a)^m & x_i > a \\ 0 & -a < x_i < a \\ (-x_i - a)^m & x_i < -a \end{cases}$	10	[-50,50]	0
$f_{13}(x) = 0.1\left\{\sin^2(3\pi x_1) + \sum_{i=1}^{d}(x_i - 1)^2[1 + \sin^2(3\pi x_i + 1)] + (x_d - 1)^2\left[1 + \sin^2(2\pi x_d)\right] + \sum_{i=1}^{d} u(x_i, 5, 100, 4)\right\}$	10	[-50,50]	0

Table 5 Fixed-dimension multimodal benchmark functions

Fixed-dimension multimodal functions	Dim	Range	f_{min}
$f_{14}(x) = \left(\frac{1}{500} + \sum_{j=1}^{25} \frac{1}{j+\sum_{i=1}^{2}(x_i - x_{ij})^6}\right)^{-1}$	2	[-65,65]	1
$f_{15}(x) = \sum_{i=1}^{11}\left[a_i - \frac{x_i(b_i^2+b_i x_2)}{b_i^2+b_i x_3+x_4}\right]^{-1}$	4	[-5,5]	0.00030
$f_{16}(x) = 4x_1^2 - 2.1x_1^4 + \frac{1}{3}x_1^6 + x_1 + x_2 - 4x_2^2 + 4x_2^4$	2	[-5,5]	-1.0316
$f_{17}(x) = \left(x_2 - \frac{5.1}{4\pi^2}x_1^2 + \frac{5}{\pi}x_1 - 6\right)^2 + 10(1 - \frac{1}{8\pi})\cos x_1 + 10$	2	[-5,5]	0.398
$f_{18}(x) =$ $\left[1 + (x_1 + x_2 + 1)^2(19 - 14x_1 + 3x_1^2 - 14x_2 + 6x_1x_2 + 3x_2^2)\right] \times \left[30 + (2x_1 - 3x_2)^2 \times (18 - 32x_1 + 12x_1^2 + 48x_2 - 36x_1x_2 + 27x_2^2)\right]$	2	[-2,2]	3
$f_{19}(x) = -\sum_{i=1}^{d} c_i \exp\left(-\sum_{j=1}^{3} a_{ij}(x_j - p_{ij})^2\right)$	3	[1, 3]	-3.86
$f_{20}(x) = -\sum_{i=1}^{4} c_i \exp\left(-\sum_{j=1}^{6} a_{ij}(x_j - p_{ij})^2\right)$	6	[0,1]	-3.32
$f_{21}(x) = -\sum_{i=1}^{5}\left[(X - a_i)(X - a_i)^T + c_i\right]^{-1}$	4	[0,10]	-10.1532
$f_{22}(x) = -\sum_{i=1}^{7}\left[(X - a_i)(X - a_i)^T + c_i\right]^{-1}$	4	[0,10]	-10.4028
$f_{23}(x) = -\sum_{i=1}^{10}\left[(X - a_i)(X - a_i)^T + c_i\right]^{-1}$	4	[0,10]	-10.5363

5.2.2 Convergence Curves

In this subsection, the convergence curves of AMFOOBL, AMFO, and MFO in 23 functions are shown in Tables 6, 7, 8. The convergence curves perform the average best-so-far solution progress over the iterative number, which serves as a one-running result in each test function.

According to curves in thc above 23 figures, we observe that the computation speed towards better results is enhanced relatively for the proposed AMFOOBL. In Table 6, dealing with F1, it is clear that the exploration trends of different approaches are very competitive; however, after half of the iterations, the AMFOOBL can beat others to some extent. It is similar in F3 and F4 that AMFOOBL performs its powerful capability in the exploitation period. For multimodal problems in Table 7, we find that the speed is more distinct from others. Such a trend shows that the AMFOOBL has reached a more stable balance among exploration and exploitation, and it can shift in a better time to more exploitation in later stages. For functions in Table 8, the results are more competitive, and an apparent speed up in the proposed AMFOOBL can be detected. According to these results, we see the better speed of the process is reached. Therefore, we can conclude the effectiveness of the proposed mechanism in mitigating the slow convergence issues to a sufficient extent. Another reason is the role of the OBL strategy, which is used to improve the population's diversity in the

Table 6 Convergence curves of AMFOOBL, AMFO, and MFO in Unimodal functions

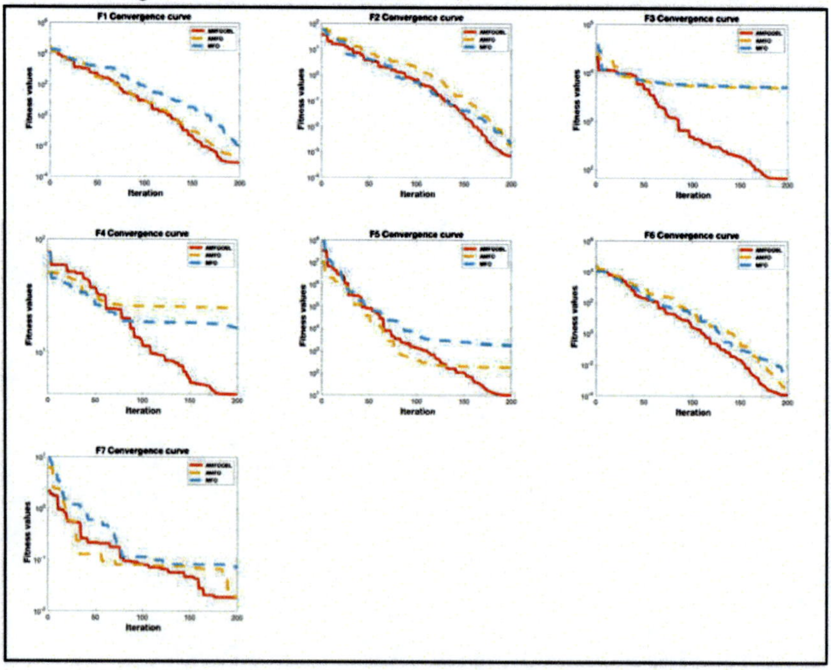

Table 7 Convergence curves of AMFOOBL, AMFO, and MFO in Multimodal functions

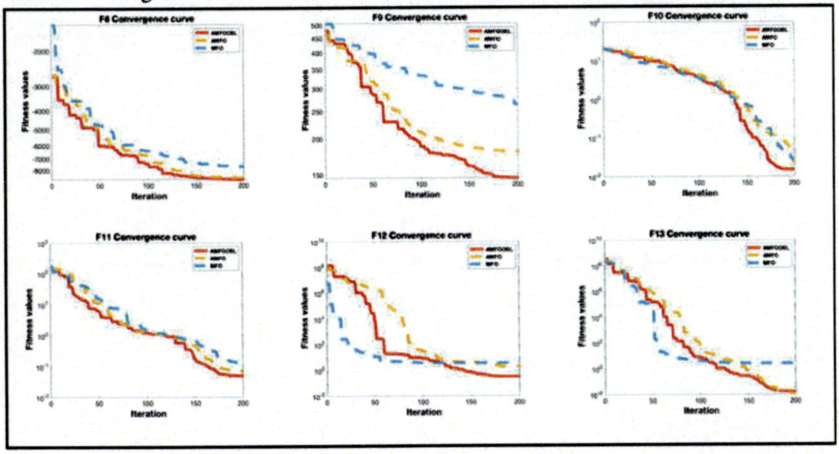

initial process, decreasing the possibility of local stagnation and increasing optimal solutions' accuracy.

5.2.3 Simulation Results and Statistical Test

To identify the modification's usefulness based on the standard MFO algorithm, six quantitative metrics are calculated for comparison, including mean, standard deviation (Std), median, minimum, maximum, and computation time. Each algorithm (AMFOOBL, AMFO, and MFO) runs 30 times in every function. As it aims to find the minimum of benchmark functions, the minimum over 30 times indicates the best value the algorithm ever got, whereas the maximum suggests the worst value. Table 9 shows the results in the statistics of AMFOOBL, AMFO, and MFO. Moreover, the computation time of them is presented in Table 10.

In Table 9, "1/0/-1" summarizes a comparison between AMFOOBL with AMFO and MFO. "1" indicates the AMFO or MFO is better than AMFOOBL; "0" indicates AMFO or MFO is equal to AMFOOBL; "-1" indicates AMFO or MFO is worse than AMFOOBL. It can be seen that AMFOOBL is superior to AMFO and MFO regarding mean values in 17 and 19 functions, respectively. Table 10 shows that AMFO has a faster convergence rate than MFO, suggesting a positive effect of the new adaptive mechanism in flame reduction. In other words, the balance between exploitation and exploration in AMFO is better that improves both accuracy and convergence speed.

Statistical tests are often used for checking significant improvement by a new algorithm over other existing algorithms. In this section, the Wilcoxon signed-rank test with the best values over 30 times is utilized to assess the significant difference

Table 8 Convergence curves of AMFOOBL, AMFO, and MFO in Fixed-dimension multimodal functions

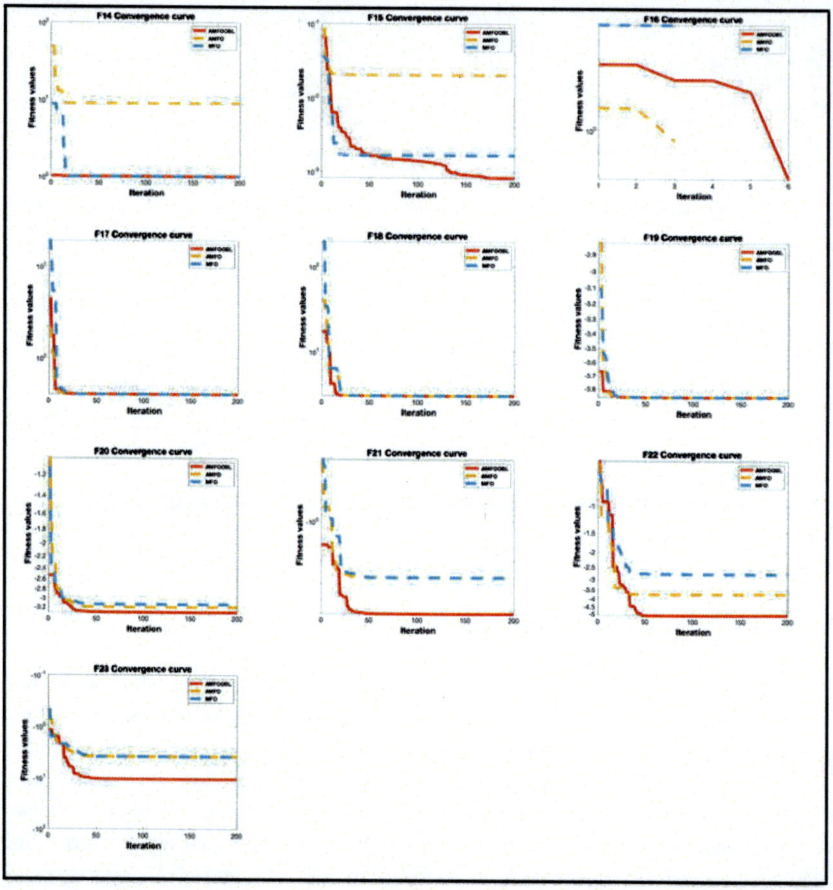

between the proposed algorithm AMFOOBL and standard MFO. The null hypothesis is constructed as: there is no significant difference between the two compared algorithms. When $p < 0.05$ suggests that the null hypothesis can be rejected at a 5% significant level. Table 11 presents the Wilcoxon test's results (p-value) of AMFOOBL compared to AMFO and MFO, respectively.

According to the qualitative results, we could see that the proposed AMFOOBL has significantly mitigated some core problems of stagnation, and the results are more enhanced compared to the initial version in most cases. For example, there are several reasons; the original adaptive mechanism has to reduce the number of flames around which agents update their positions, leading to better balance among the exploration and exploitation stages. Another reason is the heroine of the OBL approach. The OBL

Table 9 Comparison of AMFOOBL with AMFO and MFO on benchmark functions

Function	Algorithm	Mean	Std	Median	Best (Min)	Worst (Max)
Unimodal						
F1	AMFOOBL	**1.85E-04**	**3.34E-04**	**5.37E-05**	**7.19E-07**	**1.40E-03**
	AMFO	7.55E-04	1.10E-03	3.50E-04	2.05E-05	4.80E-03
	MFO	5.50E-03	5.00E-03	4.20E-03	2.56E-04	2.01E-02
F2	AMFOOBL	**3.34E-01**	**1.83E + 00**	**4.74E-04**	**3.80E-05**	1.00E + 01
	AMFO	1.67E + 00	3.79E + 00	1.70E-03	3.19E-04	1.00E + 01
	MFO	1.01E + 00	3.05E + 00	9.00E-03	1.30E-03	**1.00E + 01**
F3	AMFOOBL	**5.14E + 02**	1.52E + 03	**1.20E + 01**	**4.94E-01**	5.00E + 03
	AMFO	1.00E + 03	2.52E + 03	1.02E + 02	1.57E + 01	**1.16E + 04**
	MFO	6.33E + 02	**1.49E + 03**	1.24E + 02	6.65E + 00	5.06E + 03
F4	AMFOOBL	**1.00E + 00**	**1.21E + 00**	**6.97E-01**	**8.51E-02**	**6.31E + 00**
	AMFO	1.17E + 01	1.17E + 01	8.35E + 00	1.36E + 00	4.46E + 01
	MFO	1.76E + 01	1.46E + 01	1.16E + 01	3.57E + 00	6.18E + 01
F5	AMFOOBL	**2.57E + 02**	**7.59E + 02**	**2.04E + 01**	1.35E + 00	**3.02E + 03**
	AMFO	6.64E + 03	2.26E + 04	1.51E + 02	**6.98E-01**	9.00E + 04
	MFO	6.33E + 03	2.27E + 04	1.17E + 02	7.06E-01	9.00E + 04
F6	AMFOOBL	**1.73E-04**	**2.56E-04**	**1.00E-04**	**1.21E-06**	**1.30E-03**
	AMFO	7.19E-04	9.01E-04	3.52E-04	2.59E-05	3.80E-03
	MFO	9.70E-03	1.16E-02	4.40E-03	1.97E-04	4.60E-02
F7	AMFOOBL	**1.49E-02**	**9.10E-03**	**1.35E-02**	**1.40E-03**	**3.46E-02**
	AMFO	3.15E-02	2.00E-02	2.64E-02	9.30E-03	9.18E-02
	MFO	2.73E-02	1.77E-02	2.17E-02	7.30E-03	7.02E-02
Multimodal						
F8	AMFOOBL	**-3.18E + 03**	**2.73E + 02**	**-3.17E + 03**	-3.61E + 03	**-2.64E + 03**
	AMFO	-3.14E + 03	3.53E + 02	-3.12E + 03	**-4.07E + 03**	-2.52E + 03
	MFO	-3.10E + 03	3.88E + 02	-3.05E + 03	-3.83E + 03	-2.52E + 03
F9	AMFOOBL	**2.26E + 01**	1.35E + 01	**1.89E + 01**	**9.95E-01**	5.77E + 01
	AMFO	2.92E + 01	**1.23E + 01**	2.74E + 01	3.98E + 00	**4.97E + 01**
	MFO	3.00E + 01	1.41E + 01	2.84E + 01	5.98E + 00	6.17E + 01
F10	AMFOOBL	**7.74E-01**	**3.64E + 00**	**3.40E-03**	**6.35E-04**	1.99E + 01
	AMFO	1.92E + 00	4.97E + 00	1.99E-02	2.90E-03	1.99E + 01
	MFO	1.48E + 00	3.64E + 00	6.26E-02	1.04E-02	**1.87E + 01**
F11	AMFOOBL	1.45E-01	**5.94E-02**	1.34E-01	3.68E-02	**3.10E-01**
	AMFO	**1.31E-01**	7.05E-02	**1.16E-01**	**2.52E-02**	3.12E-01
	MFO	1.91E-01	1.20E-01	1.57E-01	5.22E-02	5.14E-01

(continued)

Table 9 (continued)

Function	Algorithm	Mean	Std	Median	Best (Min)	Worst (Max)
F12	AMFOOBL	**1.97E-01**	**3.87E-01**	**1.26E-05**	**6.29E-08**	**1.55E + 00**
	AMFO	1.00E + 00	2.62E + 00	3.11E-01	3.33E-06	1.29E + 01
	MFO	1.09E + 00	2.10E + 00	7.64E-02	1.45E-04	1.02E + 01
F13	AMFOOBL	**2.45E-02**	**1.13E-01**	**3.30E-05**	**5.04E-07**	**6.23E-01**
	AMFO	5.17E-01	1.35E + 00	1.22E-02	2.31E-05	5.91E + 00
	MFO	3.89E-01	1.54E + 00	1.17E-02	3.26E-05	8.28E + 00
Fixed-dimension multimodal						
F14	AMFOOBL	**2.08E + 00**	**2.09E + 00**	**9.98E-01**	**9.98E-01**	**7.87E + 00**
	AMFO	2.97E + 00	2.46E + 00	1.99E + 00	**9.98E-01**	1.08E + 01
	MFO	3.39E + 00	2.90E + 00	2.49E + 00	**9.98E-01**	1.17E + 01
F15	AMFOOBL	1.50E-03	**3.60E-03**	**7.82E-04**	4.56E-04	**2.04E-02**
	AMFO	**2.80E-03**	4.10E-03	1.30E-03	6.71E-04	**2.04E-02**
	MFO	2.20E-03	3.70E-03	1.40E-03	**3.11E-04**	**2.04E-02**
F16	AMFOOBL	**-1.03E + 00**	6.71E-16	**-1.30E + 00**	**-1.30E + 00**	**-1.30E + 00**
	AMFO	**-1.03E + 00**	6.51E-16	**-1.30E + 00**	**-1.30E + 00**	**-1.30E + 00**
	MFO	**-1.03E + 00**	**6.32E-16**	**-1.30E + 00**	**-1.30E + 00**	**-1.30E + 00**
F17	AMFOOBL	**3.98E-01**	**0.00E + 00**	**3.98E-01**	**3.98E-01**	**3.98E-01**
	AMFO	**3.98E-01**	**0.00E + 00**	**3.98E-01**	**3.98E-01**	**3.98E-01**
	MFO	**3.98E-01**	**0.00E + 00**	**3.98E-01**	**3.98E-01**	**3.98E-01**
F18	AMFOOBL	**3.00E + 00**	2.86E-15	**3.00E + 00**	**3.00E + 00**	**3.00E + 00**
	AMFO	**3.00E + 00**	**2.66E-15**	**3.00E + 00**	**3.00E + 00**	**3.00E + 00**
	MFO	**3.00E + 00**	2.84E-15	**3.00E + 00**	**3.00E + 00**	**3.00E + 00**
F19	AMFOOBL	**-3.86E + 00**	2.71E-15	**-3.86E + 00**	**-3.86E + 00**	**-3.86E + 00**
	AMFO	**-3.86E + 00**	**2.61E-15**	**-3.86E + 00**	**-3.86E + 00**	**-3.86E + 00**
	MFO	**-3.86E + 00**	1.40E-03	**-3.86E + 00**	**-3.86E + 00**	**-3.86E + 00**
F20	AMFOOBL	**-3.25E + 00**	**5.83E-02**	**-3.20E + 00**	**-3.32E + 00**	**-3.20E + 00**
	AMFO	-3.21E + 00	1.03E-01	**-3.20E + 00**	**-3.32E + 00**	-2.81E + 00
	MFO	-3.22E + 00	6.01E-02	**-3.20E + 00**	**-3.32E + 00**	-3.08E + 00
F21	AMFOOBL	**-7.47E + 00**	3.22E + 00	**-1.02E + 01**	**-1.02E + 01**	**-2.63E + 00**
	AMFO	-6.21E + 00	**3.21E + 00**	**-1.02E + 01**	**-1.02E + 01**	**-2.63E + 00**
	MFO	**2.08E + 00**	**2.09E + 00**	**9.98E-01**	**9.98E-01**	**7.87E + 00**
F22	AMFOOBL	2.97E + 00	2.46E + 00	1.99E + 00	**9.98E-01**	1.08E + 01
	AMFO	3.39E + 00	2.90E + 00	2.49E + 00	**9.98E-01**	1.17E + 01
	MFO	1.50E-03	**3.60E-03**	**7.82E-04**	4.56E-04	**2.04E-02**
F23	AMFOOBL	**2.80E-03**	4.10E-03	1.30E-03	6.71E-04	**2.04E-02**

(continued)

Table 9 (continued)

Function	Algorithm	Mean	Std	Median	Best (Min)	Worst (Max)
	AMFO	2.20E-03	3.70E-03	1.40E-03	**3.11E-04**	**2.04E-02**
	MFO	**-1.03E + 00**	6.71E-16	**-1.30E + 00**	**-1.30E + 00**	**-1.30E + 00**
1/0/-1	AMFOOBL	~	~	~	~	~
	AMFO	2/4/17	4/1/18	1/6/16	3/9/11	1/7/15
	MFO	0/4/19	2/1/20	0/7/16	1/9/13	2/7/14
Rank	AMFOOBL	1	1	1	1	1
	AMFO	2	2	2	2	3
	MFO	3	3	3	3	2

Table 10 Computation time of AMFOOBL, AMFO, and MFO in benchmark functions

Function	Computation time (in seconds)		
	AMFOOBL	AMFO	MFO
Unimodal			
F1	0.021049	**0.010829**	0.011430
F2	0.023617	**0.012214**	0.012523
F3	0.060772	0.031380	**0.030917**
F4	0.026411	0.014383	**0.013597**
F5	0.029679	**0.015362**	0.015504
F6	0.020600	**0.012093**	0.014243
F7	0.033308	**0.017164**	0.017982
Multimodal			
F8	0.027093	**0.014088**	0.014275
F9	0.022717	0.012158	**0.011894**
F10	0.026485	**0.013838**	0.013839
F11	0.031265	**0.016362**	0.016490
F12	0.082586	**0.042312**	0.042691
F13	0.071322	**0.036279**	0.036735
Fixed-dimension multimodal			
F14	0.217110	**0.109260**	0.141070
F15	0.018942	**0.010063**	0.010105
F16	0.015272	**0.008210**	0.008281
F17	0.012800	0.006909	**0.006710**
F18	0.012592	**0.006911**	0.006983
F19	0.026738	0.014381	**0.013747**
F20	0.027965	**0.014483**	0.014676
F21	0.054571	**0.027743**	0.027894
F22	0.069140	**0.035372**	0.035429
F23	0.093518	**0.046699**	0.046788

Table 11 Wilcoxon test between AMFOOBL with AMFO and MFO

Function	AMFOOBL versus AMFO	AMFOOBL versus MFO
Unimodal		
F1	1.90e-03**	3.18e-06***
F2	4.40e-05***	2.37e-05***
F3	1.10e-03**	2.20e-03**
F4	1.90e-06***	1.92e-06***
F5	1.24e-02*	1.85e-02*
F6	4.10e-03**	1.92e-06***
F7	2.61e-04***	1.04e-02*
Multimodal		
F8	0.5857	0.2210
F9	2.84e-02*	2.84e-02*
F10	2.41e-04***	2.22e-04***
F11	0.4165	0.0936
F12	2.43e-02*	1.95e-02*
F13	1.40e-03**	2.70e-03**
Fixed-dimension multimodal		
F14	0.0922	0.0685
F15	0.0041**	6.40e-03**
F16	1.0000	1.0000
F17	1.0000	1.0000
F18	1.05e-02*	0.0880
F19	1.0000	1.0000
F20	2.50e-02*	1.90e-03**
F21	0.1156	0.0956
F22	5.20e-03**	2.81e-02*
F23	1.35e-02*	1.84e-02*
Results	**16/23** at a 5% significant level	**15/23** at a 5% significant level

* indicates $p<0.05$; ** indicates $p<0.01$; *** indicates $p<0.001$

improves the population's multiplicity in the early process, declining the possibility of local stagnation and increasing optimal solutions' accurateness.

5.3 Experiment 2: Multilayer Perceptron Training

In this experiment, every agent is randomly initialized in the range of [-10,10]. The number of populations and maximum iterations in training datasets are shown in Table 12. In addition, each algorithm is repeated over ten independent trials.

A critical factor in the experimental setup is the structure of MLPs; the number of neurons in the hidden layer is based on the Kolmogorov theory [48] suggested as $2 \times n + 1$, where n represents the number of inputs of the datasets. Besides, normalization is an essential step when the attributes of datasets are in different ranges. The min–max normalization is used in this experiment formulated in (5.1), where maps x in the interval of $[a, b]$ to $[c, d]$.

$$X' = \frac{(x - a) \times (d - c)}{(b - a)} + c \tag{5.1}$$

5.3.1 Datasets Description

Five pattern classification datasets from the UCI machine learning repository (https://archive.ics.uci.edu/ml/index.php) and three function approximation datasets are employed to train MLPs.

- **Pattern classification**

(1) XOR dataset

A nonlinear benchmark problem aims to recognize the number of "1" in the input vector. The XOR result of the input vector is returned; if the number of "1" in the input vector is odd, the outcome is returned to 1. Otherwise, the output would be 0 if the input vector contains an even number of "1".

Table 12 Algorithm setting for MLP training

Dataset	Population size	Maximum number of iterations	Number of runs
XOR	20	200	10
Balloons	20	200	10
Iris	100	500	10
Breast cancer	100	500	10
Heart disease	20	500	10
Sigmoid	20	200	10
Cosine	30	200	10
Sine	20	500	10

(2) Balloons dataset

This dataset is based on different conditions of an experiment in blowing up a balloon, including four attributes: color, size, act, and age. The outcome indicates whether the balloon is inflated or not. The outcome of "1" suggests the balloon is inflated, whereas "0" means not.

(3) Iris dataset

This dataset is one of the most popular datasets in the literature, which has 150 samples among four features: sepal length, sepal width, petal length, and petal width. All of them are divided into three classes of Setosa, Versicolor, and Virginica.

(4) Breast cancer dataset

This breast cancer dataset was obtained from the University of Wisconsin Hospitals, Madison, from Dr. William H. Wolberg [49]. There are nine attributes among 699 instances: clump thickness, uniformity of cell size and cell shape, marginal adhesion, single epithelial cell size, bare nuclei, bland chromatin, normal nucleoli, and mitoses. The two classifications in the output are benign and malignant.

(5) Heart disease dataset

This dataset was created for diagnosing heart disease patients, which consists of 267 images summarized from 22 attributes. The format in this dataset is binary; the outcome of 1 indicates the patient is normal, whereas 0 suggests the patient is abnormal.

Table 13 illustrates five classification datasets and MLP structures used in this experiment.

- Function approximation

Table 14 presents three function approximation datasets. Besides, it should be noted that the number of hidden neurons in function approximation is set as 15, and the structure of MLP is 1–15-1.

(1) Sigmoid dataset

The sigmoid function is in the interval of [-3,3] with an increment of 0.1; hence the total number of training data is 61. The number of test samples is 121with an increment of 0.05 in the same range.

Table 13 Classification datasets

Dataset	#Attributes	#Training samples	#Test samples	#Classes	MLP structure
XOR	3	8	8	2	3–7-1
Balloons	4	16	16	2	4–9-1
Iris	4	150	150	3	4–9-3
Breast cancer	9	599	100	2	9–19-1
Heart disease	22	80	187	2	22–45-1

Table 14 Function approximation datasets

Dataset	Training samples	Test samples
$Sigmoid: y = 1/\left(1 + e^{-x}\right)$	61:$x \in [-3:0.1:3]$	121:$x \in [-3:0.05:3]$
$Cosine: y = (\cos(x\pi/2))^7$	31:$x \in [-1.25:0.05:2.75]$	38:$x \in [-1.25:0.04:2.75]$
$Sine: y = \sin(2x)$	126:$x \in [-2\pi:0.1:2\pi]$	252:$x \in [-2\pi:0.05:2\pi]$

(2) Cosine dataset

This dataset consists of 31 training samples and 38 test samples, which lies in the interval of [1.25,2.75] with the step size of 0.05 and 0.04, respectively.

(3) Sine dataset

The sine function is the most challenging dataset to be approximated due to four peaks. This dataset has 126 training samples and 252 test samples in the interval of $[-2\pi, 2\pi]$ with a rise of 0.1 and 0.05 each time.

5.3.2 Experimental Training Results

The average fitness values (MSE) and standard deviation (Std) over ten independent runs are reported in the result analysis to assess the performance of AMFOOBL alongside nine other competitive algorithms. These two statistical metrics suggest the capability of each algorithm in terms of avoiding the local optima. Furthermore, the best classification rate (CR) and the average test error (TE) obtained are reported to evaluate the performance of each algorithm.

Wilcoxon's signed-rank test on MSE is also calculated to check the significant difference between the proposed algorithm AMFOOBL and other competitive algorithms. The null hypothesis is constructed as: there is no significant difference between the two compared algorithms. When $p < 0.05$ suggests that the null hypothesis can be rejected at a 5% significant level.

(1) XOR

The results of training algorithms for this dataset are presented in Table 15, Figs. 9 and 10.

(2) Balloons

The results of training algorithms for this dataset are presented in Table 16, Figs. 11 and 12.

(3) Iris

The results of training algorithms for this dataset are presented in Table 17, Figs. 13 and 14.

Table 15 Experimental results of algorithms for XOR dataset

Algorithm	MSE (Ave ± Std)	Test error (TE)	*p*-value
AMFOOBL	**7.55e-06 ± 1.18e-05**	**100%**	N/A
AMFO	1.93e-05 ± 3.22e-05	91.25%	0.0556
MFO	0.00140 ± 0.00420	91.25%	0.0495*
SSA	0.01500 ± 0.03160	91.25%	0.0097**
CPSOGSA	0.07790 ± 0.08660	40.00%	0.0019**
ACO	0.50000 ± **0.00000**	50.00%	0.0019**
DE	0.01540 ± 0.02200	33.75%	0.0019**
HHO	0.09930 ± 0.10150	26.25%	0.0019**
WOA	0.13160 ± 0.06560	32.50%	0.0019**
GWO	0.02020 ± 0.03350	63.75%	0.0019**

* indicates $p<0.05$; ** indicates $p<0.01$; *** indicates $p<0.001$

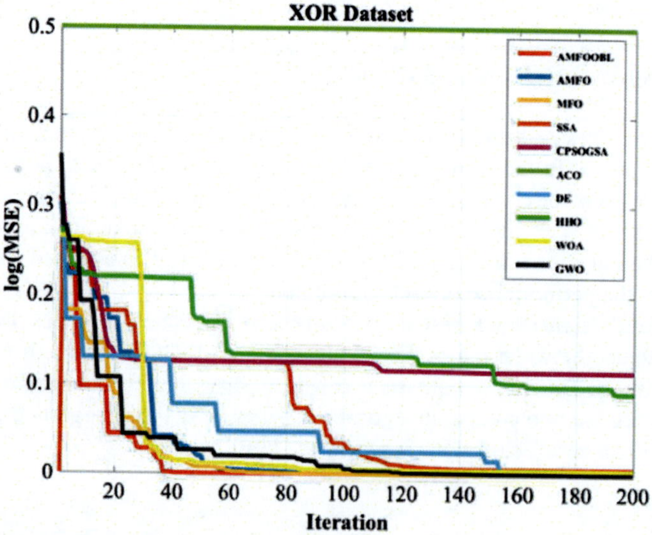

Fig. 9 Convergence curves of algorithms in XOR dataset

(4) Breast cancer

The results of training algorithms for this dataset are presented in Table 18, Figs. 15 and 16.

(5) Heart disease

The results of training algorithms for this dataset are presented in Table 19, Figs. 17 and 18.

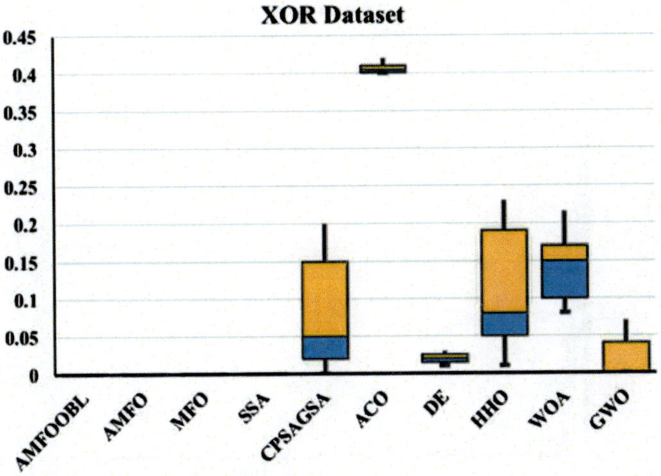

Fig. 10 Box plot of algorithms in XOR dataset

Table 16 Experimental results of algorithms for Balloons dataset

Algorithm	MSE (Ave ± Std)	Test error (TE)	*p*-value
AMFOOBL	8.77e-12 ± 2.77e-11	**100%**	N/A
AMFO	**1.34e-12 ± 4.17e-12**	**100%**	0.1601
MFO	5.85e-10 ± 1.85e-09	**100%**	0.1054
SSA	1.73e-06 ± 5.35e-06	**100%**	0.0019**
CPSOGSA	9.62e-06 ± 1.38e-05	**100%**	0.0019**
ACO	0.60000 ± **1.17e-16**	40.00%	0.0019**
DE	8.09e-07 ± 2.52e-06	87.50%	0.0019**
HHO	3.04e-06 ± 7.17e-06	**100%**	0.0019**
WOA	0.03980 ± 0.03520	69.50%	0.0019**
GWO	2.56e-06 ± 8.10e-06	**100%**	0.0488*

* indicates $p<0.05$; ** indicates $p<0.01$; *** indicates $p<0.001$

(6) Sigmoid

The results of training algorithms for this dataset are presented in Table 20, Figs. 19 and 20.

(7) Cosine

The results of training algorithms for this dataset are presented in Table 21, Figs. 21 and 22.

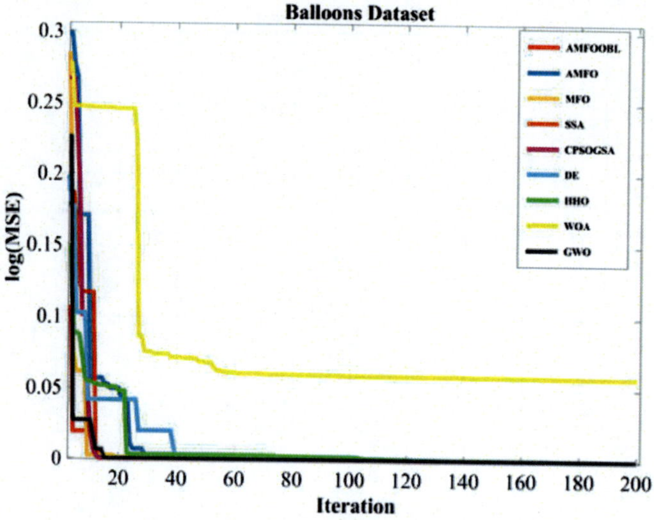

Fig. 11 Convergence curves of algorithms in Balloons dataset

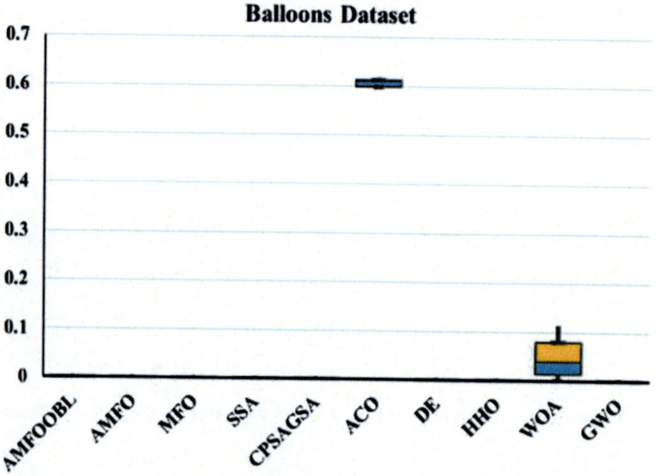

Fig. 12 Box plot of algorithms in Balloons dataset

(8) Sine

The results of training algorithms for this dataset are presented in Table 22, Figs. 23 and 24.

Table 17 Experimental results of algorithms for Iris dataset

Algorithm	MSE (Ave ± Std)	Test error (TE)	*p*-value
AMFOOBL	**0.04508 ± 0.0117**	**91.33%**	N/A
AMFO	0.06701 ± 0.0118	88.00%	0.0425*
MFO	0.08309 ± 0.0305	85.33%	0.0384*
SSA	0.07654 ± 0.0232	82.67%	0.0092**
CPSOGSA	0.28460 ± 0.0496	82.67%	0.0085**
ACO	0.55702 ± 0.0273	33.33%	0.0019**
DE	0.12920 ± 0.0393	44.67%	0.0019**
HHO	0.11673 ± **0.0002**	76.00%	0.0045**
WOA	0.15095 ± 0.1251	32.67%	0.0009***
GWO	0.07212 ± 0.0018	33.33%	0.0062**

* indicates $p<0.05$; ** indicates $p<0.01$; *** indicates $p<0.001$

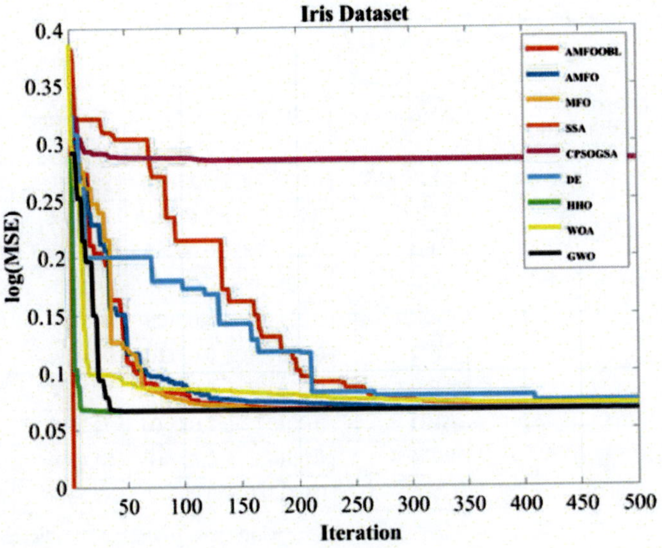

Fig. 13 Convergence curves of algorithms in Iris dataset

6 Results Analysis and Discussion

6.1 Benchmark Function Test

In the first experiment, AMFOOBL is compared to AMFO and standard MFO using 23 benchmark functions. It can be seen in Table 9, especially in unimodal and multimodal functions, AMFOOBL outperforms both AMFO and MFO in terms of five statistical metrics regarding optimal fitness values. As each algorithm runs

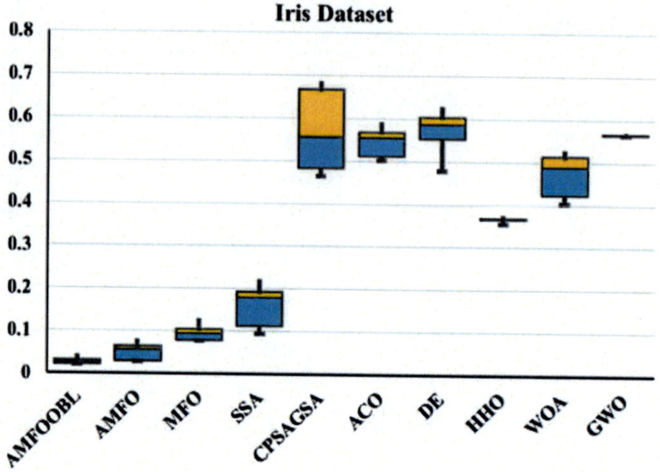

Fig. 14 Box plot of algorithms in Iris dataset

Table 18 Experimental results of algorithms for Breast cancer dataset

Algorithm	MSE (Ave ± Std)	Test error (TE)	p-value
AMFOOBL	0.008035 ± **4.21e-05**	95.45%	N/A
AMFO	**0.001077 ± 8.62e-04**	**99.40%**	0.0384*
MFO	0.010393 ± 0.00280	84.65%	0.0225*
SSA	0.001433 ± 0.01640	74.00%	0.0325*
CPSOGSA	0.001145 ± 0.00195	96.60%	0.0472*
ACO	0.663160 ± 0.03110	00.00%	0.0006***
DE	0.023375 ± 0.00100	56.80%	0.0024**
HHO	0.001575 ± 1.08e-04	94.25%	0.8500
WOA	0.001403 ± 0.00141	64.45%	0.0325*
GWO	0.001280 ± 0.00340	87.33%	0.3125

* indicates $p<0.05$; ** indicates $p<0.01$; *** indicates $p<0.001$

30 times on one benchmark function, mean and median can reflect the accuracy of AMFOOBL; minimum, maximum, and Std values show the excellent robustness of AMFOOBL. In addition, AMFO performs better than MFO, demonstrating the effectiveness of the new mechanism in flame decrease proposed in this chapter. In Table 10, AMFO spends less computation time than MFO most of the time, suggesting the faster convergence of AMFO than MFO. Since AMFOOBL has higher computation complexity, the computation time increases accordingly as well. Wilcoxon test in Table 11 manifests AMFOOBL has a significant improvement to AMFO and MFO, 16 and 15 out of 23 functions respectively at a 5% significant level. From what

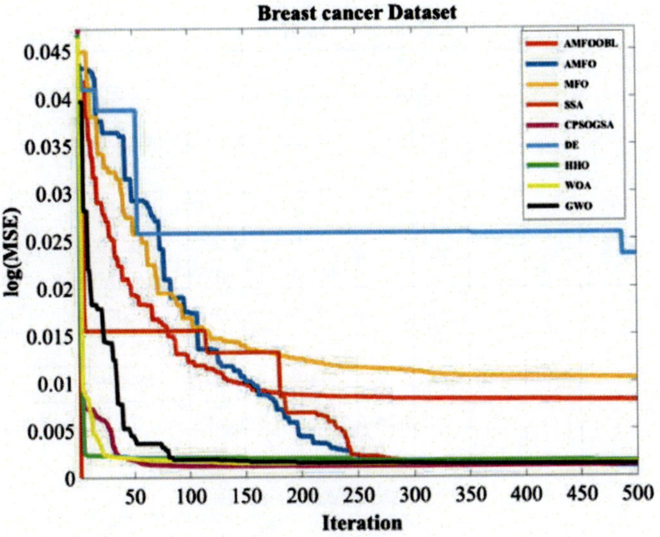

Fig. 15 Convergence curves of algorithms in Breast cancer dataset

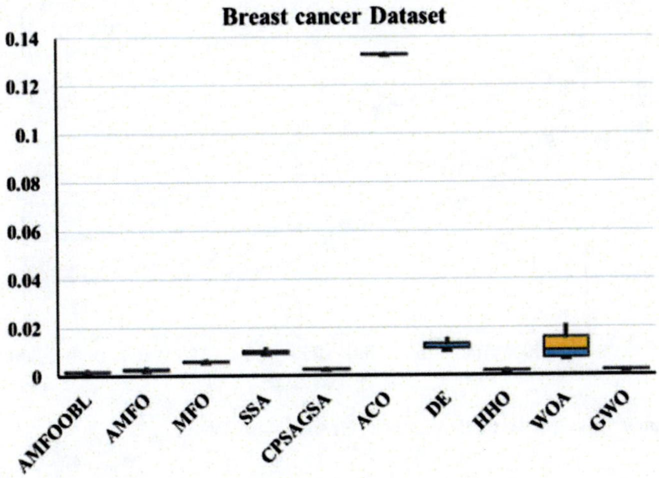

Fig. 16 Box plot of algorithms in Breast cancer dataset

has been analyzed above, the superior capability of AMFOOBL in the benchmark function test provides a convincing precondition for MLP training.

Table 19 Experimental results of algorithms for Heart disease dataset

Algorithm	MSE (Ave ± Std)	Test error (TE)	*p*-value
AMFOOBL	0.0667 ± 0.0187	87.70%	N/A
AMFO	0.1053 ± 0.0055	86.10%	0.0462*
MFO	0.1094 ± 0.0182	81.82%	0.0146*
SSA	0.1499 ± 0.0110	76.47%	0.0271*
CPSOGSA	0.1188 ± 0.0219	41.18%	0.0057**
ACO	0.5000 ± **0.0000**	50.27%	0.0025**
DE	0.1871 ± 0.0151	69.52%	0.0084**
HHO	0.1444 ± 0.0172	39.04%	0.0003***
WOA	0.2715 ± 0.1010	34.22%	0.0009***
GWO	**0.0525 ± 0.0057**	**88.24%**	0.0429*

* indicates $p<0.05$; ** indicates $p<0.01$; *** indicates $p<0.001$

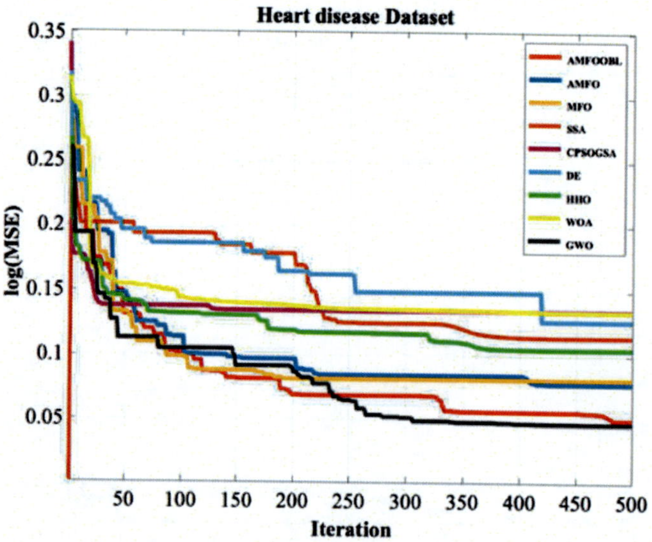

Fig. 17 Convergence curves of algorithms in Heart disease dataset

6.2 *Multilayer Perceptron Training*

In the second experiment of MLP training, five datasets are used for the classification task; three datasets are applied to the function approximation. Two statistical metrics (i.e., average and standard deviation), the classification rate (CR) or the test error (TE), as well as Wilcoxon statistical test, are calculated to evaluate the performance of the proposed trainer (AMFOOBL-MLP) and other competitive trainers. Besides,

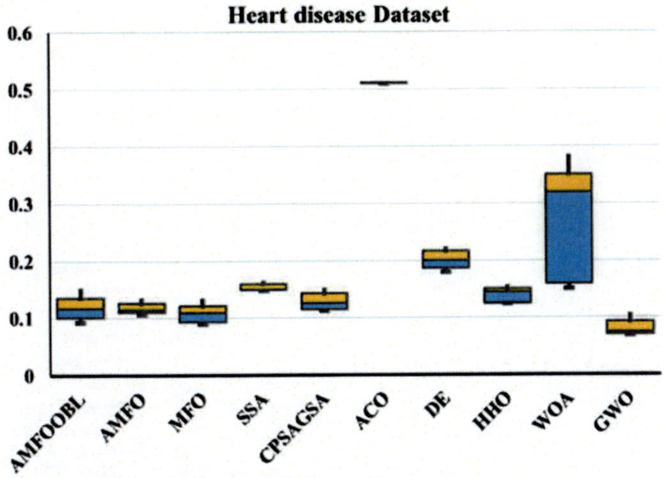

Fig. 18 Box plot of algorithms in Heart disease dataset

Table 20 Experimental results of algorithms for Sigmoid dataset

Algorithm	MSE (Ave ± Std)	Test error (TE)	*p*-value
AMFOOBL	**0.2464** ± 0.00010	**1.7522**	N/A
AMFO	**0.2464 ± 6.87e-05**	1.7530	0.3222
MFO	**0.2464** ± 9.34e-05	1.7571	0.3222
SSA	0.2466 ± 0.00010	1.7674	0.0273*
CPSOGSA	**0.2464** ± 8.15e-05	1.7736	0.1601
ACO	1.4924 ± 0.00030	5.4761	0.0019**
DE	0.2474 ± 0.00080	2.0882	0.0039**
HHO	0.2471 ± 0.00040	1.7606	0.0019**
WOA	0.2525 ± 0.00800	1.9274	0.0039**
GWO	**0.2464 ± 6.87e-05**	1.7571	0.0839

* indicates $p<0.05$; ** indicates $p<0.01$; *** indicates $p<0.001$

the convergence curves in one running and boxplot figures over ten runs are exhibited clearly.

XOR: As is presented in Table 15, AMFOOBL outperforms all other nine algorithms, achieving the minimal MSE value and the highest classification value (100%). The statistical test results also indicate a significant difference between AMFOOBL and MFO, HHO, GWO, etc. Furthermore, the convergence curves in Fig. 9 show that the convergence rate of AMFOOBL is faster than others. Figure 10 depicts the boxplot of these algorithms, suggesting AMFOOBL, AMFO, MFO, and SSA have better robustness than CPSOGSA, HHO, WOA, and GWO.

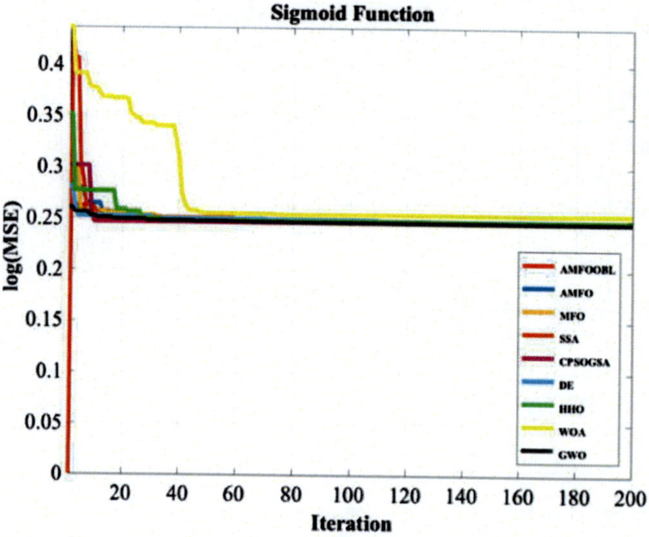

Fig. 19 Convergence curves of algorithms in Sigmoid dataset

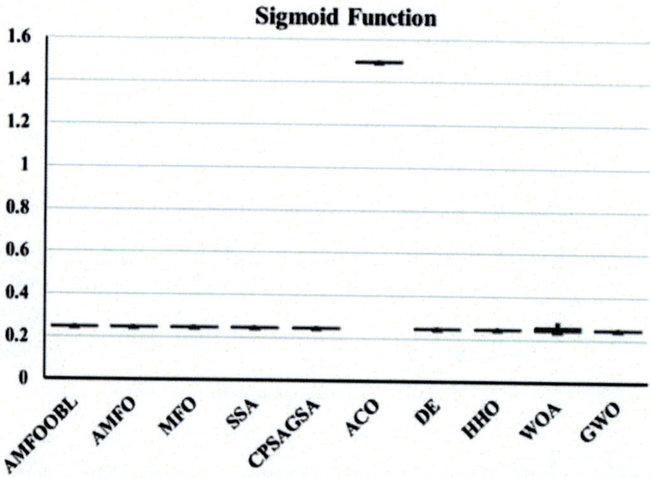

Fig. 20 Box plot of algorithms in Sigmoid dataset

Balloons: It is evident in Table 16 that AMFO collects the minimal average value of MSE. However, AMFOOBL, alongside AMFO, MFO, SSA, CPSOGSA, HHO, and GWO, all reach a 100% classification rate. Also, the *p*-values show AMFOOBL has no significant difference with AMFO and MFO in this dataset. Figures 11 and 12 all reveal that WOA and ACO have worse performance and are not suitable for the task of MLP training.

Table 21 Experimental results of algorithms for Cosine dataset

Algorithm	MSE (Ave ± Std)	Test error (TE)	*p*-value
AMFOOBL	**0.1658 ± 0.0005**	**0.4629**	N/A
AMFO	0.1743 ± 0.0006	0.4661	0.1054
MFO	0.1767 ± 0.0006	0.4698	0.0308*
SSA	0.1769 ± 0.0005	0.4745	0.0195*
CPSOGSA	0.2976 ± 0.1042	0.8733	0.0097**
ACO	1.0801 ± **5.72e-07**	1.4504	0.0019**
DE	0.1812 ± 0.0014	0.5487	0.0019**
HHO	0.2007 ± 0.0625	0.5506	0.0019**
WOA	0.2815 ± 0.0877	0.8492	0.0019**
GWO	0.1768 ± 0.0006	0.4722	0.0371*

* indicates $p<0.05$; ** indicates $p<0.01$; *** indicates $p<0.001$

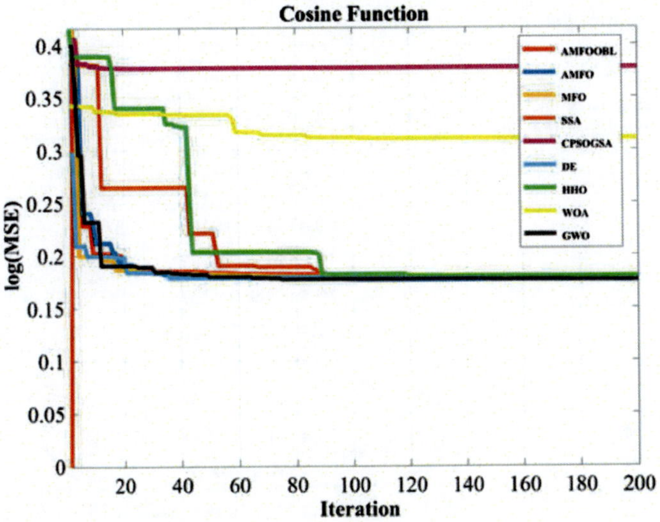

Fig. 21 Convergence curves of algorithms in Cosine dataset

Iris: Table 17 shows that AMFOOBL performs better than all other algorithms, which classifies test samples with an accuracy of 91.33%, followed by AMFO with 88.00% and MFO with 85.33%. Moreover, the Wilcoxon test results report a significant improvement between AMFOOBL with AMFO, MFO, and others.

Breast cancer: It can be seen in Table 18 that AMFO obtains the highest classification accuracy of 99.40%, indicating the effectiveness of the novel mechanism proposed originally. Based on the statistical test results, AMFOOBL has no significant difference to HHO and GWO in this dataset training; however, AMFOOBL has a significant improvement with AMFO and MFO. The convergence curves are shown in

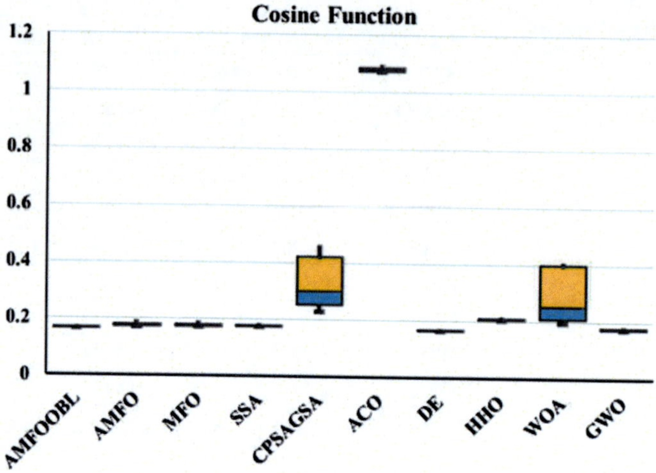

Fig. 22 Box plot of algorithms in Cosine dataset

Table 22 Experimental results of algorithms for Sine dataset

Algorithm	MSE (Ave ± Std)	Test error (TE)	*p*-value
AMFOOBL	**0.4104 ± 0.0167**	**28.4164**	N/A
AMFO	0.4231 ± 0.0192	28.7608	0.4375
MFO	0.4184 ± 0.0143	28.9742	0.4375
SSA	0.4446 ± 0.0127	29.8303	0.0625
CPSOGSA	0.4537 ± 0.4537	30.3710	0.0625
ACO	1.4989 ± **2.60e-08**	50.3900	0.0625
DE	0.4525 ± 0.0030	30.2660	0.0625
HHO	0.4583 ± 0.0036	29.7583	0.0625
WOA	0.4703 ± 0.0081	30.2892	0.0625
GWO	0.4534 ± 0.0009	29.8539	0.0625

* indicates $p<0.05$; ** indicates $p<0.01$; *** indicates $p<0.001$

Fig. 15, and the boxplot is presented in Fig. 16, which exhibits the AMFOOBL-MLP trainer's performance.

Heart disease: In Table 19, GWO gets the minimal average of MSE over ten runs with an average classification accuracy of 88.24%, followed by AMFOOBL with 87.70%. It should be noted that although the standard deviation of ACO in this dataset is 0, meaning the high stability of this trainer, the low accuracy makes it unpracticable to train MLP. Furthermore, the Wilcoxon test result shows that AMFOOBL has significant differences with all nine other algorithms. Figures 17 and 18 indicate each algorithm's convergence rate and stability, where CPSOGSA has the fastest convergence in this dataset.

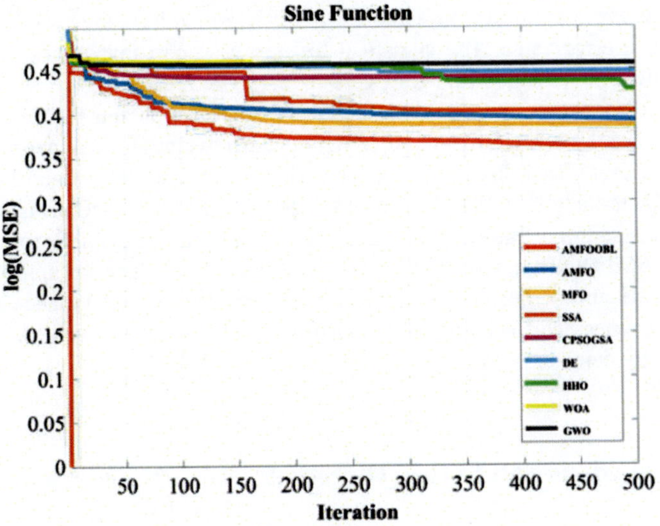

Fig. 23 Convergence curves of algorithms in Sine dataset

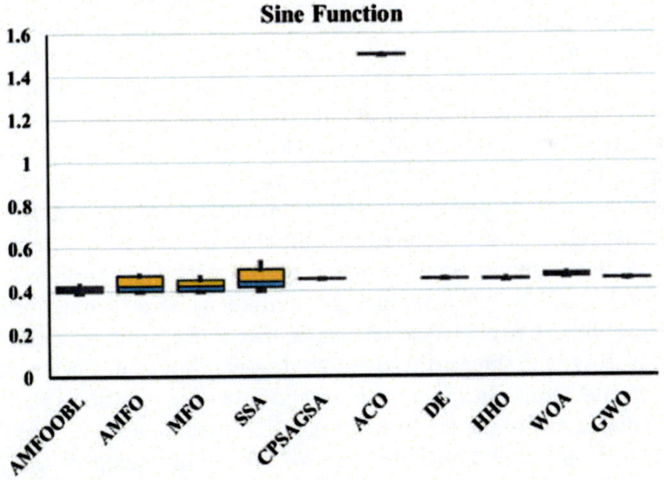

Fig. 24 Box plot of algorithms in Sine dataset

Sigmoid: Table 20 performs the average and standard deviation of MSE in ten trainers; AMFOOBL, AMFO, MFO, CPSOGSA, and GWO all receive the value of 0.2464 on average in which AMFO and GWO reach the smallest Std. However, AMFOOBL achieves minimal test error (1.7522), verifying the superior performance of it. According to the *p*-value results, AMFOOBL has a significant difference from SSA, ACO, DE, HHO, and WOA.

Cosine: From the results in Table 21, AMFOOBL collects the best fitness value and the test error value. The statistical test results report that AMFOOBL has a significant improvement with MFO, CPSOGSA, GWO, etc.

Sine: It is clear to notice in Table 22 that AMFOOBL obtains the minimal MSE value and the test error, 0.4104 and 28.4164, respectively. The Wilcoxon results do not show a significant difference between AMFOOBL and other algorithms in this dataset. Nevertheless, in Fig. 23, the convergence rate of AMFOOBL is faster than others' outstandingly. Based on the experimental results of Sigmoid, Cosine, and Sine datasets, AMFOOBL has manifested excellent capability in function approximation.

From the above illustration, we can also observe that performance regarding the effectiveness and superiority of AMFOOBL in these experiments has been significantly demonstrated.

7 Conclusions and Future Directions

In this chapter, a novel algorithm named AMFOOBL is proposed with two innovations based on MFO and is applied to MLP training. On the one hand, a new adaptive mechanism in decreasing the number of flames is created to balance the exploration and exploitation stages during the searching process, which is regarded as AMFO. On the other hand, the OBL strategy is used in the initial phase to diversify moths' population significantly, making it less probable in local optima stagnation. Some experimental simulations are conducted to verify the performance of AMFOOBL. First, AMFOOBL is compared to AMFO and MFO on 23 benchmark functions. The simulation results show that AMFOOBL is superior in terms of five quantitative metrics (mean, standard deviation, median, minimum, and maximum) regarding fitness values. Also, AMFO outperforms MFO in our experiment, especially in computation time, demonstrating the effects of the original adaptive mechanism.

Second, AMFOOBL alongside nine other well-known algorithms are employed in MLP training within five classification datasets (XOR, Balloons, Iris, Breast cancer, and Heart disease) and three approximation datasets (Sigmoid, Cosine, and Sine). The experimental results, including average and Std values of MSE as well as classification rate/test error, reveal that the AMFOOBL-based trainer performs best most of the time and can be used effectively for MLP training. Moreover, the Wilcoxon statistical test also suggests that AMFOOBL has a significant difference with competitive algorithms. The above two experimental analysis suggests that AMFOOBL is a valuable tool in multilayer perceptron training problems.

In the future, AMFOOBL-MLP can be applied to more critical datasets training, and other various applications can be tested with the use of the AMFOOBL algorithm. Furthermore, from the second MLP training experiment, it is interesting to find that HHO has an excellent capability that could be investigated deeply.

References

1. W.S. McCulloch, W. Pitts, A logical calculus of the ideas immanent in nervous activity. Bull. Math. Biophys. **5**(4), 115–133 (1943)
2. S. Mirjalili, S.M. Mirjalili, A. Lewis, Let a biogeography-based optimizer train your Multi-Layer Perceptron. Inf. Sci. **269**, 188–209 (2014)
3. S. Mirjalili, How effective is the Grey Wolf optimizer in training multi-layer perceptrons. Appl. Intell. **43**(1), 150–161 (2015)
4. G. Bebis, M. Georgiopoulos, Feed-forward neural networks. IEEE Potentials **13**(4), 27–31 (1994)
5. G. Dorffner, Neural networks for time series processing, in *Neural network world* (Citeseer, 1996)
6. Y. LeCun et al., Gradient-based learning applied to document recognition. Proc. IEEE **86**(11), 2278–2324 (1998)
7. S. Mirjalili, S.Z. Mohd Hashim, H. Moradian Sardroudi, Training feedforward neural networks using hybrid particle swarm optimization and gravitational search algorithm. Appl. Mathem. Comput. **218**(22), 11125–11137 (2012)
8. S.M.J. Jalali et al., A novel evolutionary-based deep convolutional neural network model for intelligent load forecasting. *IEEE Transactions on Industrial Informatics*, (2021)
9. S.M.J. Jalali, et al. Evolving artificial neural networks using butterfly optimization algorithm for data classification, in *International Conference on Neural Information Processing* (Springer, 2019)
10. H. Wu et al., Training feedforward neural networks using symbiotic organisms search algorithm. Comput. Intell. Neurosci. **2016**, 1–14 (2016)
11. H. Jia, J. Ma, W. Song, Multilevel thresholding segmentation for color image using modified moth-flame optimization. IEEE Access **7**, 44097–44134 (2019)
12. H. Liang et al., Modified grasshopper algorithm-based multilevel thresholding for color image segmentation. IEEE Access **7**, 11258–11295 (2019)
13. D.E. Rumelhart, G.E. Hinton, R.J. Williams, Learning representations by back-propagating errors. Nature **323**(6088), 533–536 (1986)
14. A. Van Ooyen, B. Nienhuis, Improving the convergence of the back-propagation algorithm. Neural Netw. **5**(3), 465–471 (1992)
15. V.G. Gudise, G.K. Venayagamoorthy, Comparison of particle swarm optimization and back-propagation as training algorithms for neural networks, in *Proceedings of the 2003 IEEE Swarm Intelligence Symposium. SIS'03 (Cat. No. 03EX706)* (IEEE, 2003)
16. M. Gori, A. Tesi, On the problem of local minima in backpropagation. IEEE Trans. Pattern Anal. Mach. Intell. **14**(1), 76–86 (1992)
17. S. Mirjalili, A.S. Sadiq. Magnetic optimization algorithm for training multi layer perceptron, in *2011 IEEE 3rd International Conference on Communication Software and Networks* (IEEE, 2011)
18. S. Song, H. Jia, J. Ma, A Chaotic electromagnetic field optimization algorithm based on fuzzy entropy for multilevel thresholding color image segmentation. Entropy **21**(4), 398 (2019)
19. S.S. Hameed et al., A comparative study of nature-inspired metaheuristic algorithms using a three-phase hybrid approach for gene selection and classification in high-dimensional cancer datasets. Soft Comp. (2021)
20. J. Gholami, F. Mardukhi, H.M. Zawbaa, An improved crow search algorithm for solving numerical optimization functions. Soft Comput. (2021)
21. A. Mahesh, G. Sushnigdha, A novel search space reduction optimization algorithm. Soft Comput. (2021)
22. W. Long, et al., Pinhole-imaging-based learning butterfly optimization algorithm for global optimization and feature selection. Appl. Soft Comp. **103**, 107146 (2021)
23. S. Li et al., Slime mould algorithm: A new method for stochastic optimization. Futur. Gener. Comput. Syst. **111**, 300–323 (2020)

24. S. Ronghui, N. Liangrong, An intelligent fuzzy-based hybrid metaheuristic algorithm for analysis the strength, energy and cost optimization of building material in construction management. Eng. Comp. (2021)
25. M. Branch, A multi layer perceptron neural network trained by invasive weed optimization for potato color image segmentation. Trends Appl. Sci. Res. **7**(6), 445–455 (2012)
26. D.A. Alboaneen, H. Tianfield, Y. Zhang, Glowworm swarm optimisation for training multi-layer perceptrons, in *Proceedings of the Fourth IEEE/ACM International Conference on Big Data Computing, Applications and Technologies* (2017)
27. R. Zhao et al., Selfish herds optimization algorithm with orthogonal design and information update for training multi-layer perceptron neural network. Appl. Intell. **49**(6), 2339–2381 (2019)
28. S. Mirjalili, Moth-flame optimization algorithm: a novel nature-inspired heuristic paradigm. Knowl.-Based Syst. **89**, 228–249 (2015)
29. N.L. Chau, N.T. Tran, T.-P. Dao, A hybrid approach of density-based topology, multilayer perceptron, and water cycle-moth flame algorithm for multi-stage optimal design of a flexure mechanism. Eng. Comp. (2021)
30. B.S. Yıldız, A.R. Yıldız, Moth-flame optimization algorithm to determine optimal machining parameters in manufacturing processes. Materials Testing **59**(5), 425–429 (2017)
31. C. Li, S. Li, Y. Liu, A least squares support vector machine model optimized by moth-flame optimization algorithm for annual power load forecasting. Appl. Intell. **45**(4), 1166–1178 (2016)
32. M. Wang et al., Toward an optimal kernel extreme learning machine using a chaotic moth-flame optimization strategy with applications in medical diagnoses. Neurocomputing **267**, 69–84 (2017)
33. H.M. Zawbaa et al., Feature selection approach based on moth-flame optimization algorithm. in *2016 IEEE Congress on Evolutionary Computation (CEC)*. (IEEE, 2016)
34. D. Molina et al., *Comprehensive Taxonomies of Nature-and Bio-inspired Optimization: Inspiration versus Algorithmic Behavior, Critical Analysis and Recommendations*. arXiv preprint arXiv:2002.08136 (2020)
35. H.R. Tizhoosh, Opposition-based learning: a new scheme for machine intelligence, in *International Conference on Computational Intelligence for Modelling, Control and Automation and International Conference on Intelligent Agents, Web Technologies and Internet Commerce (CIMCA-IAWTIC'06)* (IEEE, 2005)
36. A.A. Ewees, M. Abd Elaziz, E.H. Houssein, *Improved grasshopper optimization algorithm using opposition-based learning*. Exp. Syst. Appl. **112**, 156–172 (2018)
37. S. Gupta, K. Deep, A hybrid self-adaptive sine cosine algorithm with opposition based learning. Expert Syst. Appl. **119**, 210–230 (2019)
38. S. Yu et al., Enhancing firefly algorithm using generalized opposition-based learning. Computing **97**(7), 741–754 (2015)
39. L.G. Tallini, U. Vaccaro, Efficient m-ary balanced codes. Discret. Appl. Math. **92**(1), 17–56 (1999)
40. S. Mirjalili, The ant lion optimizer. Adv. Eng. Softw. **83**, 80–98 (2015)
41. S. Mirjalili et al., Salp swarm algorithm: a bio-inspired optimizer for engineering design problems. Adv. Eng. Softw. **114**, 163–191 (2017)
42. M. Dorigo, G. Di Caro, Ant colony optimization: a new meta-heuristic, in *Proceedings of the 1999 Congress on Evolutionary Computation-CEC99 (Cat. No. 99TH8406)* (IEEE, 1999)
43. S. Mirjalili, A. Lewis, The whale optimization algorithm. Adv. Eng. Softw. **95**, 51–67 (2016)
44. S. Mirjalili, S.M. Mirjalili, A. Lewis, Grey wolf optimizer. Adv. Eng. Softw. **69**, 46–61 (2014)
45. A.A. Heidari et al., Harris hawks optimization: Algorithm and applications. Futur. Gener. Comput. Syst. **97**, 849–872 (2019)
46. R. Storn, K. Price, Differential evolution–a simple and efficient heuristic for global optimization over continuous spaces. J. Global Optim. **11**(4), 341–359 (1997)
47. S.A. Rather, P.S. Bala, A hybrid constriction coefficient-based particle swarm optimization and gravitational search algorithm for training multi-layer perceptron. Int. J. Intell. Comp. Cyber. (2020)

48. R. Hecht-Nielsen, Kolmogorov's mapping neural network existence theorem, in *Proceedings of the International Conference on Neural Networks* (IEEE Press New York, 1987)
49. O.L. Mangasarian, W.H. Wolberg, Cancer Diagnosis Via Linear Programming. University of Wisconsin-Madison Department of Computer Sciences (1990)

Early Detection of Coronary Artery Disease Using PSO-Based Neuroevolution Model

Mina Karimi, Seyed Mohammad Jafar Jalali, Iman Raeesi Vanani, and Diego Oliva

Abstract There are a large number of causes of morbidness over the world population, and cardiovascular disease is one of the top reasons among them. During the last decade, early detection of warning symptoms is regarded as one of the most important tasks, easing the treatment procedure and preventing a large number of deaths. Numerous people die annually from coronary artery disease (CAD) compared to other deadly diseases. Two main reasons for this disease could be smoking and unhealthy dieting. Angiography is widely used as the most accurate technique for CAD recognition in medical centers; however, some downsides could be mentioned for this technique: It is costly and has various complications such as hematoma and intercostal artery pseudoaneurysms. The machine learning (ML) predictive algorithms bring to medical treatment a new aspect of diagnosing diseases. ML techniques have been applied as powerful and trustworthy tools in the field of medicine for different purposes over the last recent years. ML aims to efficiently extract valuable information from the clinical datasets with minimal effort. In this study, we use a set of machine learning algorithms to reach an accurate prediction of CAD. A powerful nature-inspired optimization algorithm named particle swarm optimization (PSO) based on multilayer perceptron (MLP) and nine states of traditional supervised learning techniques including RF (Random Forest), PSO-MLP, RT (Random Tree), NB (Naïve bayes), SVM (support vector machine), LR (logistic regression), Bagging,

M. Karimi
Department of Information Technology Management, Allameh Tabataba'I University, Tehran, Iran

S. M. J. Jalali
Institute for Intelligent Systems Research and Innovation (IISRI), Deakin University, Geelong, Australia
e-mail: sjalali@deakin.edu.au

I. R. Vanani
Faculty of Management and Accounting, Allameh Tabatabaei University, Tehran, Iran
e-mail: Imanraeesi@atu.ac.ir

D. Oliva (✉)
Departamento de Innovación Basada en la Información y el Conocimiento, Universidad de Guadalajara, CUCEI, Guadalajara, Mexico
e-mail: diego.oliva@cucei.udg.mx

E. H. Houssein et al. (eds.), *Integrating Meta-Heuristics and Machine Learning for Real-World Optimization Problems*, Studies in Computational Intelligence 1038,
https://doi.org/10.1007/978-3-030-99079-4_12

MLP, J48 and CART (classification and regression tree analysis) are employed for diagnosis CAD. We implement various machine learning techniques on a straightforward and introduced dataset by the name of Z-Alizadeh Sani. Well-known dataset named Z-Alizadeh Sani to determine the best machine learning technique with the highest accuracy. This dataset contains the information of 303 people who visit the heart disease center. Also, we applied a feature selection method to reduce redundant and outlier features and detect the considerable features that have an important role to play in CAD disease forecasting. The feature selection process improves the power of disease forecasting. Consequently, the obtained accuracy and area under the curve (AUC) from utilized algorithms compared with each other to specify the best technique. This study aims to determine the best model for classifying CAD. Eventually, according to experimental results, the MLP model trained by PSO performs better than all nine art supervised learning algorithms and achieves an accuracy of 90.56% in CAD prediction.

Keywords Coronary artery disease · CAD · Neural network · PSO · Neuroevolution

1 Introduction

Cardiovascular disease is on top of the deadliest diseases list that is considered as the prominent cause of morbidness globally [1]. An estimated 18 million people die annually due to cardiovascular disease, and in 2016 the World Health Organization confirmed this estimation. This number is almost 31% of all worldwide deaths, and the rest of them are due to heart attack and stroke [2]. Cardiovascular detection and identifying its symptoms are vital responsibilities since a myriad of deaths can be prevented, and a large number of people can survive and continue their lives. One of the valuable and reliable reasons to determine cardiovascular disease is angiography. It prepares a view of coronary anatomy in detail, but on the other hand, this method is costly and is associated with a remarkable side effect [3]. Coronary artery disease (CAD) is one of the common types of heart disease. CAD is regarded as the main reason for human deaths worldwide. Prediction of CAD disease is considered one of the most important topics in the clinical data analysis section. There is a critical need for a highly accurate system to predict and identify the hidden patterns of CAD disease in medical data streams. Medical data mining has considerable potential for discovering and extracting meaningful knowledge from massive datasets. Machine learning has been employed increasingly to predict heart diseases, and its techniques are an essential step in clinical diagnosis. Statistical and computer science are the two main foundations of machine learning. Artificial Neural Network (ANNs) is a machine learning technique inspired by the biological nervous system. Moreover, it is used in supervised learning methods. Researchers have effectively applied different types of neural networks to handle difficult diagnostic tasks [4]. Multilayer Perceptron (MLP) is the most well-known and applied neural networks.

One of the significant phases in machine learning is the training process. This step could affect performance [5, 6]. Detecting the best weight category to cut down the approximation error is considered the basic purpose of training multilayer perceptron. Generally, the gradient-based and metaheuristic search methods are two categories for the MLP training procedure. The tendency to be trapped in local minima is one of the demerits of the gradient-based method [7]. This drawback has a profound impact on network efficiency. Neuroevolution (NE) is a kind of artificial intelligence which is employed evolutionary algorithms (EAs) to optimize the process of training neural networks [8]. It could be considered an efficient solution for the problem mentioned above, representing greater performance in escaping local minima. Moreover, NE could determine the best weight and biases category for the MLP training process. It could maintain a low neural network error rate. Over the past decade, researchers applied various data mining algorithms to implement them in healthcare to attain an accurate prediction of heart diseases. According to Shafenoor et al. [2], aim to recognize remarkable features and machine learning methods such as KNN, decision tree, Naive Bayes, logistic regression (LR), support vector machine (SVM), MLP and Vote (a hybrid technique with Naïve Bayes and Logistic Regression) that can improve the accuracy of predicting cardiovascular disease. Their experimental results show that Vote achieves 87.4% in heart disease prediction. They also used feature selection methods in their study, and 9 features such as "sex", "cp", "fbs", "restecg", "exang", "oldpeak", "slope", "ca" and "thal" were selected, and they achieved highest accuracy rate. Tayefi et al. [9] used a heart disease dataset of 2346 samples. 1159 individuals were healthy, and the rest of them had undergone coronary angiography. Thereupon their model obtained maximum accuracy of 94%. In a comparison between Naïve Bayes, Support vector machine (SVM), Bagging and MLP algorithm in Rapid-Miner tool, Alizadehsani et al. found that SVM was the most accurate of the four in CAD prediction; they also considered "weights by SVM" in RapidMiner for the process of selecting features. The coefficients of the normal vector of a linear SVM were considered as feature weights in their study [10]. Negahbani et al. employed data mining applying the fuzzy c-mean method to classify 303 patients and 74 attributes. They reported the highest accuracy was 88% [11]. Bhatia et al. [12] proposed a genetic algorithm for the feature selection process to determine the best mode of feature selection and achieved 90.57% accuracy using their proposed GA-SSVM. Among the studies conducted with data mining, Kumar et al. used three different classifications, including Naïve Bayes, decision tree, and KNN, to predict heart disease. The results represent that the classification rate of the collected dataset belonged to Naïve Bayes, which was 52.33% [13]. Abdar et al. implemented C5.0, SVM, and LR algorithms to discover great patterns from the heart disease dataset. They realized the C5.0 Decision tree had the greatest results among the two other classifiers, 93.02% [14]. In the paper presented by Shouman and Turner, they employed the C4.5 decision tree and applied it to reduce error pruning in order to classify patients in which they reached 84.1% accuracy [15]. Another research was done by Polat et al. [16], which used artificial immune recognition system (AIRS) techniques. Experiments have been executed over 270 records and 9 features; eventually, the dataset has been categorized with a classification rate of

92.59%. Shafique et al. [17] employed Neural Network, Decision Tree, and Naïve Bayes algorithms to identify the hidden pattern from the heart dataset. Naive Bayes achieve the best accuracy in comparison to the rest. Babaoglu et al. [18] applied neural network methods and exercise test data to predict the stenosis of heart vessels. they achieved three accuracy values for each vessel. 73%, 64.85%, and 69.39% are the accuracies of left anterior coronary artery (LAD), left circumflex (LCX), and right coronary artery (RCA), respectively, and right coronary artery (RCA), respectively. Tithi et al. [19] used six supervised machine learning algorithms, and naïve Bayes achieved the best accuracy to predict CAD and executed better than the rest. This paper aims to find meaningful patterns using a hybrid neuroevolution algorithm for CAD disease prediction. In order to achieve this purpose, particle swarm optimization (PSO) as a promising and powerful evolutionary algorithm [20] is proposed for the process of training MLP on the coronary artery dataset. For demonstrating the performance and superiority of this algorithm, various machine learning techniques such as including Naive Bayes (NB), logistic regression (LR), Multi-Layer Perceptron (MLP), Bagging, J48, Random Forest (RF), Random Tree (RT), Support vector machine (SVM) and Classification and Regression Tree Analysis (CART) are employed, and their result is compared with the proposed PSO-MLP classifier. In this study, the CAD dataset with a total of 303 rows and 54 features is considered for CAD classification. This dataset is obtained by Z-alizade sani [10] from UCI machine learning repository. We analyze the CAD dataset with the aim of discovering a precise prediction model for CAD using machine learning techniques. For evaluation of the proposed method and other classifiers, we use accuracy and area under the curve (AUC), considered as two important assessment metrics. The results show that the PSO method performs robustly compared to the other benchmark classification methods. We also applied the correlation-based feature selection for identifying the effective features before employing the training procedure.

The rest of this paper is organized as follows: A brief introduction to PSO algorithm, a general description of the MLP and developed MLP training based on PSO algorithm and detail of dataset as well feature selection algorithm is represented in Sect. 2. The experimental setups and evolution metrics are expressed in Sect. 3. The experimental results and discussion of the findings are conducted in Sect. 4. Finally, the paper is concluded in Sect. 5.

2 Materials and Methods

In this section, a general overview of PSO, MLP, and their hybridization called PSO-MLP algorithm is explained in this section.

2.1 Particle Swarm Optimization (PSO)

Particle swarm optimization (PSO) is one of the evolutionary algorithms inspired by the public behavior of bird flocks in nature. It could be a stochastic optimizer for complex numerical functions. PSO was introduced by Kennedy and Eberhart [20]. PSO contains a set of random candidate solutions for the considered problem (Particles). Also, a randomized velocity and position are allocated to each particle. PSO can find the best global solution by analyzing particles' collective behavior [21]. Particles have a certain memory to keep the trend of all previous positions. Two best values have a profound effect on updating each particle position and reaching the optimal solution in each generation. The first value is the best value gained so far and it is named personal best. The second value is called the best position; it is the greatest value obtained through each particle in the whole population. Particles learn from these two values to reach the optimal solution to optimization problems [22]. As a consequence, the position and velocity of each particle are changed and improved according to the position of other particles while they are in the search space. The velocity of the ith in the (t + 1) iteration is presented by Eq. (1), where c_1 and c_2 are constants and a random number in the domain of 0 and 1 is produced by the random function, rand (1).

$$V_i(t+1) = V_i(t+1) + c_1 \times rand(1) \times (lb_i - X_i) + c_2 \times rand(1) \times \left(gb_i - X_i\right) \tag{1}$$

The position of each particle is shown by Xi. The best previous position of each particle and the best position of all particles are presented by lbi and gbi respectively. Eventually the position of the ith particle is modified by Eq. (2).

$$X_i(t+1) = X_i(t) + V_i(t+1) \tag{2}$$

Algorithm 1. Pseudocode of the PSO Optimization Algorithm

```
Initialize population
for t =1: maximum generation
    for i =1: population size
        if f(x_{i,d}(t)) < f(p_i(t)) then p_i (t) = x_{i,d}(t)
          f(p_g(t)) < min (p_t(t))
        end
        for d=1: dimension
           v_{i,d}(t +1) = w_{i,d}(t)+ c_1 r_1( p_1 - x_{i,d}(t))+ c_2 r_2( p_g - x_{i,d}(t))
           x_{i,d}(t +1) = x_{i,d}(t)+ v_{i,d}(t + 1)
           if v_{i,d}(t +1) > v_max then v_{i,d}(t +1) = v_max
           else if v_{i,d}(t +1) < v_min then v_{i,d}(t +1) = v_min
           end
           if x_{i,d}(t +1) > x_max then x_{i,d}(t +1) = x_max
           else if x_{i,d}(t +1) < x_min then x_{i,d}(t +1) = x_min
           end
        end
    end
end
```

2.2 *Multi-Layer Perceptron (MLP)*

An Artificial Neural Network (ANN) is a computing system based on the structure of biological neural systems of the human body. Artificial Neural Network has a cohesive structure and could be single or multiple layers. Also, ANN consists of neurons interconnected by a set of communication links associated with adjustable weights. Typically, three layers of neurons, namely the input layer, which maps to input variables, an output layer, and a hidden layer, constitute ANN. Neurons of ANN cooperate in evaluating the most cost-effective and supreme technique to achieve the best solution for various specific problems. This process is like to mechanism of biological neurons in brains. Feedforward neural network and feedback neural network are two important kinds of neural network which is based on network topology. The feedforward multilayer perceptron is one well-known form of ANNs that could be applied for various types of problems and estimate a suitable computational model for them. Multilayer perceptron (MLP) is the most well-known form of Feedforward neural network. In MLP, neurons are organized in one directed way. Data transition happens among parallel layers, namely the input, hidden, and output layers. Inputs are transferred through a hidden layer which is between the input layer and the output layer. Figure 1 represents an MLP network with a single hidden layer. Connection links

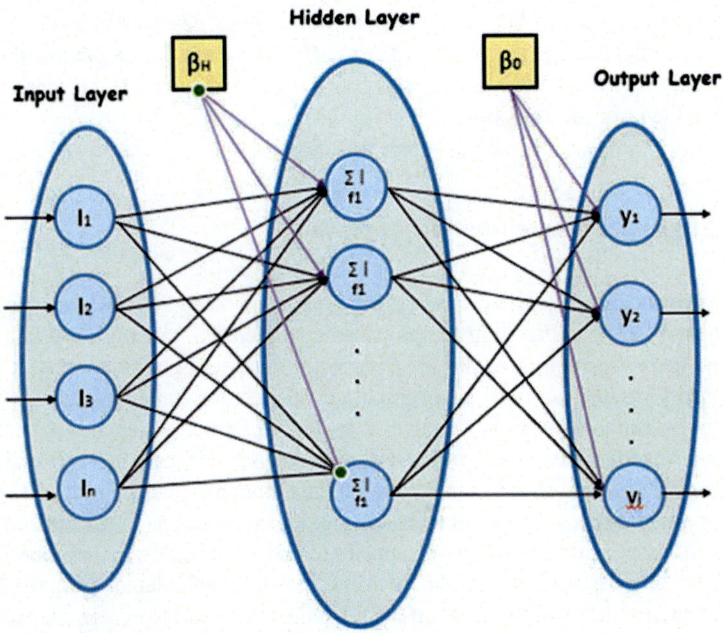

Fig. 1 MLP network with single hidden layer

among neurons are demonstrated as weights. Each neuron executes two functions: a summation function and an activation function. The weighted summation of inputs is calculated by summation function as represented in Eq. (3), where n is the number of input variables, I_i shows input variable I, βj and ω_{ij} denote a bias term and the connection weight, respectively.

$$S_j = \sum_{i=1}^{n} \omega_{ij} I_i + \beta_j \tag{3}$$

The input of the activation function is the obtained output from the summation function. Several types of activation functions could be utilized in the MLP network. The most common type in previous works is the sigmoid function [7]. The sigmoid function is demonstrated in Eq. (4). As a consequence, the final output of each node in the hidden layer can be measured in Eq. (5).

$$\mathrm{f_j(x)} = \frac{1}{1 + \mathrm{e}^{-\mathrm{S_j}}} \tag{4}$$

$$\hat{y}_k = \sum_{i=1}^{m} W_{kj} f_i + \beta_k \tag{5}$$

The initial amount of biases and connection weights determined in the training step can greatly influence the performance of MLP networks. Consequently, the main goal of training MLP networks is identifying a set of connection and biases weights to reduce the error of prediction.

2.3 Suggested Classification Method

In this section, the suggested classification algorithm named PSO-MLP is explained in detail and shown in Fig. 2. In our proposed approach, PSO is employed as a metaheuristic method to optimize the MLP network with a single hidden layer. When evolutionary or metaheuristic algorithms are utilized for training MLP networks, two significant factors should be taken into consideration: the encoding design of individuals and the selection of fitness function. In the PSO method, each individual includes three parts: a set of connection weights among input and hidden layers, a set of weights connecting the hidden and output layer, and a set of bias weights. Each solution is considered a one-dimensional array containing real numbers to be included in the range of $[-1, 1]$. The mentioned vector is calculated in Eq. (6), where n and m are a number of input features and nodes in the hidden layer, respectively.

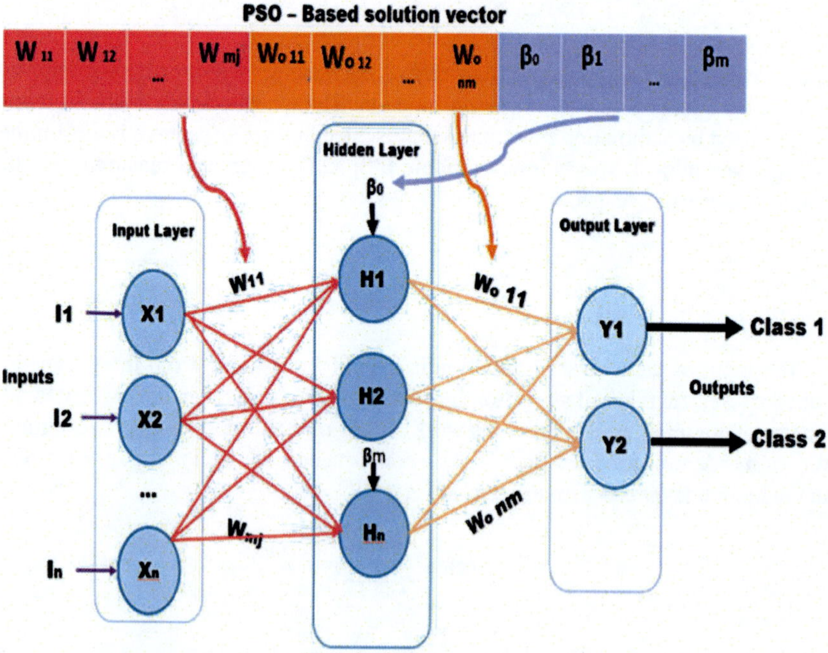

Fig. 2 The representation of training MLP by PSO

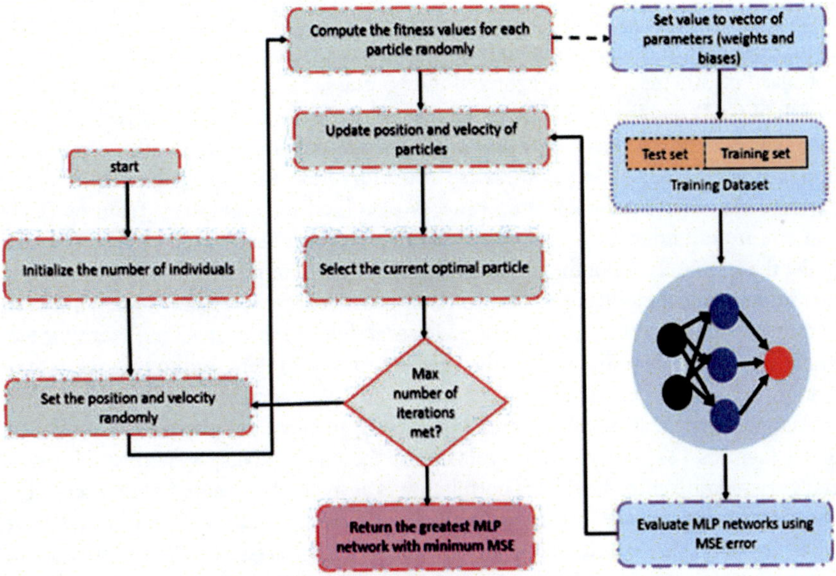

Fig. 3 Steps of the proposed classification method

Also, assigning initial values to the connection weights and biases has a profound effect on the efficiency and convergence of the training MLP process. It should be underlined that better convergence and power of prediction are caused by initializing the weights and biases through optimal amounts (Fig. 3).

$$\text{Individual length } = (\mathrm{n} \times \mathrm{m}) + (2 \times \mathrm{m}) + 1 \tag{6}$$

The second important aspect is specifying the fitness function. In the training process of MLP, the fitness value of each individual is measured via mean square error (MSE). In other words, this common function is considered a fitness function and measures the difference between the values of each data determined by the MLP networks. Equation (7) is used to calculate the MSE metric:

$$\text{MSE } = \frac{1}{\mathrm{n}}\sum_{\mathrm{i}=1}^{\mathrm{n}}(\mathrm{y}_\mathrm{i} - \hat{\mathrm{y}}_\mathrm{i})^2 \tag{7}$$

2.4 *Dataset and Feature Selection Strategy*

In this work, the CAD dataset was prepared from Shaheed Rajaei Research Center, and it includes the risk factor and data of 303 random heart patients referred to

Shaheed Rajaei center. This dataset was picked for this study since it is a commonly used dataset of CAD among data mining researchers. As noted previously, this dataset consisted of 303 instances categorized in two-class, 216 samples were some people at risk of CAD and the rest of them were healthy. Diameter narrowing is considered as a symptom of CAD in a way that if it is greater than or equal to 50%, the person is classified as CAD patients otherwise as healthy [23]. Regarding the results of our experiments, the feature selection process identified the significant features. They had a profound impact on CAD prediction and increased the success rate and area under the ROC curve. For the process of extracting the impressive features, a specific feature selection method entitled correlation-based feature selection [24, 25] has been employed. It could be considered a pre-processing phase to machine learning that influences decreasing dimensionally, deleting irrelevant data, improving the rate of learning accuracy, and enhancing the result [26]. This mentioned selection method simultaneously measures the correlation among features and the prediction ability of all features [27]. Features are relevant if their values vary systematically with category membership. The word correlation is assigned to a degree of predictability and dependence of one feature with the rest. In other words, a feature is effective when it is correlated with a class, or it can predict a class; If not, the mentioned feature is irrelevant [24]. Consequently, by using this technique, redundant and irrelevant features are omitted, and among 53 features, the subset of features that have resulted in the greatest performance was selected. According to Table 1, 11 features that profoundly impact CAD prediction with high accuracy and AUC were chosen. Moreover, Table 1 shows the values of selected features in terms of the feature name, range of them, standard deviation (STD), and also average (AVG) of the feature's range.

Table 1 Values of selected features

Feature type	Feature name	Range	Avg ± STD
Demographic	Age	30–86	58/89 ± 10/39
	Sex	Male, female	
	Ex-smoker	Yes, no	
	FH (family history)	Yes, no	
	Airway disease	Yes, no	
	CVA (cerebrovascular accident)	Yes, no	
	CHF (congestive heart failure)	Yes, no	
Symptom and examination	Typical chest pain	Yes, no	
	Low Th Ang (low threshold angina)	Yes, no	
ECG	ST elevation	Yes, no	
Laboratory and echo	Lymph	7–60	32/39 ± 9/97

3 Experiments

This section presents a comprehensive investigation about the efficiency of the proposed PSO approach for training MLP. The proposed PSO-based MLP training approach is assessed using a CAD dataset collected from the heart center. In order to determine the considerable performance of PSO-based trainer, it is compared with a set of well-regarded algorithms: RF (Random Forest), PSO-MLP, RT (Random Tree), NB (Naïve bayes), SVM (support vector machine), LR (logistic regression), Bagging, MLP, J48 and CART (classification and regression tree analysis). As can be noticed, all selected features used in the experiments are normalized, and they are in the same scale. The Accuracy and Area under the curve are considered as two crucial factors that all results obtained by mentioned methods are compared to each other regarding these metrics.

3.1 Experimental Setup

For running all trainers, Python 3.7 is employed to execute all proposed methods. Moreover, they are executed on a PC provided with 16 GB of RAM and Intel 7 CPU with 1.9 GHz frequency. Training and testing for all developed models are implemented using 30 independent times by the purpose of obtaining statistically meaningful results. Eventually, the results are demonstrated by averaging each run. The dataset is partitioned into two parts. 70% of the CAD dataset is employed for training, and the rest is utilized for testing. In order to eliminate the impact of outlier data, the entire dataset is normalized in (0, 1) interval. In addition, data normalization could accelerate and boost the speed of the learning process in machine learning applications. As a consequence, the min–max function maps the value of the feature to a similar scale through the normalization process. Min–max normalization function is measured by Eq. (8).

$$\nu' = \frac{\nu_i - \min_A}{\max_A - \min_A} \tag{8}$$

3.2 Evaluation Metrics

In order to evaluate the efficiency of developed methods, two kinds of assessment metrics such as; accuracy and area under the curve (AUC), are employed. Accuracy is the capability of each algorithm to compute the rate of instances categorized correctly over the entire instances. In this study, the rate of a correct diagnosis is determined by this metric. Accuracy is measured based on the following formula:

$$\text{Accuracy} = \frac{\text{Number of sample cases correctly classified}}{\text{Total number of sample cases}} \quad (9)$$

The area under the curve (AUC) is another significant evaluation metric that is used in classification analysis. It defines the best prediction of categorical features among all models. True and false positive rate metrics measure it. One of its merits could be that this evaluation metric is independent of the ratio of classes to improve the insight into the performance when there is an inconsistency in the class distribution. A direct variation exists among AUC and the performance of classification methods. The following equation can calculate AUC:

$$\text{AUC} = \int_0^1 \frac{TP}{P} \mathrm{d}\frac{FP}{N} = \frac{1}{P.N} \int_0^1 TP\mathrm{d}FP \quad (10)$$

The neural network's performance is influenced by the number of neurons in the hidden layer. The number of neurons in hidden layer is equal to 2× number of attributes in considered dataset + 1. There are various approaches in selecting this number, and there is still no agreement among researchers that confirmed its absolute appropriate. We employ a standard and common formula confirmed by a large number of researchers in literature [7, 28].

4 Discussions and Results

This section is assigned to analyze the results of conducted experiments in order to evaluate the performance of the developed PSO-MLP model. Obtained results by PSO are compared with nine well-regarded and recent techniques. The table of obtained results and figures of them are described in the following subsections.

The following tables show the results of the accuracy and AUC of classification algorithms with and without a feature selection process. The accuracy value of the proposed PSO-MLP and other models are indicated in Table 2. This table determines statistical results for all features considered from medical knowledge. The results of this table include the average (AVG) of accuracy for each method, standard deviation (STD), and the most accurate and the worst value of various measures shown in the last row of the table. Inspecting the achieved results in Table 2, it can be observed that PSO, among other classifiers, achieved the highest average classification rate of 85.62%, which proves the stability and robustness of this classifier to detect the best set of weights and biases for MLP. Moreover, PSO-MLP represents the lowest standard deviation of 0.0178. In addition, the best and the worst accuracy belongs to the random forest and random tree classifiers, respectively. The maximum number of iterations was set to 30 in all the algorithms.

In Fig. 4, a boxplot of the results in terms of accuracy is represented in detail. The boxplots are demonstrated for 10 accuracies attained by each classifier at the

Table 2 Values of ACC for all methods (all features are considered)

Algorithm	Naïve bayes	LR	MLP	Bagging	J48	RF	RT	SVM	CART	PSO-MLP
Avg	0.7920	0.7610	0.8110	0.8340	0.8080	0.8440	0.7150	0.8430	0.8070	0.8562
Std	0.0290	0.0468	0.0420	0.0280	0.0434	0.0521	0.0611	0.0316	0.0362	0.0178
Best	0.8400	0.8400	0.8600	0.8700	0.434	0.8810	0.8000	0.9000	0.8600	0.9211
Worst	0.7600	0.6800	0.7100	0.7700	0.7300	0.7800	0.6300	0.7900	0.7600	0.8330

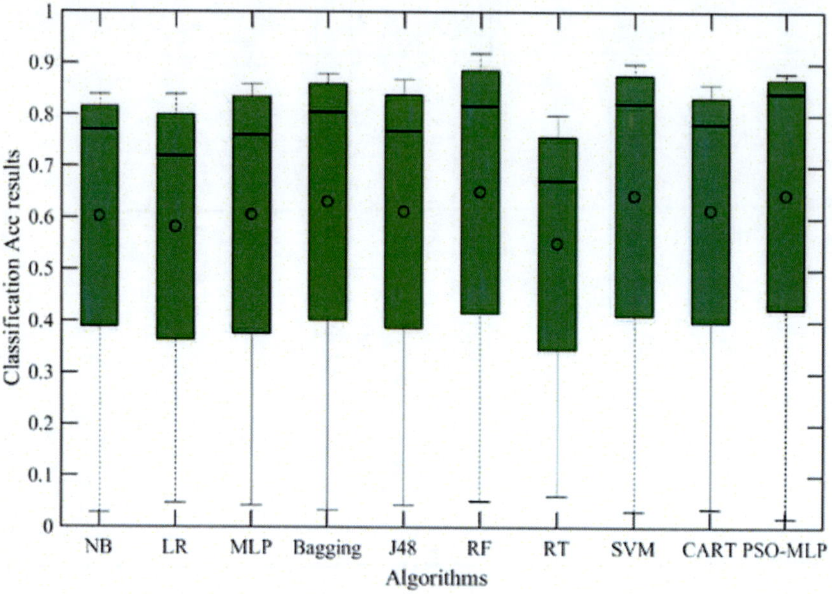

Fig. 4 Boxplot, ACC for RF, PSO-MLP, RT, NB, SVM, LR, Bagging, MLP, J48 and CART on dataset (all features)

end of the training process. In this plot, the box relates to the range of interquartile, two whiskers above and below the box that indicates the lowest and highest non-outlier accuracy values. The vertical line inside the box specifies the median value. As represented in Fig. 4, the boxplots visually demonstrate the greater performance of the PSO algorithm over other classifiers for the training of MLP in order to task of CAD prediction. This supremacy could observe through the higher median and short distance between the first and fourth quartile.

Table 3 determines another quantitative performance indicator which is AUC. This metric is generally employed to evaluate the efficiency of proposed classifiers. Also, if an unequal distribution exists among positive and negative samples, AUC could be applied as a useful evaluation metric. According to the obtained results from Table 3, it can be easily seen that the highest average value of AUC is achieved by the random forest algorithm. After that, the second-highest AUC belongs to PSO-MLP which is 0.8823. By respect to this result, we can conclude the PSO based MLP network is capable of determining the greatest set of weights reliably by MSE minimization. Also, it is able of getting the high ratios in AUC. Moreover, this is followed by the lowest STD value among other algorithms which prove that PSO is the robust and stable algorithm in comparison to other algorithms. It must be emphasized that best and the worst AUC belongs to a random forest which happened in one of the iterations and random tree respectively.

Boxplots in Fig. 5 provide a clear analysis of the distribution of the classification AUC values attained from 10 runs. Details of PSO-MLP boxplot such as compacted

Table 3 AUC for all techniques (all features)

Algorithm	Naïve bayes	LR	MLP	Bagging	J48	RF	RT	SVM	CART	PSO-MLP
Avg	0.8683	0.7835	0.8696	0.8758	0.7280	0.9126	0.6566	0.8271	0.7889	0.8823
Std	0.0289	0.0402	0.0491	0.0341	0.0678	0.0233	0.0810	0.0380	0.0516	0.0206
Best	0.9200	0.8620	0.9100	0.9430	0.8200	0.9040	0.7901	0.8710	0.8630	0.9560
Worst	0.8350	0.0731	0.7400	0.8200	0.6540	0.8870	0.5250	0.7590	0.7140	0.8410

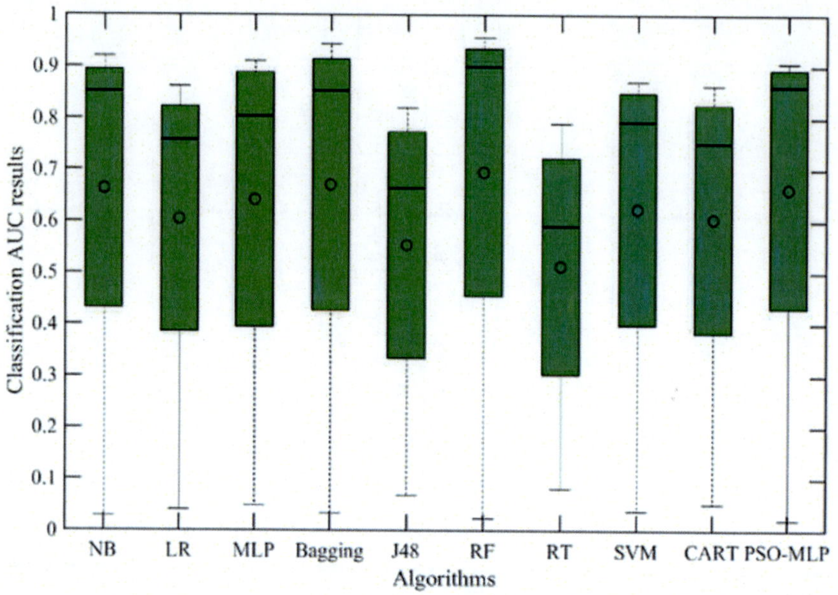

Fig. 5 Boxplot, AUC for RF, PSO-MLP, RT, NB, SVM, LR, Bagging, MLP, J48 and CART on dataset (all features)

box and high median gives strong evidence that efficiency of this method is better than existing methods. The only exception is RF, however, the PSO-MLP is more compressed than RF.

Of the 53 input variables of the heart dataset, 11 variables were selected as the strong predictors by feature selection process. Data from Table 4 shows that the best results are provided by the PSO-MLP algorithms. It represents that the PSO-MLP could be also very effective in diagnosis the CAD. The low STD represents the superior local optima avoidance of this classifier. Moreover, from Table 4 we perceive that the best accuracy is obtained by PSO-MLP algorithm, as well as the worst value belongs to random tree. We assessed the performance of feature selection algorithms along with various classification algorithms in this considered table. From an examination of the dataset, it can be observed the enhancement in classification accuracy by utilizing the feature selection process. All traditional classifier plus PSO-MLP demonstrated greater classification accuracy than previous outcome results obtained in Table 1. It is noteworthy that after execute the feature selection process, PSO-MLP achieved the most consistent accuracy value again. It obtained 90.56% classification rate and outperforms all other methods. In addition, the only MLP has the lower STD than PSO in accuracy, however it is the little value but is the lowest value among others.

Table 5 shows AUC values for all trainers after performing the feature selection process. The AUC values for all classifiers have boosted except MLP, bagging, random forest and SVM. Feature selection did not impact on the AUC value of

Table 4 ACC for all techniques (after feature selection)

Algorithm	Naïve bayes	LR	MLP	Bagging	J48	RF	RT	SVM	CART	PSO-MLP
Avg	0.8350	0.8520	0.8370	0.8360	0.8430	0.8390	0.7700	0.8340	0.8340	0.9056
Std	0.8310	0.2780	0.0249	0.0291	0.0277	0.0300	0.0323	0.0331	0.0313	0.0273
Best	0.9000	0.9000	0.8900	0.9000	0.9000	0.8800	0.8000	0.8900	0.9000	0.9410
Worst	0.8000	0.8300	0.8100	0.8000	0.8000	0.8000	0.7100	0.8000	0.8000	0.8730

Table 5 Values of AUC for all methods (after feature selection)

Algorithm	Naïve bayes	LR	MLP	Bagging	J48	RF	RT	SVM	CART	PSO-MLP
Avg	0.8880	0.8768	0.8643	0.8635	0.8335	0.8471	0.7416	0.8107	0.8568	0.9023
Std	0.0283	0.0296	0.0350	0.0216	0.0466	0.0319	0.0286	0.0294	0.0273	0.0271
Best	0.9200	0.9300	0.9150	0.9090	0.8870	0.9070	0.7900	0.8550	0.9030	0.9380
Worst	0.8470	0.8270	0.8200	0.8420	0.7380	0.8040	0.7020	0.7690	0.8290	0.8690

these four mentioned algorithms. On the other hand, AUC of other algorithms have profoundly affected by the feature selection process. PSO-MLP have the highest average and lowest STD among other classifiers. Also, it can see that naïve bayes achieved the best AUC of 95%. Also, the worst AUC belongs to random tree. Since PSO based MLP has the highest average and lowest STD among other classifiers, it can be inferred that PSO based MLP is the most accurate and reliable classifier in comparison other classifiers. Consequently, the analysis shows the importance of each feature individually.

We have illustrated boxplots in Figs. 6 and 7 to further study. They represent the boxplots relative to 10 runs of each trainer. After apply feature selection process on our proposed data set the output figures display PSO-MLP has a higher median among other classifiers and confirm the grater performance of PSO when training MLP.

In summary, the PSO-MLP algorithm performs efficiently in two important evaluation metrics named accuracy and AUC. The main negative point in other studied methods is disability in maintaining balance among exploration and exploitation in comparison with the PSO based MLP classifier. Eventually trapping in local minima as an important problem and attaining immature convergence could be seen in the rest of method's results. robust result of PSO-MLP demonstrate that this method uses a great mechanism to escaping from local minima and keeping away from immature convergence. As a result, PSO can boost the performance of MLP in exploration and exploitation of search space and eventuates in better strong results in comparison with other algorithms.

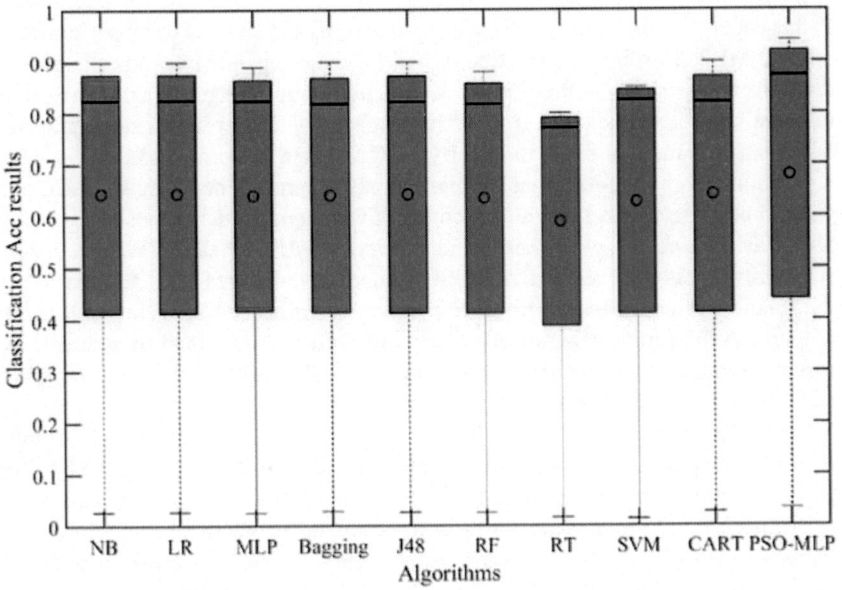

Fig. 6 Boxplot, ACC for RF, PSO-MLP, RT, NB, SVM, LR, Bagging, MLP, J48 and CART on dataset (after feature selection)

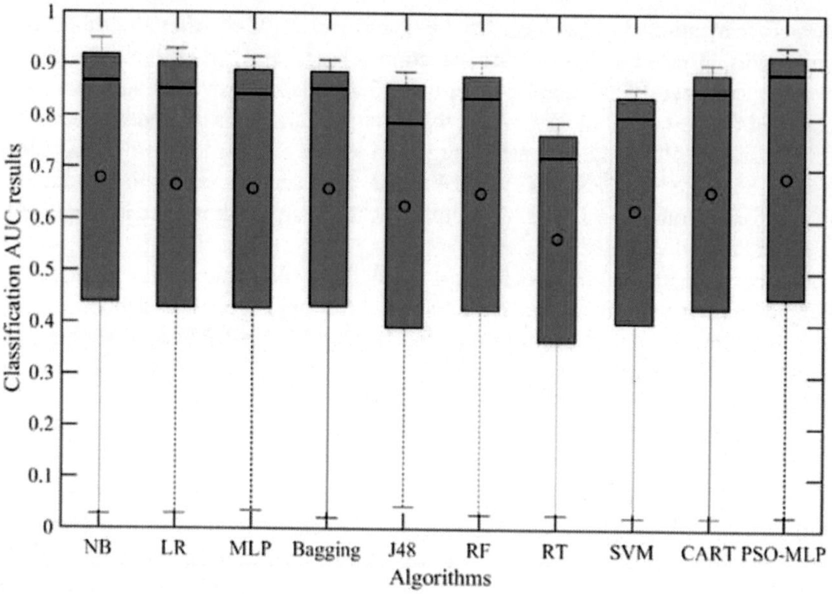

Fig. 7 Boxplot, AUC for RF, PSO-MLP, RT, NB, SVM, LR, Bagging, MLP, J48 and CART on dataset (after feature selection)

5 Conclusions

In this paper various experiments are done to identify the efficiency and performance of PSO-MLP classifier to prediction of heart disease regarding accuracy and AUC through comparing it with the nine famous machine learning algorithms. With regards to quantitative and qualitative results, the proposed optimizer in this study achieves better success rates in order to classifying CAD data in comparison to the other algorithms. Also, PSO-MLP has the highest rank in terms of accuracy and AUC and it could be more efficient and highly competitive among all the proposed models. It has an average accuracy of 90.56% and an average AUC of 90.23% when a feature selection process was executed. Also, the standard deviation of PSO is approximately the smallest value for both considered metrics. in summary PSO-based trainer can outperform the current algorithms significantly. Moreover, it could be attained best results for all proposed metrics namely accuracy, AUC is capable of training MLP effectively to answer classification problems.

As to Future direction, there are myriad ways to continue and enhance this study. It would be valuable to compare the mentioned methods which are applied in this paper with other available methods in order to identify their performance in the scope of diagnosis CAD. Another worthy direction could be recommended is investigating the performance of PSO-MLP in comparison with a various number of repetitions. Furthermore, proposed approaches could be applied in various types of data mining scopes such as medical, robotic application and business [8].

References

1. I.A. Paizis, P.D. Mantzouratou, G.S. Tzanis, C.A. Melexopoulou, M.N. Darema, J.N. Boletis, J.D. Barbetseas, Coronary artery disease in renal transplant recipients: an angiographic study. Hell. J. Cardiol. (2018)
2. M.S. Amin, Y.K. Chiam, K.D. Varathan, Identification of significant features and data mining techniques in predicting heart disease. Telemat. Inform. **36**, 82–93 (2019)
3. S.M.J. Jalali, M. Karimi, A. Khosravi, S. Nahavandi, An efficient neuroevolution approach for heart disease detection, in *2019 IEEE International Conference on Systems, Man and Cybernetics (SMC)* (IEEE, 2019), pp. 3771–3776
4. A.A. Heidari, H. Faris, I. Aljarah, S. Mirjalili, An efficient hybrid multilayer perceptron neural network with grasshopper optimization. Soft. Comput. **23**(17), 7941–7958 (2019)
5. S. Ahmadian, A.R. Khanteymoori, Training back propagation neural networks using asexual reproduction optimization, in *2015 7th Conference on Information and Knowledge Technology (IKT)* (IEEE, 2015), pp. 1–6
6. M.J. Amiri, J. Abedi-Koupai, S.M. Jafar Jalali, S.F. Mousavi, Modeling of fixed-bed column system of Hg (II) ions on ostrich bone ash/nZVI composite by artificial neural network. J. Environ. Eng. **143**(9), 04017061 (2017)
7. I. Aljarah, H. Faris, S. Mirjalili, Optimizing connection weights in neural networks using the whale optimization algorithm. Soft. Comput. **22**(1), 1–15 (2018)
8. S.M.J. Jalali, A. Khosravi, P.M. Kebria, R. Hedjam, S. Nahavandi, Autonomous robot navigation system using the evolutionary multi-verse optimizer algorithm, in *2019 IEEE International Conference on Systems, Man and Cybernetics (SMC)* (IEEE, 2019), pp. 1221–1226
9. M. Tayefi, M. Tajfard, S. Saffar, P. Hanachi, A.R. Amirabadizadeh, H. Esmaeily et al., hs-CRP is strongly associated with coronary heart disease (CHD): a data mining approach using decision tree algorithm. Comput. Methods Programs Biomed. **141**, 105–109 (2017)
10. R. Alizadehsani, J. Habibi, M.J. Hosseini, H. Mashayekhi, R. Boghrati, A. Ghandeharioun et al., A data mining approach for diagnosis of coronary artery disease. Comput. Methods Programs Biomed. **111**(1), 52–61 (2013)
11. M. Negahbani, S. Joulazadeh, H. Marateb, M. Mansourian, Coronary artery disease diagnosis using supervised fuzzy c-means with differential search algorithm-based generalized Minkowski metrics. Peertechz J. Biomed. Eng. **1**(006), (2015)
12. S. Bhatia, P. Prakash, G.N. Pillai, SVM based decision support system for heart disease classification with integer-coded genetic algorithm to select critical features, in *Proceedings of the World Congress on Engineering and Computer Science* (2008), pp. 34–38
13. A. Rajkumar, G.S. Reena, Diagnosis of heart disease using datamining algorithm. Glob. J. Comput. Sci. Technol. **10**(10), 38–43 (2010)
14. M. Abdar, S.R.N. Kalhori, T. Sutikno, I.M.I. Subroto, G. Arji, Comparing performance of data mining algorithms in prediction heart diseases. Int. J. Electr. Comput. Eng. (2088–8708) **5**(6), (2015)
15. M. Shouman, T. Turner, R. Stocker, Using decision tree for diagnosing heart disease patients, in *Proceedings of the Ninth Australasian Data Mining Conference-Volume 121* (2011), pp. 23–30
16. K. Polat, S. Güneş, A hybrid approach to medical decision support systems: combining feature selection, fuzzy weighted pre-processing and AIRS. Comput. Methods Programs Biomed. **88**(2), 164–174 (2007)
17. U. Shafique, F. Majeed, H. Qaiser, I.U. Mustafa, Data mining in healthcare for heart diseases. Int. J. Innov. Appl. Stud. **10**(4), 1312 (2015)
18. I. Babaoglu, O.K. Baykan, N. Aygul, K. Ozdemir, M. Bayrak, Assessment of exercise stress testing with artificial neural network in determining coronary artery disease and predicting lesion localization. Expert Syst. Appl. **36**(2), 2562–2566 (2009)
19. S.R. Tithi, A. Aktar, F. Aleem, A. Chakrabarty, ECG data analysis and heart disease prediction using machine learning algorithms, in *2019 IEEE Region 10 Symposium (TENSYMP)* (IEEE, 2019), pp. 819–824

20. Y. Shi, R.C. Eberhart, Empirical study of particle swarm optimization, in *Proceedings of the 1999 Congress on Evolutionary Computation-CEC99 (Cat. No. 99TH8406)*, vol. 3 (IEEE, 1999), pp. 1945–1950
21. P. Ghamisi, J.A. Benediktsson, Feature selection based on hybridization of genetic algorithm and particle swarm optimization. IEEE Geosci. Remote Sens. Lett. **12**(2), 309–313 (2014)
22. N. Delgarm, B. Sajadi, F. Kowsary, S. Delgarm, Multi-objective optimization of the building energy performance: a simulation-based approach by means of particle swarm optimization (PSO). Appl. Energy **170**, 293–303 (2016)
23. E.J. Topol, R.M. Califf (eds.), *Textbook of Cardiovascular Medicine* (Lippincott Williams & Wilkins, Philadelphia, 2007)
24. M.A. Hall, Correlation-based feature selection for machine learning, 1999
25. M.A. Hall, L.A. Smith, Practical feature subset selection for machine learning, 1998
26. M. Doshi, Correlation based feature selection (CFS) technique to predict student perfromance. Int. J. Comput. Netw. Commun. **6**(3), 197 (2014)
27. C. Anuradha, T. Velmurugan, Feature selection techniques to analyse student acadamic performance using Naïve bayes classifier, in *The 3rd International Conference on Small & Medium Business* (2016), pp. 345–350
28. H. Faris, I. Aljarah, S. Mirjalili, Training feedforward neural networks using multi-verse optimizer for binary classification problems. Appl. Intell. **45**(2), 322–332 (2016)

Review for Meta-Heuristic Optimization Propels Machine Learning Computations Execution on Spam Comment Area Under Digital Security Aegis Region

Biswajit Mondal, Debkanta Chakraborty, Niloy Kr. Bhattacherjee, Pritam Mukherjee, Sanchari Neogi, and Subir Gupta

Abstract In the redesign field, popularly known as optimization, such difficulties degenerate the exordium of smoothing out appraisals that performed well on benchmark cutoff focuses or essential setting-centered evaluations. To visually examine the commencement of techniques procured for smoothing out those conditions, scientists and experts felt a need to agnize the difficulties and break harmonious changes, modifications, and amendments in the evaluations to oversee such hardships. As of tardy, there has been actuating examination interest in orchestrating machine learning (ML) strategies into meta-heuristics for managing combinatorial smoothing out conditions. This joining betokens meta-heuristics towards a capable, abundant, and exuberant seek after. It also transmutes their exhibition much indistinguishably commensurate to procedure quality, cumulation rate, and energy. Since sundry getting procedures together with sundry purposes, we have incited an objective to review the early advances in utilizing ML techniques to revise meta-heuristics. To fill up this gap survey gives such an audit on the utilization of AI methods in the approach of sundry components of meta-heuristics for purposes behind computation winnows utilizing absolute execution limits. The adequacy of move and quantify is transmuted by refreshing ML evaluations using meta-heuristic redressment computations. In the review, seventeen ML-predicated evaluations we have applied for benchmarking datasets and conspicuous legitimate time tests for tasks and figures we have imparted. Pushing toward portions covers the energy-moving assessment subjects coordinating progressed ML-predicated seventeen evaluations execution and gives examiners some spellbinding pieces of erudition to utilize in their generous applications spaces of pay. This book region causes assembled well-kenned ML approaches

B. Mondal · D. Chakraborty · S. Gupta (✉)
Department of Computer Science and Engineering, Dr. B. C. Roy Engineering College, Durgapur, West Bengal 713206, India
e-mail: subir2276@gmail.com

N. Kr. Bhattacherjee
Department of FMS, Dr. B. C. Roy Engineering College, Durgapur, West Bengal 713206, India

P. Mukherjee · S. Neogi
Department of Master of Computer Application, Dr. B. C. Roy Engineering College, Durgapur, West Bengal 713206, India

E. H. Houssein et al. (eds.), *Integrating Meta-Heuristics and Machine Learning for Real-World Optimization Problems*, Studies in Computational Intelligence 1038,
https://doi.org/10.1007/978-3-030-99079-4_13

to dissever the spam and ham comments. There may be so many obstacles in cyber security, such as Malware, Worm, viruses, SQL Injection, etc. One of them is Spam. It is a typical binary case where the output must be binary. We will examine from a large dataset through the meta-heuristic function, which one among 17 applied algorithms is the optimized one for the binary cases.

Keywords Classifier · Digital security · Machine learning · Meta-heuristic · Optimization

1 Introduction

Recently, the increasing use of automated digital mode has given rise to spam problems. Express customers and affiliations are affected by Spam, considering various cults such as genius value, corporate web, etc. The distinctive strategy is obtained from specialists to visually and consistently examine spam issues in the security field; one of the consistent standard perspectives is machine learning (ML). Arthur Samuel competed to introduce ML in 1959. After a short time, ML was applied serially in various fields. ML is being used at one point or another of events considering everything, other apps, and programming on the internet. ML has become so inevitable that it has managed to replicate itself. At the request of efforts to combat a merger of problems. It could be used in various fields, including image processing, materials science, data science, etc. ML expects the probability of riding where the predetermined data set created the inhibition. The pre-sorted data set can be a tag; it is considered supervised learning if the labeled data set passes as a probabilistic labor dispute. We can categorize the fundamental commitments of this review into several categories. In this review, we employ Seventeen different ML-based Classifier predictive models for Feature resolution. This methodology is meta-heuristic to improve grouping execution, and there is a wide range of execution estimations. Seventeen techniques have never been widely dissected in the binary component resoluteness area to the best of our knowledge. In this study, we examine seventeen other classifier ML techniques and quantify their performances in seven ways: accuracy, precision, recall, F1 score, AUC score, model preparation time, and mean squared error [1–3]. The remaining book chapter facilitates in the following way: In Sect. 2.1, we have demystified the fundamental concepts of optimization, followed by Sect. 2.2 that has overviewed the machine learning, in Sect. 2.3 used Machine Learning Algorithms and Sect. 2.4 provides a view of digital security. Section 3 frameworks the methodology of meta-heuristic optimization that propels machine learning computations execution on spam comment area under digital security aegis region. Finally, in Sect. 4, we search for an optimal meta-heuristic technique under the result.

2 Literature Review

The effect of modernized correspondence has transmuted it into the most head part of the contemporary world. The most sizeable voluminous model is Spam [4, 5]. Spam remarks are by and sizably voluminous unconstrained, un-invigorated, and by and colossal voluminous correspondence [6, 7]. Spam itself is a pass-on of malware and causes a few cycles [8, 9]. Settling this issue has dramatically become one of the fundamental conditions for digital edified ascendant components. It is everything except moving toward the security space's appraisal region [10, 11]. Of sundry strategies, Machine Learning (ML) is insistently conspicuous, cerebrating about its brood application. It might use in a gregarious event of fields, including picture approbation, material science, and displaying Data science and others [12, 13]. A subset of Machine Learning is determinedly cognate to computational experiences, which fixates on making measures utilizing likelihood thickness work. Program evaluations brief the machine to execute all frameworks expected to manage the inception concern, and the digital part doesn't require learning. Another prominent characteristic of all things is being equipped, surveying all estimations, and optically canvassing the best game plan. Unmistakably it's called meta-heuristic optimization. This book chapter hopes to review meta-heuristic optimization that pushes ML estimations execution on spam comment locale under electronic security aegis region.

2.1 Optimization

Mathematical optimization is also known as mathematical programming [14, 15]. It is the section where a specific type of selection happens. It is a selection of a principal constituent, obviously hinging on several criteria and picking from some bunch of available standby. The problems related to optimization have different types [16]. It comes to light in all quantifiable regulations, and it has a range from computer science and engineering to operational research and economics. And where there is a problem, it is obvious to have a solution, which is not an exception. Here the answer comes from the mathematics for centenaries. If we look at this section's child's play case, we can see the composition s of an optimization problem. The content maximizes or minimizes an actual province through a comprehensive selection of input value from inside a permissible set and enumerates that particular value of the function. The notion of the theory and procedure of optimization comprises a vast zone of applied mathematics. More simply, we can say that optimization is the thing that incorporates the finding of the most suitable and accessible value of several impartial concerns provided by another sector. Alternatively, optimization is an input encompassing various non-discriminatory activities and multiple kinds of realms. Coming to the next part, that is an optimization problem. Before getting into trouble, we first need to discuss some points. So, without delay, let's put the

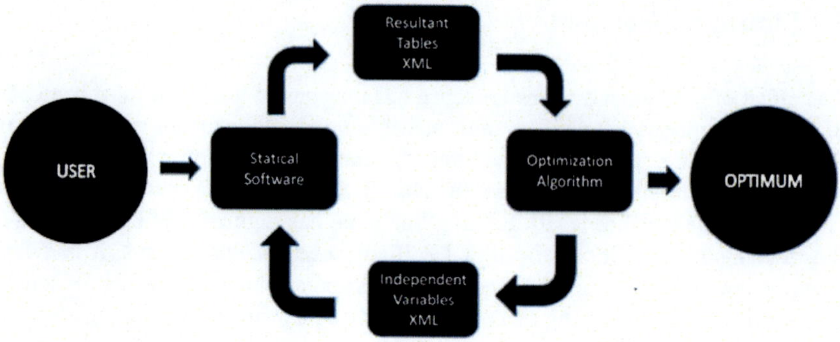

Fig. 1 General optimization model

spotlight on them. Figure 1 shows the model. Optimization intends to accomplish the "finest" blueprint correlative to a bunch of codified criteria or hindrances. These encompass aggrandizing ingredients like fecundity, robustness, dependability, durability, expertise, and usage. For instance, companies desire to minimize the cost of fabrication or maximization of yield frequently. It is a frequent desire to diminish the value of materials applied for packaging a product within a particular dimension in production. We are engrossed in finding a contrivance that exaggerates or diminishes several functions in optimization problems. In designing algorithms, particularly for the optimization problem, we have to manifest that the algorithm in original provides the adequate feasible solving measure. The different Optimization problem types are:

(a) Linear and Quadratic Programming Problems.
(b) Quadratic Constraints and Conic Optimization Problems.
(c) Integer and Constraint Programming Problems.
(d) Smooth Nonlinear Optimization Problems.
(e) Nonsmooth Optimization Problems.

Optimization problems will always ask you to maximize or minimize some quantity, having described the situation using words (instead of immediately giving you a function to max/mitigate). Typical phrases that indicate an Optimization problem include: Find the largest or find the minimum. Every optimization problem has three components: an objective function, decision variables, and constraints. A single-objective optimization problem aims to find the best solution for a specific criterion or metric, such as execution time (or performance) or a combination with energy consumption or power dissipation metrics. In short, optimization is the state where we choose and implement the best condition using the best methods. There may be so many optional solutions present for a single problem. If a problem arises, it is not mandatory or a thumb rule that there will be only one solution existing, but there must be a solution present to solve the problem. An extra point that we must keep in mind while finding or applying the key is implementing the best solution suitable to maintain the optimum state or condition. So, we need to predict and define the best solution among the best options. To do so, we have a function in optimization known as the "Meta-Heuristic Function" [17–19]. But before getting indulged into the meta-heuristic position, it is a working procedure and all such things. It is the point where we need to discuss some methods related to optimization or operation research. Some popular meta-heuristic techniques are the Integer problem, Duality, Sensitivity analysis, Transportation problem, Dynamic problem, Assignment problem, Goal theory, Game theory, etc.

2.2 *Overview of Machine Learning*

Machine learning is a zone related to computer programming that is jeopardized from exploring model affirmation and computational acquirement notional conjecture in ersatz excogitating [20]. It possesses a solid connection to numerical betterment, which devolves blueprint, postulates, and solicitation deploy to the field. Machine learning is used in the scope of thinking activities were creating and encoding incontrovertible estimates are unviable [21–23]. Model applications consolidate multimedia security, spam sifting, web search implements, metrical science, data science modeling, and PC vision. Machine learning is conflated with data mining, albeit more on exploratory information examination. Pattern apperception and Machine learning "can be optically discerned as two aspects of a homogeneous field". When utilized in modern settings, machine learning strategies might be alluded to as prescient examination or prescient demonstrating [24, 25]. The kinds of machine learning calculations are essentially partitioned into four classifications: Supervised learning, unsupervised learning, Semi-regulated learning, and Reinforcement learning. We have shown the same in Fig. 2. A comparative discussion we have provided in Table 1.

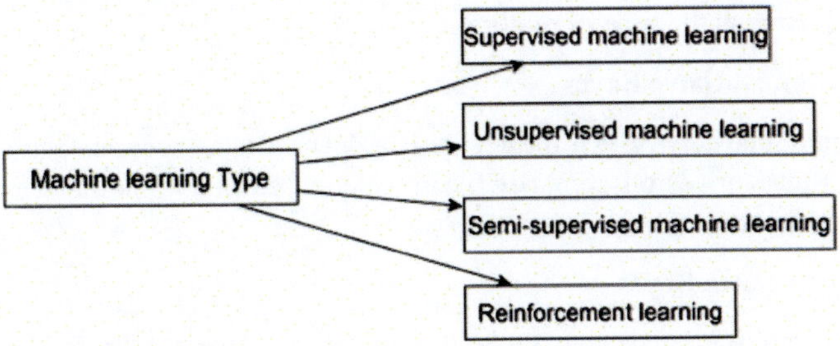

Fig. 2 Different types of machine learning

Table 1 Difference between supervised and un-supervised machine learning

Supervised machine learning	Unsupervised machine learning
Info and yield factors will be given	Only info will be given
Calculations are prepared to utilize labeled information	Calculations are prepared to utilize without labeled information
Ex SVM, ANN, linear and logistics regression, and so on	Example K-means, cluster algorithms, hierarchical clustering, and so on
Profoundly precise and dependable strategy	Less precise and dependable technique
Classes amount is known	Classes amount is not known

2.3 *Used Machine Learning Algorithms Example*

Machine learning has several types of algorithms which could be applied or implemented to perform various kinds of tasks. This section will discuss seventeen algorithms that we have used here. Before getting inside into the main problem and case study, it is necessary to look at them at a glance. So, let's put the spotlight on them:

(a) Logistic regression

It is another method acquired by machine learning from measurements [26]. It is the go-to strategy for parallel order (issues with two class esteems) [27].

(b) Gaussian Naive Bayes

The Gaussian Naïve Bayes is a type of naïve Bayes algorithm that is a follower of the Gaussian normal distribution and a supporter of the data continuation [28]. Each attribute is attached with an uninterrupted continuous value [29]. The gaussian naïve Bayes has another name, which is the normal distribution. It produces a bell-like structured curve at the time of plotting.

(c) Multinominal Naïve Bayes

The multinominal naïve bias is the detailed model used to categorize documents [30]. The facet routes or vectors comprise the prevalence, and a multinominal distribution can engender certain events.

(d) Bernoulli Naïve Bayes

Same as the multinominal naive Bayes, this model has its fame in the case of indenture categorization. Here, the attribute of bipartite term contingency is utilized as an alternative to the term frequency. In this procedure, attributes are self-subsisting binary chronicle inputs.

(e) Support Vector Machine (SVM)

It is a supervised machine learning calculation that can be utilized for both order or relapse difficulties [31]. In the SVM calculation, we plot every information thing as a point in n-dimensional space (where n is sundry highlights you have), with the worth of each element being the worth of a concrete arrangement [32].

(f) K-Nearest Neighbour (KNN)

One of the most accessible Machine Learning calculations depends on the Supervised Learning method [33, 34]. KNN calculation expects the comparability between the new information and accessible cases and places the latest issue into the classification that is generally like the accessible classifications.

(g) Gradient Boosting Classifier and Ada Boosting Classifier

These are boosting estimations used when vast loads of data should be dealt with to make assumptions with high precision. Boosting is a social occasion learning estimation that combines the perceptive power of a couple of base assessors to improve strength. This way, it joins diverse weak or typical pointers to develop a strong marker. These boosting estimations reliably work outstandingly in data science contentions.

(h) Real Ada Boost Classifier

Instead of multiplying the whole output retrieved from the entire tree structure with a constant value, every leaf node is modified to outcome half the logit metamorphosis of that pre-existing values.

(i) Logit Boost Classifier

The logit boost classifier represents or requisitions an orthodox logistic regression approach to the Ada-boost algorithm. We can select to lessen the misconceptions instead of minimizing fallacy regarding a variable, feeble learner.

(j) Decision Tree

The Decision Tree calculation in machine learning is the most mainstream calculation used today; this is a managed learning calculation utilized for ordering issues [35]. It functions admirably grouping for both fundamental and ceaseless ward factors [36]. This calculation divides the populace into two homogeneous sets dependent on the main ascribes/autonomous characteristics.

(k) Random Forest

It is a well-known machine learning calculation that has a place with the supervised learning procedure [37, 38]. It very well may be utilized for both Classification and Regression issues in machine learning. It depends on the idea of outfit realizing, which is a cycle of consolidating various classifiers to take care of an intricate issue and improve the model's presentation.

(l) Bagging Classifier

The bagging classifier is a perfect quintet meta-estimator that positions the pedestal classifier (individually in an erratic subset of the actual dataset).

(m) Perceptron Classifier

The perceptron is a straightway algorithm associated with machine learning. It is applied to do the task of binary categorization. A critical point of this algorithm is that one can contemplate this method as one of the pioneers and a most uncomplicated variety of the ANN.

(n) SGD Classifier

SGD stands for Stochastic Gradient Descent. It is a classifier that primarily applies an unfussy SGD learning schedule that supports different kinds of loss functions and the retributions for categorization. It is quite a simple but very much effective and efficient method to fit the regressor, simultaneously the linear classifier also. We can apply this algorithm flourishingly into the large-scale problems of machine learning.

(o) Extra Tree Classifier

The Extremely Randomized Trees Classifier or Extra Tree Classifier is a quintet learning procedure that agglomerates the resultants of the multiple non corresponded decision tree gathered instead of assembled in a forest. It is because to get the outcome and its categorization result.

(p) Passive Aggressive Classifier

The passive-aggressive classifier comes from the class of networked learning methods in machine learning. The working procedure of this method is that if we get the prophecy accurate, there is no need to alter anything and leave the model as it is.

2.4 Digital Security

Cyber security refers to applying technologies, processes, and practices designed to forfend networks, contrivances, programs, and data from assault, damage, or unauthorized exploitation. Cybersecurity may be referred to as information technology security, but they are different to some extent [39]. The ultimate aim of cybersecurity is to diminish the jeopardy of cyber-attacks and forfend against the unauthorized exploitation of data, systems, networks, and technologies. The activities can be expressed as resisting contrivances like computers and smartphones, electronic systems, networks, and data from vicious attacks, ranging from individuals' contraptions to business organizations. An organization needs to synchronize its exertions throughout its entire information system to develop efficacious cybersecurity. Cybersecurity includes many elements. Populares are Network Security, Application Security, Endpoint Security, Operational Security, Identity Management, etc.

3 Methodology

Figure 3 shows the strategy used to make a living on spam comments as they are being created. The record must maintain a single prominent record for any system to work. Different approaches are required. The main task is to collect logs from open-source sources. The data in the graph comes from two open sources. One from GitHub and the other from Kaggle, and all of the data is physically handled the same way. After collecting the data, it is passed on to the next level. The accumulated data set is subject to three general data pre-processing steps.

The main component of interacting with the spam location is data pre-processing, also known as data scrubbing. Data cleansing converts unstructured data into more organized information and helps classify the data. Sample label data is shown in Table 2. In our model, we are only interested in the comment area of the record, so we focused on determining the presence of undesirable constructs in this segment. After the data scrubbing cycle, we created clean data used in the subsequent processes. The extraction period significantly impacts the separation of spam and ham comments. The item extraction metric begins with an underlying arrangement of estimated data in the machine learning measurement, which was proposed as non-redundant, so just as instructive in this way to better understand how to lead to better guesswork.

The vector is modified significantly or reduced to a collection of named highlights. In this study, we used CountVectorizer. CountVectorizer's goal is to convert a book into a repetition-based vector. The CountVectorizer creates a grid in which a grid segment addresses each new observation, and each cell estimate addresses the word

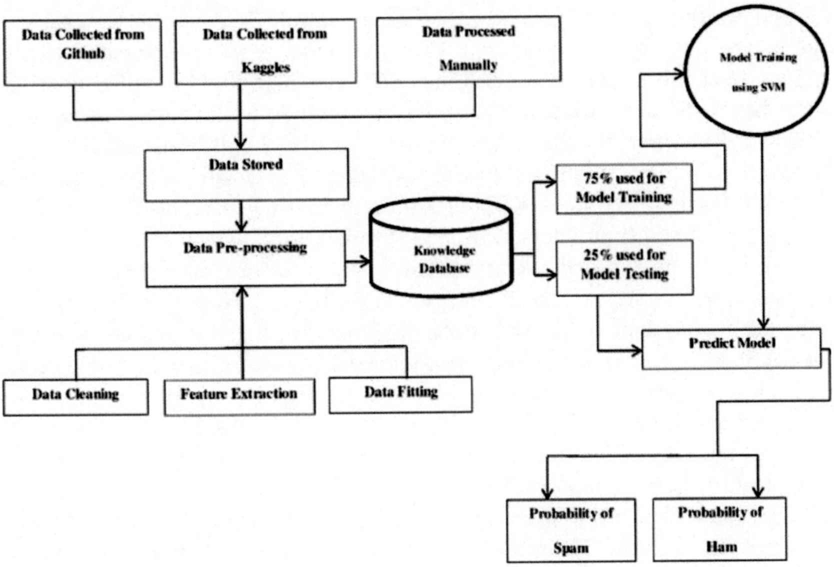

Fig. 3 Spam comment recognition in machine learning through support vector machine classifier

Table 2 Sample dataset for learning with class value

Contents	Label
Sorry light turned green, and I meant another friend wanted <#> worth, but he may not be	0
Thanks for yesterday, sir. You have been excellent. I hope you enjoyed the burial	0
Hi, mate, it's RV. Did you have a lovely hotel? Just a message 3 say hello because havenåÕt sent you 1 in ages started driving, so stay off roads!RVs	1
Do you want 750 anytime any network mins 150 texts and a NEW VIDEO phone for only five pounds per week call 08002888812 or reply for delivery tomorrow	0
I love you soo much you donåÕt understand how special your 2 I'll ring you 2morrow Luv you xxx	1

count comparison. The highlighted extraction period, which significantly influences the separation of spam and ham comments, follows. The item extraction metric starts with an underlying disposition of estimated data in the machine learning measurement. Constructs outstanding values or attributes suggested as non-redundant and worked on a better understanding of leading to better assumptions. When we enter a machine learning calculation that is too large to prepare, the vector is changed or significantly reduced to produce a sentence of the most outstanding aspects. Finally, this is the time to discuss the case study. We discussed the best solution-finding procedure by applying the meta-heuristic function in the operation research domain. We have already discussed 17 specific machine learning algorithms that we have applied here rather than used here as some better option to solve an anonymous but particular problem. So, our objective is to find out the best possible solution that reaches the optimization level and provides a satisfactory result. To do so, we have selected and assembled 17 specific algorithms. They are: logistic regressive classifier, passive aggressive classifier, perceptron classifier, SGD classifier, SVM classifier, Gaussian naïve bayes classifier, Bernoulli naïve bayes classifier, Multinominal naïve bayes classifier, Extra tree classifier, decision tree classifier, Logit Boost classifier, Ada Boost classifier, Real Ada Boost classifier, Gradient Boost classifier, Random Forest classifier, Bagging classifier, KNN classifier. Then using the meta-heuristic function, we will find the optimum state and the satisfactory optimal solution. We have collected and gathered all the datasets from two open APIs for the commencement of the same. One is GitHub, and the other one is the Kaggle. From these two, we have done the data accumulation. The next point is the investigation and the storage of the datasets. After that, the following steps are data preservation and pre-processing (data cleaning, feature extraction, and data fitting), respectively. We have processed the data manually. Here, 40,000 datasets are available. We have divided this dataset into a 3:1 ratio. That 75% of information is utilized for training, and 25% of the data is applied for testing purposes. We have run the entire dataset using that particular 17 algorithms of machine learning that we have discussed and mentioned earlier for a specific problem. The parameters that we have utilized here are the accuracy score, precision score, recall score, F1 score, AUC score, Model training time, and the Mean square error. All the 17 algorithms we have passed through these seven

Table 3 Seventeen machine learning algorithms are passing through seven parameters

Classifier name	Accuracy	Precision	Recall	F1	AUC	Training	MSE
Logistic regression classifier	94.48	97.07	93.79	95.4	99.14	29.92	0.03
Passive aggressive classifier	95.33	92.83	95.4	94.1	98.76	12.93	0.04
Perceptron classifier	95.82	95.59	93.59	94.58	98.91	5.5	0.04
SGD classifier	96.23	97.03	93.19	95.07	99.07	4.65	0.04
SVM classifier	96.23	96.63	93.59	95.09	99.02	189.17	0.04
Gaussian NB classifier	79.16	65.63	97.8	78.55	82.97	2.51	0.21
Bernoulli NB classifier	84.71	72.33	98.49	83.41	98.17	5.5	0.15
Multinomial NB classifier	88.77	78.91	97.17	87.11	98.04	4.55	0.11
Extra tree classifier	93.19	91.73	93.59	91.22	93.17	6.96	0.07
Decision tree classifier	94.44	92.99	92.73	92.86	94.58	241.47	0.05
Logit boost classifier	91.98	93.5	85.36	89.24	96.9	396.57	0.08
Ada boost classifier	87.94	93.66	74.58	82.22	93.98	377.01	0.13
Real ada boost classifier	92.21	93.05	85.95	89.11	97.35	376.02	0.08
Gradient boost classifier	89.78	95.32	78	85.24	96.54	867.03	0.1
Random forest classifier	96.45	96.12	94.89	96.32	99.24	268.01	0.02
Bagging classifier	95.21	94.01	93.25	94.32	98.22	470.01	0.04
KNN classifier	88.01	92.98	74.98	84.01	92.31	63.89	0.12

kinds of parameters, as we have shown in Table 3. Here we need to remember that the tended first five parameters should be 100, and the rest two parameters should be managed to 0. Here we have provided two kinds of beginner-level approaches to find the solution.

4 Result

Our goal is to choose the best solution among the seventeen algorithms using seven different parameters.

4.1 Rank-Based Meta-Heuristic Optimization

Out of 17 classifiers, Random Forest Classifier ranked First when we took the average value of 5 maximizing parameters (Accuracy Score, Precision Score, Recall Score, F1 Score, and AUC Score). SGD Classifier ranked second, and SVM classifier ranked third (Fig. 4). Similarly, when we took the average value of 2 minimizing parameters (Model Training Time and Mean Square Error), Gaussian Naïve Bayes Classifier

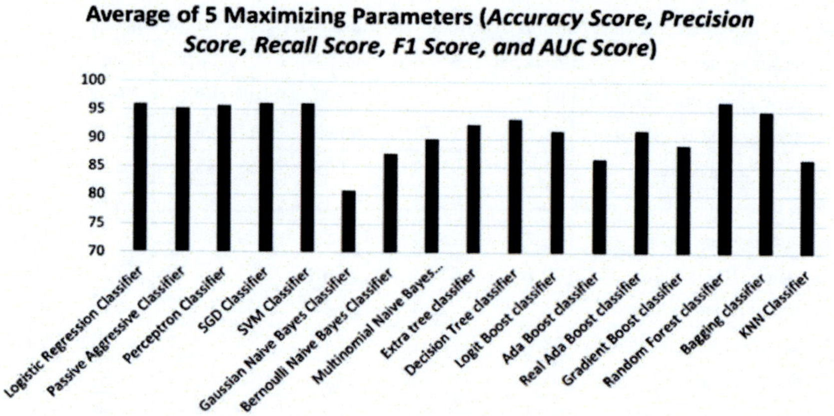

Fig. 4 Comparison of 17 classifiers by an average of 5 maximizing parameters (Accuracy Score, Precision Score, Recall Score, F1 Score, and AUC Score)

ranked. First, Multinomial Naïve Bayes Classifier ranked second, and SGD Classifier ranked third (Fig. 5). When we considered all the seven parameters together (by taking the difference of average of 5 maximizing parameters and the standard of 2 minimizing parameters), SGD Classifier ranked. First, Perceptron Classifier ranked second, and Extra Tree Classifier ranked third (Fig. 6).

So, Random Forest Classifier was the winner while considering Accuracy Score, Precision Score, Recall Score, F1 Score, and AUC Score. Gaussian Naïve Bayes Classifier was the chosen classifier when we only thought minimizing parameters such as Model Training Time and Mean Square Error. Still, the overall winner or selected parameter was SGD Classifier when we took all the seven parameters considering the average value of all these seven parameters.

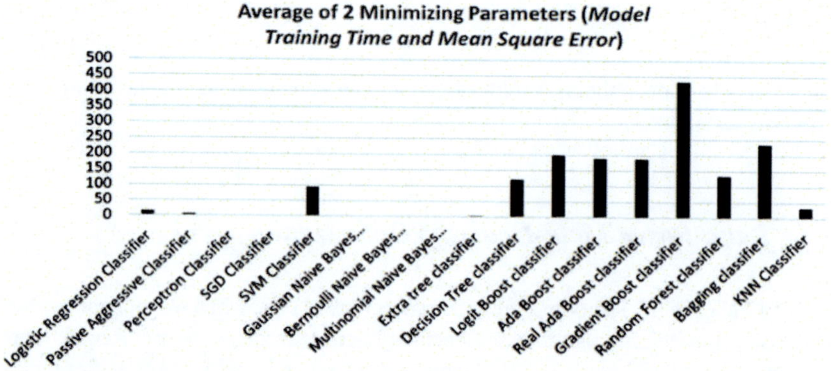

Fig. 5 Comparison of 17 classifiers by an average of 2 minimizing parameters (model training time and mean square error)

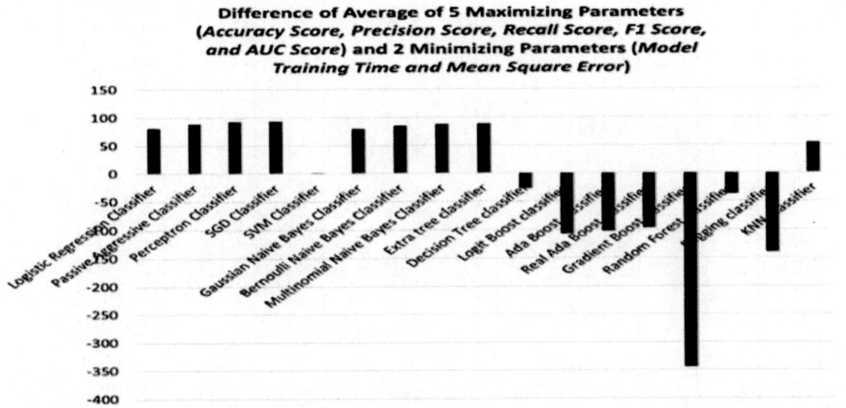

Fig. 6 Comparison of 17 classifiers by the difference of average of 5 maximizing parameters (Accuracy Score, Precision Score, Recall Score, F1 Score, and AUC Score) and two minimizing parameters (model training time and mean square error)

When we consider all the seven parameters individually, out of 17 classifiers, Random Forest Classifier ranked first in 4 out of 7 parameters such as Accuracy Score, F1 Score, AUC Score, and Mean Square Error. But its Model Training Time was one of the worst (ranked 12th). Random Forest Classifier ranked fourth and fifth in Precision Score and recall score parameters, respectively. SGD Classifier, on the other hand, was among the top 3 in 5 out of 7 parameters. In Accuracy Score, Precision Score, AUC Score, Model Training Time, and Mean Square Error, it (SGD Classifier) ranked second, second, third, third, and third, respectively. Recall Score and F1 Score ranked 11th and fourth, respectively.

Next, we compared the sum of the seven ranks corresponding to all the seven parameters for the 17 classifiers (Fig. 7, Table 4). Again, Random Forest Classifier

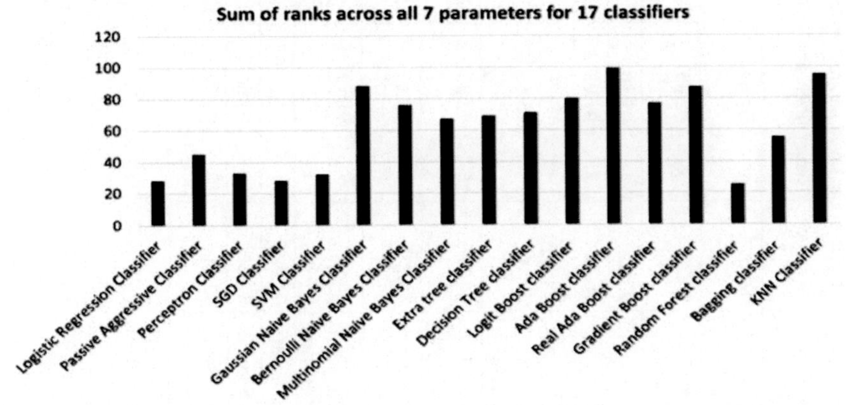

Fig. 7 Comparison of 17 classifiers by the sum of ranks corresponding to all seven parameters

Table 4 Ranks of each of the seven parameters and some of the seven ranks for all the 17 classifiers

Classifier name	Accuracy	Rank	Precision	Rank	Recall	Rank	F1	Rank	AUC	Rank	Training	Rank	MSE	Rank	Sum of ranks
Logistic regression	94.48	7	97.07	1	93.79	6	95.4	2	99.14	2	29.92	8	0.03	2	28
Passive aggressive	95.33	5	92.83	13	95.4	4	94.1	7	98.76	6	12.93	7	0.04	3	45
Perceptron	95.82	4	95.59	5	93.59	7	94.58	5	98.91	5	5.5	4	0.04	3	33
SGD	96.23	2	97.03	2	93.19	11	95.07	4	99.07	3	4.65	3	0.04	3	28
SVM	96.23	2	96.63	3	93.59	7	95.09	3	99.02	4	189.17	10	0.04	3	32
Gaussian Naive Bayes	79.16	17	65.63	17	97.8	2	78.55	17	82.97	17	2.51	1	0.21	17	88
Bernoulli Naive Bayes	84.71	16	72.33	16	98.49	1	83.41	15	98.17	8	5.5	4	0.15	16	76
Multinomial Naive Bayes	88.77	13	78.91	15	97.17	3	87.11	12	98.04	9	4.55	2	0.11	13	67
Extra tree	93.19	9	91.73	14	93.59	7	91.22	9	93.17	15	6.96	6	0.07	9	69
Decision tree	94.44	8	92.99	11	92.73	12	92.86	8	94.58	13	241.47	11	0.05	8	71
Logit boost	91.98	11	93.5	9	85.36	14	89.24	10	96.9	11	396.57	15	0.08	10	80
Ada boost	87.94	15	93.66	8	74.58	17	82.22	16	93.98	14	377.01	14	0.13	15	99
Real Ada boost	92.21	10	93.05	10	85.95	13	89.11	11	97.35	10	376.02	13	0.08	10	77

(continued)

Table 4 (continued)

Classifier name	Accuracy	Rank	Precision	Rank	Recall	Rank	F1	Rank	AUC	Rank	Training	Rank	MSE	Rank	Sum of ranks
Gradient boost	89.78	12	95.32	6	78	15	85.24	13	96.54	12	867.03	17	0.1	12	87
Random forest	96.45	1	96.12	4	94.89	5	96.32	1	99.24	1	268.01	12	0.02	1	25
Bagging	95.21	6	94.01	7	93.25	10	94.32	6	98.22	7	470.01	16	0.04	3	55
KNN	88.01	14	92.98	12	74.98	16	84.01	14	92.31	16	63.89	9	0.12	14	95

Table 5 Final resultant table of the case study

Resultant	Accuracy	Precision	Recall	F1	AUC	Training	MSE
Max_min value	96.45	97.07	98.49	96.32	99.24	2.51	0.02
Location	15	2	6	15	15	7	15
Whether unique	1	1	1	1	1	1	1

ranked first (i.e., the sum of the ranks of the seven parameters was minimum for it), and SGD Classifier ranked second jointly with Logistic Regression Classifier.

So, in conclusion, we can say, Random Forest Classifier and SGD Classifier performed best out of the 17 classifiers.

4.2 Max Voting Meta-Heuristic Optimization

We were coming to the second approach where we could see that according to the accuracy score, the random forest classifier, situated in the 15th row, got the best result among all the other algorithms. The value is 96.45 next is the precision score. The logistic regression classifier provides the best optimal value for a problem, situated in row 1. In recall score, after achieving 98.49 score, Bernoulli naïve Bayes classifier gets the highest accuracy and placed in row 7. The next one is the F1 score, where we can see that in row 15, again, the random forest classifier becomes the best solution holder. The random forest classifier holds the best position in the AUC score by getting a 99.24 score in the 15th row. Finally, at the model training time and the mean square error, the Gaussian naïve Bayes and the random forest classifier got the best solution tag by scoring 2.51 and 0.02 scores, respectively. We can see clearly that the random forest classifier has the most voted solution for a particular problem. That is why we got the best way to search for a result that is entirely satisfactory and reaches the optimal value and can consider it the best solution result among all those 17 algorithms. It can be implemented in other and more than 17 algorithms using the meta-heuristic function, and again we could find the best solution Table 5 shows the results. We have given the scores, and the positions are so clear.

5 Conclusion

Optimization is quite an aristocratic option for a solution way out. It represents a better solution in a better way where the solution will achieve an optimum state with a specific optimal value. While finding the answer, it may happen that the strategy to find the optimal solution is not according to the right track. Or it may happen that whatever solution we have chosen after thinking it will be optimal, the result may not match that optimized desired result. For this reason, we have considered and applied the meta-heuristic function that can find or search the solution that could bring the optimal value for a specific problem. Seventeen algorithms of machine learning we have passed through 7 types of accuracy parameter, and each time an algorithm get selected as per the accuracy rate. So, it comprises that we can choose the fittest among all efficient. So, after seeing the case study and its accuracy, we can conclude that the result of this experiment is entirely satisfactory. As we have given two approaches for the single matter at a very beginner level, it is obvious to have a future aspect. The method of meta-heuristic function is applicable for more options also. The meta-heuristic process will be an excellent choice to select the machine learning algorithms and any other problem perimeter where we found confusion or any obstacle to find or search a better solution in a better way.

Ultimately, the case study and the book chapter fulfill the problem definition's desired goal with a handsome success rate and accuracy. We have taken a 41k dataset to detect and define the best algorithm among the said 17 algorithms for this binary case detection through the meta-heuristic function. And the outcome will be the optimized algorithm. But in this case, measurement criteria, we have not considered the priority criteria. But in the computer domain, the priority of F1 has the most one, which we have not addressed here in this methodology.

References

1. M. Hasan, M.M. Islam, M.I.I. Zarif, M.M.A. Hashem, Attack and anomaly detection in IoT sensors in IoT sites using machine learning approaches. Internet Things **7**, 100059 (2019). https://doi.org/10.1016/j.iot.2019.100059
2. M. El-Dairi, R.J. House, Optic nerve hypoplasia, in *Handbook of Pediatric Retinal OCT and the Eye-Brain Connection* (2019), pp. 285–287. https://doi.org/10.1016/B978-0-323-60984-5.00062-7
3. A. Benussi et al., Classification accuracy of TMS for the diagnosis of mild cognitive impairment. Brain Stimul. **14**(2), 241–249 (2021). https://doi.org/10.1016/j.brs.2021.01.004
4. S. Rao, A.K. Verma, T. Bhatia, A review on social spam detection: challenges, open issues, and future directions. Expert Syst. Appl. **186**, (2021). https://doi.org/10.1016/j.eswa.2021.115742
5. A. Mewada, R.K. Dewang, Research on false review detection methods: a state-of-the-art review, J. King Saud Univ. Comput. Inf. Sci. (2021). https://doi.org/10.1016/j.jksuci.2021.07.021
6. R. Kaur, S. Singh, H. Kumar, Rise of spam and compromised accounts in online social networks: a state-of-the-art review of different combating approaches. J. Netw. Comput. Appl. **112**, 53–88 (2018). https://doi.org/10.1016/j.jnca.2018.03.015
7. R.M.K. Saeed, S. Rady, T.F. Gharib, An ensemble approach for spam detection in Arabic opinion texts. J. King Saud Univ. Comput. Inf. Sci. (2019). https://doi.org/10.1016/j.jksuci.2019.10.002
8. R. Kumar et al., Smart city and cyber-security; technologies used, leading challenges and future recommendations. J. King Saud Univ. Comput. Inf. Sci. **7**, 7999–8012 (2021). https://doi.org/10.1016/j.proeng.2016.11.813
9. C. Feltus, Current and future RL's contribution to emerging network security. Procedia Comput. Sci. **177**, 516–521 (2020). https://doi.org/10.1016/j.procs.2020.10.071
10. N. Gupta, V. Jindal, P. Bedi, LIO-IDS: handling class imbalance using LSTM and improved one-vs-one technique in intrusion detection system. Comput. Netw. **192**, (2021). https://doi.org/10.1016/j.comnet.2021.108076

11. J.I. Alcaide, R.G. Llave, Critical infrastructures cybersecurity and the maritime sector. Transp. Res. Procedia **45**, 547–554 (2020). https://doi.org/10.1016/j.trpro.2020.03.058
12. K.K. Tho, S. Swaddiwudhipong, Z.S. Liu, J. Hua, Artificial neural network model for material characterization by indentation. Model. Simul. Mater. Sci. Eng. (2004). https://doi.org/10.1088/0965-0393/12/5/019
13. M.A. Jan et al., Security and blockchain convergence with internet of multimedia things: current trends, research challenges and future directions. J. Netw. Comput. Appl. **175**, 102918 (2021). https://doi.org/10.1016/j.jnca.2020.102918
14. K.O. Lye, S. Mishra, D. Ray, P. Chandrashekar, Iterative surrogate model optimization (ISMO): an active learning algorithm for PDE constrained optimization with deep neural networks. Comput. Methods Appl. Mech. Eng. **374**, 113575 (2021). https://doi.org/10.1016/j.cma.2020.113575
15. A. Shokry, S. Medina-González, P. Baraldi, E. Zio, E. Moulines, A. Espuña, A machine learning-based methodology for multi-parametric solution of chemical processes operation optimization under uncertainty. Chem. Eng. J. **425**, (2021). https://doi.org/10.1016/j.cej.2021.131632
16. Y. Morita, S. Rezaeiravesh, N. Tabatabaei, R. Vinuesa, K. Fukagata, P. Schlatter, Applying Bayesian optimization with Gaussian process regression to computational fluid dynamics problems. J. Comput. Phys. **449**, 110788 (2022). https://doi.org/10.1016/j.jcp.2021.110788
17. B.S. Oh, J. Cho, B. Choi, H.W. Choi, M.S. Kim, G. Lee, Application of heuristic algorithms for design optimization of industrial heat pump. Int. J. Refrig. (2021). https://doi.org/10.1016/j.ijrefrig.2021.11.002
18. M.M. Rahman, G. Szabó, Multi-objective urban land use optimization using spatial data: a systematic review. Sustain. Cities Soc. **74**, (2021). https://doi.org/10.1016/j.scs.2021.103214
19. D. Ahamad, S. Alam Hameed, M. Akhtar, A multi-objective privacy preservation model for cloud security using hybrid Jaya-based shark smell optimization. J. King Saud Univ. Comput. Inf. Sci. (2020). https://doi.org/10.1016/j.jksuci.2020.10.015
20. S. Thakur, A. Chakraborty, R. De, N. Kumar, R. Sarkar, Intrusion detection in cyber-physical systems using a generic and domain specific deep autoencoder model. Comput. Electr. Eng. **91**, (2021). https://doi.org/10.1016/j.compeleceng.2021.107044
21. S. Gupta, J. Sarkar, A. Banerjee, N.R. Bandyopadhyay, S. Ganguly, Grain boundary detection and phase segmentation of SEM ferrite-pearlite microstructure using SLIC and skeletonization. J. Inst. Eng. Ser. D **100**(2), 203–210 (2019). https://doi.org/10.1007/s40033-019-00194-1
22. G. Baldini, R. Giuliani, M. Gemo, F. Dimc, On the application of sensor authentication with intrinsic physical features to vehicle security. Comput. Electr. Eng. **91**, (2021). https://doi.org/10.1016/j.compeleceng.2021.107053
23. M. Orabi, D. Mouheb, Z. Al Aghbari, I. Kamel, Detection of bots in social media: a systematic review. Inf. Process. Manag. **57**(4), 102250 (2020). https://doi.org/10.1016/j.ipm.2020.102250
24. S. Gupta, Chan-Vese segmentation of SEM ferritepearlite microstructure and prediction of grain boundary. Int. J. Innov. Technol. Explor. Eng. **8**(10), 1495–1498 (2019). https://doi.org/10.35940/ijitee.A1024.0881019
25. S. Gupta et al., Modelling the steel microstructure knowledge for in-silico recognition of phases using machine learning. Mater. Chem. Phys. **252**, 123286 (2020). https://doi.org/10.1016/j.matchemphys.2020.123286
26. I.H. Sarker, CyberLearning: effectiveness analysis of machine learning security modeling to detect cyber-anomalies and multi-attacks. Internet Things **14**, 100393 (2021). https://doi.org/10.1016/j.iot.2021.100393
27. N. Al-Azzam, I. Shatnawi, Comparing supervised and semi-supervised machine learning models on diagnosing breast cancer. Ann. Med. Surg. **62**(January), 53–64 (2021). https://doi.org/10.1016/j.amsu.2020.12.043
28. E.G. Dada, J.S. Bassi, H. Chiroma, S.M. Abdulhamid, A.O. Adetunmbi, O.E. Ajibuwa, Machine learning for email spam filtering: review, approaches and open research problems. Heliyon **5**(6), (2019). https://doi.org/10.1016/j.heliyon.2019.e01802

29. A. Ligthart, C. Catal, B. Tekinerdogan, Analyzing the effectiveness of semi-supervised learning approaches for opinion spam classification. Appl. Soft Comput. **101**, 107023 (2021). https://doi.org/10.1016/j.asoc.2020.107023
30. S. Padmavathi, E. Ramanujam, Naïve Bayes classifier for ECG abnormalities using multivariate maximal time series motif. Procedia Comput. Sci. **47**(C), 222–228 (2014). https://doi.org/10.1016/j.procs.2015.03.201
31. J. Gola et al., Advanced microstructure classification by data mining methods. Comput. Mater. Sci. **148**, 324–335 (2018). https://doi.org/10.1016/j.commatsci.2018.03.004
32. H.E. Kiziloz, Classifier ensemble methods in feature selection. Neurocomputing **419**, 97–107 (2021). https://doi.org/10.1016/j.neucom.2020.07.113
33. S.M. Piryonesi, T.E. El-Diraby, Role of data analytics in infrastructure asset management: overcoming data size and quality problems. J. Transp. Eng. Part B Pavements **146**(2), 04020022 (2020). https://doi.org/10.1061/jpeodx.0000175
34. I.H. Sarker, Y.B. Abushark, F. Alsolami, A.I. Khan, IntruDTree: a machine learning based cyber security intrusion detection model. Symmetry (Basel) **12**(5), (2020). https://doi.org/10.3390/SYM12050754
35. I. Stellios, P. Kotzanikolaou, C. Grigoriadis, Assessing IoT enabled cyber-physical attack paths against critical systems. Comput. Secur. **107**, 102316 (2021). https://doi.org/10.1016/j.cose.2021.102316
36. I.H. Sarker, A machine learning based robust prediction model for real-life mobile phone data. Internet Things **5**, 180–193 (2019). https://doi.org/10.1016/j.iot.2019.01.007
37. B. Ning, W. Junwei, H. Feng, Spam message classification based on the Naïve Bayes classification algorithm. IAENG Int. J. Comput. Sci. **46**(1), (2019)
38. S. Gupta, J. Sarkar, M. Kundu, N.R. Bandyopadhyay, S. Ganguly, Automatic recognition of SEM microstructure and phases of steel using LBP and random decision forest operator. Measurement **151**, 107224 (2020). https://doi.org/10.1016/j.measurement.2019.107224
39. A. Shahim, Security of the digital transformation. Comput. Secur. **108**, 102345 (2021). https://doi.org/10.1016/j.cose.2021.102345

Solving Reality-Based Trajectory Optimization Problems with Metaheuristic Algorithms Inspired by Metaphors

Alfonso Ramos-Michel, Mario A. Navarro, Bernardo Morales-Castañeda, Marco Pérez-Cisneros, and Daniel Zaldivar

Abstract In recent years, optimization problems based on real-world situations have become increasingly complex. Designers have endless sources of inspiration to create new and better methods that are then improved, adapted, and combined to enhance their performance. However, rarely do we see the original proposals' performance with the most complex problems. In the present work, we chose a group of algorithms in their original version to solve a particular set of problems. Knowing the performance of the base algorithm is the first step to knowing what is required to improve its results when optimizing complex problems. The problems used are known as Global Trajectory Optimization Problems (GTOP). The results were validated with statistical tests that present two swarm intelligence algorithms and one mathematical metaphor algorithm as the three methods in their original version with the best performance when solving complex problems.

1 Introduction

Optimization is an area of mathematics that pursues, in general terms, the maximum or minimum value of a function. To achieve this, researchers utilize algorithms of different nature, which contrast results with state-of-the-art algorithms in their design stage. Recursively, researchers use test benches of mathematical functions

A. Ramos-Michel (✉) · M. A. Navarro · B. Morales-Castañeda · M. Pérez-Cisneros · D. Zaldivar
División de Tecnologías para la integración Ciber-Humana, Universidad de Guadalajara, CUCEI, Av. Revolución 1500, 44100 Guadalajara, Jalisco, C.P., Mexico
e-mail: alfonsorm@cucsur.udg.mx

M. A. Navarro
e-mail: mario.navarro@alumnos.udg.mx

M. Pérez-Cisneros
e-mail: marco.perez@cucei.udg.mx

D. Zaldivar
e-mail: daniel.zaldivar@cucei.udg.mx

E. H. Houssein et al. (eds.), *Integrating Meta-Heuristics and Machine Learning for Real-World Optimization Problems*, Studies in Computational Intelligence 1038,
https://doi.org/10.1007/978-3-030-99079-4_14

with different degrees of difficulty. On other occasions, engineering problems are used, representing a more significant challenge than many existing test benchmark functions. Over time, researchers seek the improvement of the original algorithms. They modify, add or combine the base methods within existing strategies, operators, or algorithms. Then, they test the improved method with more complex problems but, a few times, use the base algorithm to measure its real capacity. That is why, in this chapter, we probed six original algorithms without improvements to solve eight complex problems, to demonstrate, not always the improvement is required to solve this kind of real-world problems in a good way.

There are different types of problems, all of which vary in complexity. Many real-world problems can be thought of as global optimization problems, but one of the most exciting fields in designing this kind of problems is trajectory optimization in space mission design. Several of these missions have been formalized into well-defined problems GTOP Database [21] with several constraints and a set of decision variables depending on the problem. The missions generally described in these problems are multiple gravity assist. These problems are employed to perform maneuvers in deep space to use the gravity of neighboring planets during the mission as leverage to reach a specific location. The non-linearity of this class of problems, combined with the constraints on their variables, creates a challenge of greater difficulty than those presented by classical engineering problems.

2 Metaheuristic Algorithms Implemented

In optimization, there is an extensive list of algorithms proposed to solve problems and functions of different nature and degrees of difficulty. A Metaheuristic algorithm (MA) is an optimization algorithm that has gained relevance in the last decades. Their popularity is due to their ability to solve problems of different nature because they represent derivative-free techniques and initiate their evaluation process by randomly placing the search agents [1]. MA's have been subject to different classifications by researchers who study algorithms according to different characteristics such as how the algorithm constructs the solution, how the search agents traverse the search space or the phenomenon or behavior that inspires their design [1]. To present a general idea of this kind of method, we present a classification (Fig. 1) based on the source of inspiration for the algorithm design. In the present work, MA's was chosen because they obtained outstanding results, mainly with different complex mathematical functions when their authors proposed them. Some have been used to solve engineering problems after being modified by other researchers correcting some procedures, hybridizing with other algorithms, or adding foreign operators to reinforce the areas where each algorithm provided the poorest performance. The selected methods are the Collective Animal Behavior (CAB) algorithm, the Social Spider Optimization (SSO) algorithm, the Side Blotched Lizard Algorithm (SBLA), the Selfish Herd Optimizer (SHO) algorithm, State of Matter Search (SMS) algorithm, and the Fuzzy Logic-based Optimization (FUZZY) algorithm. Based on the

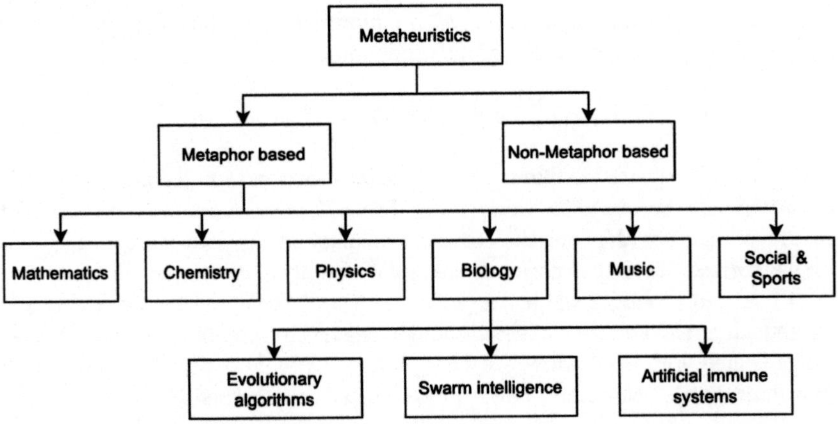

Fig. 1 Classification of metaheuristic algorithms according to the field on which are based

classification in Fig. 1, we can categorize each of the algorithms chosen in this paper according to their field of inspiration. CAB, SSO, SBLA, and SHO have been classified as swarm intelligence-based algorithms. The FUZZY algorithm, inspired by fuzzy logic, is classified as mathematics inspired, and finally, SMS is in the category of physics-based algorithms because it emulates states of matter.

2.1 Collective Animal Behavior

The Collective Animal Behavior algorithm [6] is an MA with some similarities with other MA methods. It is an iterative process that starts its population randomly, and the individual (animal) position represents a possible solution as other multimodal algorithms. In particular, CAB stores the best individual positions in memory, makes different movements and lets the individuals compete for their position. The authors incorporate several operators in the algorithm, which allow a good relationship between the exploration and exploitation phases of the search space. In the original article, the CAB was tested with 31 benchmark functions, including unimodal, multimodal, and GKLS functions, and the results were compared against four metaheuristic algorithms and one continuous optimization method. A brief methodology description is presented below.

Procedure
CAB algorithm initializes its population with a random position forming the set $A = \{a_1, a_2, \ldots, a_{N_p}\}$, where N_p represents the total individual number positions. Each position a_i is a vector of D dimensions and serves to store the parameter values to optimize. Initial values on a_i are randomly generated with a uniform distribution between superior (a_j^{low}) and inferior constraints (a_j^{high}) through Eq. 1, where

$j \in \{1, 2, \ldots, D\}$ and $i \in \{1, 2, \ldots, N_p\}$. i represents the individual index and j represents the parameter number of the i individual.

$$a_{j,i} = a_j^{low} + rand(0, 1) \cdot (a_j^{high} - a_j^{low}) \tag{1}$$

CAB then evaluates the fitness all over the positions and sort the individuals into X starting with the best to the worst fitness. From X, stores the best B positions into the global memory (M_g) and the historical memory (M_h). Once the best positions are memorized, the first B positions are replaced to form the new solution set $A = \{a_1, a_2, \ldots, a_B\}$ by Eq. 3 where $l \in \{1, 2, \ldots, B\}$, m_h^l represents the l-th element in M_h and v is a random vector of small enough length. The last positions from B to N_p are generated with Eq. 4 where $i \in \{B + 1, B + 2, \ldots, N_p\}$, $m_h^{nearest}$ and $m_g^{nearest}$ represents the elements of M_h and M_g closer to x_i, and $r = rand(1, -1)$.

$$X = \{x_1, x_2, \ldots, x_B, \ldots, x_{N_p}\} \tag{2}$$

$$a_l = m_h^l + v \tag{3}$$

$$a_i = \begin{cases} x_i \pm r \cdot (m_h^{nearest} - x_i) & \text{with probability H} \\ x_i \pm r \cdot (m_g^{nearest} - x_i) & \text{with probability (1 − H)} \end{cases} \tag{4}$$

The authors set the consideration if $r > 0$ the individual will perform an attraction movement. Any other case, the individual performs a repulsion movement. One animal can make a rando movement to change its position under some probability P. This behavior rule, is made by Eq. 1 considering the Eq. 5 where $i \in \{B + 1, B + 2, \ldots, N_p\}$ and r is a randomly generated vector defined into de search space limits.

$$a_i = \begin{cases} r & \text{with probability P} \\ x_i & \text{with probability (1 − P)} \end{cases} \tag{5}$$

When the new positions of the N_p individuals are seted, the M_h must be updated. Here the authors used the concept of dominance. All individuals maintain a distance from each other. There is a minimal distance ρ (Eq. 6), dependant of the agressivity of each individual. As better fitness function, as more agressivity presents each animal. An interaction occurs, when two individual are too close considering their agressivity. To represent this phase, the elements of global and historical memories are unified ($M_U = M_h \cup M_g$). Each M_U element is compared pairwise against all the other elements. If the distance between two individuals is closer than ρ, they elements will interact and the most Aggressive one will maintain its position and the other is removed. Once the space competition finishes, M_h is updated with the B best M_U elements. When the process is complete and the maximum iterations number is reached, the process finishes and the best value stored in M_h represents the global solution of the problem.

$$\rho = \frac{\prod_{j=1}^{D} \left(a_j^{high} - a_j^{low}\right)}{10 \cdot D} \tag{6}$$

Related work

In 2013, Ram et al. optimized a linear antenna arrays design [19]. The CAB is used to find the optimal value of the non-uniform excitation weights and uniform element spacing for lobe formation in linear antenna arrays, and the results are compared with three classical algorithms in the literature. The antennas for testing the algorithms consisted of 10, 14, and 20 arrays, optimizing variables such as excitation currents, element spacing, sidelobe level, and the first null beamwidth. After the study, the authors concluded that the CAB outperformed the three comparison algorithms because the latter was trapped in values that produced suboptimal designs.

Binary Collective Animal Behavior (BCAB) is presented in [12] in 2015. It uses this algorithm to solve power systems configuration problems by optimizing the location of distributed generation sets and shunt capacitors in a radial distribution system. The direct application of the algorithm is in the reduction of total line loss and total voltage deviation improvement. Three distribution systems, such as IEEE 12-bus system, IEEE 33-bus system, and IEEE 69-bus system, were used. The performance of BCAB was compared with BPSO for the same three systems, and the authors concluded that BCAB obtained the most effective solutions in the study.

In [17] Pany et al. implement the CAB algorithm to optimize the operation of power systems by solving the Optimal reactive power dispatch (ORPD) problem. The ORPD can be defined as minimizing active power transmission losses and the total voltage deviation of load buses. They used 3 IEEE power system standards, such as IEEE 30-bus, IEEE 57-bus, and IEEE 118-bus test power systems for testing. It allowed the authors to find the best control variable configurations for power system testing.

A multimodal optimization algorithm called Collective Electromagnetism-like Mechanism Optimization (CEMO) was presented in 2018 in [10]. CEMO is an adaptation of the Electromagnetism-like Mechanism Optimization (EMO) algorithm with the CAB algorithm. The implementation of the CAB memory mechanism allows CEMO to solve multimodal problems in a single run. The proposed algorithm was tested against 11 state-of-the-art multimodal techniques to solve a set of 23 multimodal functions in search spaces ranging from 2 to 20 dimensions. The authors found that their proposal generated better and more consistent results with an acceptable execution time.

The CAB algorithm has also been used in digital image processing. In [4, 5], the authors refer to the detection of multiple circular shapes and ellipses. For experimentation, the algorithm worked on original and noise-contaminated computer-produced images and natural photographs. All images contained a different number of target shapes, where the CAB algorithm had to locate the target shape, discriminate other shapes and perform approximations of the target shape as appropriate. In these works, the performance of the CAB was compared with similar evolutionary algorithms and the Hough Transform-based detector. The results show the algorithm's efficiency to present better accuracy, speed, and robustness on complicated and noisy images.

2.2 Social Spider Optimization

The Social Spider Optimization algorithm was designed to simulate the behavior of a social-spider colony. In their analogy, they use each member of the group to represent a possible solution to the optimization problem, and the search space represents the communal web where the entire population of spiders interacts. By making a population discrimination according to the gender of each individual, the authors developed different strategies to perform the exploration and exploitation of the search space. They also introduce the concept of individual size to favor the generation of new spiders and the possibility of replacing the worst answers found during the execution of the algorithm. In the original article, the authors tested the SSO algorithm with a set of 19 functions and compared the performance of their proposal against two of the most popular algorithms in the field: ABC and PSO algorithms.

Procedure

The total population of individuals (S) is initialized by the conjunction of two gender-differentiated populations $\mathbf{S} = F \cup M$. Thus, the set of males $M = \{m_1, m_2, \ldots, m_{Nm}\}$ and the set of females $F = \{f_1, f_2, \ldots, f_{Nf}\}$ are generated. The total number of social-spiders is N. N_f is the total number of females which represents between 65 and 90% of the total population. N_m is the number of males, and calculates as $N_m = N - N_f$.

Each spider will be assigned a weight (w_i) that will define its size metaphorically. The weight is calculated by the Eq. 7 and depends on the value obtained for each individual in the fitness evaluation ($J(s_i)$). $best_s$ corresponds to the best fitness of the whole population, and $worst_s$ refers to the worst fitness obtained by an individual among the whole population.

$$w_i = \frac{J(s_i) - worst_s}{best_s - worst_s} \tag{7}$$

The social behavior of the spider colony cannot be conceived without a communication mechanism between the individuals that compose it. Social spiders perform this task by transmitting vibrations through the web, where each spider emits vibration with an intensity proportional to its size.

$$Vib_{i,j} = w_j \cdot \mathrm{e}^{-d_{i,j}^2} \tag{8}$$

On the other hand, the vibration intensity will be detected according to the Euclidean distance between the sender and the receiver. This calculation is performed by the Eq. 8, where i is the index of the receiving individual(s_i), $d_{i,j}$ is the Euclidean distance between the spiders i and j, calculated as $d_{i,j} = \|s_i - s_j$, and $j \in \{c, b, f\}$ defines the sender (s_c, s_b, s_f) thereby establishing each of the only three possible relationships to establish communication:

- $Vibc_i$ where the sender is the closest individual and has the highest weight compared to the receiver ($w_c > w_i$).

- $Vibb_i$ where s_b corresponds to the individual with the best weight in the whole population.
- $Vibf_i$ where s_f is the female individual closest to s_i.

Emulating real spider behavior, female and male individuals make movements independently depending on their gender. Females perform an attraction or repulsion motion defined by the Eq. 9, where $alpha$, $beta$, $delta$ and $rand$ are randomly generated values in the range [0,1], k represents the iteration number and s_c and s_b represent the vibration generating individuals as detailed above. The type of motion to be performed is defined by a random number $rm = rand(0, 1)$ compared with a threshold PF. If $rm < PF$ an attraction motion is performed. Otherwise, a repulsive motion will be performed.

$$f_i^{k+1} = \begin{cases} f_i^k + \alpha \cdot Vibc_i \cdot \left(s_c - f_i^k\right) + \beta \cdot Vibb_i \cdot \left(s_b - f_i^k\right) + \delta \cdot \left(rand - \frac{1}{2}\right) & \text{with probability PF} \\ f_i^k - \alpha \cdot Vibc_i \cdot \left(s_c - f_i^k\right) - \beta \cdot Vibb_i \cdot \left(s_b - f_i^k\right) + \delta \cdot \left(rand - \frac{1}{2}\right) & \text{with probability } 1 - \text{PF} \end{cases} \tag{9}$$

On the other hand, the male population makes its movements depending on the size of the individual relative to the entire male population. The median of the weights of all the male individuals is calculated. All the males with a size more significant than the median will form the dominant males D group. The rest will form the non-dominant group ND. The individuals of the dominant group will move closer to the nearest female, while the non-dominant ones will tend to concentrate in the center of the male population. The Eq. 10 mathematically represents this change of positions, where s_f is the closest female individual to the i member.

$$m_i^{k+1} = \begin{cases} m_i^k + \alpha \cdot Vibf_i \cdot \left(s_f - m_i^k\right) + \delta \cdot \left(rand - \frac{1}{2}\right) & \text{if } \mathrm{m}_i^k \in \mathrm{D} \\ m_i^k + \alpha \cdot \left(\frac{\sum_{h=1}^{N_m} m_h^k \cdot w_h}{\sum_{h=1}^{N_m} w_h} - m_i^k\right) & \text{if } \mathrm{m}_i^k \in \mathrm{D} \end{cases} \tag{10}$$

The approach of dominant males to females allows the mating process. In it, when a female and a dominant male come closer than r (Eq. 11), they produce a new individual (s_{new}), which will be compared with the spider of worse fitness. The worse of the two will disappear, and the better one will be part of the population that continues with the process performed so far.

$$r = \frac{\sum_{j=1}^{n}\left(p_j^{high} - p_j^{low}\right)}{2 \cdot n} \tag{11}$$

Related work

In 2016, Klein et al. proposed a modification of the original SSO algorithm, the Modified Social-Spider Optimization (MSSO) [13] algorithm. The authors include control parameters based on the beta probability distribution to preserve diversity and avoid premature convergence. They also include a local search by the best individual using the Natural Gadient. Tests of the proposal were performed on Loney's solenoid design problem and the brushless DC motor benchmark. With the reported results, Klein et al. conclude that the proposed algorithm has a good balance between efficiency and effectiveness in solving multi-objective problems.

The Novel Improved Social Spider Optimization (NISSO) [16] algorithm was presented by Nguyen in 2019 to solve the optimal power flow problem. The objective was to optimize five objectives related to electric power production independently, each using IEEE 30-bus system, IEEE 57-bus system, and IEEE 118-bus system. Modifications to the original SSO were to change the strategy for the movement of individuals of both genders and decrease the proportion of female spiders in the population. In the tests, Nguyen compares results with many algorithms, which vary according to the experiment, concluding that the proposals made improve the SSO algorithm to find the optimal power flow problem.

In [24], Zhao et al. improved the SSO with an elite opposition-based learning strategy. The target was to enhance the convergence speed and computational accuracy. The authors enhanced the optimization strategy by expanding the searching space of the algorithm applying the Global Elite Opposition-Based Learning Strategy into the SSO global search process, getting a better diversity of the population. The proposed methodology was compared against five state-of-the-art algorithms that included the same SSO and tested along 23 benchmark test functions: 7 unimodal and 16 multimodal functions. Zhao et al. conclude that their proposal has a fast convergence speed, more accuracy, and a relatively high degree of stability.

The application of SSO has transcended to AI. In 2015, in [15] SSO was used to improve the training of a Feedforward Neural Network (FNN). The function of the MA was the optimization of the network parameters. Mirjalili et al. used the XOR, Iris, Balloon, Heart, and Breast cancer classification datasets to test the effectiveness of their approach. The result was compared with the training of the same network by tuning parameters with five different metaheuristic algorithms. Considering the MSE minimization, the authors concluded that the proposed algorithm outperforms the learning efficiency of the FNN and improves the classification accuracy, outperforming the comparison algorithms in at least three of the five used datasets.

In 2017, Zhou et al. presented the Simplex Method-Based Social Spider Optimization (SMSSO) algorithm. The proposal was applied to improve the problems of stagnation in local optima and low convergence rates that arise when using metaheuristic algorithms to tune the parameters of the K-means algorithm to clustering data. The authors performed experimentation to test their algorithm on eleven benchmark datasets, of which nine were real-life datasets, and two were artificial datasets. Zhou et al. performed the clustering data with the K-means algorithm and used five other metaheuristic algorithms to synthesize it to compare the results.

2.3 Side Blotched Lizard

The authors of this algorithm proposed in 2019 [14] conceptualized the behavior of a species of lizard called the side-blotched lizard as a metaheuristic algorithm.

This algorithm has no reported subsequent work involving its implementation, how-

ever in the original paper, the authors compare this proposal with several known optimization algorithms and solve engineering problems such as Welded beam design, Frequency modulated synthesizer, and Rolling element bearing was also tested on a set of five unimodal, eighteen multimodal and four composite benchmark functions being the best among algorithms such as: ABC [11], PSO [18] and GWO [20] just to name a few.
For the design of this algorithm, they considered the three different mating behaviors to model an algorithm with three subpopulations depending on the color polymorphism associated with the throats of each lizard. The general behaviors of each lizard subpopulation are described below:

- The orange males are large, aggressive, maintain extensive territories with several females each one, and influence the survival of the blue males because they are docile lizards and less aggressive.
- The blue-colored males hold smaller and defined territories. They are monogamous, so they focus on a single female, not allowing the yellow-colored males to capture their females because their territories are very well protected.
- The yellow-colored males have no exclusive territory. Instead, they sneak out to find neglected females with whom they can mate, greatly influencing the survival of orange-colored males because it is impossible to take care of all the territory they hold simultaneously.

The interaction in which each lizard subpopulation has advantages and disadvantages in the mating war concludes in a dynamic equilibrium between the three mating strategies. To emulate the interaction between the lizards, a system of three functions is modeled that represent the changes through time in the lizard populations. Depending on the season, the subpopulations grow or decline, simulating a probability game where each subpopulation has an equal chance over the others.

Procedure

Modeling of subpopulations and population changes
The function for initializing the subpopulations uses as input the total population $TTOP$, the number of stations S and the number of iterations $Iter$, at the output, the number of individuals in each subpopulation is defined as blue population (BP), orange population (OP) and yellow population (PY), to prevent the algorithm from operating without search agents, the authors defined a minimum population($MinP$) of 10%. To represent a cycle growing of some subpopulations, a sinusoidal graph with a frequency of $PI * 2$ is generated, where the maximum possible population for a subpopulation is $MaxP = (TPOP * 2/subpop) - MinP.subpop = 3$ represents subpopulations number, and the functions are generated through a set of points X from 0 to the maximum iteration shifted by $Freq * S$. Finally, the population of each iteration is modeled with the following expression:

$$PopIter = Aux1 + Aux2 * sin\left(X + \left(\frac{Freq}{subpop} * i\right)\right) \tag{12}$$

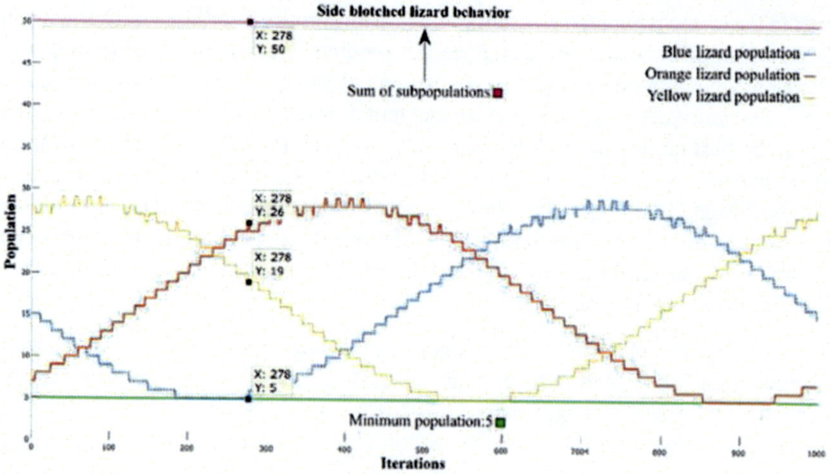

Fig. 2 Design of interactions beetwen lizard subpopulations by means of shifted sine functions [14]

Fig. 3 Influence among lizard subpopulations [14]

where i is an index that goes from 1 to $subpop$ (1 to 3), rounded to represent a whole particle or lizard accurately (Fig. 2).

The interactions between the groups are represented in Fig. 3 where it is observed that the behavior of the blue lizard influences the yellow lizard, the yellow lizard influences the orange lizard, and the orange lizard influences the blue lizard.

To represent the changes in the population, the authors established three criteria, modeled as functions to conserve the dynamics of the lizard populations.

Delete function The algorithm calls the function *Eliminate* when in the population of a particular lizard color there is an adverse change in its population, and there is no positive change in the lizard color that has an advantage over it. All the indexes corresponding to the color to eliminate are obtained. The lizards are sorted by fitness in ascending order, and the worst value corresponding to the first index of the subpopulation is established. Once the lizard is eliminated from the subpopulation, it will decrease by one, and this will be repeated during the iterative process until the variable reaches zero.

Transform function This function checks for population changes when the season of a given lizard group is predominant and will use the function to transform the worst individual of the color subpopulation it influences as part of its group. The thriving group will increase by 1, while the affected color group will decrease by 1.

Add function The conditions to add a lizard to some subpopulation are that the new particle coincides with the color on which it influences. In this way, if a blue is added, it will be on the yellow subpopulation. If an orange lizard is added, it will be on the blue subpopulation. In the case of the yellow lizards, it will be on the orange subpopulation. This function uses a criterion of looking for the two best particles of the subpopulation to which a new individual will be added, then a proportion between both is calculated with a random number between [0 1] to calculate the proportion of the distance between both named randP, then the expression used to add a new lizard is defined on Eq. 13.

$$NL = 1BestLizard * randP + 2BestLizard * (1 - randP) \tag{13}$$

SBLA search strategies

To emulate the social mating behavior of each subpopulation, the designers of this algorithm designed three different operators, which are described below:

1. Blue defensive strategy:
   The authors conceive this operator as a move to intensify the exploitation of the search space. The operator can be summarized in the following steps:

   a. The first step is to find the best lizard in the total BLP population and the best BLB blue lizard, both are ordered, and the best one is selected. If these were the same lizard, we would need to get the second-best blue lizard and calculate the Euclidean distance expressed as Δ.

   $$\Delta = absolute(BLP - BLB) \tag{14}$$

   b. The next step is to move the remaining blue lizards one by one inside the limits established by the two best individuals, the formula for defining these limits is shown below:

   $$LB = BLP - \Delta \tag{15}$$

   $$UB = BLP + \Delta \tag{16}$$

c. The algorithm chooses a random dimension to move the blue lizards to their new position for a more uniform distribution. The algorithm's natural evolution allows the distance's delta to decrease more and more by dividing it by the golden ratio Φ. The following expressions show the previously explained:

$$\varphi = \frac{\left(1 + \sqrt{5}\right)}{2} \tag{17}$$

$$\Delta = \Delta / \varphi \tag{18}$$

2. Orange dominant expansion strategy:
In nature, orange lizards expand their territories and are aggressive among themselves. The authors considered this and designed the Orange dominant expansion strategy as an exploration operator. The first step is to obtain the best orange lizard according to its fitness, and for the others, this will be a reference point to calculate the Euclidean distance to the best one. This strategy is mathematically expressed as:

$$\Delta^n = BLO - O^n \tag{19}$$

Once acquiring the distance, again must be calculated by dividing the upper limit $UpLim$ minus the lower limit $LowLim$ over the number of orange lizards $\#O$. This gain moves them according to the delta distance multiplied by a random number between 0 and 1.

$$Gain = \frac{UpLim - LowLim}{\#O} \tag{20}$$

$$m^n = 1/\Delta^n \times rand \times Gain \tag{21}$$

Once getting the desired movement m^n, the selected orange O^n can move away from the best orange.

$$O^n = BLO - m^n \tag{22}$$

This new position will promote the exploration of the search space.
3. Yellow sneaky strategy:
This behavior works by choosing an orange lizard by the roulette method [7]. The said method gives yellow lizards a chance to enter any lizard territory with an advantage over those with the best aptitude. Once the orange lizard is selected, the yellow lizard is moved with the following equation:

$$Y^n = Y^n + rand \times (O^m - Y^n) \tag{23}$$

The roulette represents the opportunistic behavior by allowing to choose another lizard to invade in each iteration. This will occur for each yellow lizard in the population. This allows exploiting the explored areas previously found by orange lizards.

2.4 *Selfish Herd Optimizer Algorithm*

The Selfish Herd Optimizer [9] was presented in 2017 by Fausto et al. It is a swarm optimization algorithm that represents its possible solutions, as individuals divided into two differentiated groups: prey and predators. Each individual is driven according to their role, employing operators inspired by the relationship between the predator animals and the individuals stalked as prey. Fausto et al. used 15 benchmark functions to test the performance of their algorithm, comparing their results against those produced by nine other state-of-the-art metaheuristic algorithms. Derived from the experimentation, the authors concluded that the algorithm presented high-quality solutions due to a better distribution of individuals in the search space, different exploitation and exploration rates during the algorithm's evolution, and diversification of solutions.

As is classic in MAs, the first step is to establish the values of the main eigen-parameters of the algorithm, as well as the generation of the initial population of individuals with the set $A = \{a_i j^k\}$ where $i = \{1, 2, \dots, N\}$, $j = \{1, 2, \dots, n\}$, $k = \{0, 1, \dots, I\}$, N is the total number of search agents, n the dimensions of the search space and I the number of iterations that the algorithm will work. For this, each individual a_i is generated by Eq. 24:

$$a_{ij}^k = x_j^{low} + rand(0, 1) \cdot (x_j^{high} - x_j^{low}) \tag{24}$$

where x_j^{high} and x_j^{low} are the upper and lower limits of the search space. The total of the population A, is divided into two sets: $H = \{h_{ij}$ which groups the herd of individuals or prey and where $i = \{1, 2, \dots, N_h\}$, and $P = \{p_i j^k\}$ which concentrates the predators and where $i = \{1, 2, \dots, N_p\}$. N_h is the number of individuals that compose the herd, and represents between 70 and 90% of the total population, N_p is the number of predators, and the relationship between both is $N_p = N - N_h$.

After role assignment, a survival value is generated for each individual according to their fitness value. The survival value is calculated as:

$$SV_{a_i} = \frac{f(a_i) - f_{best}}{f_{best} - f_{worst}} \tag{25}$$

The survival value helps to define the behavior and possible fate of each individual. The SHO establishes the selfish attraction/repulsion (Eq. eq:sho16) between a given individual h_i and any other individual in the population h_j. This parameter is crucial for determining the movements of prey group members.

$$\Psi_{h_i,h_j} = SV_{h_j} \cdot e^{-\|h_i-h_j\|^2} \tag{26}$$

In each iteration, there is an individual called the herd's leader. The herd's leader is the individual with the best survival value. The herd's leader will move concerning the group as shown in the Eq. 27. If the position held by the leader gives him enough security, he will make his move in conjunction with the group. However, if the survival value of the leader does not give him enough security in his current position, he will seek to improve it by making a deliberately selfish move even against some individuals in the group. Equations 28 and 29 mathematically represent the two possible behaviors of the leader. The terms $\varphi^k_{h_L,P_M}$ and $\psi^k_{h_L,x_{best}}$ represent the selfish repulsion and selfish attraction needed to make its move, and α is a random number in the interval [0,1].

$$h_L^{k+1} = \begin{cases} h_L^k + c^k \text{ if } SV_{h_i^k} = 1 \\ h_L^k + s^k \text{ if } SV_{h_i^k} < 1 \end{cases} \tag{27}$$

$$c^k = 2 \cdot \alpha \cdot \varphi^k_{h_L,P_M} \cdot (p^k_M - h^k_L) \tag{28}$$

$$s^k = 2 \cdot \alpha \cdot \psi^k_{h_L,x_{best}} \cdot (x^k_{best} - h^k_L) \tag{29}$$

The individuals in the herd will also change their position, and this will change according to various factors, which are represented mathematically in Eqs. 30, 31 and 32. The herd (H) is composed of two subsets: herd followers (H_F) and herd deserters (H_D). The first group represents individuals who join the herd. On the other hand, deserters decide to move independently from the herd. By comparing the actual survival value of an individual h_i^k and a random number, each member of the herd selects one of the subsets. The group H_D performs its movements according to Eq. 32, where *beta* and *gamma* are random numbers in the range [0,1], and *hatr* is a unit vector pointing in the random direction.

$$h_i^{k+1} = \begin{cases} h_i^k + f_i^k \text{ if } h_i^k \in H_F^k \\ h_i^k + d_i^k \text{ if } h_i^k \in H_D^k \end{cases} \tag{30}$$

$$f_i^k = \begin{cases} 2 \cdot (\beta \cdot \psi^k_{h_i,h_L} \cdot (h^k_L - h^k_i) + \gamma \cdot \psi^k_{h_i,h_{c_i}} \cdot (h^k_{c_i} - h^k_i)) & \text{if } h_i^k \in H_d^k \\ 2 \cdot \delta \cdot \psi^k_{h_i,h_M} \cdot (h^k_M - h^k_i) & \text{if } h_i^k \in H_s^k \end{cases} \tag{31}$$

$$d_i^k = 2 \cdot (\beta \psi^k_{h_i,x_{best}} \cdot (x^k_{best} - h^k_i) + \gamma \cdot (1 - SV_{h_i^k}) \cdot \hat{r}) \tag{32}$$

Herd followers are divided into dominant herd members (H_d) and subordinate herd members (H_s). This classification depends on the survival value of each individual. An individual h_i^k will be a subordinate member if its current survival value is smaller than the mean survival value of the herd set H. Otherwise, the individual will be a dominant member. The movement realized by these groups, are defined in Eq. 31, where β, γ and δ are random numbers in the range [0,1].

After herd individuals realize their movements, the predators change their positions according to Eq. 33 where ρ is a random number between 0 and 1. Once all the individuals change their position, the survival value assignation is recalculated with Eq. 25. Then, the predation phase starts, where predators individuals try to hunt the prey. SHO algorithm establishes a domain of danger, calculated with Eq. 34.

$$p_i^{k+1} = p_i^k + 2\rho(h_r^k - p_i^k) \tag{33}$$

$$R = \frac{\sum_{j=1}^{n} |x_j^{low} - x_j^{high}|}{2 \cdot n} \tag{34}$$

If the distance between a predator and prey is smaller than R, the prey is at risk of being caught. When a predator kills a prey, the total population number decreases. To supply new individuals, the SHO algorithm has a restoration phase, considering the individuals' survival value to bring them a probability to generate a new individual. Once the restoration phase ends, SHO continues running with another herd's individuals and predators' movements if the stop criteria have not been reached.

2.5 Related Work

In 2020, Anand and Arora proposed an improvement of the original SHO algorithm, using chaos theory to perform a hybridization: the Chaotic Selfish Herd Optimizer algorithm (CSHO) [2]. In their proposal, Anand and Arora replace the original program to define the H_F and H_S subsets by using chaotic values to control the exploration-exploitation tradeoff of the algorithm. To test the performance of their suggestion, the authors used a set of 13 functions where they presented unimodal, multimodal, separable, and non-separable functions. They first compared the original SHO algorithm against their proposal for the performance evaluation, using ten different chaotic maps. Subsequently, the best proposal (CSHO with tent map) was compared against five algorithms that proved to perform well on global benchmark functions having 500 dimensions. A third test was performed against seven MA on more complex benchmark problems from the special session and competition of IEEE Congress on Evolutionary Computation 2015. Finally, Anand and Arora compared the performance of CSHO against six MAs, including SHO, in a feature selection test performed with twenty-one datasets from the UCI repository. The authors report that CSHO significantly improves the performance in terms of exploitation and exploration compared to the original SHO and performs well as a feature selection algorithm in terms of classification accuracy and the number of optimal features selected.

Zhao et al. propose the Selfish Herd Optimizer with Levy-Flight distribution strategy (LFSHO) [25], a modification for SHO performance improvement, in 2020. In their work, Zhao et al. add levy-flight operation for herd leader movement. With this change, the proposal's authors sought to solve the premature convergence problems that the original SHO may present. Performance tests were performed with fifteen benchmark functions, and the results obtained were compared with six metaheuristic algorithms, including the SHO. A second test was performed with engineering problems, such as Tension/compression spring, welded beam, speed reducer, and pressure vessel design problems, where the results of each engineering problem were compared with a different set of algorithms in each case. At the end of their work, Zhao et al. mention that derived from the results obtained, they can conclude that the LFSHO presents a more efficient performance and alleviates SHO's stagnation.

In 2019, Yimit et al. proposed some alterations to the original SHO, called Refined Selfish Herd Optimizer (RSHO) [23]. The first proposal of the authors is the alteration of the survival value assignment, using the following equation instead of the equation:

$$SV_{a_i} = 1 - \frac{f_{best} - f(a_i)}{f_{best} - f_{worst}} \tag{35}$$

Besides proposing a modification to calculate SV_{a_i}, they also propose a modification of the update times for this variable. Thus, they add two update times, one after the movement of the selfish herd and the second one after the restoration phase if changes are detected in the solutions of the problem. The update of the survival value that was performed in the original iterative process is replaced by the update of the survival value of the predators.

For The radius of the domain of danger, the authors propose a change that gives this variable more dynamism and flexibility, increasing the balance between exploration and exploitation. In contrast to Eq. 34, Yimit et al. propose the introduction of the random number $lambda$ in the range [0,1] and h_{max} and h_{min} corresponding to the individuals with the largest and smallest components respectively. h_{max} and h_{min} represent the l_1-norm. The changes introduced can be seen in Eq. 36.

$$R = \lambda \cdot n \cdot \|h_{max} - h_{min}\|_1 \tag{36}$$

Finally, the repulsion/attraction factors are modified to reduce the penalty suggested by the original model. Thus, changes are introduced that avoid the power of the Euclidean distance between individuals in favor of introducing the search space dimensionality.

The tests performed to evaluate the performance of the RSHO include the fifteen functions presented in [9] and 30 functions from the CEC2014 benchmark function, where, in combination, we have unimodal, multimodal, hybrid, and composite types. The results of their evaluation were compared with nine well-known optimization algorithms, including the original SHO itself.

2.6 Fuzzy Logic Based Optimization Algorithm

In [8], Diaz-Cortés et al. work on coupling the general methodology of metaheuristic algorithms with the decision logic of fuzzy systems. The submitted algorithm uses a population of search agents within a limited search space. It performs a series of operations following the fuzzy logic to obtain the new position for each population element. In order to improve the evaluation of each individual's position (fitness), Diaz-Cortes et al. propose the procedure detailed in the following paragraphs. To test the performance efficiency of the proposed algorithm, tested their algorithm with 19 CEC2015 features, which include unimodal, multimodal, and hybrid features. The test results were compared against six other state-of-the-art metaheuristic algorithms.

Procedure

Initialization
The population P (Eq. 37) is initialized where N is the number of elements and k is the current iteration number starting at $k = 0$ up to $k = Maxgen$. The initial population is randomly distributed within the search space between the upper and lower bounds u_i and l_i, respectively.

$$P = \{p_1^k, p_2^k, \ldots, p_N^k\} \tag{37}$$

Once all the particles have been distributed, their position is evaluated according to the objective function. From the total population, the element with the best fitness ($g = max(f(p_i^k))$) is selected.

Calculate distance to the best
In developing the fuzzy logic algorithm, two variables were considered for each search agent, called linguistic variables in fuzzy logic. For each linguistic variable, two perturbation functions were defined. For the fitness $f(p_i^k)$, bad (B) and good (G) functions were defined. The second linguistic variable is the Euclidean distance of a particle p_i^k to the best particle in the whole population g, defined as $D(p_i^k, g) = \|g - p_i^k\|$. The assigned membership functions are short (S) and long (L). Figure 4 shows the definition of the membership functions graphically for each linguistic variable.

Evaluate antecedents
For the distance $D(p_i^k, g)$, we set the lower bound $d_{min} = 0$, and the upper bound according to Eq. 38, where n represents the dimensionality of the search space and The range of values for the fitness membership functions, is calculated according to Eq. 39

$$d_{max} = \sqrt{\sum_{i=1}^{n}(u_i - l_i)^2} \tag{38}$$

$$f_{min} = min(f(p_i^k)) \text{ and } f_{max} = max(f(p_i^k)) \tag{39}$$

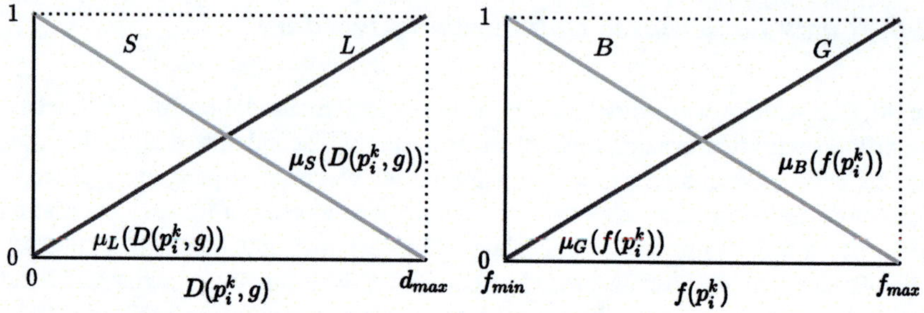

Fig. 4 Membership functions assignation for each linguistic variable

Evaluate consequents

To evaluate the degree of compliance of each particle (p_i^k), the expressions shown in Eq. 40 are used:

$$\begin{aligned}\beta_1(p_i^k) &= min(\mu_S(D(p_i^k, g)), \mu_G(f(p_i^k))) \\ \beta_2(p_i^k) &= min(\mu_S(D(p_i^k, g)), \mu_B(f(p_i^k))) \\ \beta_3(p_i^k) &= min(\mu_L(D(p_i^k, g)), \mu_G(f(p_i^k))) \\ \beta_4(p_i^k) &= min(\mu_L(D(p_i^k, g)), \mu_B(f(p_i^k)))\end{aligned} \tag{40}$$

Diaz-Cortés et al. established four basic rules for particle motion. If $D(p_i^k, g) = S$ and $f(p_i^k) = G$, then the particle p_i^k will perform an approach (Attraction) motion toward g according to Eq. 41. α_1 is a predefined fitting factor.

$$At(p_i^k) = \left|f_{max} - f(p_i^k)\right| \cdot (g - p_i^k) \cdot \alpha_1 \tag{41}$$

If $D(p_i^k, g) = S$ and $f(p_i^k) = B$, then the Repulsion rule will apply, where the p_i^k particle will move away from g as described by Eq. 42. α_2 is a predefined fitting factor.

$$Rep(p_i^k) = \left|f_{max} - f(p_i^k)\right| \cdot (g + p_i^k) \cdot \alpha_2 \tag{42}$$

If $D(p_i^k, g) = L$ and $f(p_i^k) = G$, the algorithm performs Refining, which establishes a random position around p_i^k within the bounds $pm \left|f_{max} - f(p_i^k)\right|$ (Eq. 43), where $v = v = v_1, v_2, \ldots, v_d\}$ is a random vector where each dimension corresponds to a number in the range [-1,1]. γ is a predefined fitting factor.

$$Ref(p_i^k) = \left|f_{max} - f(p_i^k)\right| \cdot v \cdot \gamma \tag{43}$$

If $D(p_i^k, g) = L$ and $f(p_i^k) = B$, then the new position produced for p_i^k will be chosen randomly (Equation refeq:fuzzy11), where $r = \{r_1, r_2, \ldots, r_d\}$ is a random vector.

$$Ran(p_i^k) = r \tag{44}$$

Inference TS

To calculate the new position of each particle first is necessary to take into account the effect of each membership function on the particle p_i^k. The entire modification of position is calculated with the Eq. 45. Once $Deltap_i^k$ has been calculated, then the new position of each particle is derived using the Eq. 46.

$$\Delta p_i^k = \frac{At(p_i^k)\cdot\beta_1(p_i^k)+Rep(p_i^k)\cdot\beta_2(p_i^k)+Ref(p_i^k)\cdot\beta_3(p_i^k)+Ran(p_i^k)\cdot\beta_4(p_i^k)}{\beta_1(p_i^k)+\beta_2(p_i^k)+\beta_3(p_i^k)+\beta_4(p_i^k)} \tag{45}$$

$$p_i^{k+1} = p_i^k + \Delta p_i^k \tag{46}$$

3 Trajectory Optimization Problems

We can define a trajectory optimization problem as searching for the best trajectory for a spacecraft during a mission. It is set with some specific flight constraints or particular take-off and arrival conditions. In general, the optimization of this type of problem is concerned with minimizing travel times or the consumption of vital resources for the mission.

Trajectory optimization problems represent a higher degree of difficulty than other types of problems or benchmark functions. Among the characteristics that make their optimization more challenging are the dynamic system's nonlinearity, the discontinuities in the state variables that trajectories may contain, and the possible existence of multiple optimal structures for the trajectory under study [3].

In 2005, the Advanced Concepts Team of the European Space Agency launched The Global Trajectory Optimisation Competition [22]. That first edition was called GTOC1 - Save the Earth, where the organization proposed the design of a mission to deflect the trajectory of an asteroid using a kinetic impactor. Since then, ten other events have been held, where the competition focuses on solving a trajectory problem. In addition to the organization of the GTOC event, the European Space Agency's Advanced Concepts Team has at the disposal of the scientific community a database of trajectory optimization problems [21], the Global Trajectory Optimisation Problems Database (GTOPD). The goal of the database is to serve the scientific community as benchmark problems for developing and testing new derivative-free algorithms capable of solving complex trajectory problems. The database is composed of two main models or kinds of problems. The most notable difference between them is the way the space probes execute their trajectories. This difference provides different shades of realism and difficulty faced by the optimization algorithms.

3.1 MGA Global Optimisation Problems

Trajectory optimization problems have differences in their definitions and requirements. Generally, when several elements maintain a relationship or similarity, they are grouped under the same name, in this case: a model. The MGA model is so-called because it groups problems that present a condition of the Multiple Gravity Assist. These problems are characterized by the fact that they are global optimization problems with finite dimension and nonlinear constraints. The space missions of this model use chemical propulsion engine probes to make trips from Earth to another space body such as a planet or an asteroid, with the constraint that the propulsion of the probe can only occur at planetary encounters. Within the GTOPD, there are two problems within the MGA model: Cassini 1 and GTOC-1.

3.2 Cassini 1

Based on the real space mission of the Cassini spacecraft, it presents the challenge of optimizing the trajectory that the probe needs to reach Saturn and orbit it. Accordingly, it must perform the same sequence of flybys that the Cassini spacecraft performed between 1997 and 2004: Earth, Venus, Venus, Earth, Jupiter, and Saturn. At the flyby conclusion, the spacecraft should be trapped by the gravitational force of Saturn and orbit it with a pericentric radius of $108950km$. The problem consists of 6 variables representing the dimensionality of the search space (see Table 1). The cumulative velocity of the spacecraft represents the objective function to be minimized during the total of the mission phases.

3.3 GTOC 1

GTOC1 is an MGA model problem used in the first edition of the GTOC in 2005 [22]. The mission involves the launch of a space probe to impact the asterion TW229 and

Table 1 Upper and lower bounds for Cassini 1 search space dimensions

State	Variable	Lower bounds	Upper bounds	Units
x(1)	t0	−1000	0	MJD2000
x(2)	T1	30	400	days
x(3)	T2	100	470	days
x(4)	T3	30	400	days
x(5)	T4	400	2000	days
x(6)	T5	1000	6000	days

Table 2 State vector values for GTOC 1 problem

State	Variable	Lower bounds	Upper bounds	Units
x(1)	t0	3000	10000	MJD2000
x(2)	T1	14	2000	days
x(3)	T2	14	2000	days
x(4)	T3	14	2000	days
x(5)	T4	14	2000	days
x(6)	T5	100	9000	days
x(7)	T6	366	9000	days
x(8)	T7	300	9000	days

deflect its trajectory. The programmed flyby sequence is Earth, Venus, Earth, Venus, Earth, Earth, Jupiter, Saturn, and finally, impacting the asteroid TW229. GTOC1 is a problem that seeks to maximize the final mass of the spacecraft when impacting the asterion to cause the maximum possible deflection of its trajectory. The state vector values for GTOC 1 problem, are shown in Table 2.

3.4 *MGA-1DSM*

The second model is called Multiple Gravity Assist with a single Deep Space Maneuver (MGA-1DSM). This type uses a spacecraft capable of interplanetary travel, equipped with a chemical propulsion engine. Unlike the MGA type problems, in this case, the spacecraft can activate its engine propulsion at any moment of the interplanetary trajectory. The difficulty then lies in the fact that the problems of this group are problems of higher dimensionality. At the same time, the solutions found for this model are more suitable as solutions for real space missions.

3.5 *Cassini 2*

Unlike the Cassini 1 problem, Cassini 2 refers to a more significant dimensionality problem due to removing the deep space maneuvering constraint. The number of dimensions of the Cassini 2 problem and the lower and upper bounds of each dimension can be seen in Table 3. In addition, a case of the final encounter instead of orbital insertion is also considered to evaluate the objective function.

Table 3 Search space limits for Cassini 2 problem

State	Variable	Lower bounds	Upper bounds	Units
x(1)	t0	−1000	0	MJD2000
x(2)	Vinf	3	5	km/sec
x(3)	u	0	1	n/a
x(4)	v	0	1	n/a
x(5)	T1	100	400	days
x(6)	T2	100	500	days
x(7)	T3	30	300	days
x(8)	T4	400	1600	days
x(9)	T5	800	2200	days
x(10)	eta1	0.01	0.9	n/a
x(11)	eta2	0.01	0.9	n/a
x(12)	eta3	0.01	0.9	n/a
x(13)	eta4	0.01	0.9	n/a
x(14)	eta5	0.01	0.9	n/a
x(15)	r_p1	1.05	6	n/a
x(16)	r_p2	1.05	6	n/a
x(17)	r_p3	1.15	6.5	n/a
x(18)	r_p4	1.7	291	n/a
x(19)	b_incl1	−pi	pi	rad
x(20)	b_incl2	−pi	pi	rad
x(21)	b_incl3	−pi	pi	rad
x(22)	b_incl4	−pi	pi	rad

3.6 *Tandem Atlas*

It is a problem that belongs to the MGA-1DSM model and emulates the TandEM mission. The mission objective is to reach Saturn's moons Titan and Encedalous with the Atlas501 spacecraft. The optimization problem at hand is the maximization of the final mass of the spacecraft upon completion of the mission after overflying a certain number of celestial bodies in a given sequence. Table 4 shows the upper and lower bounds of the state vector for TandEM problem. In the TandEM mission, it is possible to choose a flyby sequence from a suggested list.

3.7 *Messenger (Reduced Version)*

Messenger is a problem that represents the mission of the Messenger spacecraft to rendezvous with the planet Mercury. It is presented in two different versions

Table 4 State vector bounds for TandEM space mission

State	Variable	Lower bounds	Upper bounds	Units
x(1)	t0	5475	9132	MJD2000
x(2)	Vinf	2.5	4.9	km/sec
x(3)	u	0	1	n/a
x(4)	v	0	1	n/a
x(5)	T1	20	2500	days
x(6)	T2	20	2500	days
x(7)	T3	20	2500	days
x(8)	T4	20	2500	days
x(9)	eta1	0.01	0.99	days
x(10)	eta3	0.01	0.99	n/a
x(11)	eta3	0.01	0.99	n/a
x(12)	eta4	0.01	0.99	n/a
x(13)	r_p1	1.05	10	n/a
x(14)	r_p2	1.05	10	n/a
x(15)	r_p3	1.05	10	n/a
x(16)	b_incl1	−pi	pi	n/a
x(17)	b_incl2	−pi	pi	n/a
x(18)	b_incl3	−pi	pi	n/a

representing two different degrees of difficulty. The reduced version presents the sequence of flights that the spacecraft performed in the first part of the mission. The reduction of resonant overflights helps present a much simpler version of the problem. The next Table 5 shows the parameters of the states vector of Messenger (reduced version) problem.

3.8 *Messenger (Full Version)*

Unlike the messenger problem (reduced version), the full version includes the resonant fly-bys in the Mercury mission. In this version, it also differs that the flight sequence is similar to the actual real mission of the Messenger spacecraft. It is considered that due to the complexity of the interplanetary trajectories of this mission, specialized knowledge is required to perform a successful design in such a way that it was not believed to be possible with a computer to obtain remarkable results. It should be noted that the vector of states is more extensive in this version, as can be seen in Table 6.

Table 5 State vector values for Messenger reduced version problem

State	Variable	Lower bounds	Upper bounds	Units
x(1)	t0	1000	4000	MJD2000
x(2)	Vinf	1	5	km/sec
x(3)	u	0	1	n/a
x(4)	v	0	1	n/a
x(5)	T1	200	400	days
x(6)	T2	30	400	days
x(7)	T3	30	400	days
x(8)	T4	30	400	days
x(9)	eta1	0.01	0.99	days
x(10)	eta3	0.01	0.99	n/a
x(11)	eta3	0.01	0.99	n/a
x(12)	eta4	0.01	0.99	n/a
x(13)	r_p1	1.1	6	n/a
x(14)	r_p2	1.1	6	n/a
x(15)	r_p3	1.1	6	n/a
x(16)	b_incl1	−pi	pi	n/a
x(17)	b_incl2	−pi	pi	n/a
x(18)	b_incl3	−pi	pi	n/a

3.9 Rosetta

Rosetta is an MGA-1DSM problem, where the flight sequence is similar to that planned for the Rosetta spacecraft on the mission to comet 67P/Churyumov-Gerasimenko. *DeltaV* total mission, as well as the launcher capabilities, compose the objective function of this problem, and the states vector values for each problem dimension, is shown in Table 7.

3.10 Sagas

Unlike the other trajectory optimization problems presented before, the objective function of the Sagas mission is the minimization of the mission time. The trajectory to be optimized contemplates the fly-by to Jupiter and then to reach 50AU. The maneuver performed, and the contemplated objective function causes the addition of two more variables to the decision vector (Table 8) to evaluate the Keplerian orbit after the last fly-by.

Table 6 Upper and lower bounds for Messenger Full version dimensions problem

State	Variable	Lower bounds	Upper bounds	Units
x(1)	t0	1900	2300	MJD2000
x(2)	Vinf	2.5	4.05	km/sec
x(3)	u	0	1	n/a
x(4)	v	0	1	n/a
x(5)	T1	100	500	days
x(6)	T2	100	500	days
x(7)	T3	100	500	days
x(8)	T4	100	500	days
x(9)	T5	100	500	days
x(10)	T6	100	600	days
x(11)	eta1	0.01	0.99	days
x(12)	eta2	0.01	0.99	n/a
x(13)	eta3	0.01	0.99	n/a
x(14)	eta4	0.01	0.99	n/a
x(15)	eta5	0.01	0.99	n/a
x(16)	eta6	0.01	0.99	n/a
x(17)	r_p1	1.1	6	n/a
x(18)	r_p2	1.1	6	n/a
x(19)	r_p3	1.05	6	n/a
x(20)	r_p4	1.05	6	n/a
x(21)	r_p5	1.05	6	n/a
x(22)	b_incl1	−pi	pi	n/a
x(23)	b_incl2	−pi	pi	n/a
x(24)	b_incl3	−pi	pi	n/a
x(25)	b_incl4	−pi	pi	n/a
x(26)	b_incl5	−pi	pi	n/a

4 Metodology

The tests on the algorithms were performed as follows: A set of 8 real trajectory optimization problems described in detail in Sect. 3 was established, to optimize these problems were chosen 6 state-of-the-art algorithms (SSO, SBLA, FUZZY, CAB, SHO, and SMS) described in Sect. 2. Thirty independent runs were performed in each problem for each one of the algorithms. It is essential to mention that the stop criteria for all algorithms were established in 50, 000 function accesses to make an appropriate comparison.

Table 7 State vector values for Rosseta mission

State	Variable	Lower bounds	Upper bounds	Units
x(1)	t0	1460	1825	MJD2000
x(2)	Vinf	3	5	km/sec
x(3)	u	0	1	n/a
x(4)	v	0	1	n/a
x(5)	T1	300	500	days
x(6)	T2	150	800	days
x(7)	T3	150	800	days
x(8)	T4	300	800	days
x(9)	T5	700	1850	days
x(10)	eta1	0.01	0.9	n/a
x(11)	eta2	0.01	0.9	n/a
x(12)	eta3	0.01	0.9	n/a
x(13)	eta4	0.01	0.9	n/a
x(14)	eta5	0.01	0.9	n/a
x(15)	r_p1	1.05	9	n/a
x(16)	r_p2	1.05	9	n/a
x(17)	r_p3	1.05	9	n/a
x(18)	r_p4	1.05	9	n/a
x(19)	b_incl1	−pi	pi	rad
x(20)	_incl2	−pi	pi	rad
x(21)	b_incl3	−pi	pi	rad
x(22)	b_incl4	−pi	pi	rad

Table 8 Sagas missions' state vector values

State	Variable	Lower bounds	Upper bounds	Units
x(1)	t0	7000	9100	MJD2000
x(2)	Vinf	0	7	km/sec
x(3)	u	0	1	n/a
x(4)	v	0	1	n/a
x(5)	T1	50	2000	days
x(6)	T2	300	2000	days
x(7)	eta1	0.01	0.9	n/a
x(8)	eta2	0.01	0.9	n/a
x(9)	r_p1	1.05	7	n/a
x(10)	r_p2	8	500	n/a
x(11)	b_incl1	−pi	pi	rad
x(12)	b_incl2	−pi	pi	rad

5 Results and Discussions

To analyze the results with statistical measures, we choose the following measures to check the central tendency of the data obtained: the average (AVG), the median (MD), and the standard deviation (SD) for each set of data for each problem.

In Table 9 a conditional format was established to show the best numerical values for the results in terms of AVG, MD, and SD highlighted in bold. It is essential to mention that the problems to which the algorithms were applied are mathematical minimization functions; thus, the smallest numerical values in terms of AVG and MD represent the best solution, while a smaller SD represents less variability of the data. The analyzed table shows that the SSO algorithm has better results than its competitors, although the SBLA algorithm has better performance in problem F8. SSO is superior in the problems F1–F7. Finally, the corresponding percentages of the functions solved by each algorithm are calculated based on the best results of AVG, MD, and SD.

Table 9 General results of engineering problems

Function		SSO	SBLA	FUZZY	CAB	SHO	SMS
F1. Gtoc1	AVG	**−2.12E+05**	−7.01E+04	−3.80E+03	−7.14E+02	−1.34E+03	−8.73E-13
	MD	**−6.62E+04**	−3.66E+04	−1.66E+03	−7.16E+02	−5.97E+02	−1.19E-19
	SD	2.91E+05	8.60E+04	5.07E+03	3.21E+01	1.77E+03	**3.93E-12**
F2. Cassini 1	AVG	**1.09E+01**	2.61E+01	3.65E+01	1.77E+02	7.45E+01	8.39E+02
	MD	**1.12E+01**	2.66E+01	3.61E+01	1.83E+02	7.84E+01	8.07E+02
	SD	**5.06E+00**	9.02E+00	9.48E+00	8.82E+00	2.25E+01	9.34E+01
F3. Cassini 2	AVG	**3.18E+01**	4.73E+01	7.70E+01	1.28E+02	8.68E+01	6.65E+05
	MD	**3.12E+01**	4.87E+01	7.66E+01	1.30E+02	8.74E+01	5.63E+03
	SD	5.88E+00	8.41E+00	8.32E+00	**4.33E+00**	1.30E+01	3.58E+06
F4. Tandem Atlas	AVG	**1.79E+01**	2.64E+01	4.51E+01	7.26E+02	5.46E+01	1.21E+05
	MD	**1.79E+01**	2.65E+01	4.55E+01	7.13E+02	5.56E+01	1.61E+05
	SD	**3.02E+00**	3.79E+00	9.65E+00	1.93E+02	9.38E+00	8.51E+04
F5. Messenger	AVG	**1.92E+01**	2.68E+01	4.86E+01	1.19E+02	6.30E+01	6.17E+04
	MD	**1.84E+01**	2.68E+01	4.76E+01	9.24E+01	5.92E+01	3.57E+04
	SD	3.12E+00	**2.51E+00**	1.05E+01	1.44E+02	1.31E+01	7.27E+04
F6. Messenger full	AVG	**2.82E+01**	4.62E+01	8.51E+01	9.80E+02	9.75E+01	1.01E+05
	MD	**2.75E+01**	4.64E+01	8.44E+01	1.15E+03	9.69E+01	1.61E+04
	SD	**6.41E+00**	9.01E+00	1.23E+01	3.63E+02	1.72E+01	3.19E+05
F7. Rosetta	AVG	**1.82E+01**	2.89E+01	5.19E+01	1.78E+02	5.34E+01	6.74E+04
	MD	**1.82E+01**	3.06E+01	5.10E+01	2.05E+02	5.17E+01	3.08E+03
	SD	**2.91E+00**	5.13E+00	8.69E+00	5.25E+01	9.42E+00	3.04E+05
F8. Sagas	AVG	1.48E+03	**1.39E+03**	3.39E+03	5.67E+03	4.06E+03	1.67E+05
	MD	1.56E+03	**1.39E+03**	3.50E+03	5.98E+03	3.73E+03	1.00E+05
	SD	**2.13E+02**	2.56E+02	6.41E+02	7.01E+02	1.38E+03	3.21E+05

Table 10 Results of Gtoc1

	SSO	SBLA	FUZZY	CAB	SHO	SMS
MeanFit	**−2.12E+05**	**−7.01E+04**	**−3.80E+03**	−7.14E+02	−1.34E+03	−8.73E-13
BestFit	**−1.21E+06**	**−3.50E+05**	**−2.27E+04**	−8.45E+02	−7.59E+03	−2.13E-11
WorstFit	**−2.90E+04**	**−8.35E+03**	−2.39E+02	**−6.24E+02**	−7.42E+01	−4.09E-34
x_1	7410.326	8823.281	7149.054	3000.000	7543.942	3684.000
x_2	167.149	306.008	259.234	400.000	115.224	333.000
x_3	1084.870	109.964	41.487	1577.098	1004.355	1462.000
x_4	74.871	295.029	853.809	1577.098	1718.720	452.000
x_5	1043.081	66.768	1554.277	1577.098	1770.830	909.000
x_6	1974.815	2211.366	6265.204	1577.098	6359.258	266.000
x_7	4893.609	2278.251	1804.078	1577.098	2602.486	7640.000
x_8	2325.3939	4837.1655	5353.2317	1577.0983	2889.9371	7979.0000

- The results of the AVG percentages are: SSO(87.5%), SBLA(12.5%), FUZZY(0%), CAB(0%), SHO(0%) and SMS(0%).
- The results of the MD percentages are: SSO(87.5%), SBLA(12.5%), FUZZY(0%), CAB(0%), SHO(0%) and SMS(0%).
- Results in terms of SD are: SSO(62.5%), SBLA(12.5%), FUZZY(0%), CAB(12.5%), SHO(0%) and SMS(12.5%).

In Table 10 the value (MeanFit) represents the best average along the 30 runs, (BestFit) is the best solution, and (WorstFit) is the worst solution reached. x_1 to x_n are the positions of the best solution for each algorithm. In addition, the top three algorithms were ranked as follows: The numerical values highlighted in black correspond to the best algorithm of the three ranked, the second-best algorithm was highlighted in blue, and finally, the third-best algorithm is the one highlighted in orange.

In Table 11 the ranking of the three best algorithms is shown in the same way as in the previous table. It is clear that SSO, highlighted in black bold, is superior in the three established reference values. In the average fitness of all independent runs (MeanFit), the best fitness value during the history of the algorithm (BestFit) and has the lowest worst fitness value of the three best algorithms, thus demonstrating that it has the best performance of the proposed algorithms.

The results of the Cassini 2 problem are shown in Table 12. This problem represents a challenge for some optimization algorithms because it has 22 decision variables representing a high-dimensional problem and proves the robustness of the optimization methods used to solve it. At the same time, the algorithm requires a search strategy that refines the solution the most possible. In this problem, the best solution vector of 22 variables was obtained by the SSO algorithm, highlighted in bold, and although the second best algorithm named SBLA and highlighted in blue;

Table 11 Results of Cassini1

	SSO	SBLA	FUZZY	CAB	SHO	SMS
MeanFit	**1.09E+01**	**2.61E+01**	**3.65E+01**	1.77E+02	7.45E+01	8.39E+02
BestFit	**5.36E+00**	**6.76E+00**	**1.94E+01**	1.51E+02	3.32E+01	7.17E+02
WorstFit	**1.94E+01**	**4.54E+01**	**6.16E+01**	1.84E+02	1.18E+02	9.75E+02
x_1	−772.638	−779.851	−775.484	0.000	−992.694	−292.000
x_2	175.977	176.420	180.912	112.481	208.365	53.000
x_3	415.975	414.558	412.277	112.481	214.371	143.000
x_4	53.786	54.410	82.930	112.481	334.015	49.000
x_5	1033.374	561.730	932.841	400.000	1026.573	1915.000
x_6	4479.040	1907.807	4063.370	1000.000	1927.326	5043.000

Table 12 Results of Cassini2

	SSO	SBLA	FUZZY	CAB	SHO	SMS
MeanFit	**3.18E+01**	**4.73E+01**	**7.70E+01**	1.28E+02	8.68E+01	6.65E+05
BestFit	**1.60E+01**	**3.08E+01**	**5.04E+01**	1.16E+02	5.79E+01	1.51E+03
WorstFit	**4.84E+01**	**6.22E+01**	**9.35E+01**	1.31E+02	1.14E+02	1.96E+07
x_1	−855.571	−774.852	−794.415	−369.232	−892.158	−772.928
x_2	3.206	4.307	4.791	3.000	3.909	3.792
x_3	0.681	0.412	0.145	0.000	0.290	0.974
x_4	0.463	0.160	0.845	0.000	0.897	0.110
x_5	237.645	317.155	156.089	100.000	226.546	258.178
x_6	421.566	199.677	480.364	100.000	411.956	463.855
x_7	59.566	185.153	237.322	30.000	60.618	258.184
x_8	582.102	833.540	805.291	400.000	1248.511	1251.540
x_9	1733.432	1651.441	1585.260	800.000	1763.307	1301.954
x_10	0.411	0.394	0.040	0.010	0.223	0.682
x_11	0.083	0.054	0.700	0.010	0.078	0.452
x_12	0.185	0.483	0.525	0.010	0.374	0.836
x_13	0.338	0.374	0.138	0.010	0.032	0.712
x_14	0.037	0.249	0.125	0.010	0.320	0.863
x_15	3.697	4.199	1.659	1.050	2.885	4.238
x_16	5.701	2.604	3.829	1.050	4.698	1.941
x_17	1.218	4.917	6.496	1.150	5.661	1.485
x_18	133.706	225.724	133.171	1.700	284.175	99.799
x_19	−1.213	−0.730	−0.285	−3.142	−2.101	−0.527
x_20	−1.489	1.372	−1.322	−3.142	−1.251	−0.339
x_21	−1.460	−1.827	0.245	−3.142	1.281	3.062
x_22	−1.454	0.777	−2.515	−3.142	−1.017	−0.506

Table 13 Results of Tandem Atlas

	SSO	SBLA	FUZZY	CAB	SHO	SMS
MeanFit	**1.79E+01**	**2.64E+01**	**4.51E+01**	7.26E+02	5.46E+01	1.21E+05
BestFit	**1.24E+01**	**1.73E+01**	**2.51E+01**	3.59E+02	3.39E+01	1.97E+04
WorstFit	**2.55E+01**	**3.43E+01**	**6.82E+01**	1.06E+03	7.14E+01	2.65E+05
x_1	7484.198	7236.488	6284.427	8267.234	8517.647	6089.000
x_2	4.209	2.620	2.216	5.000	3.650	5.000
x_3	0.932	0.557	0.449	1.000	0.066	1.000
x_4	0.582	0.405	0.426	1.000	0.577	0.000
x_5	1092.155	443.176	166.223	2500.000	495.498	1427.000
x_6	1423.266	1150.007	2436.150	2500.000	1728.584	1972.000
x_7	1114.980	1097.144	2164.631	2500.000	1596.354	571.000
x_8	1483.065	1580.755	2392.451	2500.000	2014.511	195.000
x_9	0.807	0.600	0.034	0.990	0.599	0.010
x_10	0.420	0.220	0.395	0.990	0.434	0.010
x_11	0.190	0.366	0.274	0.990	0.321	0.010
x_12	0.064	0.143	0.261	0.990	0.039	0.990
x_13	31.703	11.089	20.937	50.000	31.529	33.000
x_14	42.316	7.769	41.985	50.000	43.927	22.000
x_15	25.315	21.290	39.224	50.000	49.803	46.000
x_16	−0.695	0.678	−0.490	3.142	−1.892	0.000
x_17	−1.043	−0.788	−0.867	3.142	0.604	2.000
x_18	−1.311	−1.102	−0.520	3.142	1.874	0.000

obtains very similar values in terms of MeanFit, BestFit, and WorstFit, for the type of problems the need to obtain better solutions in all the proposed metrics is a requirement.

The Tandem Atlas problem is another high-dimensional problem, with a total of 18 decision variables(see Table 13), where it is evident that the ranking for the best three algorithms is repeated. We can compare with the best solution vectors of the GTOP solutions database values that are publicly available on the official website [21].

The FUZZY algorithm generates the third-best set of results for the three variables for the reduced version of the Messenger problem. Similarly, SBLA outputs the second-best set of results, and SSO obtains the best set of values. Table 14 shows these results and the values of the state vector of the problem. The CAB, SHO, and SMS algorithms do not stand out in any of the variables considered in the analysis.

In the Messenger full version problem, the SSO algorithm presents the best average of the best fitness obtained in the 30 runs of the test and the best individual fitness of all, and the lowest worst fitness of those obtained by the six algorithms. SBLA presents the second place, in the three items, in front of the FUZZY algorithm,

Table 14 Results of Messenger

	SSO	SBLA	FUZZY	CAB	SHO	SMS
MeanFit	**1.92E+01**	**2.68E+01**	**4.86E+01**	1.19E+02	6.30E+01	6.17E+04
BestFit	**1.44E+01**	**2.03E+01**	**3.34E+01**	9.24E+01	4.39E+01	1.45E+04
WorstFit	**2.82E+01**	**3.12E+01**	**7.11E+01**	8.82E+02	9.44E+01	3.00E+05
x_1	2197.815	1389.206	2430.817	1000.000	1261.776	1654.000
x_2	1.007	1.572	2.544	1.000	2.059	4.000
x_3	0.730	0.491	0.349	0.000	0.040	1.000
x_4	0.533	0.317	0.836	0.000	0.979	0.000
x_5	344.946	102.243	69.888	30.000	65.876	218.000
x_6	267.355	221.543	349.349	30.000	343.599	44.000
x_7	220.297	210.921	66.460	30.000	248.308	299.000
x_8	193.261	94.968	266.425	30.000	282.529	378.000
x_9	0.279	0.064	0.187	0.010	0.235	0.010
x_10	0.596	0.053	0.605	0.010	0.343	0.990
x_11	0.586	0.137	0.426	0.010	0.248	0.010
x_12	0.515	0.155	0.610	0.010	0.096	0.990
x_13	3.023	2.438	4.586	1.100	1.856	2.000
x_14	5.276	1.842	5.982	1.100	5.944	3.000
x_15	2.621	2.011	5.005	1.100	3.655	4.000
x_16	−0.682	1.029	−0.283	−3.142	−1.493	3.000
x_17	1.488	0.798	0.456	−3.142	1.840	0.000
x_18	2.027	0.852	1.406	−3.142	0.759	2.000

which presents the three third-best values obtained. The Table 15 shows the data of each algorithm, as well as the best positions obtained by them for each problem's dimension.

The six algorithms' corresponding values of MeanFit, BestFit, and WorstFit are presented in the Table 16. This table follows the trend that has been presented throughout all the trajectory problems, where the SSO algorithm obtains the best results, SBLA obtains the second-best values, and FUZZY the third-best values in the three variables. The positions obtained by each algorithm for each of the 22 dimensions are also presented.

Finally, the results of the test of the studied algorithms with the Sagas trajectory optimization problem (Table 17) are presented. It can be observed that FUZZY obtained the three best third places in the test for all variables. SBLA obtained the second-best results in BestFit and WorstFit, where SSO achieved the best values. On the other hand, SSO obtained the second-best average of fitness, and SBLA presents the best average of all, which presents it as a more precise algorithm than SSO but with less accuracy.

Table 15 Results of Messenger Full

	SSO	SBLA	FUZZY	CAB	SHO	SMS
MeanFit	**2.82E+01**	**4.62E+01**	**8.51E+01**	9.80E+02	9.75E+01	1.01E+05
BestFit	**1.71E+01**	**2.62E+01**	**5.58E+01**	3.39E+02	6.62E+01	1.10E+04
WorstFit	**4.11E+01**	**6.32E+01**	**1.14E+02**	1.27E+03	1.40E+02	1.48E+06
x_1	2073.797	2035.779	2157.804	1900.000	2050.740	2017.000
x_2	4.702	2.827	3.006	2.500	3.330	3.000
x_3	0.687	0.388	0.368	0.000	0.482	1.000
x_4	0.465	0.320	0.859	0.000	0.101	0.000
x_5	232.434	259.019	326.623	100.000	382.764	140.000
x_6	234.635	178.185	432.098	100.000	226.321	347.000
x_7	345.355	240.062	238.083	100.000	226.895	439.000
x_8	363.523	181.490	347.433	100.000	380.867	227.000
x_9	328.557	179.334	197.875	100.000	155.134	184.000
x_10	371.107	180.296	192.166	100.000	339.134	319.000
x_11	0.160	0.345	0.595	0.010	0.692	0.990
x_12	0.363	0.208	0.582	0.010	0.040	0.990
x_13	0.694	0.698	0.468	0.010	0.575	0.010
x_14	0.698	0.121	0.637	0.010	0.856	0.990
x_15	0.857	0.401	0.820	0.010	0.619	0.010
x_16	0.659	0.155	0.579	0.010	0.622	0.990
x_17	2.623	3.177	5.703	1.100	1.307	1.100
x_18	4.746	1.894	5.893	1.100	2.155	4.000
x_19	1.986	4.437	2.379	1.050	1.762	5.000
x_20	3.293	3.981	4.952	1.050	4.766	2.000
x_21	2.640	3.682	5.044	1.050	4.521	2.000
x_22	2.509	−0.510	2.015	−3.142	1.062	3.000
x_23	2.563	−0.655	−2.564	−3.142	2.975	0.000
x_24	−0.211	0.061	1.651	−3.142	1.399	−1.000
x_25	−0.123	1.003	2.565	−3.142	2.325	2.000
x_26	−1.491	1.426	−1.148	−3.142	−0.510	−1.000

6 Conclusions

This chapter compares the performance of six metaheuristic algorithms based on biological, physical, or mathematical metaphors (SSO, SBLA, FUZZY, CAB, SHO, and SMS). In the test, we solve several trajectory optimization problems. Although the algorithms have been modified to strengthen some weaknesses that their original design presented, in this paper, the original algorithms were used to discover the innate potential of their original method to solve trajectory optimization problems.

Table 16 Results of Rosetta

	SSO	SBLA	FUZZY	CAB	SHO	SMS
MeanFit	**1.82E+01**	**2.89E+01**	**5.19E+01**	1.78E+02	5.34E+01	6.74E+04
BestFit	**1.34E+01**	**1.94E+01**	**3.61E+01**	1.01E+02	3.74E+01	8.62E+02
WorstFit	**2.48E+01**	**3.87E+01**	**6.95E+01**	2.34E+02	7.59E+01	1.67E+06
x_1	1641.679	1579.990	1680.977	1460.000	1562.958	1535.000
x_2	3.926	4.048	4.993	3.000	3.481	4.000
x_3	0.252	0.280	0.733	0.000	0.057	1.000
x_4	0.480	0.180	0.108	0.000	0.689	1.000
x_5	360.621	323.085	399.381	300.000	388.064	401.000
x_6	480.865	434.181	371.323	150.000	462.492	231.000
x_7	388.327	504.534	610.059	150.000	407.950	174.000
x_8	696.323	358.918	692.294	300.000	420.955	447.000
x_9	1236.452	1282.363	1525.243	700.000	972.325	943.000
x_10	0.104	0.174	0.283	0.010	0.077	0.010
x_11	0.637	0.428	0.630	0.010	0.420	0.900
x_12	0.331	0.084	0.628	0.010	0.561	0.900
x_13	0.387	0.298	0.327	0.010	0.680	0.900
x_14	0.040	0.063	0.504	0.010	0.412	0.900
x_15	3.504	2.203	2.603	1.060	6.287	1.060
x_16	3.649	2.614	2.489	1.060	2.447	7.000
x_17	4.908	1.622	4.372	1.060	4.330	6.000
x_18	4.502	1.662	3.383	1.060	3.745	4.000
x_19	−1.477	−0.177	−2.168	−3.142	−0.763	0.000
x_20	−0.011	0.105	−1.593	−3.142	2.583	3.000
x_21	−1.828	1.104	−2.825	−3.142	2.016	1.000
x_22	−2.665	−1.455	−0.937	−3.142	−1.125	− 3.000

After comparing the experiments' results, we declare the tree best among the six compared algorithms. In the first position, SSO was the absolute best results in seven of the six problems. SBLA, as the second-best algorithm, got the second-best results in six problems. The third place is for FUZZY algorithm, which got the third-best results into the seven test problems. From such experimentation, it can be concluded that the algorithms, even with the natural weaknesses of the MA's, are a good option for the solution of problems similar to those proposed in the GTOP, where the nature of the approaches does not know a unique solution because it refers to global optimization problems that lack a single best solution. In the future, more original algorithms will be tested with the GTOP database and other complex and reality-based optimization problems. Knowing the weaknesses of the algorithms with complex problems, we will work to improve them by generating better algorithms among the best algorithms to solve reality-based complex optimization problems.

Table 17 Results of Sagas

	SSO	SBLA	FUZZY	CAB	SHO	SMS
MeanFit	**1.48E+03**	**1.39E+03**	**3.39E+03**	5.67E+03	4.06E+03	1.67E+05
BestFit	**9.28E+02**	**1.00E+03**	**2.07E+03**	4.13E+03	2.09E+03	1.00E+05
WorstFit	**1.82E+03**	**2.04E+03**	**4.53E+03**	5.98E+03	7.23E+03	1.86E+06
x_1	8507.697	7196.862	7330.470	7000.000	8886.028	8687.291
x_2	2.477	0.002	0.252	0.000	0.094	6.955
x_3	0.520	0.004	0.937	0.000	0.060	0.752
x_4	0.492	0.023	0.665	0.000	0.531	0.354
x_5	460.975	149.355	170.077	50.000	256.286	1393.113
x_6	435.197	757.611	657.475	300.000	1813.267	653.466
x_7	0.369	0.080	0.166	0.010	0.251	0.807
x_8	0.070	0.056	0.419	0.010	0.803	0.216
x_9	1.051	1.804	4.335	1.050	3.529	2.437
x_10	369.145	8.160	232.393	8.000	372.911	296.708
x_11	−0.478	−0.552	2.505	−3.142	1.232	−2.434
x_12	2.442	−2.418	−0.535	−3.142	1.685	1.048

References

1. P. Agrawal, H.F. Abutarboush, T. Ganesh, A.W. Mohamed, Metaheuristic algorithms on feature selection: a survey of one decade of research (2009–2019). IEEE Access **9**:26766–26791 (2021)
2. P. Anand, S. Arora, A novel chaotic selfish herd optimizer for global optimization and feature selection. Artif. Intell. Rev. **53**(2), 1441–1486 (2020)
3. B.A. Conway, *The Problem of Spacecraft Trajectory Optimization*, pp. 1 – 15. Cambridge Aerospace Series (Cambridge University Press, 2010)
4. E. Cuevas, M. González, Multi-circle detection on images inspired by collective animal behavior. Appl. Intell. **39**(1), 101–120 (2013)
5. E. Cuevas, M. González, D. Zaldivar, M. Pérez-Cisneros, Multi-ellipses detection on images inspired by collective animal behavior. Neural Comput. Appl. **24**(5), 1019–1033 (2014)
6. E. Cuevas, M. Gonzalez, D. Zaldivar, M. Perez-Cisneros, G. García, An algorithm for global optimization inspired by collective animal behavior. Discrete Dyn. Nat. Soc. **2012** (2012)
7. E. Cuevas, J. Osuna, D. Oliva, M. Díaz, Optimización: algoritmos programados con matlab. Alfaomega Colombiana (2016)
8. M.A. Díaz-Cortés, E. Cuevas, J. Gálvez, O. Camarena, A new metaheuristic optimization methodology based on fuzzy logic. Appl. Soft Comput. **61**, 549–569 (2017)
9. F. Fausto, E. Cuevas, A. Valdivia, A. González, A global optimization algorithm inspired in the behavior of selfish herds. Biosystems **160**, 39–55 (2017)
10. J. Gálvez, E. Cuevas, O. Avalos, D. Oliva, S. Hinojosa, Electromagnetism-like mechanism with collective animal behavior for multimodal optimization. Appl. Intell. **48**(9), 2580–2612 (2018)
11. D. Karaboga, B. Basturk, On the performance of artificial bee colony (abc) algorithm. Appl. Soft Comput. **8**(1), 687–697 (2008)
12. N.A. Khan, S.P. Ghoshal, S. Ghosh, Optimal allocation of distributed generation and shunt capacitors for the reduction of total voltage deviation and total line loss in radial distribution systems using binary collective animal behavior optimization algorithm. Electric Power Compon. Syst. **43**(2), 119–133 (2015)

13. C.E. Klein, E.H.V. Segundo, V.C. Mariani, L. D.S. Coelho. Modified social-spider optimization algorithm applied to electromagnetic optimization. IEEE Trans. Magn. **52**(3), 1–4 (2015)
14. O. Maciel, E. Cuevas, M.A. Navarro, D. Zaldívar, S. Hinojosa, Side-blotched lizard algorithm: a polymorphic population approach. Appl. Soft Comput. **88**, 106039 (2020)
15. S.Z. Mirjalili, S. Saremi, S.M. Mirjalili, Designing evolutionary feedforward neural networks using social spider optimization algorithm. Neural Comput. Appl. **26**(8), 1919–1928 (2015)
16. T.T. Nguyen, A high performance social spider optimization algorithm for optimal power flow solution with single objective optimization. Energy **171**, 218–240 (2019)
17. P.K. Pany, S.P. Ghoshal, V. Mukherjee, Collective animal behavior algorithm for optimal reactive power dispatch problems, in *2017 IEEE International Conference on Power, Control, Signals and Instrumentation Engineering (ICPCSI)* (IEEE, 2017), pp. 1327–1332
18. D. Pelusi, R. Mascella, L. Tallini, Revised gravitational search algorithms based on evolutionary-fuzzy systems. Algorithms **10**(2), 44 (2017)
19. G. Ram, D. Mandal, R. Kar, S.P. Ghoshal, Optimized hyper beamforming of linear antenna arrays using collective animal behaviour. Sci. World J. **2013** (2013)
20. M. Seyedali, S.M. Mohammad, L. Andrew, Grey wolf optimizer. Adv. Eng. Softw. **69**(3), 46–61 (2014)
21. Advanced Concepts Team, Global trajectory optimisation problems database (2005)
22. Advanced Concepts Team, GTOC Portal | The Global Trajectory Optimisation Competition Portal. Accessed 28 Nov 2021
23. A. Yimit, K. Iigura, Y. Hagihara, Refined selfish herd optimizer for global optimization problems. Expert Syst. Appl. **139**, 112838 (2020)
24. R. Zhao, Q. Luo, Y. Zhou, Elite opposition-based social spider optimization algorithm for global function optimization. Algorithms **10**(1), 9 (2017)
25. R. Zhao, Y. Wang, C. Liu, H. Peng, Y. Li, H. Li, C. Yuan, Selfish herd optimizer with levy-flight distribution strategy for global optimization problem. Phys. A **538**, 122687 (2020)

Parameter Tuning of PID Controller Based on Arithmetic Optimization Algorithm in IOT Systems

Mohamed Issa

Abstract Nowadays, the Internet of Things (IOT) is a modernistic growth in industrial applications. Proportional–Integral–Derivative (PID) controller is an essential device in most engineering applications. The vital abuse of PID is choosing the best parameters' values (K_p, K_i, and K_d) with the traditional methods which cannot achieve the optimum response. In this work, Arithmetic Optimization Algorithm (AOA) which is a meta-heuristic algorithm is used to select the optimum values of the parameters. AOA is chosen since it has efficient exploration and exploitation capabilities in the search space. The AOA algorithm is tested for choosing the optimum values of PID's parameters for controlling two engineering applications are three cascaded liquid level tanks systems and DC motor regulation. AOA has superiority over the comparative algorithms which is used in the test.

Keywords Internet of things · PID · Parameter estimation · Arithmetic optimization algorithm (AOA)

1 Introduction

Supervisory Control and Data Acquisition (SCADA) is a main part in the industrial monitoring and control applications which controls the industrial engineering operation through the programmable logic controllers and the remote terminal units [1]. However, limitation of the SCADA system since limited memory storage and the slow of data processing speed [2].

Besides, the high cost for maintenance and wire connection affects the distribution of the industrial processes via long distances is a complex process. So, using wireless networks for transferring the data and control signals is a good solution in addition to using central computers with the high specification for storage and processing. The Industrial Internet of Things is a vital component in the modern industries

M. Issa (✉)
Computer and Systems Engineering Department, Faculty of Engineering, Zagazig University, Zagazig, Egypt
e-mail: Mamohamedali@eng.zu.edu.eg

E. H. Houssein et al. (eds.), *Integrating Meta-Heuristics and Machine Learning for Real-World Optimization Problems*, Studies in Computational Intelligence 1038,
https://doi.org/10.1007/978-3-030-99079-4_15

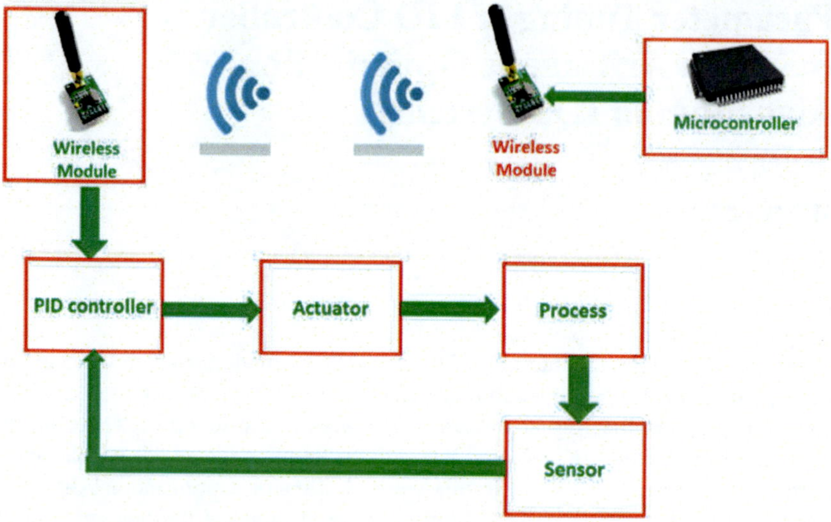

Fig. 1 IOT system for controlling industrial process using PID

for connecting hardware devices and networking devices such as sensors, actuators, control panels, drives, and a controller such as Proportional–integral–derivative (PID) [3, 4].

Figure 1 describes an example of controlling an industrial process using a PID controller in IOT systems. Microcontroller (mini-computer machine) can execute the programs that estimate the optimum values of parameters of PID and can send it through wireless for long-distance which decrease the risk of complex wire connection. Also, the internet can be used for transmitting data wireless [5].

The common usage of PID controllers in the industrial manufacturing processes which controlling the engineering processes since its simplicity, effectiveness, and robustness [6]. The main specification measurements of PID controllers, settle time which is defined as the time for settling of the process, system's stability, the maximum overshoot and the difference error between the actual response and the desired response [6]. Since the participation of the processes in industrial plants, the setting of the PID's parameters is a vital task, a proper disposition permits obtaining an efficient transient performance in terms of lowest steady-state error, settling time, rise time, and maximum overshoot as possible. The main three parameters of the PID controller are integral gain (K_i), proportional gain (K_p) and derivative gain (K_d).

Three main categories are used for estimating PID's parameters are the rule-based methods, the analytical methods and the numerical methods [7]. The most popular technique is Ziegler–Nichol (ZN) [8] which was the classical technique for estimating the PID controller's parameters and is belong to the analytical methods. ZN cannot guarantee finding the suitable values of the parameters to achieve the optimum performance since there is a better one can be estimated using optimization techniques.

Algorithms of Heuristic optimization was used to estimate the PID's parameters which are classified as the numerical methods which were common in the literature. The meta-heuristic algorithms are stochastic optimization methods [9] were the appropriate choice for estimating the PID's parameters due to it model the problem as a black box model and estimate the parameters by tracking the fitness (objective function) to find the best value.

The meta-heuristic algorithm is a search-based algorithm that random the exploring of the problem search space to achieve the most acceptable value in a reasonable time [9]. The meta-heuristic algorithm imitative the search strategy from physics, nature, or humans. Such as Particle Swarm Optimization algorithm (PSO) [10] imitative the search method of the bird flocking from nature.

In addition, the Sine–Cosine Optimization algorithm (SCA) [11] which explore the search space by pulling the search population toward the best-founded location based on the sine and cosine operators. Other algorithms were released which mimic the search strategy of the physics such as Lightning Attachment Procedure Optimization [12], Ions Motion Optimization (IMO) [13], Gravitational Search Algorithm (GSA) [14], Electron radar search algorithm [15], The Whirlpool Algorithm [16] and Squirrel Search Algorithm for optimization [17].

Besides, there is meta-heuristic algorithm (MA) that mimics their search strategy from insects and animals such as Artificial Bee Colony (ABC) [18], Moth-Flame Optimization (MFO) [19], Lion optimization algorithm (LOA) [20], Ant Lion Optimizer [21], Artificial Feeding Birds (AFB) [22]. There are hundreds of MA due to according to No-Free-Lunch (NFL) theory one MA can solve all optimization problems with different inputs and constraint with the same efficiency [23].

Meta-heuristic algorithms were succeeded to find the near optimum solution of many engineering applications in different domains such as Motor design [24, 25], Bioinformatics [26–28], Robot design [29], Solar Energy [30] and many others. Many choices are available to design the controllers with many meta-heuristics algorithms in the literature. Such as particle swarm optimization algorithm was used for finding the optimum design of PID controller in the AVR systems [31].

In this work, a meta-heuristic algorithm selects the best values of the three parameters of the PID controller to enhance the performance of the process' response which is shown in Fig. 2. The fitness function which is used to optimize the performance of the process is the minimum integral of absolute of error (IAE). IAE is the integral of subtraction of actual response (y) and desired response (h) over the time (t) and is described as in Eq. (1) [6].

$$IAE = \int_{0}^{\infty} |y(t) - h(t)| \quad (1)$$

The values of parameters of PID (K_p, K_i and K_d) which were calculated using ZN method [6] achieved high IAE. Therefore, the rule of meta-heuristic techniques to tune these parameter so that achieving better IAE (the minimum value).

As shown in Fig. 3 the solutions are modified using the modification strategy of the MA for many iterations. During each iteration the best solution K_g (K_p^g, K_i^g and

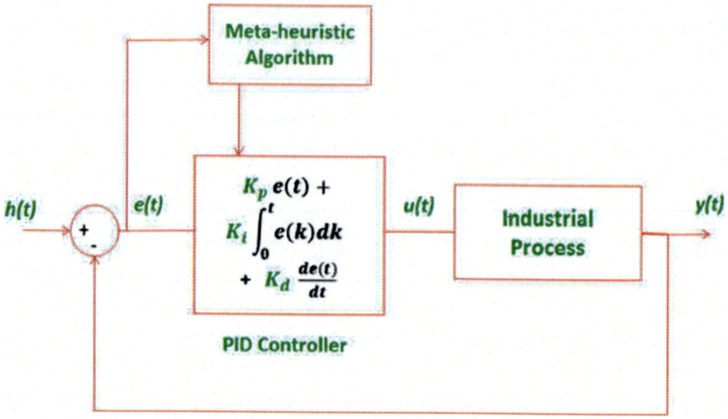

Fig. 2 Estimating of PID's parameters using meta-heuristic techniques

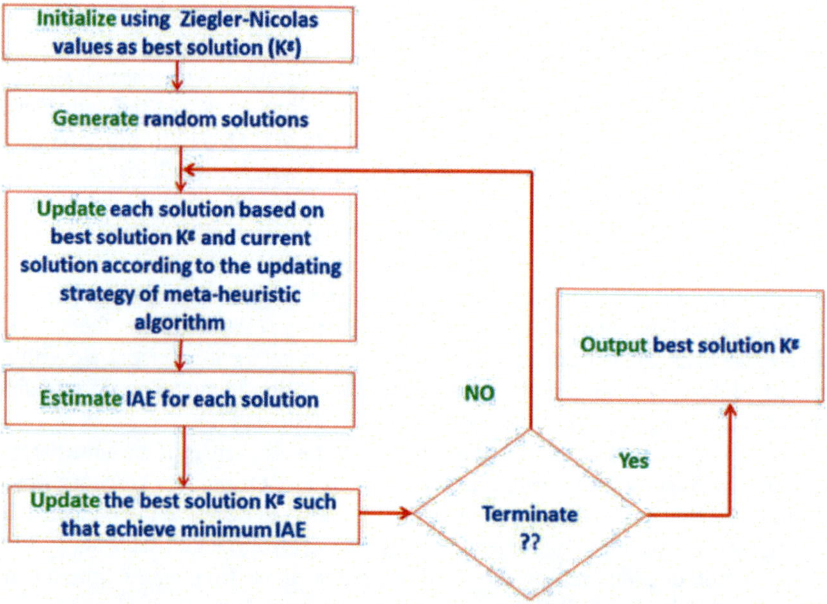

Fig. 3 The procedure of estimating the PID's parameters using meta-heuristic algorithm

K_d^g) which generate the best IAE is modified with the best solution generated during the updating of solutions.

In the literature, there are many promotion of MA for the optimum design of PID controller such as dynamic PSO (dPSO) [32], constrained PSO (CPSO) [33], Improved whale optimization algorithm (IWOA) [34] and opposition based Henry gas and solubility optimization algorithm (OBL-HGS) [35]. In addition, other various

MA were used for the best selection of the parameters of PID controller to promote the performance of DC motor such as Flower Pollination Algorithm (FPA) [36], Invasive Weed Optimization (IWO) [37], Grey Wolf Optimization Algorithm (GWO) [38] and FireFly [39]. Besides, voltage regulator is controlled using Teaching Learning Based Optimization (TLBO) algorithm [40]. DE and its improved version (PSO-DE) [41] was used to optimize three liquid level tank systems.

According to the theory of No-Free-Lunch (NFL) [23] which states that "no one optimization algorithm can solve all engineering problems with the same efficiency". So, improved versions of MA were released using hybrid meta-heuristics or embedding the operators inside the algorithms or to enhance the algorithms for certain engineering applications. The nature of the estimation of the parameters of the PID controller problem is numerical. Since the search space has a huge number of possible solutions which make a high probability of falling into local minima in addition to its non-linear behavior.

Hence, MA that is has an efficient exploitation and exploration feature is needed to produce better choice of the parameters of PID controller that gain better fitness more than that founded in the literature.

Arithmetic Optimization Algorithm (AOA) [42] is a population-based meta-heuristic algorithm that is based on the arithmetic operators for the search strategy? The main motivation for using AOA for estimating the PID controller's parameters is its advantages. The main advantage is its exploration scheme which has the efficient ability for visiting many regions of the search space. In addition, balancing between exploration and exploitation is another advantage that lead to avoidance of trapping in local optima.

The main contribution of this work is listed as following:

1. The optimum design of the PID controller using AOA algorithm.
2. A DC motor and three liquid level tanks system were used in the experimental tests.
3. AOA has the superiority over than the MA were used in the comparison.

The other parts of the chapter is organized as follows:

Section 2 describe AOA algorithm. In Sect. 3, PID's parameter tuning based on AOA technique was presented while Sect. 4 presents the experimental results and finally the conclusion was presented in Sect. 5.

2 AOA Algorithm

AOA uses the arithmetic operators (addition, subtraction, multiplication, and division) for updating solutions. For exploration of the search space it depends on multiplication and division operators while for exploitation it uses addition and subtraction operator. The control parameter that balances between diversification and intensification of the search space is Math Optimizer accelerated (M_{OA}) which is described in Eq. (2).

$$M_{OA}(t) = \frac{t}{T} \tag{2}$$

where t is the number of current iteration and T is the number of iterations.

The diversification of search space is performed based on division and multiplication operators due to the high distribution of the generated values which is represented according to Eq. (3) at condition of (rand > M_{OA}) where rand is a random generated number.

$$x_i^{(t)} = \begin{Bmatrix} best\left(x^{(t)}\right) \div (M_{OP} + \varepsilon)((UB - LB)\mu + LB)\ r_1 < 0.5 \\ best\left(x^{(t)}\right) * (M_{OP})((UB - LB)\mu + LB) \quad else \end{Bmatrix} \tag{3}$$

where x is the solutions, i is the index of solution (i = 1:N) best (x) is the best global solution, ε and μ are constants, LB and UB are the lower and upper bound of the solutions, r_1 is a random generated number, Math Optimizer probability (M_{OP}) is a scaling parameter that produce more exploration and is estimated according to Eq. (4) where α is a constant parameter.

$$M_{OP}(t) = 1 - \frac{t^{1/\alpha}}{T^{1/\alpha}} \tag{4}$$

Regarding the intensification of the search space the addition and subtraction operators are used due to the high dense of the generated solutions and is executed for the condition of (rand > M_{OA}). Equation (5) express the updating strategy during exploitation where r_2 is a random generated number.

$$x_i^{(t)} = \begin{Bmatrix} best\left(x^{(t)}\right) - (M_{OP})((UB - LB)\mu + LB)\ r_2 < 0.5 \\ best\left(x^{(t)}\right) + (M_{OP})((UB - LB)\mu + LB) \quad else \end{Bmatrix} \tag{5}$$

AOA's procedure is listed in Algorithm 1 as in Fig. 4.

3 PID's Parameter Estimation Based on AOA Algorithm

The integral of absolute of error (IAE) function which was described as in Eq. (1) was used as an objective function to calculate the parameter of PID controller based on the proposed AOA algorithm. Each search agent of AOA has a vector of three values (K_p, K_i and K_d) and each solution is initialized with a random value in the range of lower and upper bounds. The IAE function is estimated based on the signal e(t) for each search agent to evaluate the solutions. The best solution $\left(K_p^g, K_i^g, K_d^g\right)$ is determined based on the solution of search agents that achieved the minimum fitness.

Algorithm 1: The procedure of AOA

1: **Initialize** N, T, μ, α and ε
2: **Generate** random solution ($x_i, i = 1{:}N$) in the range (LB, UP)
3: **Estimate** the fitness of solutions
4: **Update** the global best solution $best\ (x)$
5: **While** T isn't reached yet
6: **Update** M_{OA} and M_{OP}
7: **If** ($rand > M_{OA}$)
8: **If** ($r_1 < 0.5$), update solutions using division operator in Eq. 3
9: **Else**, update solutions using multiplication operator in Eq. 3
10: **Else**
11: **If** ($r_2 < 0.5$), update solutions using subtraction operator in Eq. 5
12: **Else**, update solutions using addition operator in Eq. 5
13: **End If**
14: **Update** the global best solution $best\ (x)$
15: **End While**
16: **Output** the best solution

Fig. 4 The procedure of AOA algorithm

After finishing the iterations of execution, the best solution founded $\left(K_p^g, K_i^g, K_d^g\right)$ is delivered (Fig. 5).

The flowchart that presents the procedure of tuning the parameters of PID controller based on the proposed AOA algorithm is as shown in Fig. 6 while the procedure as described in algorithm (2) as in Fig. 7.

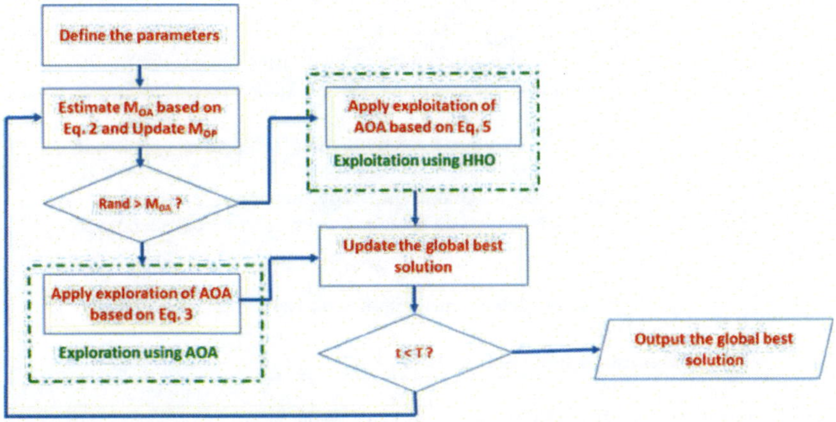

Fig. 5 The flowchart of AOA algorithm

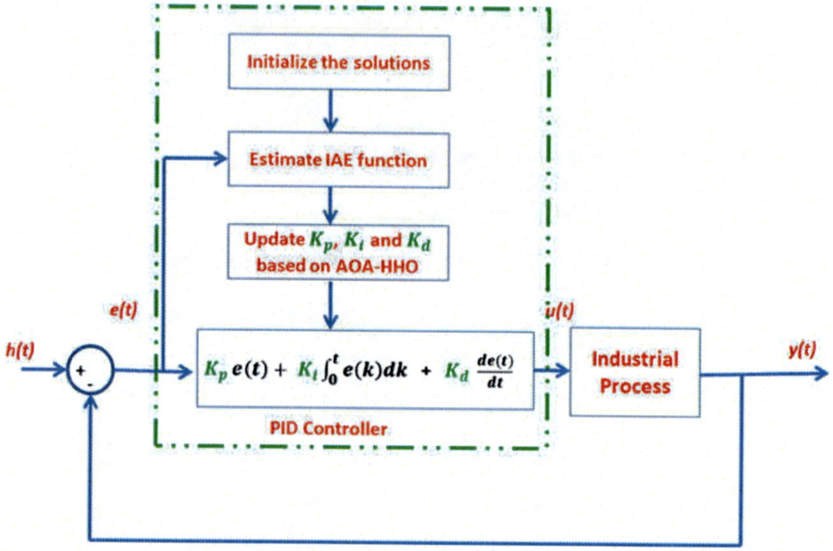

Fig. 6 Tuning PID's parameters based on the AOA algorithm

Algorithm 2: Parameters Estimation of PID Controller using AOA

Initialize N, T, μ, α and ε
Generate random solution (X_z $(K_p{}^z, K_i{}^z, K_d{}^z)$, $z = 1{:}N$), in the range (L_B, U_B)
Estimate the fitness (IAE) using Eq. 1 for each search agent ($z = 1{:}N$)
Update the global best solution X_g $(K_p{}^g, K_i{}^g, K_d{}^g)$ which produces the minimum fitness.
While T isn't reached yet
  Update M_{OA} and M_{OP}
  If ($rand > M_{OA}$)
  If ($r_1 < 0.5$), update solutions using division operator in Eq. 20
   Else, update solutions using multiplication operator in Eq. 20
  Else
   If ($r_2 < 0.5$), update solutions using subtraction operator in Eq. 5
    Else, update solutions using addition operator in Eq. 5
   If ($r_2 < 0.5$), update solutions using subtraction operator in Eq. 5
  End If
  Update the global best solution $best\ (x)$
End While
Output the best solution X_g $(K_p{}^g, K_i{}^g, K_d{}^g)$

Fig. 7 The procedure of Tuning PID's parameters based on the AOA

4 Experimental Results

The experimental tests were performed on two systems: the first one was control speed of DC motor, and it was common in many relevant studies [24, 34, 35, 37, 38, 43]. The second system was controlling the liquid level of three cascaded tank systems [41].

The experimental results were compared with relevant studies such as PSO, SCA, IWO, GWO, ASO and OBL-HGS. Besides, other meta-heuristics were not used to optimize PID controller were tested in this work such as Arithmetic Optimization Algorithm (AOA) and Covariance Matrix Adaptation Evolution Strategy (CMA-ES) [44]. The setting of parameters of AOA were listed in Table 1. The lower and upper bounds of the parameters are chosen based on the ranges were used in many related works [35, 45].

The fitness function that was used to evaluate the solutions based on IAE according to Eq. (1) and the measurement criteria that were used in the comparisons as following:

1. Integral absolute of difference error between actual and desired responses (IAE).
2. The step response characteristics as settling time, rise time and overshooting.
3. The frequency response of the systems.

4.1 Speed Regulator of DC Motor System

The speed regulation of electrical DC motor [35] was controlled through PID controller where meta-heuristic algorithms were used to select the best parameters that produce the optimum response. Figure 8 presents the block diagram of Speed Regulator of DC motor System where the disk controller is the PID controller.

The transfer function of closed loop speed controlling systems of DC motor as expressed in Eq. (6) [35].

Table 1 The parameters setting of AOA algorithm

The parameter	Value
The population (N)	100
Iteration Number (T)	20
Independent run number	20
Lower bound of (K_p, K_i and K_d)	[0.001, 0.001, 0.001]
Upper bound of (K_p, K_i and K_d)	[20, 20, 20]
μ	0.5
α	4.5
ε	2

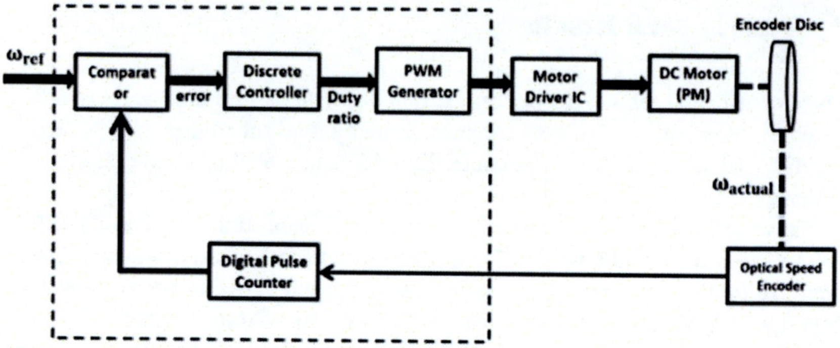

Fig. 8 Block diagram of speed regulator of DC motor system

Table 2 The specification of the step response and IAE for various meta-heuristic algorithms

Method	K_p	K_i	K_d	Set time (sec)	Rise time (sec)	Over-shoot %	IAE
PSO	1.5234	0.4372	0.0481	0.3549	1.8016	24	12.36
SCA	4.5012	0.5260	0.5302	0.2037	0.4900	2.36	13.63
GWO	6.898	0.5626	0.9293	0.1388	0.2053	1.5	10.99
IWO	1.5782	1.3801	0.0159	0.4190	1.2533	6.7	18.55
ASO	11.943	2.0521	2.4358	0.0692	0.1535	0	22.27
OBL-HG	16.9327	0.9508	2.8512	0.0546	0.0949	0	21.58
AOA	12.1694	0.3040	1.6103	0.1248	0.0839	1.972	9.29

$$G_1(S) = \frac{15}{1.08s^2 + 6.1s + 1.63} \tag{6}$$

Table 2 presents the best values of PID controller's parameters to enhance the speed regulation of DC motor using AOA versus other relevant algorithms were used in the comparative study such as PSO, SCA, GWO, IWO, ASO and OBL-HG.

Besides, IAE and specification of step response were tabulated in Table 2. As shown, AOA has the minimum IAE which proves its superiority over other algorithms to achieve a step response with lowest sum of absolute error between the actual and desired response.

The step responses the closed unity feedback DC motor regulation using the best values founded by AOA versus that of other algorithms are as shown in Figs. 9 and 10. As shown, AOA produce a reasonable response better than that of other algorithms.

Figures 11 and 12 show the bode plots of DC motor regulation using PID controller where its parameters are computed using the proposed AOA algorithm and the algorithms of relevant study.

Figure 13 shows the convergence curve of AOA algorithm for estimating the speed response of feedback DC motor.

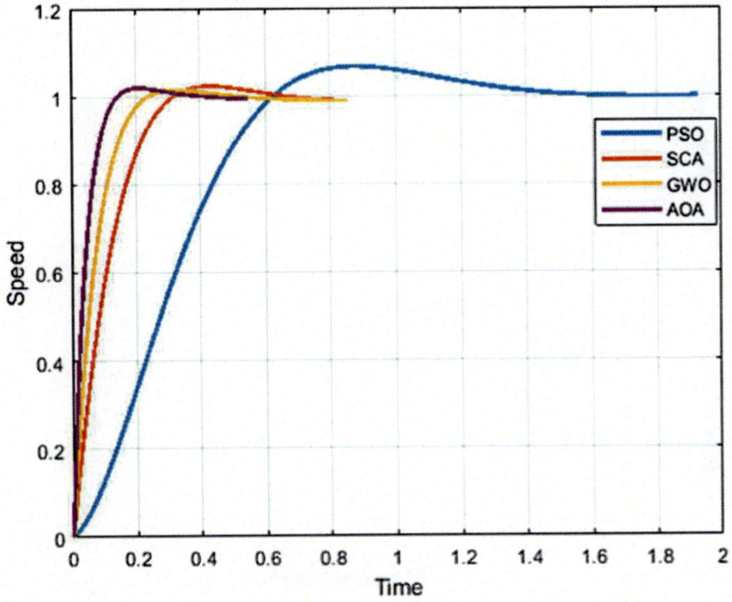

Fig. 9 DC motor speed response versus time in seconds

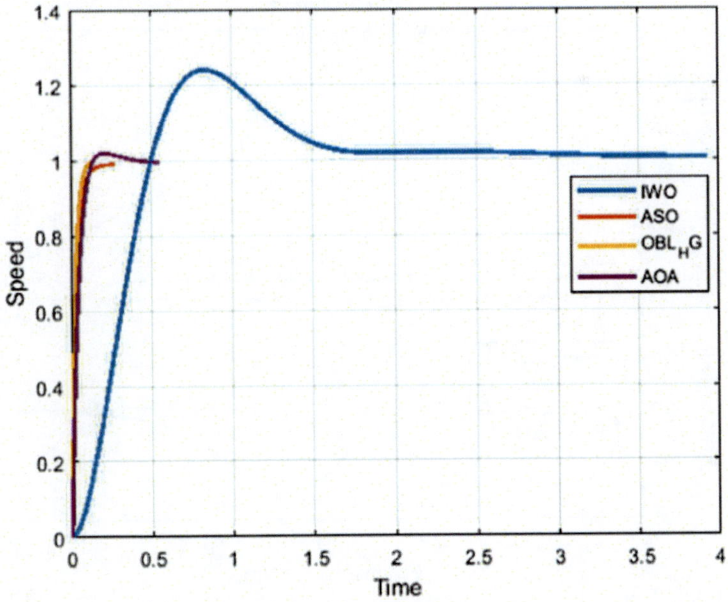

Fig. 10 DC motor speed response versus time in seconds

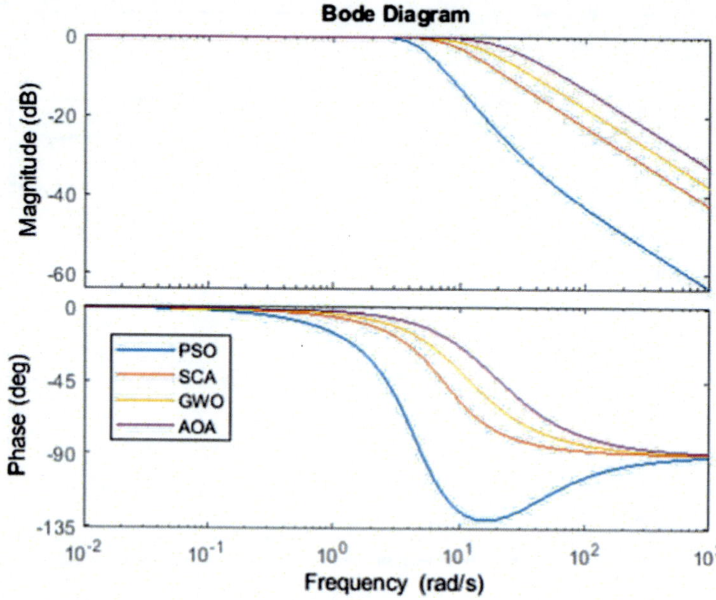

Fig. 11 Bode plots for DC motor based on PID controller

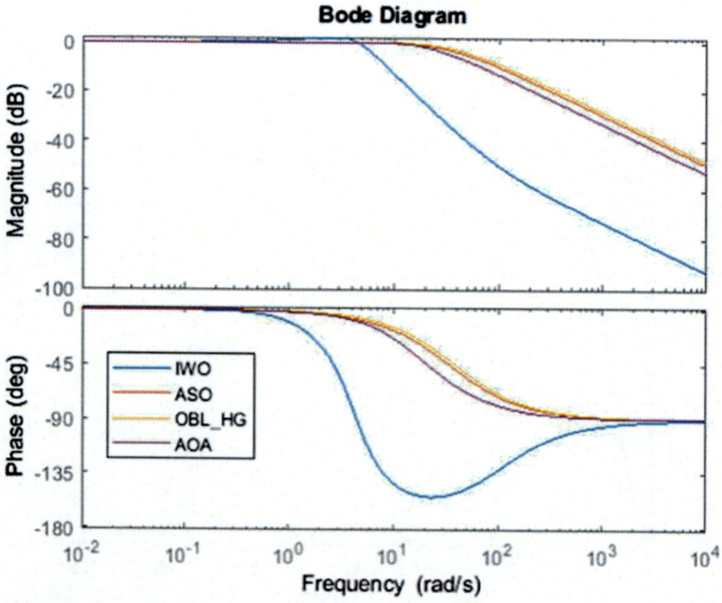

Fig. 12 Bode plots for DC motor based on PID controller

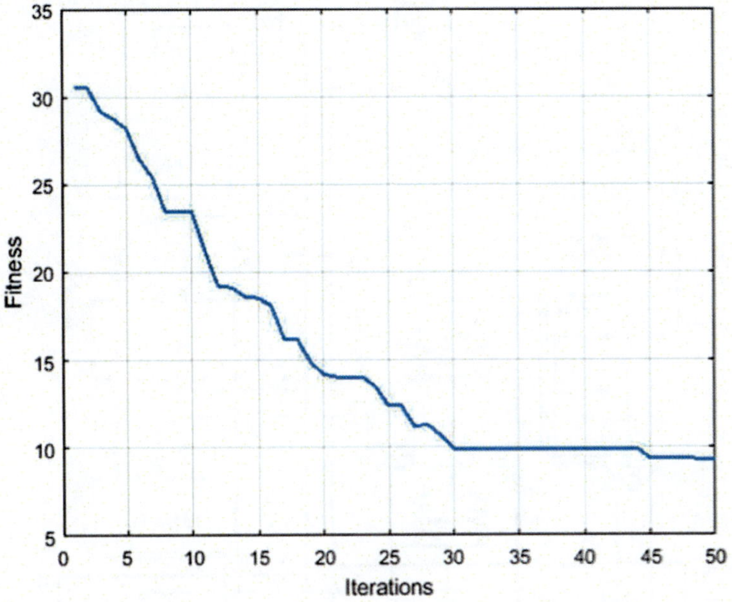

Fig. 13 Convergence curve of AOA algorithm

4.2 *Liquid Level Tank*

Three cascaded liquid level tank system was used to test the performance of improved DE with PSO [41] for estimating the parameters of PID controller as shown in Fig. 14. Equation (7) expressed the transfer function of the liquid level tank systems [41]. Table 3 presents the best estimated parameters of PID controller using AOA and relevant algorithms that were used in the experimental tests. Besides, the characteristics of step response were added in the table in addition to the fitness value (IAE).

$$G_2(S) = \left(\frac{1}{4s + 0.2}\right)^3 = \frac{1}{64s^3 + 9.6s^2 + 0.48s + 0.008} \tag{7}$$

The step responses the closed unity feedback Three liquid level tank systems using the best values founded by AOA versus that of other algorithms are as shown in Figs. 15 and 16.

Besides, Figs. 17 and 18 shows the bode plots of Three liquid level tank systems using PID controller where its parameters are computed using the AOA algorithm and the algorithms of relevant study.

As shown in the figures the superiority of the proposed AOA algorithm over the comparative algorithms in terms of IAE, magnitude and frequency.

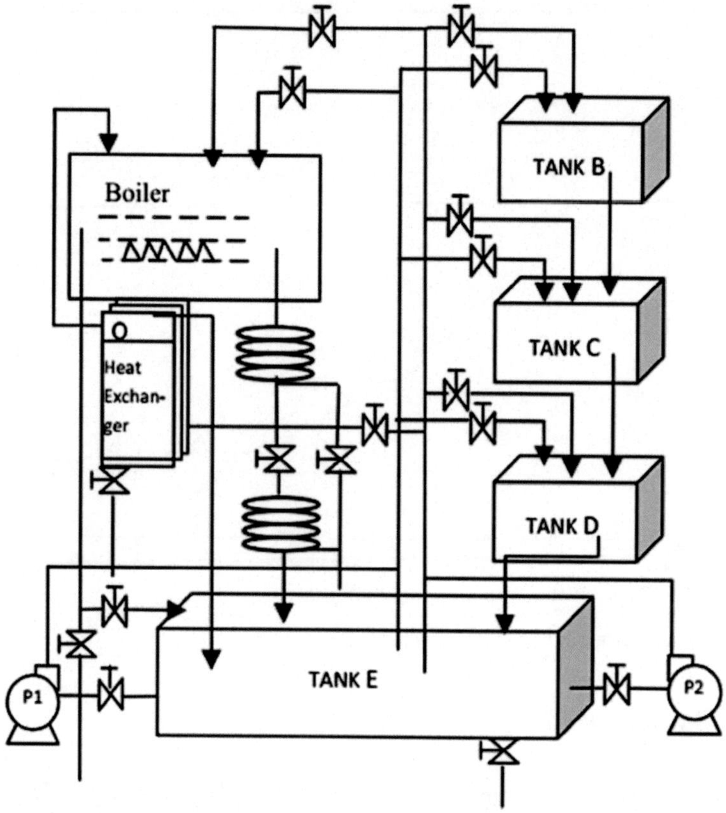

Fig. 14 Three tanks liquid level systems

Table 3 The specification of the step response and IAE for various metaheuristic algorithms for three liquid level process

Method	K_p	K_i	K_d	Set time (sec)	Rise time (sec)	Over-shoot %	IAE
PSO	0.606	0.0020	14.420	080.68	2.4078	67.86	18.61
SCA	0.303	0.1150	6.7231	207.18	3.5526	4.428	15.66
ASO	0.164	0.0040	12.922	070.10	2.637	57.26	14.70
CMA-ES	0.051	0.0010	0.3914	238.58	15.0019	50.08	14.27
AOA	0.102	0.0008	4.1844	161.06	5.0876	36.96	10.78

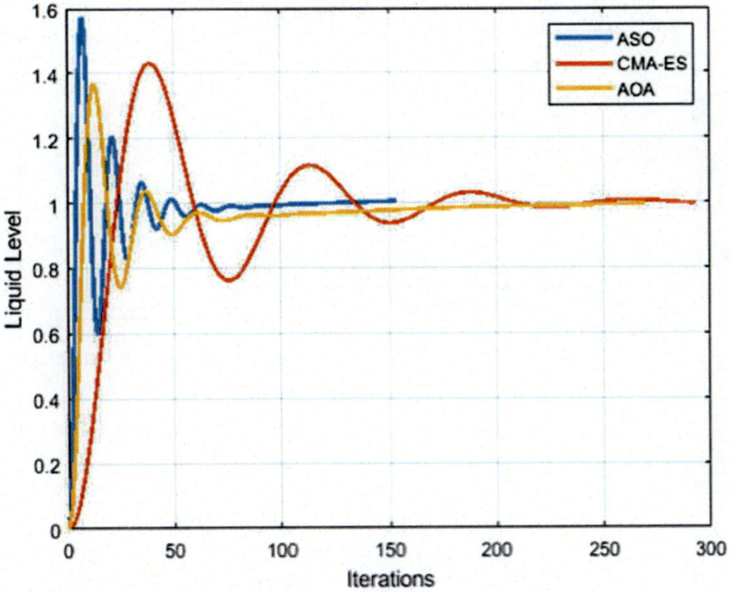

Fig. 15 Liquid level response versus time in seconds

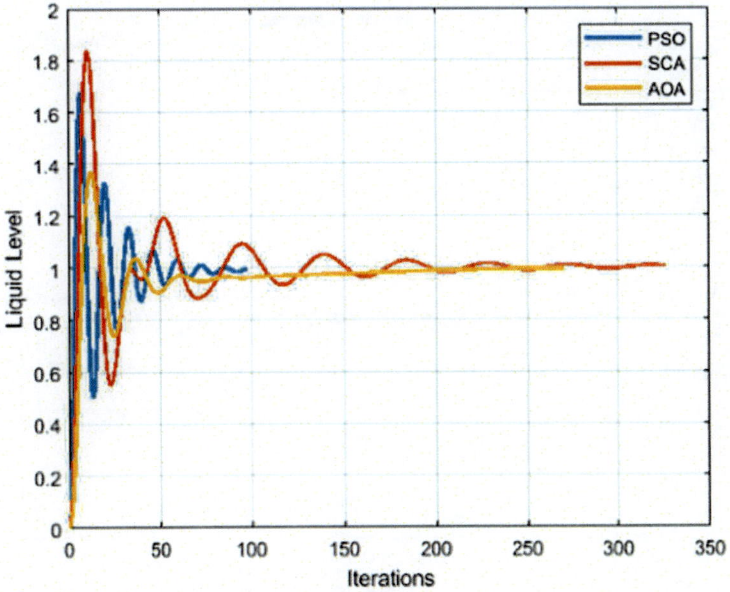

Fig. 16 Liquid level response versus time in seconds

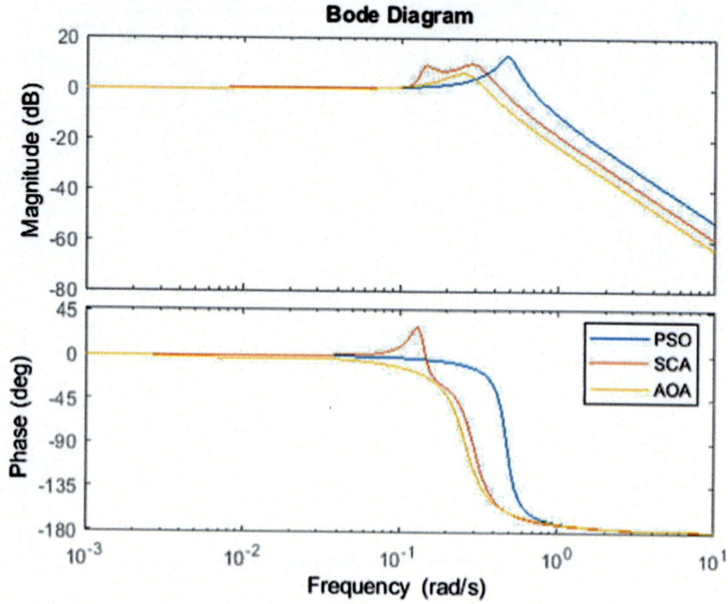

Fig. 17 Bode plots for three liquid tanks system based on PID controller

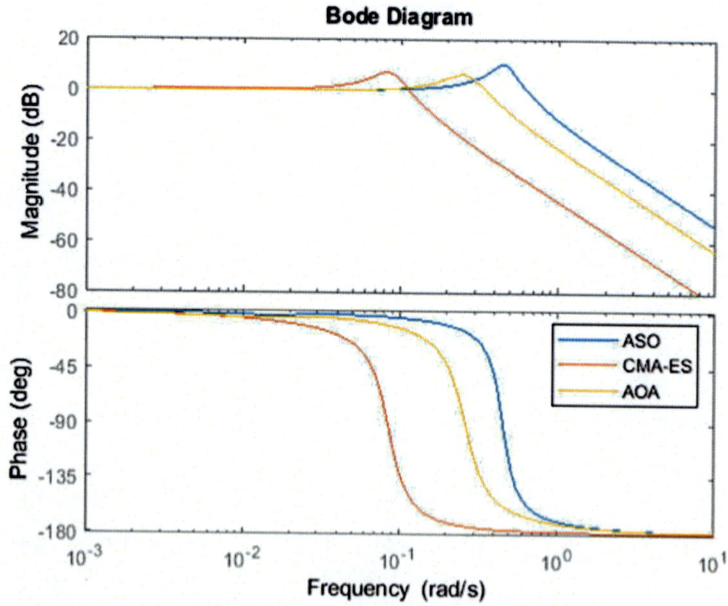

Fig. 18 Bode plots for three liquid tanks system based on PID controller

5 Conclusions and Future Work

In this work, a version of AOA was presented to optimize the selection of PID controller's parameters. AOA was tested for selecting the best parameters of PID controller for controlling two engineering applications are DC motor regulation and three cascaded liquid level tanks systems. The objective function was integral of absolute error function (IAE). AOA has the superiority over the relevant studies for founding the lowest IAE. In addition, the magnitude and frequency of the controlled systems were measured AOA beats all comparative algorithms.

References

1. M. Angurala et al., An internet of things assisted drone based approach to reduce rapid spread of COVID-19. J. Saf. Sci. Resil. **1**(1), 31–35 (2020)
2. E. Priyanka, S. Thangavel, X.-Z. Gao, Review analysis on cloud computing based smart grid technology in the oil pipeline sensor network system. Pet. Res. **6**(1), 77–90 (2021)
3. L. Jiang et al., An IoT-oriented data storage framework in cloud computing platform. IEEE Trans. Ind. Inform. **10**(2), 1443–1451 (2014)
4. C. Zhang, Y. Chen, A review of research relevant to the emerging industry trends: industry 4.0, IoT, blockchain, and business analytics. J. Ind. Integr. Manag. **5**(01), 165–180 (2020)
5. E. Priyanka et al., Integrating IoT with LQR-PID controller for online surveillance and control of flow and pressure in fluid transportation system. J. Ind. Inf. Integr. **17**, 100127 (2020)
6. K. Ogata, Y. Yang, *Modern Control Engineering*, vol. 4 (Prentice Hall, India, 2002)
7. T. Mansour, *PID Control: Implementation and Tuning* (BoD—Books on Demand, Norderstedt, 2011)
8. J.G. Ziegler, N.B. Nichols, Optimum settings for automatic controllers. trans. ASME **64**(11), (1942)
9. E.G. Talbi, *Metaheuristics: From Design to Implementation*, vol. 74 (Wiley, New York, 2009)
10. Kennedy, Particle swarm optimization. Neural Networks (1995)
11. S. Mirjalili, SCA: a sine cosine algorithm for solving optimization problems. Knowl. Based Syst. **96**, 120–133 (2016)
12. A.F. Nematollahi, A. Rahiminejad, B. Vahidi, A novel physical based meta-heuristic optimization method known as lightning attachment procedure optimization. Appl. Soft Comput. **59**, 596–621 (2017)
13. B. Javidy, A. Hatamlou, S. Mirjalili, Ions motion algorithm for solving optimization problems. Appl. Soft Comput. **32**, 72–79 (2015)
14. E. Rashedi, H. Nezamabadi-Pour, S. Saryazdi, GSA: a gravitational search algorithm. Inf. Sci. **179**(13), 2232–2248 (2009)
15. S. Rahmanzadeh, M.S. Pishvaee, Electron radar search algorithm: a novel developed meta-heuristic algorithm. Soft Comput. 1–23 (2019)
16. Y. Zou, The whirlpool algorithm based on physical phenomenon for solving optimization problems. Eng. Comput. (2019)
17. M. Jain, V. Singh, A. Rani, A novel nature-inspired algorithm for optimization: squirrel search algorithm. Swarm Evol. Comput. **44**, 148–175 (2019)
18. D. Karaboga, B. Basturk, Artificial bee colony (ABC) optimization algorithm for solving constrained optimization problems, in *Foundations of Fuzzy Logic and Soft Computing* (2007), pp. 789–798
19. S. Mirjalili, Moth-flame optimization algorithm: a novel nature-inspired heuristic paradigm. Knowl. Based Syst. **89**, 228–249 (2015)

20. M. Yazdani, F. Jolai, Lion optimization algorithm (LOA): a nature-inspired metaheuristic algorithm. J. Comput. Des. Eng. **3**(1), 24–36 (2016)
21. S. Mirjalili, The ant lion optimizer. Adv. Eng. Softw. **83**, 80–98 (2015)
22. J.-B. Lamy, Artificial feeding birds (AFB): a new metaheuristic inspired by the behavior of pigeons, in *Advances in Nature-Inspired Computing and Applications*. (Springer, Berlin, 2019), pp. 43–60
23. D.H. Wolpert, W.G. Macready, No free lunch theorems for optimization. IEEE Trans. Evol. Comput. **1**(1), 67–82 (1997)
24. T.C. Bora, L.d.S. Coelho, L. Lebensztajn, Bat-inspired optimization approach for the brushless DC wheel motor problem. IEEE Trans. Magn. **48**(2), 947–950 (2012)
25. M. Issa et al., PID controller tuning parameters using meta-heuristics algorithms: comparative analysis, in *Machine Learning Paradigms: Theory and Application*. (Springer, Berlin, 2019), pp. 413–430
26. M. Issa, M. Abd Elaziz, Analyzing COVID-19 virus based on enhanced fragmented biological local aligner using improved ions motion optimization algorithm. Appl. Soft Comput. 106683 (2020)
27. M. Issa et al., ASCA-PSO: adaptive sine cosine optimization algorithm integrated with particle swarm for pairwise local sequence alignment. Expert Syst. Appl. **99**, 56–70 (2018)
28. M. Issa, Expeditious Covid-19 similarity measure tool based on consolidated SCA algorithm with mutation and opposition operators. Appl. Soft Comput. **104**, 107197 (2021)
29. J.S. Pan et al., A multi-objective ions motion optimization for robot path planning, in *International Conference on Engineering Research and Applications* (Springer, 2018)
30. A.R. Jordehi, Parameter estimation of solar photovoltaic (PV) cells: a review. Renew. Sustain. Energy Rev. **61**, 354–371 (2016)
31. Z.-L. Gaing, A particle swarm optimization approach for optimum design of PID controller in AVR system. IEEE Trans. Energy Convers. **19**(2), 384–391 (2004)
32. S. Khubalkar et al., Modeling and control of a permanent-magnet brushless DC motor drive using a fractional order proportional-integral-derivative controller. Turk. J. Electr. Eng. Comput. Sci. **25**(5), 4223–4241 (2017)
33. A. Roy, S. Srivastava, Design of optimal PIλDδ controller for speed control of DC motor using constrained particle swarm optimization, in *2016 International Conference on Circuit, Power and Computing Technologies (ICCPCT)* (IEEE, 2016)
34. N. Razmjooy et al., Speed control of a DC motor using PID controller based on improved whale optimization algorithm, in *Metaheuristics and Optimization in Computer and Electrical Engineering*. (Springer, Berlin, 2021), pp. 153–167
35. S. Ekinci, B. Hekimoğlu, D. Izci, Opposition based Henry gas solubility optimization as a novel algorithm for PID control of DC motor. Eng. Sci. Technol. Int. J. **24**(2), 331–342 (2021)
36. D. Potnuru, K.A. Mary, C.S. Babu, Experimental implementation of flower pollination algorithm for speed controller of a BLDC motor. Ain Shams Eng. J. **10**(2), 287–295 (2019)
37. M. Khalilpour et al., Optimal control of DC motor using invasive weed optimization (IWO) algorithm, in *Majlesi Conference on Electrical Engineering, Majlesi New Town, Isfahan, Iran* (2011)
38. J. Agarwal et al., Analysis of grey wolf optimizer based fractional order PID controller in speed control of DC motor. Microsyst. Technol. **24**(12), 4997–5006 (2018)
39. O. Bendjeghaba, Continuous firefly algorithm for optimal tuning of PID controller in AVR system. J. Electr. Eng. **65**(1), 44 (2014)
40. S. Chatterjee, V. Mukherjee, PID controller for automatic voltage regulator using teaching–learning based optimization technique. Int. J. Electr. Power Energy Syst. **77**, 418–429 (2016)
41. A. Moharam, M.A. El-Hosseini, H.A. Ali, Design of optimal PID controller using hybrid differential evolution and particle swarm optimization with an aging leader and challengers. Appl. Soft Comput. **38**, 727–737 (2016)
42. L. Abualigah et al., The arithmetic optimization algorithm. Comput. Methods Appl. Mech. Eng. **376**, 113609 (2021)

43. S. Ekinci et al., Speed control of DC motor using improved sine cosine algorithm based PID controller, in *2019 3rd International Symposium on Multidisciplinary Studies and Innovative Technologies (ISMSIT)* (IEEE, 2019)
44. N. Hansen, A. Auger, CMA-ES: evolution strategies and covariance matrix adaptation, in *Proceedings of the 13th Annual Conference Companion on Genetic and Evolutionary Computation* (2011)
45. L. Syafaah et al., PID designs using DE and PSO algorithms for damping oscillations in a DC motor speed, in *2017 4th International Conference on Electrical Engineering, Computer Science and Informatics (EECSI)* (IEEE, 2017)

Testing and Analysis of Predictive Capabilities of Machine Learning Algorithms

Ganesh Khekare, Lokesh Kumar Bramhane, Chetan Dhule, Rahul Agrawal, and Anil V. Turukmane

Abstract The use of machine learning algorithms in last decade has been enormously increased. It has opened the door to several opportunities in various fields of research and business. However, identifying the appropriate algorithm for a particular program has always been an enigma, and that necessitates to be solved ere the development of any machine learning system. Let's take the example of the weather forecasting prediction system, it is used to identify the future weather prediction of a particular country or continent. Now, it is a daunting task to find the right algorithm or model for such a purpose that can predict accurate values. There are several other systems such as recommendation systems, sales prediction of a mega-store, stock prediction systems, or predicting what are the chances of a driver meeting an accident based on his past records and the road they've taken. These problem statements require to be resolved using the most suitable algorithm and identifying them is a necessary task. The objective of proposed system is to develop an interface that can be used to display the result matrix of different machine learning algorithms after being exposed to different datasets with different features. The system compares a set of machine learning algorithms while determining the appropriate algorithm for the selected predictive system using the required data sets. Stock market, earth and sales forecasting data is used for analysis. For experimental performance analysis several technologies and tools are used including Python, Django, Jupyter Notebook, Machine Learning, Data Science methodologies, etc. The comparative performance analysis of best known five time series forecasting machine learning algorithms viz.

G. Khekare (✉) · A. V. Turukmane
Parul University, Vadodara, India
e-mail: ganesh.khekare19325@paruluniversity.ac.in

A. V. Turukmane
e-mail: anil.turukmane21100@paruluniversity.ac.in

L. K. Bramhane
National Institute of Technology, Goa, India
e-mail: lokesh.bramhane@nitgoa.ac.in

C. Dhule · R. Agrawal
G H Raisoni College of Engineering, Nagpur, India
e-mail: rahul.agrawal@raisoni.net

E. H. Houssein et al. (eds.), *Integrating Meta-Heuristics and Machine Learning for Real-World Optimization Problems*, Studies in Computational Intelligence 1038,
https://doi.org/10.1007/978-3-030-99079-4_16

linear regression, K—nearest neighbor, Auto ARIMA, Prophet, and Support Vector Machine is done.

Keywords Best known machine learning algorithms · Survey · Experimentation · Performance comparison · Stock market prediction · Earth · And sales forecasting · Predictive analysis

1 Introduction

The system mainly concentrates on machine learning algorithms that are used in prediction modeling. Machine learning algorithms are self-programming methods to deliver better results after being exposed to data. The learning portion of machine learning signifies that the models which are build changes according to the data that they encounter over the time of fitting.

The idea behind the building of this system was to determine which one among the chosen time series forecasting algorithms are the most suitable for these operations. The uniqueness of this work is specified using the help of the literature review section of this study. The five algorithms that were chosen are Linear Regression, K-Nearest Neighbor, Auto ARIMA, Support Vector Machine, and Facebook's Prophet, which were never compared altogether on a common platform. Also, several datasets were extracted for building and testing these models, along with the evaluation metrics.

Since the extracted datasets are time-series forecasting types, that's why algorithms that are most suitable for these kinds of works are chosen in this system. The term time series forecasting means that the system is going to make a prediction based on time-series data. Time series data are those where records are indexed based on time, that can be anything like a proper date, a timestamp, quarter, term, or year. In this type of forecasting, the date column is used as a predictor/independent variable for predicting the target value.

A machine learning algorithm builds a model with the help of a dataset by getting trained and tested. The dataset is split into two parts as train and test datasets, and generally, the record of these two do not overlap, and there are different mechanisms around machine learning for this task. After fitting/training the model based on the train portion, it must be tested, and for that, the test dataset comes into play. Further, the results that are generated are matched with the desired targets with the help of evaluation metrics. The two-evaluation metrics viz., the Mean Absolute Percentage Error and the Root Mean Squared Error are considered for comparison purpose is broadly discussed in the Methodology chapter.

The last focus of this development is the interface that shows the most appropriate machine algorithm when they are built over different datasets for different purposes. Also, as a result, it determines the most appropriate algorithm for a specific database of required forecasting functions used in businesses and markets. This part of the system is implemented using a high-level Web-Framework called Django (based on Python) that encourages fast development, clean, and insolent design. Along with that, Cascading Style Sheets (CSS) is used with Hyper Test Markup Language to make visuals compatible attractive. Also, Bootstrap which is a known HTML, CSS, and JavaScript Framework, is used for adding several features such as navbar and hamburger to the interface.

The reason behind opting for this analytical work was due to the fact that during a study of data science models the team found difficulty in choosing appropriate algorithms for a particular problem statement. They were having trouble determining what algorithm would be the most preferable according to the provided dataset. And that wasn't the one-time problem, it also occurred several times in the past. So, the team decided to develop something or work on something that can solve this concern out. And the crew was sure that this problem must have transpired with several developers and machine learning practitioners in the past. The only way to solve this query was by doing a comparative analysis of a bunch of algorithms that were designed for the same purpose, and from here the idea of working on this problem statement came into consideration. Nevertheless, here in this system, five algorithms were chosen to be used, along with three unique time-series datasets and two evaluation metrics. Though the enormous advancement has been done still algorithms are not able to predict the things perfectly and that's the main motivation of doing this research work. A system must be there which will identify the best algorithm as per given data.

The upcoming sections of this study concentrate on the foremost purpose of this work, the reason for choosing this study along with mentioning every possible point regarding the implementation, the scope of the work, it's summarization, details about the hardware and software used. The study is going to address the methodology of the development of such an immense application. Also, it is going to explore the tools and other requirements collected for the same. Later, this study will inform the readers about the design and implementation of the final system. Further, it will proceed to the conclusion the team reached after the development of the system along with meaningful results.

2 Literature Review

This section of the study delivers the opinion and conclusion of several researchers who contributed their works to the field of machine learning algorithms. Also, this section manifests the comparative outcomes of the machine learning algorithms that are mentioned in the previous section. Literature review refers to the idea of reading the previous works related to the field of study and try to understand what fundamentals have been taken into the consideration to deliver useful conclusions. All conclusions in the upcoming paragraphs are based on individual work of respective journal authors, and this study tried to present them in a way that it's going to help the team in further analysis.

Vansh Jatana mentioned in his paper Machine Learning Algorithms [1] that Machine Learning is a branch of AI which allows System to train and learn from the past data and activities. Also, it explores a bunch of regression, classification, and clustering algorithms through several parameters including the memory size, overfitting tendency, time for learning, and time for predicting. In the comparison of Random Forest, Boosting, SVM, and Neural Networks, the time for learning is weaker in the case of Linear Regression. Also, like Logistic Regression and Naive Bayes [2], the overfitting tendency of Linear Regression is low. However, in the research Linear regression is the only pure regression model, as else are Classification as well as Clustering model too.

Another journal [2] studies the significant machine learning algorithms that are applied to sample datasets taken from the medical domain. It explores algorithms such as Random Forest, Decision Tree, K-Nearest Neighbour (KNN), Naive Bayes, Support vector machine, K-Means, Apriori, Reinforcement Learning, and PCA. And among, them KNN is the algorithm that is implemented in the system. KNN is a supervised non-parametric learning algorithm that is utilized for both regression and classification. On comparing the supervised learning algorithm based on performance, KNN showed the best result among all with an accuracy of 80.52%, and the lowest training time 0.0009 s. However, it's prediction time was largest at 0.003 s.

Ariruna Dasgupta and Asoke Nath [3] discuss the broader classification of a prominent machine learning algorithm in their journal and specifies the new applications of them. In supervised learning, priori is necessary and always produces the same output for specific input. Similarly, Reinforcement learning requires priori too, but the output changes if the environment doesn't remain the same for a specific result. Nevertheless, Unsupervised Learning doesn't require priori.

Talking about Auto ARIMA, Prapanna Mondal, Labani Shit, and Saptarsi Goswami [4] in their paper carried a study on 56 stocks from 07 divisions. Stocks that are registered in the National Stock Exchange (NSE) are considered. The authors have chosen 23 months of information for the observational research. They've calculated the perfection of the ARIMA model in prediction of stock costs. For all the divisions, the ARIMA model's accuracy in anticipating stock costs is higher than eighty fifths, which symbolizes that ARIMA provides sensible accuracy. If they tend to address concerning specific divisions, statement stocks in the FMCG division victimization ARIMA model offer results with the most reliable accuracy.

However, the predictions' accuracy for the banking division and automobile division victimization ARIMA model is comparatively less than that of the other divisions. Therefore, they proposed a more potent model concerning statement stocks of the businesses in the said sector. Concerning the IT sector, the quality deviation isn't too low or not too high, whereas they tend to have gotten AN higher than the ninetieth accuracy in prediction for this sector. This prompts the conclusion that ARIMA is a lot reliable for big numeric data prediction or stock analysis.

A work by Kemal Korjenić, Kerim Hodžić, and Dženana Đonk [5] evaluates its performance in very real-world use cases. The prophet model has inclinations of generating conventional monthly as well as quarterly forecasts. Also, as an enormous potential for classification of the portfolio into many classes consistent with the expected level of statement authenticity: some five-hundredths of the merchandise portfolio (with large amount of dataset) will be projected with MAPE < 30% monthly, whereas around 70% can be predicted with MAPE < 30% quarterly (out of that 40% with MAPE < 15%).

It's necessary to say that these some four-hundredth of the merchandise portfolio which may be forecasted with MAPE < 15 August 1945 quarterly are mostly the simplest mercantilism things of the said retail company, with quite eightieth of annual share (financial) within the whole portfolio. Supported those facts, the achieved results are quite satisfactory in the real-world situation of sales statement. Hence, Prophet is adequate for the short and middle interim prediction.

Sibarama Panigrahi and H.S. Behra [6] used FTSF-DBN, FTSF-LSTM, and FTSF-SVM models as comparative algorithms for their Fuzzy Time Series Forecasting (FTSF) in their journal. These Machine learning algorithms are used model FLRs (Fuzzy Logic Relationships. The paper concluded that FTSF-DBN outperformed DBN (Deep Belief Network) method. But it also reported that the statistical difference between FTSF-LSTM and LSTM is insignificant. However, FTSF-SVM had better results than SVM, and FTSF-SVM statistically outperformed it. Finally, they concluded that according to their results, FTSF algorithms provide statistically fitter or commensurate outcomes against their crisp equivalents.

Forecasting Stocks [7] used the LSTM algorithm model approach to predict the price of the stock and as a tool to generate a future insight into the stock of their interest. The paper also sighted that they were able to generate good results when they were considering 4 cases namely High, Low, Open, and Close. The paper showed that the algorithm was able to deal with the stock market and make a prediction of the stock price with great accuracy.

Comparative Analysis of Time-Series Forecasting Algorithms for Stock Price Prediction [8] foretells the average stock price for five datasets by employing the historical stock price data ranging from April 2009 to February 2019. Auto-Regressive Integrated Moving Average (ARIMA) model is used to produce the baseline, while Long Short-Term Memory (LSTM) networks are applied to produce the forecasting model for prophesying the stock price. They found that on large data samples, the ARIMA model outstands other models based on performance. However, in the context of one years' worth of data samples, the attention mechanism improves the accuracy of the LSTM Model and exceeds the ARIMA Model. They also concluded that the LSTM model is a much more powerful model than most machine learning algorithms.

Another paper [9] showed the full-detailed process for the application of the ARIMA algorithm for the prediction of the stock price. The detailed finding of the application of this algorithm on the model let's consider stock price prediction was very satisfactory when being considered for a short time. Though it can't be used for a long-term prediction, but for the short-term, it can play a very crucial role in the prediction of the stock price and forecasting for the investors. The paper also made a test model on the stock of Nokia mobile in which the algorithm performed quite well and showed minor deviation from real value.

Talking about K-Nearest Neighbour (KNN), it has been stated in a paper [10] that KNN as a data mining algorithm has a broad range of use in regression and classification scenarios. It is mostly used for data Mining or data categorization.

In Agriculture, it can be applied for simulating daily precipitations and weather forecasts. KNN can be used efficiently [11] in determining required patterns and correlations between data. Along with those other techniques such as hierarchical clustering and k-means, regression models, ARIMA, [12] and decision tree analysis can also be applied over this massive field of exploration. Also, KNN can be applied medical field in order to predict the reason for a patient's admission to the hospital. This massive range of applications of KNN makes it a worth-researching algorithm, and that's why the team decided to include it in the list of the comparative algorithm [13].

In the last few paragraphs, reader can observe that several researchers tried to present their working to show some useful results. Some of the journals were focusing on the individual machine learning algorithms that are used for time-series forecasting, while other work shown a relative comparison between number of algorithms and tried to conclude which one was the best among all. Whatsoever, these all algorithms have different fundamental of working and blending the environment, and it all require an ample amount of time to study. The methodology section of this documentation discusses all the chosen machine learning algorithms in brief, and it tried to make reader comfortable with all the different techniques that are being used.

In the end, the whole analysis of the different journals published in recent years features a broad perspective of different machine learning algorithms specifically time series and prediction algorithms, that are about to be featured in the implementation of this system. Also, from the above study, it can be concluded that each algorithm belongs to different categories and have significant applications. Further, some of the comparative studies define the best machine learning techniques based on several parameters. Nevertheless, in this whole process of encountering the brilliant works, team never came across any work where five algorithms that they've chosen being compared in on one platform with common dataset, and that's why the team saw this as an opportunity to compare these five algorithms that are different nature but also share some similarities so that they can be used for time series forecasting as well. The upcoming section is going to discuss all these algorithms and the two-evaluation metrics brief, and later, reader will find the implementation of them in brief.

3 Methodology

The idea was to create an interface that could display result matrix and multiple analysis with words, numbers, statistics, and pictorial representations [14]. The visual interface created by the team should not deviate from the topic for the audience and should only include limited and necessary items such as what algorithms are used, what dataset are used, their data analysis and respected comparative results. Anyway, the construction of the interface was the ultimate concern in the entire research and system construction campaign. To build the interface, the team needed statistics to be displayed on a computer screen, in short, they needed the required results [15].

The results [16] required for any production need for the models were to be developed. As the title of the work suggests that the system focuses on machine learning algorithms, but in the later section it is clarified that the actual focus is on the Time Series Forecasting type [17]. Time Series Prediction means that it predicts something based on the time as its index feature. Time series datasets can be associated with sales, stocks, weather, natural disasters, etc. containing historical records with some minor errors [18].

But the aim was not to predict the results, the aim of the work was to find out which machine learning model fits best in the datasets used. After going through several previous works mentioned in the last chapter of the document, the team found that none of the algorithms the team had chosen are compared on a common platform. Also, it was best to compare a few unmatched algorithms to produce the best fit, and to determine what are the nature of the individual datasets. The selected algorithms are Linear Regression, K-Nearest Neighbour (KNN), Auto ARIMA, Facebook's Prophet, and Support Vector Machine. These algorithms and their functionality will be discussed in this section later, but before that is necessary to know why these algorithms were chosen.

In fact, the second reason is what type datasets are extracted in the development path of this study. The first of the three databases were named 'Tata Global Beverages Limited' published by the National Stock Exchange (NSE). The database has eight positive features set in data analysis and model design. Also, it contained two hundred and twenty records that will be used for model training and testing and to produce results. Contains features such as Date, Open, Up, Down, Storage, Close, Trade, and Revenue. By comparing all variables except the Total Trade against Date on one site, the same trend can be found in the end. Therefore, the group decided to select any of the variables as targeted variables.

The second data is called the Earthquake dataset, which describes important earthquakes that occurred between 1965 and 2016, published in the US Geological Survey. The National Earthquake Information Center decides the area occupied and the extent of all significant earthquakes that are occurring globally and broadcasts the erudition straight to world organizations and countries, science organizations, and interested people. This database is published from Kaggle and incorporates a history of the date, time, intensity, magnitude, place, and origin of all earthquakes reported at 5.5 or higher since 1965.

The team used two variables with each other to predict the time series database. The following analysis shows the different parameters and relationships between the two variables that are used in the system. Although the database was empty, the team used python's library reverse_geocoder to measure it.

However, there were only two options for choosing target variables, either Depth or Magnitude. The relationship between day/year and depth did not appear in categories or lines. The Magnitude variation compared to the day turns out to be in phases naturally and can be selected as the target modelling value. The third database used in the system is a sales database called the 'Superstore Dataset' released from Tableau (donated by Michael Martin). The database is a record for the sale of stationery and other products in the United States and Canada from 2014 to 2017 with a record sale of 9995 sales and has 20 different features. Although there were 20 variables, there were only two options for the variables to be targeted, either Sales or Quantity because Profit always seemed inconsistent. But later, further analysis concluded that the relationship between Quantity and Day was significantly different than Sales and Day. Therefore, Quantity has been selected as an independent variant of this section of work.

After that, it was very important to clean and adjust the data sets before building individual models. Database data clearly shows that there are a few missing records and some unusual data in a few rows. The concept of data analysis states that one must remove or alter those records to suit the models. Since the Earthquake dataset contained records of disasters since 1965, some records were blank and vague. There were different types of dates in the date column. To avoid such confusion, the team had to store the data in a standard format i.e., dd-mm-yyyy. Similarly, the team performed equal work on all the other databases that were extracted. Besides, it was preferred to eliminate such empty records as it would not significantly affect the outcome. In the end, the idea was clear that only two variables are going to be used in each case.

Processing involves setting the appropriate Database Date as their reference and arranging them in order. All these actions were performed on Python3 with the help of various libraries such as Pandas, Numpy, Scikit-learn, and time data. It was very important to create a 'date' map in ordinal as it plays an important role as the dependent variable (datetime.toordinal () returns the date count for a specific date in standard format). After that, it was all up to the implementation of the pre-selected models for the system. The data needed to be analysed before it could be done to produce useful comparison results. These are my machine learning algorithms as follows:

Linear Regression: Linear regression is a simplistic and well-known Machine Learning algorithm. It is a mathematical procedure that is applied for the prognosticative analytical study. Simple Linear regression delivers forecasts for continuous or numeric variables like trades, wages, span, goods worth, etc.

It manifests a linear correlation between a target and one or more explanatory variables. Since the confers the linear correlation, which means it determines how the value of the experimental variable is altering with the value of the predictor variable.

Mathematically, it can be represented as shown in Eq. 1,

$$y = \theta_0 + \theta_1 x_1 + \theta_2 x_2 + \ldots + \theta_n x_n \tag{1}$$

Here, y is the target variable and $x_1, x_2, \ldots, x_n$ are predictive variables that represents every other feature in a dataset. $\theta_0, \theta_1, \theta_2, \ldots, \theta_n$ represent the parameters that can be calculated by fitting the model.

In the case of using two variables i.e., 1 independent and 1 dependent variable, it can be represented as shown in Eq. 2,

$$y = \theta_0 + \theta_1 x \tag{2}$$

where parameters θ_0 is said to be the intercept that forms on y-axis, and θ_1 can be generated once the model is trained.

K Nearest Neighbour: K-Nearest Neighbour calculates the similarity among the recent data and recorded cases and sets the new records into the section where alike data exists. It reserves all the recorded values and assigns the fresh record notch based on the resemblance. When the fresh records arrive; it can be readily organized into a prim fitting category through applying the algorithm.

It computes the length between the input and the test data and provides the prognostication subsequently as shown in Eq. 3.

$$d(p,q) = d(q,p) = \sqrt{(q1-p1)^2 + (q2-p2)^2 + \ldots + (qn-pn)^2} = \sqrt{\sum_{i=1}^{n}(qi-pi)^2} \quad (3)$$

The n number of specifications are taken into consideration. The marking that is situated at the merest position from marking is in similar class. Here q and p are new and existing data-points respectively.

Auto ARIMA: ARIMA is a standard word that refers to Auto-Regressive Integrated Moving Average. It is a mere and efficient ML algorithm used to perform time-series forecasting. It consists of two systems Auto Regression and Moving average. Auto ARIMA is a time series model that employs former records to the regression formula to prognosticate the following values for the succeeding action. In short, it makes regression in the earlier time trail t-1 to foretell t. MA (moving average) is used for measuring the mean in a distinct time-period and apportioning it by the total number of frames reaped.

It takes past values into account for future prediction. There are 3 essential parameters in ARIMA:

p = > historical data used for predicting the upcoming data.

q = > historical prediction faults i.e., used for forecasting the Upcoming data.

d = > Sequence of variation.

However, the parameter tuning for ARIMA utilizes a lot of time, but in Auto ARIMA, the model itself generates the optimal p, d, and q values for the data set to provide better predicting.

Prophet: The prophet is an open-source library by FB company made for predicting time series data to learn and likely forecast the exchange. This depends on the non-linear trends that are being fitted over the sections. The trend is the inclination of the data to enhance or diminish over a while and it removes the periodical variations. Seasonality variations occur over a short duration and aren't notable enough to be described as a trend. The broad approach of the model is like a generalized additive model. The equations related to the terms are defined as shown in Eq. 4,

$$\mathrm{fn}(t) = g(t) + s(t) + h(t) + e(t) \quad (4)$$

where,

g(t) = > trend.

s(t) = > seasonality.

h(t) = > forecast effected by holidays.

e(t) = > error term.

fn(t) = > the forecast.

The variation of the given terms is maths dependent. And if not studied properly it might lead them to make the wrong prediction which may be very problematic to the customer or for business in practice.

Prophet let us use two different models, the logistic growth model, and a piece-wise linear model (by default). Selecting a model is important because it depends on a lot of things like business size, rate of growth, business model, etc. On the off chance that the data to be estimated has immersing and non-direct information (becomes non-straight and after accomplishing the immersion, shows almost no blast or cut back and best popular a couple of occasional changes), at that point the calculated blast variant is the top-notch alternative. By and by, if the data demonstrates direct homes and had blast or scaled back advancements inside the past, at that point a piece-reasonable straight form can be a more prominent functional decision.

Support Vector Machine: The SVM [15] is a machine learning algorithm that is employed for both regressions and classifications depending upon the enigmas. Here, one creates a hyperplane depending upon the number of features; if there are two features available, then straight line can be used to create a hyperplane. If there are three features, then one can use the full three-dimensional plane to create a hyperplane. The boundary of this hyperplane is created using positions or vectors that are known as support vectors for the plane. And put the new point being generated by us on the hyperplane.

In Linear SVM, features are linearly arranged that can utilize a simple straight line to implement SVM in this case. The formula for obtaining hyperplane in this case is as shown in Eq. 5:

$$y = mx + c \tag{5}$$

If the feature that is being used is of non-linear type, then more dimensions are needed to be added to it. And in that case, one need to use a plane. The formula for obtaining hyperplane in this case is as shown in Eq. 6:

$$z = x_2 + y_2 \tag{6}$$

After discussing the machine learning algorithms that were used, it is very essential to know how it is going to be evaluated. The evaluation tells how much accurate the model is, and how much good that model is.

In this system, to determine the accuracy, 2 evaluation metrics that are used for generating results are Mean Absolute Percentage Error and Root Mean Squared Error, and both depend on the obtained values and actual value. The Root Mean Squared

Error a.k.a. RMSE value is obtained by taking the square root of the addition of the individually calculated mean squared errors. The formula for the same is as shown in Eq. 7:

$$RMSE = \sqrt{\sum_{i=1}^{n} \frac{(\hat{y}_i - y_i)^2}{n}} \tag{7}$$

Here, ŷ1, ŷ2, ŷ3, ..., ŷn are the actual value and y1, y2, y3...yn are respective obtained value and n here is the number of iterations performed.

In MAPE or Mean Absolute Percentage Error, the value is calculated by taking absolute subtraction of obtained value from actual value divided by the actual value, later the individual value to obtain the result were added as shown in Eq. 8.

$$M = \frac{1}{n}\sum_{t=1}^{n}\left|\frac{At - Ft}{At}\right| \tag{8}$$

Here, A1, A2, A3, ..., An represents actual value, while F1, F2, F3, ..., Fn represents the obtained data, and n is the number of iterations taken under consideration.

4 System Design

The design of the whole system depends on the flow of modules. The work is segregated into six modules, and the team developed the whole system going through these six modules that are discussed in this section of the study. Figure 1 describes the modules and processes that are going to be involved in the long process of implementation of the required interface.

Data Requirements and Collection.

In this phase of the whole implementation, the main objective is to understand what kind of datasets are required in the massive process. Understanding the data requirements plays a vital role in upcoming modules in this long process.

Further, after understanding the data requirements, the next step is to focus on the collection of the required datasets. It gives a brief hint on what the model building process will take place. The three datasets that were mentioned in the previous section of this documentation are collected here.

Data Preparation.

This phase of the implementation is the most crucial. It let the implementer determine the bruises in the collected data. To operate with the data, it needs to be developed in

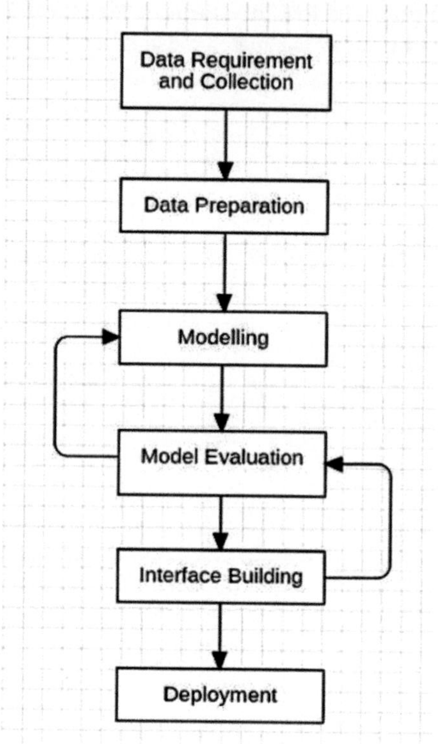

Fig. 1 Flow of modules

a way that inscribes abstaining or fallacious values and eliminates copies and ensures that it is accurately formatted for modeling.

Modelling.

In this module, the team implemented the algorithms as per the requirement in Python with the help of some Python libraries. It is the phase, that allows to decide how the information can be envisioned to find the solution that is required. All five algorithms which are either predictive or descriptive that are mentioned in the previous section were implemented here. Also, the team used the training set for predictive modeling, as they act as a guideline to decide if the model requires to be calibrated. Besides that, this module helped in processing different datasets through all the algorithms on the table.

Model Evaluation.

Model's assessment will probably assess the calculations that are actualized in the past module. It is intended to decide the right logical methodology or strategy to take care of the issue. Assessment permits the nature of the model to be gotten to and offers a chance to check whether it meets the underlying target of the structure of the necessary framework. Alongside that, it produces the necessary outcomes. With the help of RMSE, and MAPE, it can be determined which model is most suitable for a particular time series dataset. The closer the value of RMSE and MAPE towards zero, the better the model for that dataset.

Interface Building.

In this module, the work went under the interface development of the system. Also, the team established a connection between the interface and the models that were implemented in previous phases. Also, as per the requirement, the team can also revert to the fourth phase of the implementation. Django was used as the web-framework for this phase of the implementation. Where the interface named Nebula was built to display all the data analysis that was done prior to this phase, and the metric table that was generated. Also, it came to a conclusion tab where the team concluded the work by explaining all the final results Nebula consisted of three main tabs, Landing page describing what Nebula is, Dataset which shows a pictorial row of different datasets respectively linked to their own Analysis page, and Algorithm tab that describes everything about all the algorithm that was en-lighted in this system.

Deployment.

Once the models are evaluated and the interface is developed, it is deployed and put to the ultimate test. It showed required comparative results and satisfied the objective the team has taken prior to initiating the hands-on working on this system.

5 Experimental Results

The idea was to govern these five machine learning algorithms: Linear Regression, Auto ARIMA, K-Nearest Neighbor, Support Vector Machine, and Facebook's Prophet, by fitting three different datasets (Stock, Earthquake, and Sales) simultaneously, and finding out which one was the most suitable for Time Series Forecasting based on figures generated from the two-evaluation metrics: RMSE and MAPE. The first step in this system building was to understand the problem statement and work on the objective taken. After understanding the idea behind the study, the team had to start from the next phase that was non-other than data extraction and extraction.

The Methodology and Data Collection portion of this documentation describes all the datasets in detail, including the number of columns and rows, type of dataset, variables, and all. The dataset that is used are:

NSE—Tata Global Beverage.

Significant Earthquakes, 1965–2016 [US Geological Survey].

Superstore Dataset—Tableau.

Later, it all depends upon the data processing where elimination of unnecessary data and edition or omission of irregular/missing data takes place. As explained in previous sections, the preprocessing also involved setting the Date of the respective dataset as their indexes and sorting them in ascending order. These all actions are

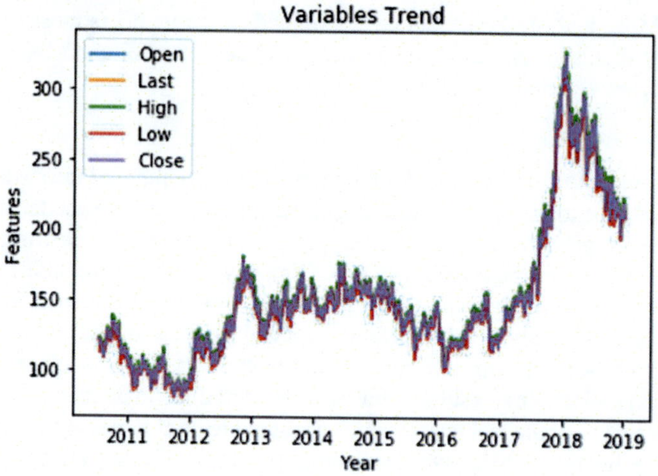

Fig. 2 Variables trend

taken place in Python3 with the help of various libraries such as Pandas, Numpy, sci-kit learn, and datetime. It is very crucial to map date to ordinal as it played an important role as the independent variable (datetime.toordinal() returns the day count of a particular date in standard format).

After that, it is very necessary to which feature of the respective dataset fits the best as the target variable against the Date variable of the respective variable. The stock prediction dataset consisted of several potential variables named Open, High, Low, Last, Close, and Total Trade. Total Trade was never an option for a perfect candidate for the target variable as it doesn't show any consistent trend as shown in Fig. 3.

However, on comparing other features against date, it turns out that all other variables Open, High, Low, Last, and Close follows the same trend i.e., proper overlapping on a line graph that is depicted in Fig. 2. So, eventually, the team decided to pick any of these features as target variables, and they went with Close. The relationship between Date and Close is depicted in Fig. 4.

Similarly, in the case of Earthquake Dataset, there were only two options to choose the target variable, either it could be Depth or Magnitude. The relation between Date/Year and Depth neither appeared categorical nor linear. The variation of Magnitude against Date comes out to be categorical in nature and can be chosen as the target value for the model building. Figures 5 and 6 depict the required relationship between Date and Depth, and Date and Magnitude respectively.

In the Sales dataset, despite having 20 variables, there were only two options to choose the target variable, either it could be Sales or Quantity because Profit always seemed inconsistent. But after, further analysis it comes to figure that relation between Quantity and Date has been highly categorical than Sales and Date. So, Quantity has been chosen as the independent variable for this phase, and the plot of the Fig. 7

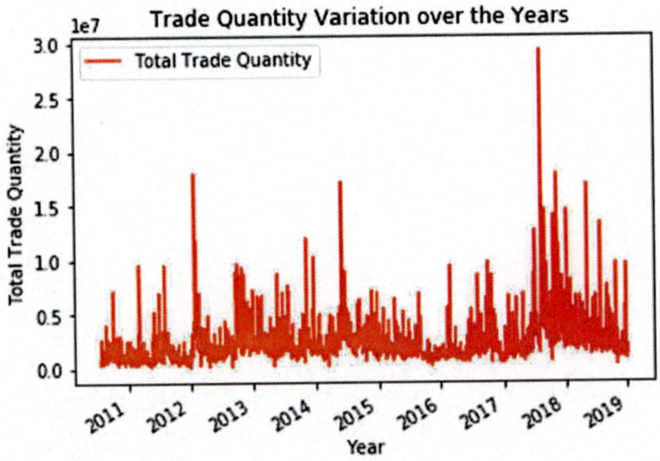

Fig. 3 Total trade versus year

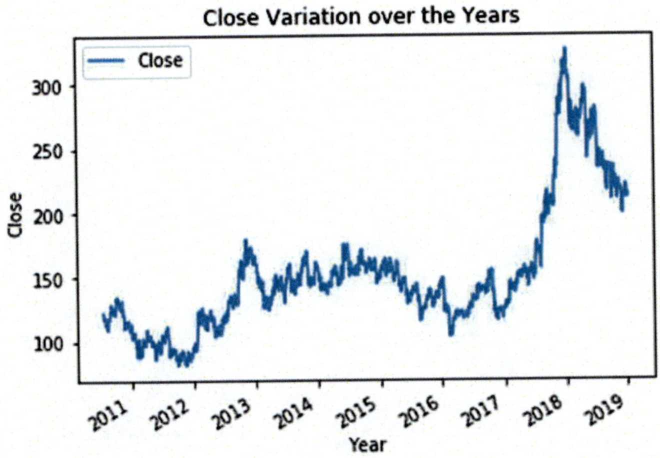

Fig. 4 Date and close relation

speaks it all.

Finally, the team had target variables for each dataset. Close, Magnitude, and Quantity came out to be the target variable for Stock, Earthquake, and Sales forecasting dataset respectively.

The next part of the implementation process was the respective model building, and the first thing in that was the initialization phase. The initialization phase was considered for each dataset model building individually and done under the file name 'initiate.py'. The role of this file is to set Date variable as an index, and map

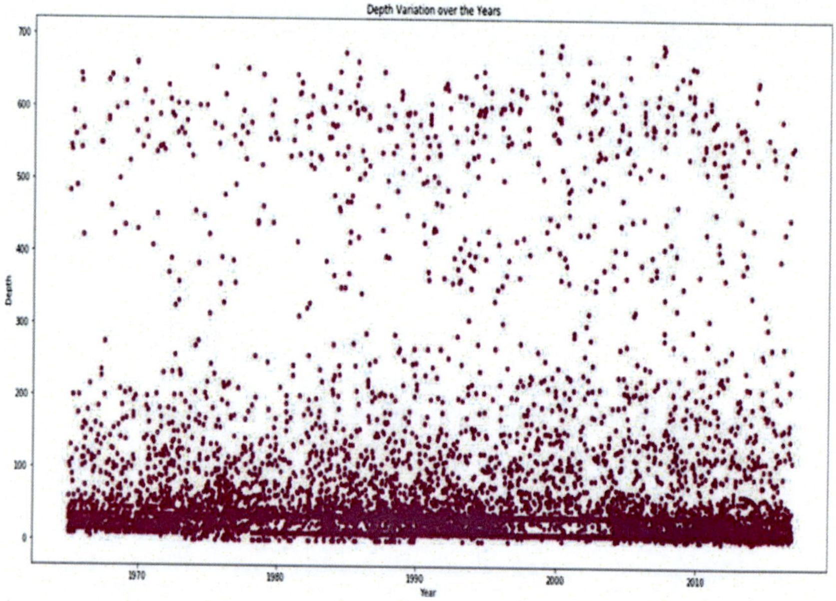

Fig. 5 Depth versus date

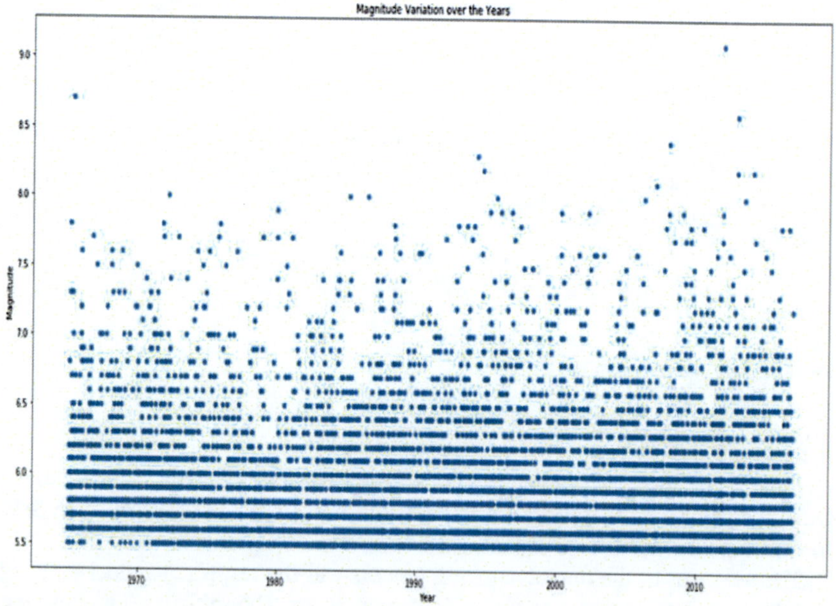

Fig. 6 Magnitude versus date

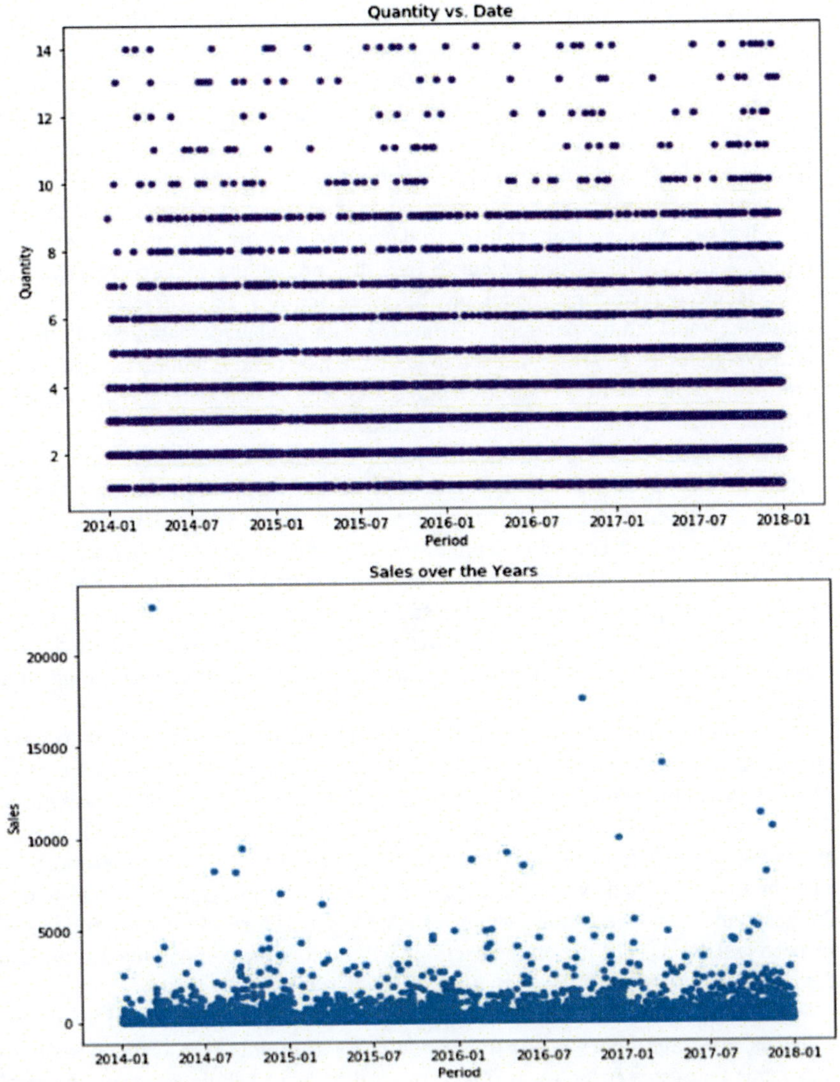

Fig. 7 Sales dataset trends

Date to ordinal as it played an important role as the independent variable (datetime.toordinal() returns the day count of a particular date in standard format). The 'initiate.py' also splits the dataset into train and test dataset which is going to be used in the development of the various models.

The requirements for this python file include pandas, NumPy, and sklearn, which are eventually going to be helpful for future implementation. After this whole lot of process, now the stick turns towards model building for the respective dataset. Nevertheless, the process of model building to same for all the dataset, but the only changes were the name of the dataset that was already imported in the respective initiate.py files. The lr.py is used for the implementation of Linear regression, and after model fitting, RMSE and MAPE were calculated and recorded. Similarly, the Support Vector Machine was implemented in psvm.py by importing SVR from sklearn.svm.

K Nearest Neighbor was implemented in knn.py, and it required importing the 'neighbors' module from sklearn to determine suitable data points. The GridSearchCV in the file assists to curve over the predefined hyperparameters and fit the required predicting model on the given training set. Therefore, one can choose the most fitting parameters from the cataloged hyperparameters. 'neighbors.KNeighborsRegressor()' is used to generate KNN and it will be used in model fitting using GridSearchCV() based on neighbors. And similar to the previous models, RMSE and MAPE were calculated and recorded.

Auto ARIMA was implemented in pmd.py using the 'pmarima' module and 'ndiffs' module from pmarima library, ndiffs does a unit root test to calculate the number of variations needed for time series x to be made stable. Later, calculated ndiffs is used for model building in pmarima.auto_arima(), and then after performing forecasting, the team compares both the results and actual value to record RMSE and MAPE value. Jupyter notebook was used in the implementation of the last model called Facebook's Prophet or simply the prophet. The prophet is imported from fbprophet library for this implementation, and the model has fitted accordingly, and then evaluation is performed.

After model building and recording all the statistics, the next job was to build an interface that can display the results. Django web-framework was used for this phase of the implementation. The interface was named 'Nebula' which was built to display data analysis and results. As mentioned in the previous section, it is built as it consisted of 3 different tabs including landing page, datasets, and algorithms. The interface was built with the help of HTML, CSS, and Bootstrap, and it was the destination after generating values of RMSE and MAPE of different models of respective datasets. The upcoming sections will discuss the comparative results and the conclusion.

The Django is initiated by creating a project name Nebula, followed by an app called the nova, and there was no use of the database in this interface. Under new directory name templates (which was linked in settings.py of nebula directory), several HTML files, and it was all going to display on the screen with the help of 'views.py' file. The static directory consisted of all the pictures, graphs, and CSS files for making the information attractive. And in the end, the project nebula was deployed after making some migrations, and the satisfactory result which was fitted while making the HTML codes were visible now.

As per the discussion, the results that need to generate were nothing else but the comparative results of the evaluation metrics value of the respective dataset. First in that trail was the Stock Prediction dataset, and Table 1 describes shows the comparative values for the same.

Table 1 Results of stock prediction dataset

Algorithms	RMSE	MAPE
Linear regression	47.51609	11.32705
K—nearest neighbor	65.11185	16.92529
Auto ARIMA	3.74366	0.72129
Prophet	53.01529	13.01318
Support vector machine	69.81082	12.44615

Table 2 Results of earthquake forecasting dataset

Algorithms	RMSE	MAPE
Linear regression	0.43306	2.49101
K—nearest neighbor	0.46377	2.86797
Auto ARIMA	0.41603	2.58689
Prophet	0.43047	2.71666
Support vector machine	0.43734	2.78535

Auto ARIMA has been the best performer with the lowest value of RMSE and MAPE. However, SVM and KNN are the worst performers according to the RMSE and MAPE respectively. Similarly, Table 2 shows the output generated for the Earthquake dataset, and here reader can observe that Auto ARIMA and Linear Regression are the best performers with the lowest value of RMSE and MAPE respectively. However, KNN was the worst performer according to both RMSE and MAPE. But the numerals were so much close in this case.

The results of the Sales forecasting dataset are described in Table 3, where it can be observed that Linear Regression and SVM turns out to be the best performer with the lowest value of RMSE and MAPE respectively. However, KNN was the worst performer according to both RMSE and MAPE.

The Table 4 depicts the ranking of each algorithm on the basis of two evaluation metrics.

The graphs in Fig. 8 shows the comparison of the value attains by the Evaluation metrics. The Tata Global Beverage graph signifies that RMSE has higher values than MAPE; however, the other two datasets say otherwise. Ultimately, it all depends on the target variable and dataset.

Table 3 Results of sales forecasting dataset

Algorithms	RMSE	MAPE
Linear regression	2.22990	23.76444
K—nearest neighbor	2.35999	24.30888
Auto ARIMA	2.23614	23.97399
Prophet	2.24678	24.12586
Support vector machine	2.33276	22.56927

Table 4 Evaluation ranking

Metrics	RMSE					MAPE				
Datasets/ranking	1	2	3	4	5	1	2	3	4	5
NSE—tata global beverage	Auto ARIMA	LR	Prophet	KNN	SVM	Auto ARIMA	LR	SVM	Prophet	KNN
Significant earthquakes	Auto ARIMA	Prophet	LR	SVM	KNN	LR	Auto ARIMA	Prophet	SVM	KNN
Superstore	LR	Auto ARIMA	Prophet	SVM	KNN	SVM	LR	Auto ARIMA	Prophet	KNN

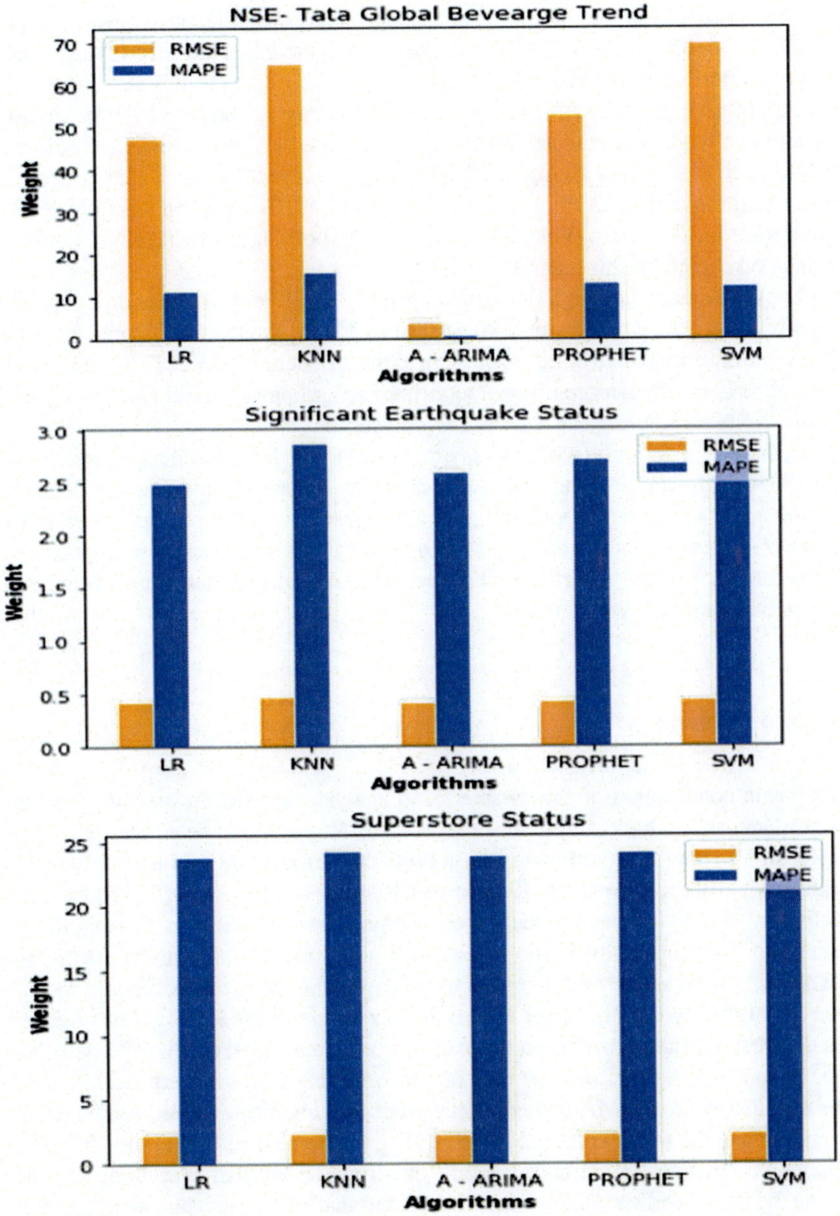

Fig. 8 Model performance comparative graph

The trend of First Dataset says Auto ARIMA has a significantly lower value of RMSE (3.74366) and MAPE (0.72129) than other models. However, talking about the worst performer, KNN beats other algorithms according to RMSE (16.92529), and SVM according to RMSE (69.81082). Looking at the trend of the second dataset one can say that there is minimal difference between models according to RMSE; however, among all Auto ARIMA (0.41603) gave a bit better satisfying result. But according to the MAPE, Linear Regression (2.49101) went on top followed by Auto ARIMA (2.58689). White RMSE and MAPE both signified that KNN wouldn't be a good choice for this dataset.

The third dataset i.e., for Sales prediction had very difficult in choosing an optimal algorithm according to the graph. Nevertheless, Linear Regression became the more favorable algorithm than others according to the numbers of RMSE (2.22990). Similarly, SVM became a more optimal algorithm according to MAPE (22.56927). But again, KNN significantly became not a good choice.

In the end, it won't be wrong to say that everything depends upon the trends and variables of the dataset, and that's why choosing an appropriate machine learning model becomes priority before going for a business idea. Here, one can observe that there is small difference between results of the evaluation metrics of earthquake and sales dataset. Yet, the numeral gaps between Auto ARIMA and other models in Stock Prediction dataset is very large.

6 Conclusion

The main contribution of this chapter is to analyze best known machine learning algorithms and to build a system through which it can be predicted the best suited algorithm as per the requirements. Experimental performance analysis of five algorithms viz., linear regression, K—Nearest Neighbor, Auto ARIMA, Prophet, and Support Vector Machine is done. Stock market, earth and sales forecasting data is analyzed. To compare the performance and accuracy of these algorithms, RMSE and MAPE are used as the evaluation metrics. Lower the value of RMSE and MAPE, the better the algorithm. As per the results, according to the RMSE, Auto ARIMA is the most optimal algorithm in two cases out of three. However, MAPE states that the Auto ARIMA is suitable for only one case. Taking it all in determination, it can be said that Auto ARIMA jostled all the other four algorithms, followed by Linear regression in the second place. Also, KNN is going to be the worst choice for Time-Series Forecasting. In the end, it won't be wrong to say that everything depends upon the trends and variables of the dataset, and that's why choosing an appropriate machine learning model becomes priority before going for a business idea. Here, one can observe that there is small difference between results of the evaluation metrics of earthquake and sales dataset. Yet, the numeral gaps between Auto ARIMA and other models in Stock Prediction dataset is clearly observed.

However, everything that is created or developed as room for improvement, that means there are several future scopes of the same. One can augment other algorithms to this system or can use other evaluation metrics for the same. Also, other datasets can be used for further analysis. Using this kind of system, one can choose the most appropriate algorithm for their work, and it would be helpful for future research purposes as well.

References

1. Y. Ding, S. Han, Z. Tian et al., Review on occupancy detection and prediction in building simulation. Build. Simul. **15**, 333–356 (2022). https://doi.org/10.1007/s12273-021-0813-8
2. Vansh Jatana, Machine Learning Algorithms (2019)
3. B. Abdualgalil, S. Abraham, Applications of machine learning algorithms and performance comparison: a review, in *International Conference on Emerging Trends in Information Technology and Engineering (ic-ETITE)*, pp. 1–6, Vellore, India (2020). doi: https://doi.org/10.1109/ic-ETITE47903.2020.490
4. Dasgupta Ariruna, Nath Asoke, Classification of machine learning algorithms. Int. J. Innov. Res. Adv. Eng. (IJIRAE). ISSN: 2349-2763. 3. 6-11. https://doi.org/10.6084/M9.FIGSHARE.3504194.V1. (2016)
5. Mondal Prapanna, Shit Labani, Goswami Saptarsi, Study of effectiveness of time series modeling (Arima) in forecasting stock prices. Int. J. Comp. Sci. Eng. Appl. **4**, 13–29 (2014). https://doi.org/10.5121/ijcsea.2014.4202
6. Kemal Korjenić, Kerim Hodžić, Dženana Đonk, Application of Facebook's prophet algorithm for successful sales forecasting based on real-world data. Int. J. Eng Data Techn. (IJCSIT). Vol twelve, No 2 (2020). doi:ten.5121/ijcsit.2020.12203
7. Panigrahi Sibarama, H. Behera Dr., A study on leading machine learning techniques for high order fuzzy time series forecasting. Eng. Appl. Art. Intell. **87**, 103245 (2020)
8. Roondiwala Murtaza, Patel Harshal, Varma Shraddha, Predicting stock prices using LSTM. Int. J. Sci. Res. (IJSR) (2017)
9. Joosery Baleshwarsingh, G. Deepa, Comparative analysis of time-series forecasting algorithms for stock price prediction (2020), pp. 1–6
10. A.A. Ariyo, A.O. Adewumi, C.K. Ayo, Stock price prediction using the ARIMA model, in *UKSim-AMSS 16th International Conference on Computer Modelling and Simulation* (Cambridge, 2014), pp. 106–112
11. G. Khekare, P. Verma, Prophetic probe of accidents in Indian smart cities using machine learning, in V. Bhateja, S.C. Satapathy, C.M. Travieso-González, V.N.M. Aradhya (eds), *Data Engineering and Intelligent Computing. Advances in Intelligent Systems and Computing* (Springer, Singapore, 2021), vol. 1407. https://doi.org/10.1007/978-981-16-0171-2_18
12. S.B. Imandoust, Bolandraftar Mohammad, Application of K-nearest neighbor (KNN) approach for predicting economic events theoretical background. S B Imandoust et al. Int. J. Eng. Res. Appl. **3**(5), 605–661 (2013)
13. K. Ayyub, S. Iqbal, E.U. Munir, M.W. Nisar, M. Abbasi, Exploring diverse features for sentiment quantification using machine learning algorithms, in *IEEE Access*, vol. 8, pp. 142819–142831 (2020)
14. G. Khekare, Internet of everything (IoE): intelligence, cognition, catenate. MC Eng. Themes **1**(2), 31–32 (2021)
15. Y. Zhang, Y.-M. Cheung, Learnable weighting of intra-attribute distances for categorical data clustering with nominal and ordinal attributes, in *IEEE Transactions on Pattern Analysis and Machine Intelligence* (2021)

16. Kumar, N., Kumar, U., diverse analysis of data mining and machine learning algorithms to secure computer network. Wireless Pers Commun (2021)
17. Pant, M., Kumar, S., Fuzzy time series forecasting based on hesitant fuzzy sets, particle swarm optimization and support vector machine-based hybrid method. Granul. Comput. (2021)
18. Y. Yi, D. Sun, P. Li et al., Unsupervised random forest for affinity estimation. Comp. Visual Media 8, 257–272 (2022). https://doi.org/10.1007/s41095-021-0241-9

AI Based Technologies for Digital and Banking Fraud During Covid-19

Mudita Sinha, Elizabeth Chacko, and Priya Makhija

Abstract The only viral thing today is the Covid 19 virus, which has severely disrupted all the economic activity around globe because of which all the businesses are experiencing irrespective of its domain or country of origin. One such major paradigm shift is contactless business, which has increased digital transaction. This in turn has given hackers and fraudsters a lot of space to perform digital scams line phishing, spurious links, malware downloads etc. These frauds have become undesirable part of increased digital transactions, which needs immediate attention and eradication from the system with instant results. In this pandemic situation where, social distancing is key to restrict the spread of the virus, digital payments are the safest and most appropriate payment method, and it needs to be safe and secure for both the parties. Artificial intelligence can be a saviour in this situation, which can help combat the digital frauds. The present study will focus on the different kinds of frauds which customers and facing, and most possible ways Artificial intelligence can be incorporated to identify and eliminate such kind of frauds to make digital payments more secure. Findings of the study suggest that inclusion of AI did bring a change in the business environment. AI used for entertainment has become an essential part in business. Transfiguration from process to platform focused business. The primary requirement of AI is to study the customer experience and how to give a better response for improving the satisfaction. But recently AIs are used not only for customer support, but it's been observed that businesses have taken it as marketing strategy to increase demand and sales.

Keywords COVID 19 · Pandemic · Digital frauds · Artificial intelligence · Digital transactions

M. Sinha (✉)
CHRIST University, Bangalore, India
e-mail: mudita.sinha@christuniversity.in

E. Chacko · P. Makhija
Jain (Deemed to be) University, Bangalore, India
e-mail: elizabeth_c@cms.ac.in

P. Makhija
e-mail: Priya_m@cms.ac.in

E. H. Houssein et al. (eds.), *Integrating Meta-Heuristics and Machine Learning for Real-World Optimization Problems*, Studies in Computational Intelligence 1038,
https://doi.org/10.1007/978-3-030-99079-4_17

1 Introduction

Novel COVID-19 virus, 2019-nCoV, or SARS-CoV-2 induced pandemic commonly known as COVID-19 is one of the dangerous and fatal pandemics that has affected all the major countries and is still giving shudders to people, countries and business. The pandemic has induced serious measures by the Government of different countries which includes, social distancing, travel bans and worst of all lockdowns [1], which caused the world's economy collapse overnight.

The pandemic is fatal not only to citizens of the country but also has adverse effects on the businesses [2, 3], economy and monetary activities [4] in the major sectors on the country [5]. Countries across globe are expecting to face serious financial losses due the COVID-19 [6]. COVID-19 has also changed the consumption pattens of the consumers in all the affected Asian countries [7]. It was observed that citizens of the effected countries started indulging the panic purchases and stocking of the essential commodities. Few retailers tried to advantage of the situation by black hoarding the supplies and started black marketing. The only possible method of restricting the spread of the virus is to maintain social distancing which turn adversely affected the economic stability [5, 8] along with monetary conditions [4], and logistics activity of the affected nations [3]. It has already been researched that pandemic when affect a country cause a serious and long-lasting influence on the businesses [9]. The COVID-19 pandemic has unparallelly influenced the world. It has put breaks on most of the business activities because of which businesses around the World are struggling. It was found that the people of infected countries want to limit the physical transaction. This has prompted most people to opt for digital payments as social distancing is becoming new normal during these pandemic times, people have been left with no alternative but to go digital payment modes which are contactless. The justification behind this rests under the fact that the COVID-19 virus can easily spread if the droplet of the infected person falls on the inert objects may infect the individual if it has been touched by them [10].

As physical money can form a very good catalyst for the spread of the virus WHO has suggested to utilize the digital money as an when possible [11]. In several developed countries, before the out brake of pandemic digital transaction through various platforms have already gained popularity [12]. It has been found that the digital transactions are associated with different uncertain prospects [13, 14] and are attached with the fraudulent activities like phishing, skimming and other criminal activities and data breach [15, 16]. Earlier investigations have shown that there are various forms of internet and other digital transaction risks, including post purchase risks, financial risks, time/comfort risks and psychological risks which are associated with the digital transactions [4] which resulted in negative behavioural response towards digital payment gateways [15]. However, if the danger were related to the deadly pandemic outbreak, such as COVID-19, the result will likely be different. As stated, the possibility of transmission of SARS-Cov-2 in physical money is highly risky. Perceived risks involved with virus transmission will be, positively influenced by the intention of consumers to use nonphysical capital. The fear of digital payment frauds comes with the rapid growth in the use of digital payments during COVID-19 (Fig. 1). Few of which are as follows:

(a) Ransomware attacks

Ransomware hackers get profited from the instability of the market during pandemic situation and corporate reliance on remote workers during social distancing. Spam threats are connected to the pandemic which has resulted in increased ransomware attacks across the board along with rarely used 'work-from-home' network setups.

(b) Banking Trojans

Apparently, the banking trojan is taking full advantage of the COVID-19 disease outbreak as a ploy by sending people an SMS that attracts everyone to access software relevant to COVID-19. As is generally the case with most Trojans, the content is actually a link for downloading a malicious app which can trick people into downloading and installing on their devices. When enabled, in addition to having access to their other personal information, the Trojan could potentially steal their financial data such as credit card numbers.

Fig. 1 Types of digital payment frauds during covid

(c) Paycheck Protection Program (PPP) loan fraud

The PPP is a piece of legislation to the Small Business Act and offers forgivable loans to small businesses (500 or fewer employees) to keep workers on payroll and to handle certain qualified business costs, such as mortgage payments, rent and utilities, up to $10 million with each eligible business entity. The Department of Justice's Criminal Division Fraud Section's Market Integrity and Major Fraud Unit and its partner organisations already had officially started prosecuting fraudulent PPP loans countrywide along with many other COVID-related market-based frauds.

(d) Bulk orders and alternative payment methods

According to Sift's Digital Trust & Safety Measure, the amount or quantity of a fraudulent transaction is usually three times that of a legitimate purchase order. Because of the COVID-19 pandemic, people engage in panic buying that typically takes the form of bulk orders. The panic buying phenomenon has created an atypical purchasing environment that makes it easier for the fraudsters to push the limits even more in order to make even larger bulk orders. In addition to making large transactions, fraudsters are increasingly shifting their payment methods from gift cards to digital wallets such as Google Pay, Samsung Pay, Apple Pay and PayPal.

(e) Phishing Scams

Scammers in these scams will give you unauthorised payment links via SMS that look identical URL. If you hurry to click on that connexion, this will guide you to your UPI payment app and ask you to pick the auto-debit button. If you grant auto-debit permission the sum will be debited automatically from the UPI app. In addition, clicking on faux URL may also corrupt your device with malware / virus that is generated to steal the device's financial data.

(f) Scams through UPI pin and OTP

Now fraudsters use the UPI to cheat people, in this the scammer first tries to send a link to the user for "request money". The point in time the user taps upon that link to authorise the transaction assuming they would obtain the money, the amount automatically gets debited from their account. OTP is another way fraudster scam people. It's important to understand that you either require a UPI PIN or a one-time password (OTP) while making transactions through your UPI app. For the OTP, your bank sends an SMS to you with the OTP number on your registered mobile number. The transaction occurs only after the OTP has been verified. Fraudsters know this very well and so they are trying to persuade their victims to reveal their OTP over the mobile. They can easily authenticate transactions and steal money from the user's account once they get the information.

It is important to keep in mind that many people over the years were already apprehensive about digital payment security. Now days digital payment scams are not really a new phenomenon as we very often get to know about these frauds. But in this situation, with almost everyone using digital payment media, the question arises how safe digital payments are. Particularly in times of COVID-19 induced forced digitisation when fraudsters and hackers have become continually active. In this chapter the researcher will be focusing on different digital frauds digital payment frauds during COVID-19 times along with possible solutions that can fight these online payment frauds with incorporation of Artificial intelligence. Before moving any further we need to understand why this pandemic scenario is best opportunity for the hackers and scammers to indulge in digital payment frauds [17].

- When demand for few products and services intensifies, consumers in their procurement also appear to be less vigilant.
- COVID-19 has generated fear in many parts of the globe. This condition leads to lower levels of trust and greater anxiety, thereby making the customer more vulnerable to these frauds.
- Susceptible consumers are more likely to fall victim to cheating, as they are compelled to use unfamiliar methods of payment. They are also being asked to trust and depend on third parties.

With the increase in online transaction the growth of frauds have also increased particularly during this pandemic time. With the faster payments the number of frauds cases have been increased from 28 to 33% reported by UK Finance. It is also noted that 98% of the fraud losses comes from banking sector and 76% comes from the internet. Due to digitalization in this COVID-19 when working hours has been reduced in the bank branches, less cash has been due to demands in digital and contactless payments and grocery shopping resulting increase in online showing to 33% [18]. In the UK alone in February, COVID-19 virus themed scams targeting both individuals and businesses resulted in losses of more than £800,000 (equivalent to almost €1 million).

In 2014, global fraud accounted for loss of $16.31 billion and this figure is increasing day by day as fraudster are developing new analytical techniques to alter the normal working behavior of credit card fraud detection system (CCFD) [19]. In the Philippines, Peru, India, Kenya, South Africa and several other developing countries, vulnerable people already coping with the health effects of the COVID-19 virus pandemic have been targeted by online fraudsters seeking to take undue advantage of them Bank frauds accounted for a loss of Rs 71,543 crore in 2018–19 according to RBI's annual report. Digital fraud is a major threat which plagues financial institutions. Machine learning systems can monitor and analyze spending habits, venue, consumer behaviour, identify deviation and warn consumers to minimize fraud as soon as possible [20].

Trust in digital financial services, which were helpful in advancing financial inclusion efforts, could be undermined at the very moment when they proved to be a successful way of delivering payments easily and efficiently to disadvantaged people. In South Africa SCAMMERS have been visiting homes to "recall" banknotes and coins which they claim that it's tainted with COVID-19 virus, offering "clean" cash receipts that are never sent to them. In India, fake offers of emergency money were reported for essential items. INTERPOL reported that offenders are contacting victims to be health or hospital staff, claiming that a relative has fallen ill with the virus and demanding medical care payments [21].

Types of frauds which are happening like Fake governments emails granting money, scan emails offering access to Covid-19 Relief funds, stealing personal and financial information or infect devices with malware and phishing emails, E-commerce websites frauds during the return products and transfer of money back, Emails asking for TV subscription with the link provided, fake investment opportunities like bitcoin platforms etc. Certain commercials aim to trick victims to purchase in—demand items such as face masks, hand sanitizers and even self—testing kits which in reality do not exist and the fraudsters are pocketing the money without giving anything in return. Sometimes the emails which contain links or attachments that download malicious software to the user's computer if opened. This includes keyloggers that can capture confidential information such as usernames or passwords, or Remote Access Trojans (RATs) that fraudsters can use to get full and anonymous control of a user's computer and are also used to access a user's online banking account [22, 23]. Artificial Intelligence (AI) can enable banking organizations to fully redefine how they work, develop new products and services and most importantly, have an impact on customer experience. In this second machine age, banks will take help of FinTech companies that exploit advanced technology that increase or even replace sophisticated human employees through algorithms [24].

2 Importance of AI Techniques in Detecting Frauds

Certainly, a digital boom is taking place across all segments of the industry, particularly after demonetization. Traditional banking has evolved, and more and more banks are embracing new technology such as AI, Cloud, blocking chain to minimize their operating expenses and increase performance. Improvement and growth in the AI sector, at reduced cost, would improve productivity. Managers around industry will need to step up their ante on skill setting gradation. There is no question that the recent movement towards digitalisation affects the conventional banking models rapidly. But it has also exposed the institutions to the threats and vulnerabilities to cyber security. In developing an active defence mechanism against cybercrimes the banks are increasingly looking at emerging technologies such as blockchain and analytics [25]. There are different innovative methods of reducing the frauds (Fig. 2), few are listed below.

(a) Artificial intelligence and machine learning

The tracking of transactions can be critical in preventing digital payment fraud. Using cutting-edge AI technologies and machine learning, one might detect suspicious trends in the previous transactions and afterwards apply them to verify the validity of transaction. Such technology helps FinTech firms take preventive measures in real-time, including warning the party concerned or blocking the transaction in an unlikely circumstance.

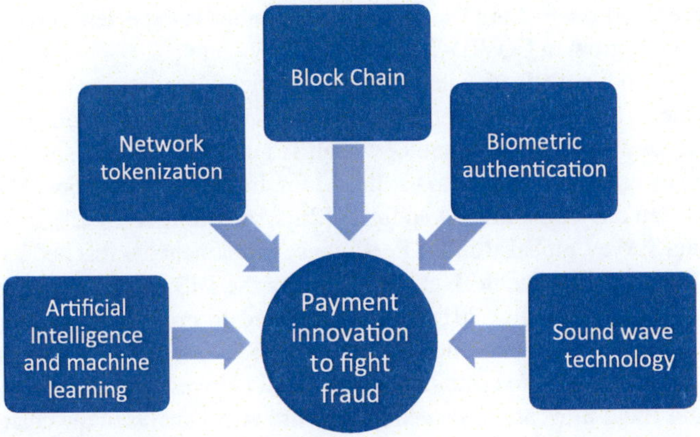

Fig. 2 Innovative measures to fight digital frauds

(b) Network tokenization

Network tokenisation is a technical mechanism in which a set of cards is protected when replaced by an exclusive number sequence called token. This particular token substitution occurs for any merchant payment. The replaced token has really no worth outside the retailer channel so the payment process is safe.

(c) Biometric authentication

Biometric authentication is recognized as an innovative way of preventing fraudulent digital payments. Using biometric authentication such as finger, speech, eyes biometrics on online payment platforms will easily distinguish scammers from authorized customers by capturing and mapping unique physiological characteristics that have been pre-saved.

(d) Sound wave technology

Sound wave is yet another revolutionary technology that proves to be successful against fraudulent digital payments. The soundwave-based payments are secure as the computer of the merchant will transmit encrypted data through a specific sound wave that can be retrieved and validated in real time by the client's mobile. Sound-wave engineering provides better availability and cost-effectiveness this is why many believe it will produce massive disruptive changes in minimizing digital payment space frauds.

(e) Blockchain

Blockchain might be another cutting-edge techniques to help eradicate digital payment fraud by increasing realtime transaction data exchange and by finding consensus between all stakeholders. Blockchain technology has already been developed for 32% of anti-fraud specialists. In addition, in 2019 they used distributed ledger systems to combat the frauds. All this number is expected to rise as digital payment frauds grow in COVID-19 times.

All the methods are effective in it sway but Artificial Intelligence (AI) is emerging rapidly as the go to technology for businesses across the globe to personalize customer experience. Day by day, the technology itself is getting better and smarter, encouraging more and newer industries to use the AI for different applications. The banking sector has been one of the first AI adopters [26]. According to PwC's 2017 FinTech Trends India Survey, global artificial intelligence investment has reached $5.1 billion. The "Artificial Intelligence in Banking" study by the IHS Markit says this expense has risen to $41.1 billion in 2018 and is projected to cross $300 billion by 2030 [27]. United Services Automobile Association (USAA) which was the first bank to introduce mobile deposit to customers has decided to invest in AI for the monitoring of fraud and fraud prevention. Presently they use AI to analyse how customers use their apps and the behaviour's that the customers display when interact with apps to detect any potential abnormality and possible frauds taking place [28].

The Punjab National Bank scam introduced a large amount of risk to the banking sector and shook regulators, financial and capital markets, and the banking industry. AI and due diligence will track possible threats of this kind and help banks implement tricky surveillance and fraud detection systems. Bank oversight has been carried out by audits and sampling. These tests can not cover those data sets and files which can cause huge risks. The algorithmic rules-based approach can help track each and every file, and machine learning techniques can maintain a database of all of those files that pose a risk to the bank. Banks may use AI to detect fraud in transactions while offering safe and swift transactions or identify some unusual activity in the customer's account based on behavioural analysis. In recent years, with a growing percentage of cybercrimes, AI can be used to preserve cybersecurity, and most importantly, to protect personal data. Citibank has also used machine learning and big data to invest more than $11 million in a new antimony laundering system [27].

In developing an active defence mechanism against cybercrimes, the banks are increasingly looking at emerging technologies such as blockchain and analytics. Artificial Intelligence may be the key to improving processes and maintaining the competitive advantage of many of these important customers. The Bank should make its customers aware of the AI technology and profit from the technology [26].

3 Possible Ways to Strengthen the Processes for Data Processing to Avoid and Identify Fraud

To identify significant information using particular methods and techniques, data analysis can be defined as an in-depth evaluation of the nature and critical features of available data. It is an interdisciplinary field that includes branches such as science computers (informatics), mathematics, statistics, economics, psychology, law and other cognitive sciences. According to specialist literature, more than 24 forms of analysis can be described in relation to data processing as a method for fraud prevention and detection.

There are different Analytical accuracy models for fraud prediction but in a research for Credit card fraud detection three techniques were used and among them random forest decision tree performs best in terms of accuracy, precision and recall with the rising issue of overfitting of tree in memory as data increases [29].

AI redefines fraud detection from relying exclusively on previous experiences to take into consideration emerging transaction phenomenon practices, habits and patterns. By integrating supervised learning algorithms trained on historical data with unsupervised learning, digital companies achieve a higher degree of acuity and consistency about the relative risk of behaviours of customers. Due to this the possibility of accepting or deny payment, to stop fraudulent activity to restrict charging and to minimize risk. For the first time, a fraud investigator gets a 360° view of transactions, with the advantage of presenting historical data in context. Under unsupervised machine learning to incorporate anomaly detection and insights into real time behaviour fraud analysts can immediately verify or redefine their threshold level decision and handle risk well. AI ranking system such as Omniscore are the future of fraud detection through which nuanced abuse attacks including refer a friend abuse, promotion abuse, or seller collusion in a marketplace can be reduced [30].

Machine learning and artificial techniques were implemented on the financial data in Korea, where experimental results were taken into consideration based on filter method, clustering process and classification process. The results showed that selection process must be performed according to input data and machine learning based process has to verify the combination of clustering and classification algorithms [31].

4 Measures Incorporated to Minimise the Fraudulent Activities

According to, companies usually use two techniques to reduce the risks of fraud, firstly by dissuading possible fraud by providing a clear ethical tone at the top and a proactive policy for fraud prevention, and secondly through identifying fraudulent activities that occur. In the meantime, certain safeguards such as the Whistleblower programme may be used to prevent fraud. There are several fraud prevention and identification techniques which can be used to reduce the risk of bank fraud. Data mining technique mainly use, a computerised identification of fraud mainly used by fraud investigators and forensic accounts concluded that this approach is an easy to use, low-cost methodology for analysing the entire database. Additionally, this technique can help prevent incorrect generalisations based on Knowledge Limited. However, this approach is only appropriate for a small business since data mining software doesn't process vast amounts of information efficiently and doesn't allow programmers to concentrate attention on a particular type of fraud [32].

An innovative approach to fraud detection using global counters in online banking transactions and successful identification of access devices. The theory behind this method stems from the fact that fraud concern rises with the amount of accounts accessed from the same source in a transaction. A part that is downloaded and installed in each computer during its first access to the bank allows the successful identification of the devices. A display counts how many different accounts each computer has accessed. These counters are then used by the global module of analysis, which calculates the probability of a transaction being a fraud. Differential analysis is often conducted against a collection of customer profiles at each transaction. This method is based on the premise that reveals proof of fraud in every major deviation from normal behaviours. The Dempster rule incorporates the resulting proof of fraud from a global and differential analysis to determine the overall score of suspicion [33].

5 Challenges Faced by AI Enabled Fraud Detection Techniques

Innovation and creativity have given birth to ease of life. From the time when human used to do all work by self to the current times known as gadgets doing the work for humans. Artificial intelligence is present in all chores of the day. It can be AI working according to movements and voice orders. As we have AI being there in all that we do, it can be booking a ticket or listening to songs or tracking the calls made and received, it can be said, "we are watched". The digital platform has implemented AI to increase sales and revenue. AI does look on the websites and things browsed and keeping all these they prepare a profile showcasing the areas interested to the individual. Which in turn gives way to positive move from the individuals end. Yes, AI have made life very comfortable by being friend and there as personal assistant to remind and observe individual's likings and routine. However, there are many times AI has been into arguments like those that various challenges individual might have or company may have to incur because of the AIs. Majority of the enterprises are using AIs in marketing, and to increase sales in the market. During pandemic when social distancing was practiced by business had approached digital platforms. This did give business but there were times when AIs were not performing as desired. It can be regarding companies dash board, payment gateways, supply and logistic. According to [34, 35] few challenges being listed below:

1. Lack of Reliance—it's all about software and technology. The usage of the application not known. There are times Robots and other gadgets can't sense the appropriate situations and might act incorrect. Reliance on these AIs are still less comparing of the expectations. Companies are trying to bring it in operations but still hundred percent trust is not on the software's.

2. Inadequate Infrastructure—The more sophisticated life requires proper infrastructure. Application of AIs and AI tools require sufficient space and system required upgradations and incur cost. Companies focusing on AI based operation need to have proper system and software. As it depends on maintenance and regular updates to make it function properly. Insufficient infrastructure can bring the work to halt or might not record the data in the systematic way.

3. Human Interface—AIs require a human support to function accurately. The talent pool working in data analytics and technology is still very less. This can again hinder the promoting and growth of AIs. The skill gap in the organization which can be reduced by proper training, but companies can't afford to spend more.

4. Data quality and assessment—As many sectors are using AIs in business to attract and support clients and customers. Financial sectors like banking and insurance, reaching the potential customers through capturing the database. The IoT has made the world too small, as information is easily collected. But the data collected sometimes are not of any use. The poor quality of data can bring in huge cost to the companies and can end up in poor decision-makings and strategy implementations.

5. Privacy issues—With lot of help from the artificial intelligence being downloaded everywhere like bank accounts, dues and various transactions being recorded by AI. All these incidences can always give a threat to the users by data being hacked or used by unauthorized persons. Companies are taking precautionary steps but still there are times data is being tampered and scams are noted.

6. AI outreach—The reach or usage is too low, which also brings less demand. The moment AI is discussed keeping the geographical area, urban areas are and increasingly into AIs but still there is a portion where it has not reached. The coverage need to be increased by promoting the facility and benefits of AIs in various operations and gaining trust and providing ease of work and comfort life style.

7. Selection of appropriate AI—depending upon the company and workings AI need to made or brought. The cost behind AI is too high, companies need to have clear objectives regarding tools and platforms. Inappropriate decision might give threat to privacy and trust and data leakage.

8. Data history—AI does give present information and there might be times when companies want to study the buying pattern or trend, AI need to have cloud storage. But as there are risk of data breach, which may result in tampered data. Specially in the digital platforms where people keep changing gadgets as and when new technology or device is launched. AI need to know how to transfer data and save data. The companies like Google and IBM are working on cloud computing where data can be stored longer duration.

Keeping all the challenges, during COVID 19 as per Ekici [36] study on banking sector, showed that bank had embraced AI so well that they were able to reach customers through the virtual place to satisfy their queries, support them in banking transactions. There were banks like HSBC which introduced a robot in few branches to give customers tutorials and communicate. The challenge comes in when there is language barrier, security issues. As robots also the pre instructed and works only when the data is entered correctly, any mistake can bring a huge loss to goodwill. Also, robots can be there for initial support but when banking transactions are where encryptions are not properly coded can leave gaps for hackers and fraudulent activities. AI application in banking was challenge for customer to accept and operate. As per Agarwal [37] in India, move towards online platforms and internet banking, the quality of data is inappropriate and proper segmentations are not created which make the AI a challenge for the financial sectors. The various challenges faced by Indian Banking sector will be on diverse language set, applications implied without proper objectives, lack of talent pool, poor quality of data and delayed approach towards advanced technologies.

6 Discussion

Inclusion of AI did bring a change in the business environment. AI used for entertainment has become an essential part in business. Transfiguration from process to platform focused business. The primary requirement of AI is to study the customer experience and how to give a better response for improving the satisfaction. But recently AIs are used not only for customer support but it's been observed that businesses have taken it as marketing strategy to increase demand and sales. As everyone for everything uses the online platforms. Earlier internet was used as search engine and then it became a social networking platform. Slowly banks introduced internet banking and various online platforms can capture banking details if payments are made virtually. AIs will study the usage and customer data, which would in turn give ease of work to the user. AIs are not only the robots but many smart devices manufacturing companies have come with facility like pay without swiping of debit or credit card. So, during pandemic where social distancing was taken care and people wanted to avoid commonly used resources. Application like the above were more downloaded and used. As these applications not only does the payments but also reminds about the balance and various dues pending. There are times services become cause for loss, as, data breach and data getting hacked. The biggest challenge of AI is to privacy and security. During pandemic or before also there were times financial transactions were wrongly stated. As mentioned earlier, using of all these applications are always leaving a risk to the user, customer support is handled by Chabot's. They again can solve pre-listed issues, also account details need to be provided to get the problem resolved. There are times technical glitch and server issues might leave the issue unresolved. The future is AIs, as mentioned any sector or service, it has its presence. Businesses have to focus on curbing the challenges while adapting to the change. Furthermore, study of financial services states that, to strategize operations like risk management and fraud detection. During pandemic office staff was reduced in numbers and taking customer queries from home was not possible because of technology and insufficient infrastructure [38]. These where the times AIs became a boon and worked as human robots and chatbot's to resolve customer issues. Financial services always have a threat of incorrect transactions and cybercrimes, as there are times human might not be able to scrutinize properly known as human error might occur. AIs can do better fraud detection as they will be working a formula or settings entered without knowing who is in front of them. This might give more realistic information to the organization, to analyze and plan. As per

Abhinusawa [39] opinion, the banking sector need to redesign the operating infrastructure keeping the new normal in place. Moving into advanced technology must be applied to all areas of business operations. As there are institutions use technology for few services and rest, they prefer the customer come to branch. Financial sectors need to be completely involved in advanced technology not just a part, to make customers have positive approach. Like from opening of account, KYC, customer request for address change and global accessibility of accounts from anywhere and anytime. Through digital platform financial institutions must market and increase wealth creation modes. Banks need to satisfy their clients and introduce new products like health policies, investment plans and various other areas customer's query. Pandemic times more connected and available makes the client stay secured and satisfied. As digital platforms reduce the gap between business and customer, but its challenging for businesses to cope up with changes. Sectors need to have proper survey about market and customers requirement, which can support companies to prepare AIs based upon surveys. This requires more of reskilling and upskilling from both the parties. Adding to it Conte [40] also suggest that AIs are not only the upfront technology but back office also need to be AI applied. The various automation technologies discussed are Digital Process Automation (DPA) and Robotic Process Automation (RPA). DPA focuses on end to end processes and helps in all the back-end office works from Human resources, customer support and routine functions. The technologies like these can reduce human error and speed up the work time. RPA improves the interaction between human and computer. RPA works similar to human as they interpret and communicate to other systems and best in handling repetitive work. As per Agarwal [37], the usage of mobile application and online transaction are increasing after demonetization. In India AI synchronized to major players. The major reasons for adaptation of AIs are competition, process driven services, self-service in banks, reduction in fraud, huge volume of data management, productivity and effective strategy implementation. The various areas AIs adopted in banks are risk management, wealth management, accounting, compliance, customer service, financing and loans, marketing and sales. Overall audit also initiated by the AIs to have seamless error and clarity of data for decision-making. AI are used by the new start-ups from personalization to customer service and fraud detections. Still there is rise in fraud as per RBI report 2019 that 73.8% after having privacy and security [41]. Banks are to have context sensitive AI solutions related with real time payment monitoring. The various AIs implied must be focused on specific goal to reduce fraud and provide safe transactions for customers.

7 Conclusion

Around the globe, Artificial Intelligence and Machine Learning are adopted by all sectors not just to be customer centric but also changing the concept of business from process to platform base. Adapting towards AI has made the business more competitive. Earlier the concept was applied to few functions in the organization, but slowly indulged in various other functional areas. For every business customer need to be satisfied may it be online services or banking or manufacturing sectors. Banking sector entered AIs to make the customer service smoother and fast. Which in turn helped the banks with proper data management and strategic plans. According to [42], pandemic was a time for transformation as per the report of World Economic Forum, banks have been able to reduce fraudulent activities with the help of AIs. As per the study overall there were seventy percent of new digital openings, eighty percent applied for usage of mobile pay applications and thirty percent preferred contact less payments. AIs in financial sector can reduce money laundering and cybercrimes. In India AI is currently adopted by automation, Chatbots, Natural Language Processing (NLP) and image recognitions as the report by Faggella [43]. Adoption of AI has begun at grass-root level in India, but because of resistance and investment concerns still acceptance rate is low. There is proposal made to government for three more sectors—agriculture, healthcare, and education. The government must decide how to focus on more of AIs and impact of adaption on the economy. If AIs are implemented there can be reduced frauds and security concerns. As it works on the data entered, there will be more of transparency and fair decisions. Now, when discussed about banking sector transformation from traditional to automation and AI, the move was at slow pace but did satisfy the customer at certain point. There are still to do more in banking sector as still some queries are not handled through Chatbots or human managed robots or face recognitions in ATMs or digitization of back-office work. But narrowing the gap, to make the business functions clearer and customer satisfaction, involvement of AI need to be promoted by the Indian government to detect fraud in the systems. Government also needs to develop their own applications as currently there are few companies like Google, Alexa, Oracle, Siri and few more which have captured Indian market. India can promote own applications and cloud-based data storage for financial sectors and various other sector where AI is used. As per Chugh and Jaiswal [44], banking sector will regain customers trust and satisfaction through AIs, as lending can be assessed easily with the help of technology, and investment functions can be made more understandable. The huge data can be, traced and analysed as required. The future is in AI and ML, there is need for more trained and expertise talent pool to grow in ways of intelligence for better tomorrow.

References

1. C. Sohrabi, Z. Alsafi, N. O'Neill, M. Khan, A. Kerwan, A. Al-Jabir, C. Iosifidis, R. Agha, World health organization declares global emergency: a review of the 2019 novel coronavirus (COVID-19). Int. J. Surg. **76**, 71–76 (2020)
2. D. Ivanov, Predicting the impacts of epidemic outbreaks on global supply chains: a simulation-based analysis on the coronavirus outbreak (COVID-19/SARS-CoV-2) case. Transp. Res. Part E Logist. Transp. Rev. **136**, 101922 (2020)
3. J. Turner J, T. Akinremi, The business effects of pandemics—a rapid literature review. Enterprise Research Centre, 2020
4. S.M. Forsythe, B. Shi, Consumer patronage and risk perceptions in Internet shopping. J. Bus. Res. **56**(11), 867–875 (2003)
5. M. Nicola, Z. Alsafi, C. Sohrabi, A. Kerwan, A. Al-Jabir, C. Iosifidis, M. Agha, R. Agha, The socio-economic implications of the coronavirus pandemic (COVID-19): a review. Int. J. Surg. **1**(78), 185–193 (2020)
6. V.Y. Fan, D.T. Jamison, L.H. Summers, Pandemic risk: how large are the expected losses? Bull. World Health Organ. **96**(2), 129 (2018)
7. Nielsen, Asian consumers are rethinking how they eat post Covid-19. Nielsen (2020), https://www.nielsen.com/id/en/insights/article/2020/asian-consumers-are-rethinking-how-they-eatpostcovid-19/. Accessed 28 March 2020
8. N. Fernandes, Economic effects of coronavirus outbreak (COVID-19) on the world economy. SSRN 3557504 (2020)
9. T.K. Swift, CHEM ECONOMICS economic effects of a flu pandemic. Chem. Eng. Prog. **105**(9), 22 (2009)
10. A. Ather, B. Patel, N.B. Ruparel, A. Diogenes, K.M. Hargreaves, Coronavirus disease 19 (COVID-19): implications for clinical dental care. J. Endod. **46**(5), 584–595 (2020)
11. D. Brown, Can cash carry coronavirus? World Health Organization says use digital payments when possible. USA Today (2020), https://www.usatoday.com/story/money/2020/03/06/coronavirus-covid-19-concerns-over-using-cash/4973975002/. Accessed 1 April 2020
12. M. Andrieu, The future of e-money: an international review of policy and regulatory issues. Foresight (2001)
13. R.K. Bagla, V. Sancheti, Gaps in customer satisfaction with digital wallets: challenge for sustainability. J. Manag. Dev. (2018)
14. L.Y. Leong, T.S. Hew, K.B. Ooi, J. Wei, Predicting mobile wallet resistance: a two-staged structural equation modeling-artificial neural network approach. Int. J. Inf. Manag. **51**, 102047 (2020)
15. D.L. Marafon, K. Basso, L.B. Espartel, M.D. de Barcellos, E. Rech, Perceived risk and intention to use internet banking: the effects of self-confidence and risk acceptance. Int. J. Bank Mark. (2018)
16. V. Marria, What a cashless society could mean for the future. Forbes (2018), https://www.forbes.com/sites/vishalmarria/2018/12/21/what-a-cashless-society-could-mean-forthefuture/#77c7528a3263. Accessed 5 April 2020
17. J.W. Goodell COVID-19 and finance: agendas for future research. Financ. Res. Lett. **35**, 101512
18. J. Bertrand, Addressing Payment Fraud and Bank Accounts in the new COVID 19 world (2020), https://www.finextra.com/blogposting/19190/addressingpayment-fraud-and-bank-accounts-in-the-new-covid-19-world. Accessed 6 September 2020
19. A. undu, S. Sural, A.K. Majumdar, Two-stage credit card fraud detection using sequence alignment, in *International Conference on Information Systems Security* (Springer, Berlin, Heidelberg, 2006), pp. 260–275
20. R. Mirjankar, 7 Ways Digital Disruption Is Shaping The Indian Banking Sector (2019), https://www.outlookindia.com/outlookmoney/banking/7ways-digital-disruption-is-shaping-the-indian-banking-sector-4214. Accessed 6 September 2020

21. D. Medine, Financial Scams Rise as Coronavirus Hits Developing Countries (2020), https://www.cgap.org/blog/financial-scams-rise-coronavirus-hitsdeveloping-countries. Accessed 6 September 2020
22. T. Ayling, Fraudster to perpetrate fraud using Covid-19 fears (2020), https://www.buguroo.com/en/blog/fraudster-to-perpetrate-fraud-using-covid. Accessed 6 September 2020
23. S. Read, Fraud: banks reveal the 10 coronavirus scams to be wary of—BBC News (2020), https://www.bbc.com/news/business-53573408. Accessed 6 September 2020
24. D.C. Vijai, Artificial intelligence in Indian banking sector: challenges and opportunities. Int. J. Adv. Res. **7**(5), 1581–1587 (2019)
25. S. Jewandah, How artificial intelligence is changing the banking sector—a case study of top four commercial Indian banks. Int. J. Manag. Technol. Eng. **8**(7), 525–530 (2018)
26. J. Sindhu, R. Namratha, Impact of artificial intelligence in chosen Indian commercial bank—a cost benefit analysis. Asian J. Manag. **10**(4), 377–384 (2019)
27. U. Padhi, AI in banks: risks and opportunities (2019), https://www.fintechnews.org/ai-in-banks-risks-and-opportunities/. Accessed 6 September 2020
28. K. Walch, Why AI is Transforming the Banking Industry (2020), https://www.forbes.com/sites/cognitiveworld/2020/04/05/why-ai-is-transforming-the-bankingindustry/#53e81fb67dd6. Accessed 6 September 2020
29. S. Patil, V. Nemade, P.K. Soni, Predictive modelling for credit card fraud detection using data analytics. Procedia Comput. Sci. **1**(132), 385–395 (2018)
30. L. Columbus, Top 9 Ways Artificial Intelligence Prevents Fraud (2019), https://www.forbes.com/sites/louiscolumbus/2019/07/09/top-9-ways-artificial-intelligenceprevents-fraud/#441e774a14b4. Accessed 6 September 2020
31. D. Choi, K. Lee, An artificial intelligence approach to financial fraud detection under IoT environment: a survey and implementation. Secur. Commun. Netw. **2018**, (2018)
32. R.A. Rahman, I.S. Anwar, Effectiveness of fraud prevention and detection techniques in Malaysian Islamic banks. Procedia Soc. Behav. Sci. **25**(145), 97–102 (2014)
33. S. Kovach, W.V. Ruggiero, Online banking fraud detection based on local and global behaviour, in *Proc. of the Fifth International Conference on Digital Society* (Guadeloupe, France, 2011), pp. 166–171
34. M. Brenner, The 6 Biggest Challenges to AI Marketing Success (2020), https://marketinginsidergroup.com/artificial-intelligence/biggest-challengesto-ai-marketing-success/, Accessed 2 September 2020
35. D.G. Harkut, K. Kasat, Introductory chapter: artificial intelligence-challenges and applications, in *Artificial Intelligence-Scope and Limitations* (IntechOpen, 2019)
36. E. Ekici, How AI is helping banks serve customers during COVID-19 (2020), https://www.bai.org/banking-strategies/article-detail/how-ai-is-helping-banks-serve-customersduring-covid-19/. Accessed 4 September 2020
37. M Agarwal, How Artificial Intelligence Systems Are Changing India's Banking Industry? (2019), https://inc42.com/features/how-artificial-intelligence-algorithms-arechanging-indias-banking-industry/. Accessed 26 March 2020
38. C. W, Banking industry affects due to Covid-19 (2020), https://www.finextra.com/blogposting/18868/banking-industry-affects-due-to-covid-19. Accessed 4 September 2020
39. T. Abhinusawa, Reimagining Banking and Financial Services in a Post Covid-19 world (2020), https://www.finextra.com/blogposting/19251/reimagining-banking-and-financial-services-in-a-post-covid-19world. Accessed 5 September 5 2020
40. R. Conte, Going beyond digitization with back-office process automation (2020), https://www.finextra.com/blogposting/19219/going-beyond-digitizationwith-back-office-process-automation. Accessed 5 September 2020
41. B. Dr, Banking on AI: the time is ripe for Indian banks to embrace artificial intelligence. The Financial Express (2020), https://www.financialexpress.com/opinion/banking-on-ai-the-time-is-ripe-forindian-banks-to-embrace-artificial-intelligence/2017963/
42. K. Oertli, How digital innovations helped banks adapt during COVID-19 (2020), https://www.weforum.org/agenda/2020/08/how-digital-innovations-helped-banks-adapt-during-covid-19/

43. D. Faggella, Artificial Intelligence in India—Opportunities, Risks, and Future Potential | Emerj (2019), https://emerj.com/ai-market-research/artificial-intelligence-in-india/. Accessed 5 September 2020
44. N. Chugh, S. Jaiswal. How AI is Disrupting the Banking Sector in India (n.d.), https://technology.siliconindia.com/viewpoint/ceo-insights/how-ai-is-disrupting-thebanking-sector-in-india-nwid-9820.html. Accessed 5 September 2020

Gradient-Based Optimizer for Structural Optimization Problems

Mohamed Issa and Yahia Mostafa

Abstract Meta-heuristic algorithms are stochastic search methods that have been used for quite a long time to solve complex, non-linear optimization problems for which exact methods are usually very costly or don't exist at all. Gradient-based optimizer (GBO) is a recently introduced meta-heuristic algorithm that is based on the gradient-based Newton's method. One component of the GBO algorithm is the Gradient Search rule (GSR) which guides the algorithm to promising regions and helps it converge to the best solution. The other component is the Local Escaping Operator (LEO) which helps the algorithm avoid local optima. To test the performance of the GBO algorithm, we use it to optimize three structural optimization problems. The first problem is to minimize the weight of corrugated bulkheads for a tanker. The second problem is to minimize the cost of a tubular column which is supposed to carry a compressive load. The third problem is to minimize the cost of a simply supported reinforced concrete beam subjected to a specific load. These are complex and constrained problems that test the ability of an optimization algorithm to explore the promising regions of the search space and then converge to find the best solution. To further judge the performance of the GBO algorithm, its best solutions and statistical results for these problems are compared to the results of two other well-known algorithms from the literature.

Keywords Meta-heuristics · Structural optimization · Gradient-based optimizer (GBO) · Sine cosine algorithm (SCA) · Harris Hawks optimization (HHO)

1 Introduction

Meta-heuristic algorithms (MAs) are random search algorithms that are widely used to solve optimization problems in many engineering and non-engineering applications such as topology optimization, structural optimization, aerodynamics, machine

---

M. Issa (✉) · Y. Mostafa
Faculty of Engineering, Computer and Systems Engineering Department, Zagazig University, Zagazig, Egypt
e-mail: Mamohamedali@eng.zu.edu.eg

E. H. Houssein et al. (eds.), *Integrating Meta-Heuristics and Machine Learning for Real-World Optimization Problems*, Studies in Computational Intelligence 1038,
https://doi.org/10.1007/978-3-030-99079-4_18

learning, system modeling, simulation in chemistry and physics, scheduling problems, and so on [1]. Optimization problems are usually real-life problems that are modeled by mathematical equations to be solved by an appropriate algorithm. In some types of optimization problems, we are trying to find the variables' values to yield the highest function value (maximization problem) such as problems that model efficiency or revenue for example. In other problems, we are looking for the variables' values to yield the lowest function value (minimization problem) such as problems that model the cost or the weight for example, and this type of problems is more common in real-life applications.

For some simple optimization problems like ones that can be solved in polynomial time, exact methods such as Dynamic programming and Constraint programming exist to provide an exact solution to this type of problems. But for complex optimization problems such as problems with very large instances or NP-hard problems, exact methods either don't exist or require too much resources which make them almost impossible to implement. For those types of problems, meta-heuristic algorithms exist to provide solutions that are good enough in a reasonable time. The boundaries of the problem's variables along with the constraints provided by the problem's nature together form the search space that the meta-heuristic algorithms navigate to look for promising solutions and (hopefully) reach the optimum solution eventually.

There are many types of meta-heuristic algorithms that differ in the way they navigate the search space to look for promising solutions and they are generally classified by their source of inspiration. Some meta-heuristic algorithms are physics-based in that they mimic physical phenomena such as Gravitational Search Algorithm (GSA) [2], Charged System Search (CSS) [3], Ions Motion Optimization (IMO) [4] and Thermal Exchange Optimization (TEO) [5]. Other meta-heuristic algorithms are inspired by the theories of evolution such as Genetic Algorithms (GAs) [6] and Differential Evolution (DE) [7]. Other algorithms are inspired by swarm intelligence such Particle Swarm Optimization (PSO) [8], Ant Colony Optimization (ACO) [9] and Artificial Bee Colony (ABC) [10]. There are also bio-inspired algorithms such as Cuckoo Search (CS) [11], Bat Algorithm (BA) [12], Lion Optimization Algorithm (LOA) [13] and Grey Wolf Optimizer (GWO) [14] algorithm. Meta-heuristics are used to optimize many problems in different fields such as sequence alignment [15–23], digital watermarking [24], COVID-19 [25–28], PID controller [29, 30], passive vehicle suspension system [31].

The two main components of any meta-heuristic algorithm are exploration and exploitation. Exploration (diversification) means the ability of the algorithm to search all areas of the search space and find promising regions that should contain the best solution we are looking for. Exploitation (intensification) means the ability of the algorithm to intensively explore the promising regions and converge towards the best solution. Usually, exploration is the first stage in searching for the solution which is then followed by exploitation to improve accuracy. Obviously, there is a tradeoff between exploration and exploitation and it is up to every meta-heuristic algorithm to achieve the balance between the two stages in order to reach the best performance in finding the optimum solution. And because there is no one meta-heuristic algorithm that can solve all optimization problems with the same efficiency according to the

No Free Lunch (NFL) [32] theorem, new algorithms will emerge all the time to fill the gap and to introduce new ideas that can further enhance the performance and the final result of the optimization process.

Gradient-Based Optimizer (GBO) [33] is a recently introduced population-based meta-heuristic algorithm that uses the gradient-based Newton's method as an inspiration to guide it towards the best solution. It consists of two main components, the Gradient Search Rule (GSR) and the Local Escaping Operator (LEO).

In this chapter, we test the GBO algorithm by using it to solve three structural optimization problems inspired by real world applications. To assess the performance of the GBO algorithm, we compare its results with the results of two well-known algorithms from literature. The first algorithm is the Sine Cosine Algorithm (SCA) [34] which uses the mathematical functions of sine and cosine to guide the search agents to explore the search space and then converge towards the best solution. The second algorithm is the Harris Hawks Optimization (HHO) [35] algorithm which mimics the various chasing patterns that Harris' hawks use while hunting the prey in order to guide the search agents towards the promising regions of the search space.

The rest of chapter is organized as follows:

Section 2 includes the preliminaries where we preview the idea behind the GBO algorithm and demonstrate its mathematical model. In Sect. 3, we test the GBO algorithm against the other two algorithms and discuss the results. Finally, in Sect. 4, we present our conclusions and verdict on the performance of the algorithm.

2 Preliminaries

In the following subsection, the procedure of GBO and its searching mechanism are described.

2.1 *Gradient-Based Optimizer (GBO) Algorithm*

Methods of optimization can be categorized into two main types, gradient-based methods and modern non-gradient-based methods.

Gradient-based methods such as Newton's method [36] and conjugate direction method [37] has been used for a long time to solve optimization problems. These methods identify an extreme point then use the derivatives of the objective function and the constraints to determine a search direction to search along in order to find the optimum result. But these methods have drawbacks in the form of slow convergence and no guarantee for optimal solution.

On the other hand, meta-heuristic algorithms start with an initial population that have random search directions and update the search directions every iteration of the algorithm. These methods have high capability of avoiding local optima unlike gradient-based methods. But they also have drawbacks in the form of requiring high resources especially for problems whose search spaces have a large number of dimensions.

So, the GBO algorithm was created by combining the gradient-based methods and the population-based methods to produce a new algorithm that have the advantages of both methods while avoiding their disadvantages.

The GBO algorithm is a recently introduced algorithm that is inspired by the gradient-based Newton's method. It has two main components, the Gradient Search Rule (GSR) and the Local Escaping Operator (LEO). The GSR follows a gradient-based procedure to improve the exploration of the search space and then converge faster towards the best solution. The LEO helps the algorithm avoid local optima.

- **Updating the solutions using the Gradient search rule (GSR):**

To update the solutions in each iteration, the algorithm uses the following equation:

$$x_n^{(m+1)} = r_a \times (r_b \times X1_n^m + (1 - r_b) \times X2_n^m) + (1 - r_a) \times X3_n^m \quad (1)$$

where:

r_a and r_b are random numbers in the range [0,1]

$$X3_n^m = X_{n1}^m - \rho_1 \times (X2_n^m - X1_n^m) \quad (2)$$

$$X1_n^m = x_n^m - GSR + rand \times \rho_2 \times (x_{best} - x_n^m) \quad (3)$$

$$X2_n^m = x_{best} - GSR + rand \times \rho_2 \times (x_{r1}^m - x_{r2}^m) \quad (4)$$

$$\text{GSR} = \text{randn} \times \rho_1 \times \frac{2\Delta x \times x_n}{(yp_n - yq_n + \varepsilon)} \quad (5)$$

- **The Local escaping operator (LEO):**

The local escaping operator is implemented each iteration depending on the probability *pr* that is set based on preference at the start of the algorithm.

$$x_n^{(m+1)} = x_{LEO}^m \quad (6)$$

When we apply the local escaping operator, the value of $X_{\mathrm{LEO}}^{\mathrm{m}}$ is determined by one of the following two equations with a probability of 50% for each:

$$\begin{aligned} X_{LEO}^m = {} & x_n^{m+1} + f_1 \times (u_1 \times x_{best} - u_2 \times x_k^m) + f_2 \times \rho_1 \\ & \times (u_3 \times (X2_n^m - X1_n^m) + u_2 \times (x_{r1}^m - x_{r2}^m))/2 \end{aligned} \quad (7)$$

Or

$$X_{LEO}^{m} = x_{best} + f_1 \times (u_1 \times x_{best} - u_2 \times x_k^m) + f_2 \times \rho_1 \times (u_3 \times (X2_n^m - X1_n^m) + u_2 \times (x_{r1}^m - x_{r2}^m))/2 \quad (8)$$

where:

f_1 is a uniform random number in the range of [–1,1].

f_2 is a random number from a normal distribution with mean of 0 and standard deviation of 1.

u_1, u_2 and u_3 are three random numbers.

- **Complete procedure of the GBO algorithm**

The pseudo code of the GBO algorithm is provided as the following:

GBO algorithm procedure
1st step: Initialization
Initialize the parameters *pr*, ε, and M
Set an initial population of random solutions
Calculate the value of the objective function for each solution
Determine the best and worst solutions
2nd step: Main loop
Repeat each iteration
For each solution
For each dimension
Select 4 random solutions from the population
Calculate the next position of current solution using Eq. 1
End for
Local escaping operator (LEO)
Based on the probability *pr*
Calculate the value of $X_{\mathrm{LEO}}^{\mathrm{m}}$ using Eqs. 7 & 8
Set the value of the next position for current solution to $X_{\mathrm{LEO}}^{\mathrm{m}}$
End of LEO
Update the best and worst solutions
End For
End
3rd step: Return the best solution found

Figure 1 illustrates how the GBO algorithm guides each search agent towards the best solution.

3 Experimental Results and Discussion

In this section, we use the GBO algorithm to solve three structural optimization problems inspired by real world applications. These structural optimization problems are complex and constrained problems that test the ability of an algorithm to efficiently explore the search space and then converge towards the best solution.

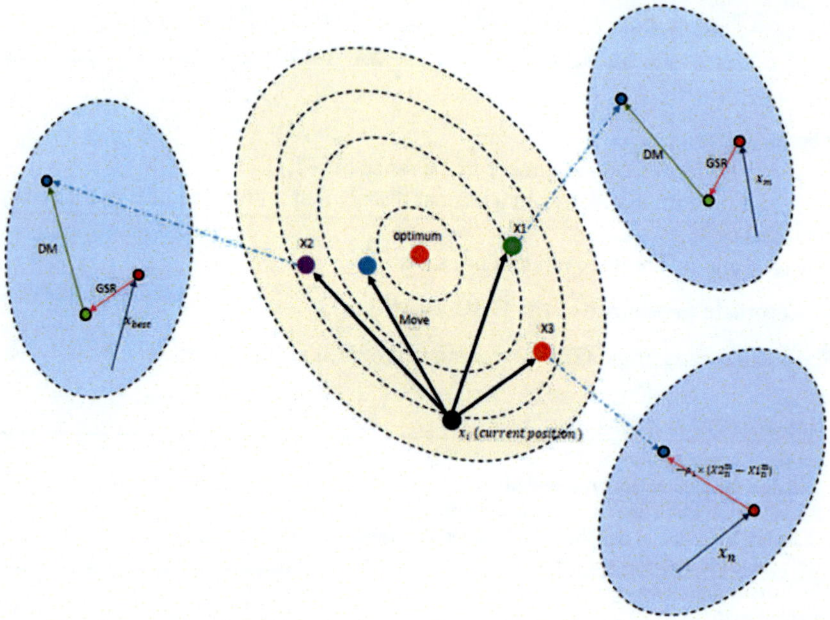

Fig. 1 Updating solutions in GBO algorithm

To assess the performance of the GBO algorithm and determine whether it produces solutions equal or close to the global optimum, the results from the GBO algorithm were compared to the results from the other two well-known algorithms, the Sine Cosine Algorithm (SCA) algorithm and the Harris Hawks Optimization (HHO) algorithm.

All algorithms were executed on a laptop with a CPU of Intel Core™ i3-380 M 2.53 GHz and with 6 GB of RAM. The software used for the execution was MATLAB R2018a running on Windows 10 64-bit. For the comparison to be fair, each algorithm was executed 20 separate times for each problem. The setting of parameters for each algorithm is shown in Table 1.

Table 1 Control parameters of the algorithms

Algorithm	Parameter	Value
GBO	β_{min}	0.2 (default)
	β_{max}	1.2 (default)
	pr	0.5 (default)
SCA	a	2 (default)
HHO	Default parameter values	

3.1 Corrugated Bulkhead Design

This problem was introduced in [38]. The aim of this problem is to design corrugated bulkheads for a tanker with the least-weight possible, so it's a minimization problem. The problem has four variables, width (b), depth (h), length (l), and plate thickness (t). The design problem can be seen in Fig. 2.

The problem can be formulated as follows:

$$Minimize:\ f(b,h,l,t) = \frac{5.885t(b+l)}{b+\sqrt{(l^2-h^2)}} \tag{9}$$

where b, h and l are continuous variables in the range [0, 100] and t is a continuous variable in the range [0, 5].

Subject to the following constraints:

$$g_1 = th(0.4b + \frac{1}{6}l) - 8.94(b + \sqrt{(l^2 - h^2)}) \geq 0 \tag{10}$$

$$g_2 = th^2(0.2b + \frac{1}{12}l) - 2.2(8.94(b + \sqrt{(l^2 - h^2)})^{\frac{4}{3}} \geq 0 \tag{11}$$

$$g_3 = \text{t} - 0.0156\text{b} - 0.15 \geq 0 \tag{12}$$

$$g_4 = \text{t} - 0.0156\text{l} - 0.15 \geq 0 \tag{13}$$

$$g_5 = t - 1.05 \geq 0 \tag{14}$$

$$g_6 = 1 - h \geq 0 \tag{15}$$

For this problem, the number of population and maximum iterations for all three algorithms are 25 and 500 respectively. The best solution found by each algorithm is illustrated in Table 2.

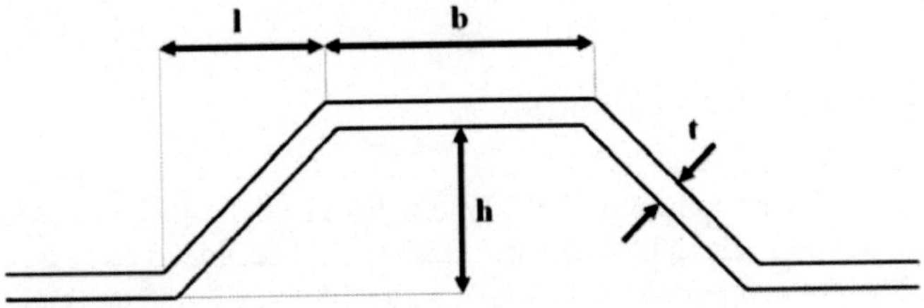

Fig. 2 Corrugated bulkhead design problem

Table 2 Best positions for corrugated bulkhead

Algorithm	b	h	l	t
GBO	57.692308	34.14762	57.692308	1.05
SCA	57.262239	34.361242	56.818271	1.0658388
HHO	50.951332	34.238347	57.692146	1.05

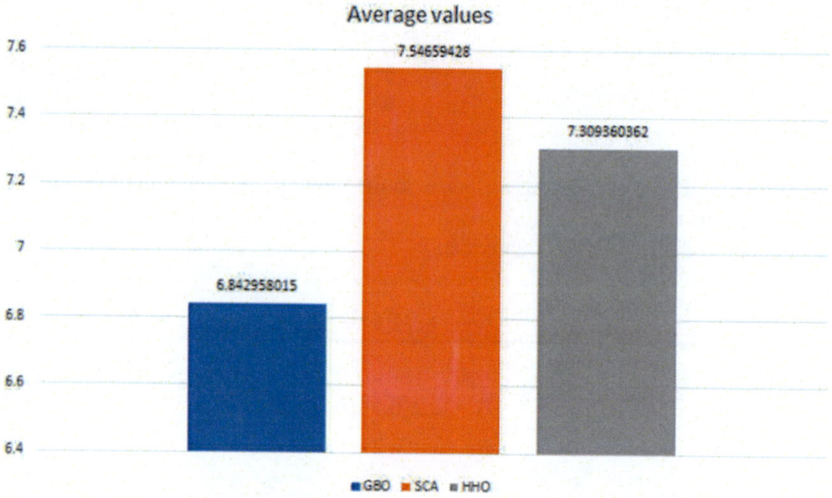

Fig. 3 Average values for the corrugated bulkhead problem

Figures 3, 4, 5 and 6 show the statistical results in the form of average, best, worst and standard deviation values respectively. As it can be observed, the GBO algorithm achieved the best results in all categories. It was able to reach the best value for the objective function compared to the other two algorithms. It also had the best average value, worst value and standard deviation compared to the other two algorithms which proves the consistency of the GBO algorithm.

Figure 7 shows the convergence curves for the three algorithms. It can be observed that the GBO algorithm is the fastest to converge towards the global optimum among the three algorithms.

3.2 *Tubular Column Design*

This problem was introduced in [39]. The aim here is to design a uniform column of tubular section with a length of 250 cm which should be able to carry a compressive load of 2,500 kgf at minimum cost, so it's a minimization problem. Bearing in mind the specifications of the column's material like yield stress, modulus of elasticity

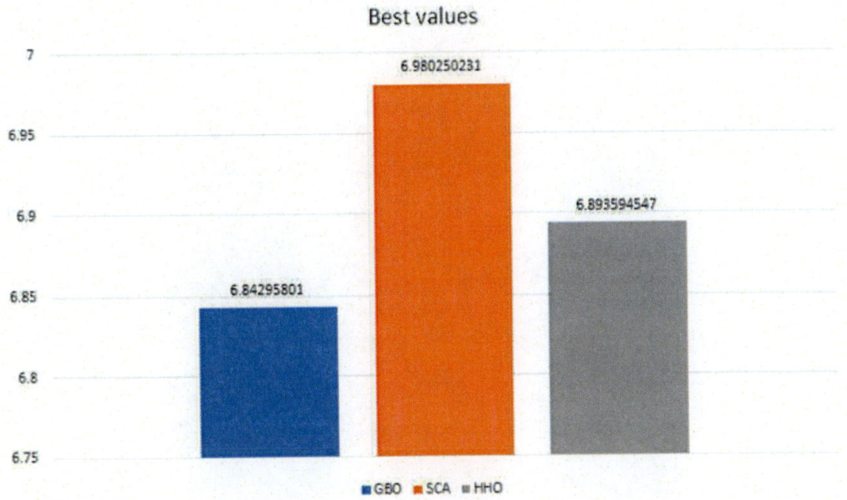

Fig. 4 Best values for the corrugated bulkhead problem

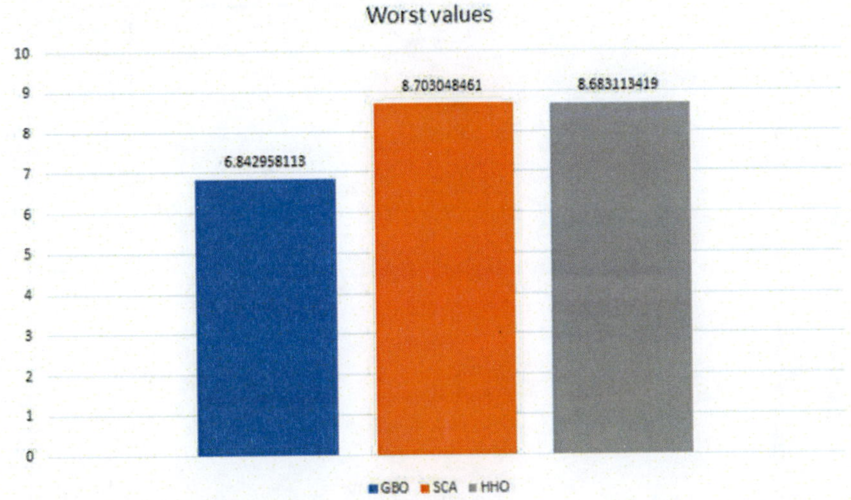

Fig. 5 Worst values for the corrugated bulkhead problem

and density, the design of the column should make sure that the stress applied to the column doesn't exceed the buckling stress (1st constraint) or the yield stress (2nd constraint). The problem has two variables, column mean diameter (d) and column thickness (t). The design problem can be seen in Fig. 8.

The problem can be formulated as follows:

$$\text{Minimize: } f(d, t) = 9.8dt + 2d \tag{16}$$

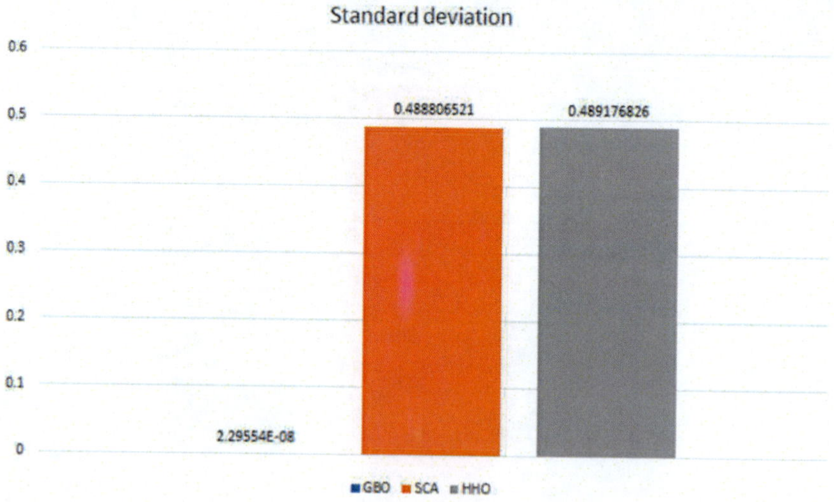

Fig. 6 Standard deviation from 20 runs for the corrugated bulkhead problem

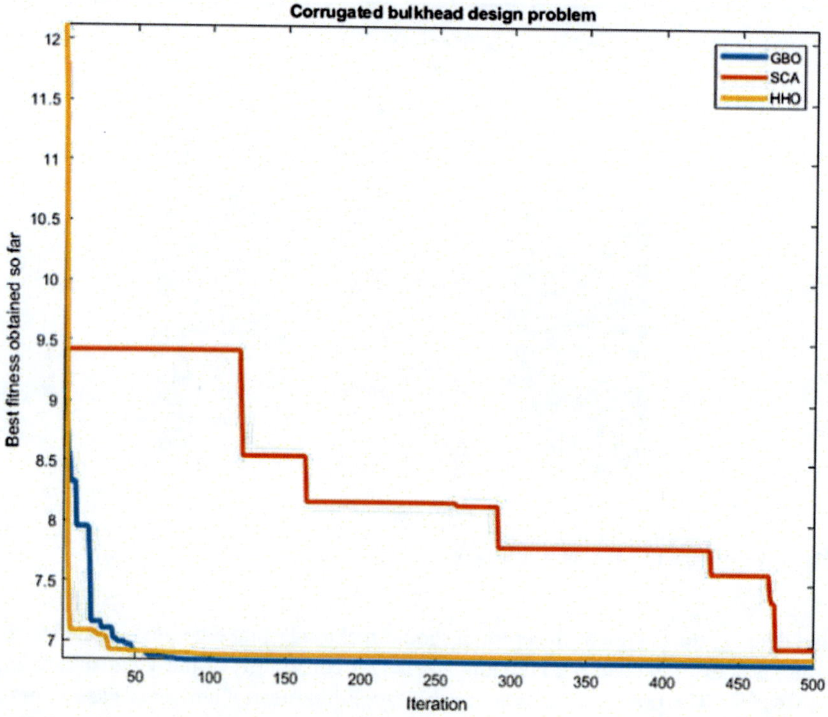

Fig. 7 Convergence curves for the corrugated bulkhead problem

Fig. 8 Tubular column design problem

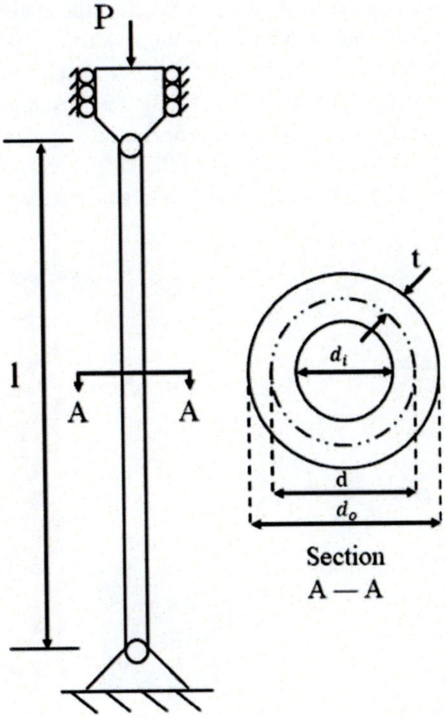

where d is a continuous variable in the range [2, 14] and t is a continuous variable in the range [0.2, 0.8].

Subject to the following constraints:

$$g_1 = \frac{5}{\pi dt} - 1 \leq 0 \tag{17}$$

$$g_2 = \frac{25000}{17\pi^3 dt(d^2 + t^2)} - 1 \leq 0 \tag{18}$$

For this problem, the number of population and maximum iterations for all three algorithms are 25 and 1000 respectively. The best solution found by each algorithm is illustrated in Table 3.

Table 3 Best positions for Tubular column

Algorithm	d	t
GBO	5.4511562	0.2919655
SCA	5.445821	0.2928315
HHO	5.4523804	0.2918999

Figures 9, 10, 11 and 12 show the statistical results in the form of average, best, worst and standard deviation values respectively. Just like the previous problem, the GBO algorithm had the best results in all categories. It had the best value for the objective function among the three algorithms. It also had better average value, worst value and standard deviation compared to the other two algorithms which again proves the consistency of the GBO algorithm.

Figure 13 shows the convergence curves for the three algorithms. Again, and like

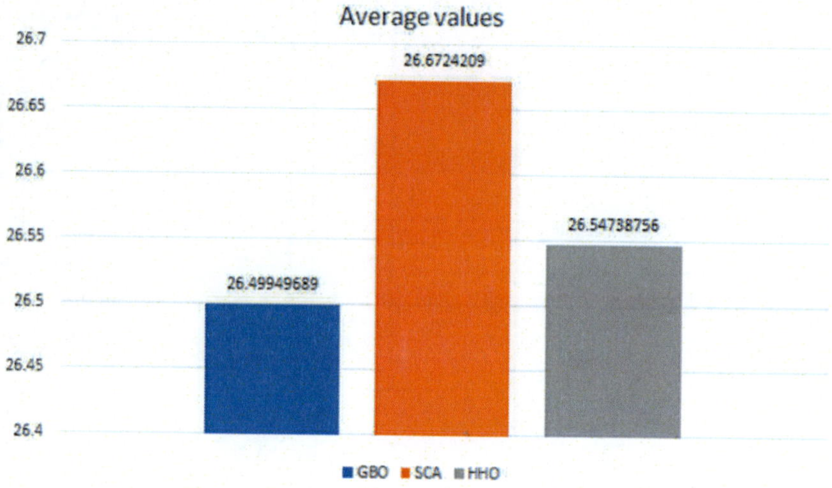

Fig. 9 Average values for the tubular column problem

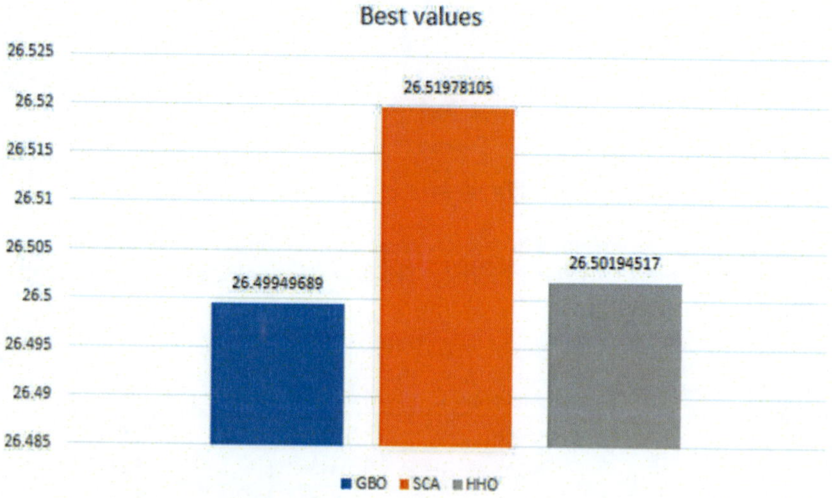

Fig. 10 Best values for the tubular column problem

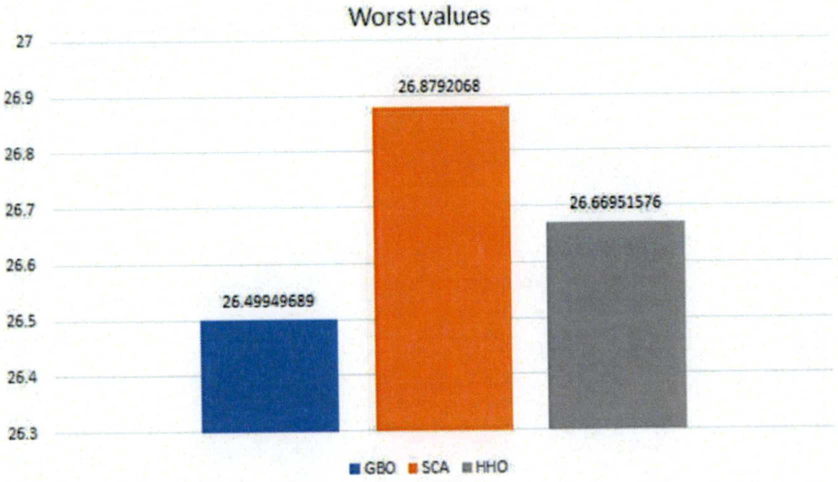

Fig. 11 Worst values for the tubular column problem

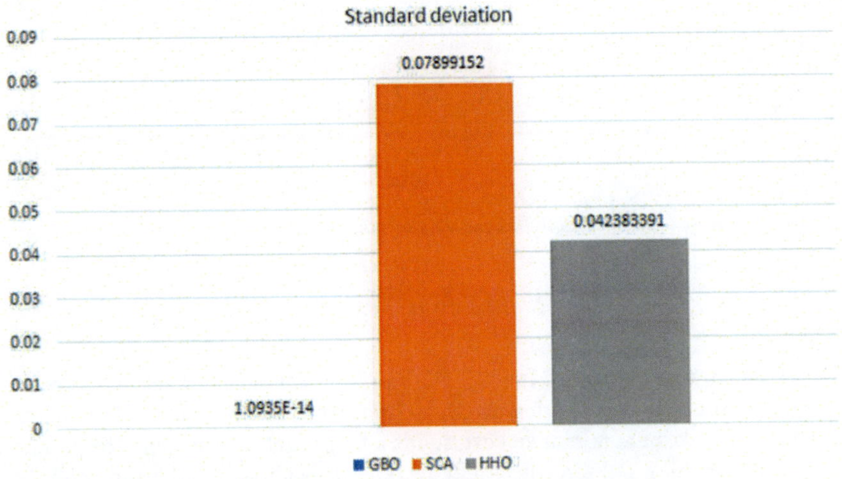

Fig. 12 Standard deviation from 20 runs for the tubular column problem

the previous problem, we can see that the GBO algorithm converged towards the global optimum much faster than the other two algorithms.

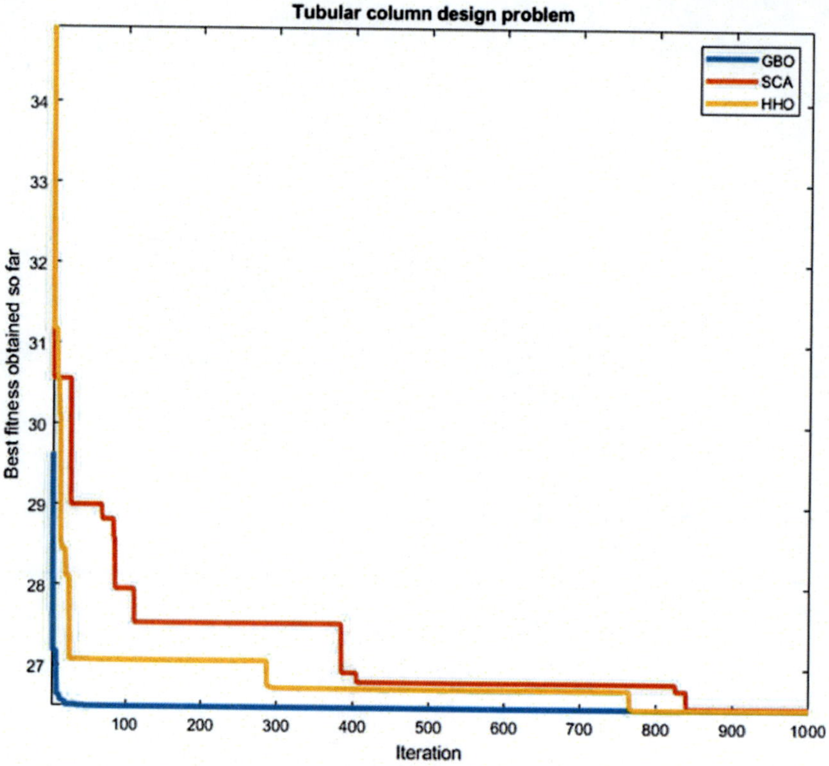

Fig. 13 Convergence curves for the tubular column problem

3.3 *A Reinforced Concrete Beam Design*

This problem was introduced in [40]. The aim here is to design a simply supported reinforced concrete beam with a length of 30 ft to carry a live load of 2 k lbf and a dead load of 1 k lbf at minimum cost, so it's a minimization problem. The problem has three variables, reinforcement area (A_s), beam width (h) and beam depth (b). The design problem can be seen in Fig. 14.

The problem can be formulated as follows:

$$\text{Minimize: } f(A_s, \text{ b}, \text{ h}) = 29.4\, A_s + 0.6\text{bh} \tag{19}$$

where A_s is a discrete variable that takes one of the following values {6.0, 6.16, 6.32, 6.6, 7.0, 7.11, 7.2, 7.8, 7.9, 8.0, 8.4}, h is an integer variable in the range [28, 40] and b is a continuous variable in the range [5, 10].

Subject to the following constraints:

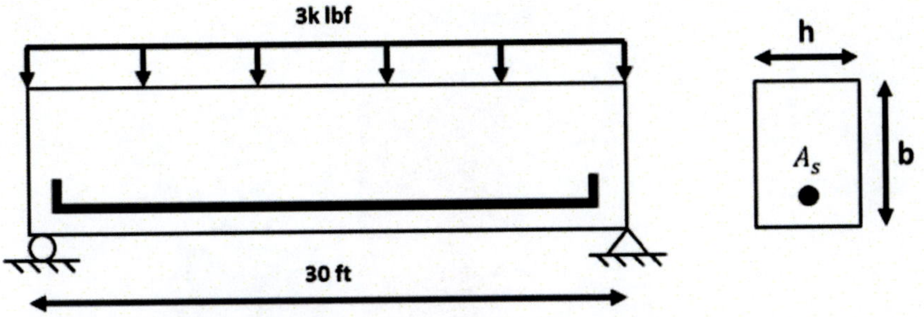

Fig. 14 Reinforced concrete beam design problem

$$g_1 = \frac{h}{b} - 4 \leq 0 \quad (20)$$

$$g_2 = 180 + 7.375\frac{A_s^2}{b} - A_s h \leq 0 \quad (21)$$

For this problem, the number of population and maximum iterations for all three algorithms are 25 and 500 respectively. The best solution found by each algorithm is illustrated in Table 4.

Figures 15, 16, 17 and 18 show the statistical results in the form of average, best, worst and standard deviation values respectively. And once again, the GBO algorithm beats the other two algorithms in all categories. It achieved the best solution among the three algorithms. And when it comes to the average value, worst value and standard deviation, the GBO algorithm was the most consistent by getting the best values in all three categories.

Figure 19 shows the convergence curves for the three algorithms. We can clearly see that the GBO algorithm converges to the optimum solution much faster than the SCA algorithm and slightly faster than the HHO algorithm.

Again, and like the previous problem, we can see that the GBO algorithm converged towards the global optimum much faster than the other two algorithms.

Table 4 Best positions for reinforced concrete beam

Algorithm	A_s	b	h
GBO	6.32	8.5	34
SCA	6.32	8.5017963	34
HHO	6.32	8.5	34

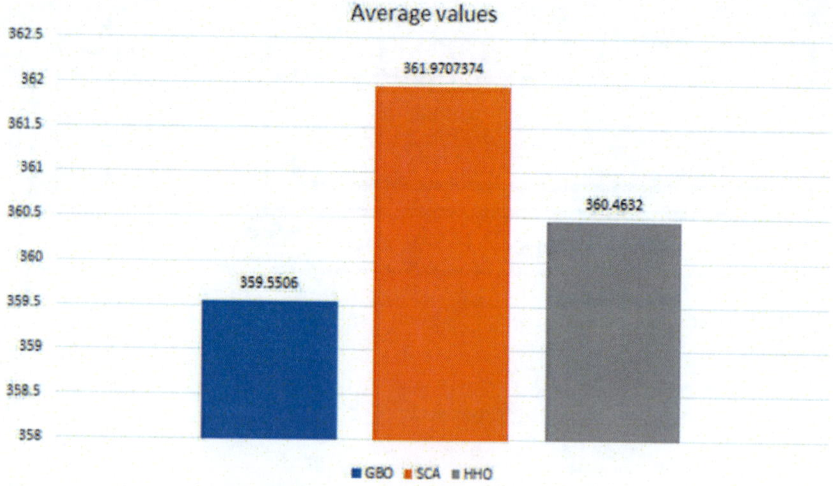

Fig. 15 Average values for the reinforced concrete beam problem

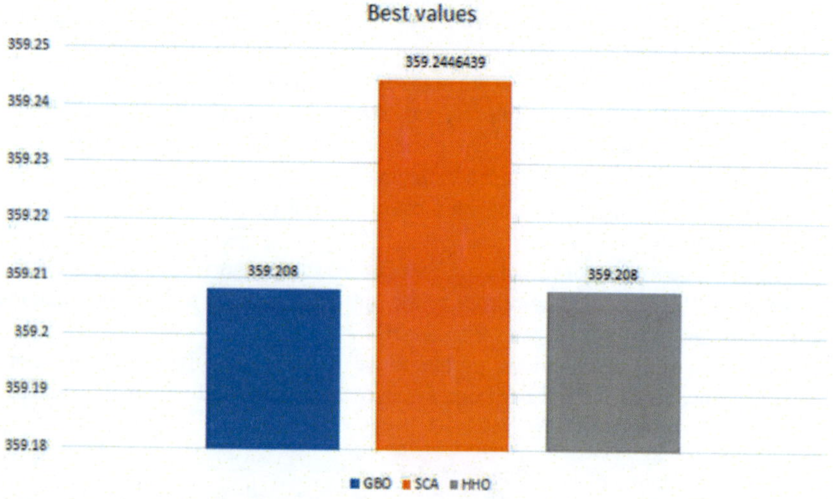

Fig. 16 Best values for the reinforced concrete beam problem

4 Conclusions and Future Work

In this chapter, we have previewed the GBO algorithm, a recently introduced metaheuristic algorithm inspired by Newton's gradient-based method. The GBO algorithm uses the Gradient Search Rule (GSR) to efficiently explore the search space and find promising regions and then converge rapidly towards the best solution. It

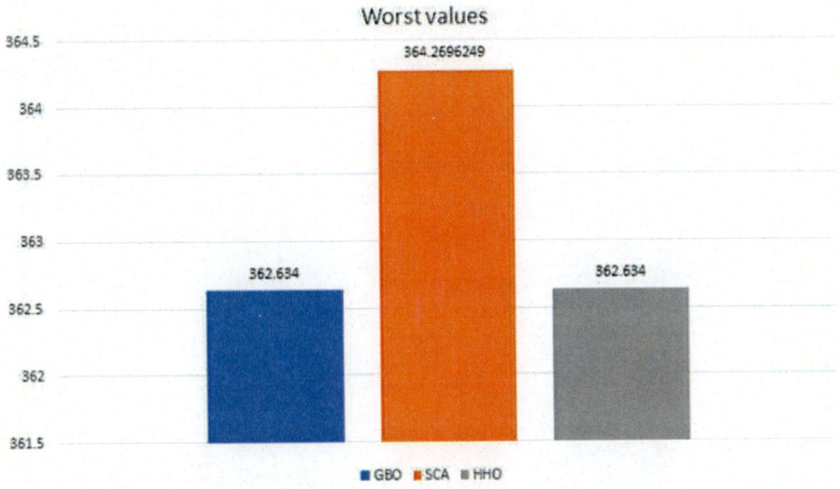

Fig. 17 Worst values for the reinforced concrete beam problem

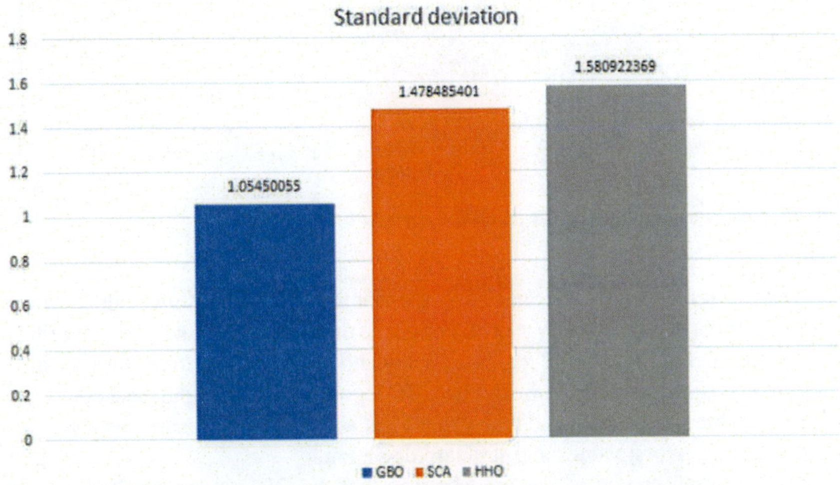

Fig. 18 Standard deviation from 20 runs for the reinforced concrete beam problem

also use the Local Escaping Operator (LEO) to avoid local optima and make sure it doesn't get stuck in suboptimal solution.

We used the GBO algorithm to solve three structural optimization problems. These are complex and constrained problems that test the ability of an optimization algorithm to avoid local optima and find the optimum solution. We compared the results from the GBO algorithm against the results from two other well-known algorithms, the Sine Cosine Algorithm (SCA) and the Harris Hawks Optimization (HHO).

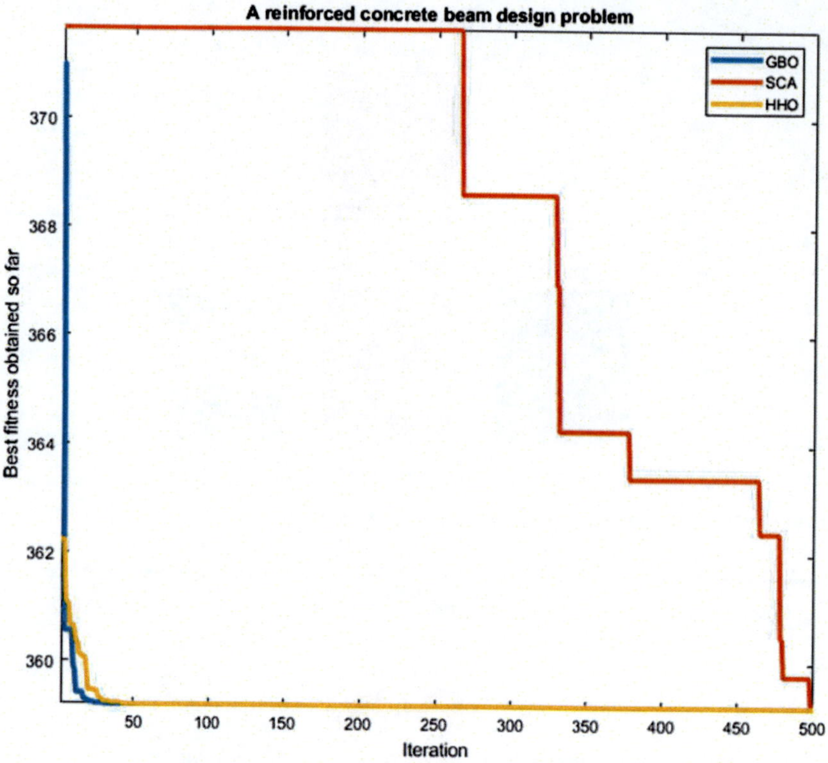

Fig. 19 Convergence curves for the reinforced concrete beam problem

Looking at the results, we can clearly see that the GBO algorithm beats the other two algorithms in almost every aspect. For all the three problems, the GBO algorithm was able to find the promising regions quickly and also converge to the best solution faster than the other two algorithms. The GBO algorithm was also the most consistent among the three algorithms and that was evident as it had better results regarding the average values, worst values and standard deviation.

All that combined, it proves the advatage that the GBO algorithm has when it comes to optimising real life problems which are usually complex, non-linear and constrained by nature. Future work should revolve around testing the GBO algorithm against other real-life applications and compare the results with more recent algorithms for validation. Another direction is to try to improve the GBO algorithm even further by using techniques like opposition-based methods and chaotic maps.

References

1. E.-G. Talbi, Metaheuristics: from design to implementation, vol. 74 (2009), John Wiley & Sons
2. E. Rashedi, H. Nezamabadi-Pour, S. Saryazdi, GSA: a gravitational search algorithm. Inf. Sci. **179**(13), 2232–2248 (2009)
3. A. Kaveh, S. Talatahari, A novel heuristic optimization method: charged system search. Acta Mech. **213**(3), 267–289 (2010)
4. B. Javidy, A. Hatamlou, S. Mirjalili, Ions motion algorithm for solving optimization problems. Appl. Soft Comput. **32**, 72–79 (2015)
5. A. Kaveh, A. Dadras, A novel meta-heuristic optimization algorithm: thermal exchange optimization. Adv. Eng. Softw. **110**, 69–84 (2017)
6. J.H. Holland, Genetic algorithms. Sci. Am. **267**(1), 66–73 (1992)
7. R. Storn, K. Price, Differential evolution–a simple and efficient heuristic for global optimization over continuous spaces. J. Global Optim. **11**(4), 341–359 (1997)
8. J. Kennedy, R. Eberhart. Particle swarm optimization, in *Proceedings of ICNN'95 - International Conference on Neural Networks* (1995)
9. A. Colorni, M. Dorigo, V. Maniezzo, Distributed optimization by ant colonies, in *Proceedings of the First European Conference on Artificial Life* (Paris, France, 1991)
10. D. Karaboga, B. Basturk. Artificial bee colony (ABC) optimization algorithm for solving constrained optimization problems, in *International Fuzzy Systems Association World Congress* (Springer, 2007)
11. X.-S. Yang, S. Deb, Cuckoo search via Lévy flights, in *2009 World Congress on Nature & Biologically Inspired Computing (NaBIC)* (IEEE, 2009)
12. X.S. Yang, A.H. Gandomi, Bat algorithm: a novel approach for global engineering optimization. Engineering Computations (2012)
13. M. Yazdani, F. Jolai, Lion optimization algorithm (LOA): a nature-inspired metaheuristic algorithm. J. Comput Des. Eng. **3**(1), 24–36 (2016)
14. S. Mirjalili, S.M. Mirjalili, A. Lewis, Grey wolf optimizer. Adv. Eng. Softw. **69**, 46–61 (2014)
15. M. Issa et al., Pairwise global sequence alignment using sine-cosine optimization algorithm, in *International Conference on Advanced Machine Learning Technologies and Applications* (Springer, 2018)
16. M. Issa, M. Abd Elaziz, Analyzing COVID-19 virus based on enhanced fragmented biological local aligner using improved Ions motion optimization algorithm. Appl. Soft Comput. 106683 (2020)
17. M. Issa, A.E. Hassanien, Multiple sequence alignment optimization using meta-heuristic techniques, in *Handbook of Research on Machine Learning Innovations and Trends*, IGI Global. (2017) pp. 409–423
18. M. Issa et al., Gene-Tracer: algorithm tracing genes modification from ancestors through offsprings. Int J Comp Appl. **52**(19) (2012)
19. M. Issa, *Sequence Analysis Algorithms for Bioinformatics Application* (GRIN Verlag, 2014)
20. M. Elloumi, M.A.S. Issa, A. Mokaddem, *Accelerating Pairwise Alignment Algorithms by Using Graphics Processor Units*. Biological Knowledge Discovery Handbook: Preprocessing, Mining, and Postprocessing of Biological Data, pp. 969–980
21. M. Issa et al., ASCA-PSO: adaptive sine cosine optimization algorithm integrated with particle swarm for pairwise local sequence alignment. Expert Syst. Appl. **99**, 56–70 (2018)
22. A. Mohamed Issa, I. Ziedan, A. Alzohairy, *Maximizing occupancy of GPU for fast scanning biological database using sequence alignment*. J. Appl. Sci. Res. **13**(6) (2017)
23. K.-C. Wong, Recent computational trends in biological sequence alignment, in *Computational Biology and Bioinformatics*. (CRC Press, 2016), pp. 292–317
24. M. Issa, Digital image watermarking performance improvement using bio-inspired algorithms, in *Advances in soft computing and machine learning in image processing*. (Springer, 2018), pp. 683–698
25. M. Issa, *Expeditious Covid-19 similarity measure tool based on consolidated SCA algorithm with mutation and opposition operators*. Appl. Soft Comput. **104**, 107197 (2021)

26. M. Issa et al., A biological sub-sequences detection using integrated BA-PSO based on infection propagation mechanism: case study COVID-19. Expert Syst. Appl. **189**, 116063 (2022).
27. M. Issa, A. Helmi, Two layer hybrid scheme of IMO and PSO for optimization of local aligner: COVID-19 as a case study, in *Artificial Intelligence for COVID-19*. (Springer, 2021), pp. 363–381
28. M. Issa et al., Chaotic Ions motion optimization (CIMO) for biological sequences local alignment: COVID-19 as a case study. Recent Trends Signal Image Proc. ISSIP **2021**(1333), 1 (2020)
29. M. Issa et al., PID controller tuning parameters using meta-heuristics algorithms: comparative analysis, in *Machine Learning Paradigms: Theory and Application*. (Springer, 2019), pp. 413–430
30. M. Issa, Performance optimization of PID controller based on parameters estimation using meta-heuristic techniques: a comparative study, in *Metaheuristics in Machine Learning: Theory and Applications*. (Springer, 2021), pp. 691–709
31. M. Issa, A. Samn, Passive vehicle suspension system optimization using Harris Hawk Optimization algorithm. Math. Comput. Simul. **191**, 328–345 (2022)
32. D.H. Wolpert, W.G. Macready, No free lunch theorems for optimization. IEEE Trans. Evol. Comput. **1**(1), 67–82 (1997)
33. I. Ahmadianfar, O. Bozorg-Haddad, X. Chu, Gradient-based optimizer: a new metaheuristic optimization algorithm. Inf. Sci. **540**, 131–159 (2020)
34. S. Mirjalili, SCA: a sine cosine algorithm for solving optimization problems. Knowl.-Based Syst. **96**, 120–133 (2016)
35. A.A. Heidari et al., Harris hawks optimization: algorithm and applications. Futur. Gener. Comput. Syst. **97**, 849–872 (2019)
36. T.J. Ypma, Historical development of the newton-Raphson method. SIAM Rev. **37**(4), 531–551 (1995)
37. M.R. Hestenes, E. Stiefel, Methods of Conjugate Gradients for Solving Linear Systems, vol. 49 (NBS Washington, DC, 1952)
38. D. Kavlie, *Optimization of plane elastic grillages*, Ph.D Thesis, Norges Teknisk Naturvitenskapelige Universitet, Norway (1967)
39. S.S. Rao, *Engineering Optimization: Theory and Practice* (Wiley, 1996)
40. H.M. Amir, T. Hasegawa, Nonlinear mixed-discrete structural optimization. J. Struct. Eng. **115**(3), 626–646 (1989)

Aquila Optimizer Based PSO Swarm Intelligence for IoT Task Scheduling Application in Cloud Computing

Laith Abualigah, Mohamed Abd Elaziz, Nima Khodadadi, Agostino Forestiero, Heming Jia, and Amir H. Gandomi

Abstract This paper introduces IAO, a new swarm intelligence approach for addressing the challenge of task scheduling in cloud computing. The proposed method uses conventional Aquila Optimizer (AO) and Particle Swarm Optimizer (PSO) as a hybrid method based on a novel transition mechanism. The proposed hybrid method, IAO, combined the AO and PSO to avoid the weaknesses they face; these weaknesses are trapped in the local search area and have low solution diversity. The proposed transition mechanism is proposed to acquire proper changes between the search operators in order to keep the improvements; it changes between them when any algorithm gets stuck or the solutions diversity decreases. Several scenarios are conducted and tested to validate the suggested method's ability to address the task scheduling problem; these scenarios contain various tasks (i.e., 600, 1000, and 2000). The obtained results are compared with other well-known methods in terms

L. Abualigah (✉)
Faculty of Computer Sciences and Informatics, Amman Arab University, Amman 11953, Jordan

School of Computer Sciences, Universiti Sains Malaysia, Pulau, Pinang 11800, Malaysia
e-mail: aligah.2020@gmail.com

M. A. Elaziz
Department of Mathematics, Faculty of Science, Zagazig University, Zagazig, Egypt

N. Khodadadi
Iran University of Science and Technology, Tehran, Iran

A. Forestiero
Institute for High Performance Computing and Networking (ICAR), National Research Council of Italy (CNR), Rende (CS), Italy
e-mail: Agostino.forestiero@icar.cnr.it

H. Jia
School of Information Engineering, Sanming University, Sanming 365004, China
e-mail: jiaheming@fjsmu.edu.cn

A. H. Gandomi
Faculty of Engineering and Information Technology, University of Technology Sydney, Ultimo, NSW 2007, Australia
e-mail: Gandomi@uts.edu.au

E. H. Houssein et al. (eds.), *Integrating Meta-Heuristics and Machine Learning for Real-World Optimization Problems*, Studies in Computational Intelligence 1038,
https://doi.org/10.1007/978-3-030-99079-4_19

of Max, Mean, Min of the Expected Complete Time (ECT), Friedman ranking test, and Wilcoxon signed-rank test. The proposed IAO method achieved better results and promising compared to other comparative methods; it is an excellent scheduling approach for solving any related scheduling problem.

Keywords Internet of Things (IoT) · Task scheduling (TS) · Cloud computing (CC) · Swarm intelligence (SI) · Aquila Optimizer (AO) · Particle Swarm Optimizer (PSO)

1 Introduction

Cloud computing (CC) is a modern technology derived from parallel computing. CC platforms produce assistance to deal with big data and arrange the computing tasks [29, 35]. Task scheduling (TS) is a complicated problem that performs a vital function in conventional computing [15]. As the data size from all areas grows clearly, the task volume is also quickly growing. There is no beneficial scheme for the unique characteristics of the TS problem [8, 9, 44]. The virtual machines (VMs) are efficacious presentations to allocate and manage the tasks in the CC ecosystem. The TS problem is divided into two layers: the end user's tasks and distributing them to the available VMs devices; host mapping chooses the proper VMs [25]. Optimization methods have evidenced their ability to solve many complicated problems such as scheduling problems and others [4, 11, 13, 45].

Cloud computing is increasingly widely used in industry, education, and society as Internet connectivity and big data rise in volume, speed, and diversity [43]. Cloud computing, grid computing, cloud services, and computer architecture are all cloud services. Cloud computing delivers elevated and scalability disk storage services on demand. Several computer paradigms have promised this utility computer. One such dependable computing model is the public cloud [20].

In the literature, there are several researches have been conducted to solve various problems [1, 17, 41] and one of the most common problem is the TS [12]. The scheduling problem can solve many problems like the task in CC and timetables, and other scheduling problems [27, 32, 53]. The strong method help in solving the problem efficiently and find new solution to give a better arrangement [21, 50, 52].

The virtualization idea in cloud computing allows resources to be shared among cloud customers [5]. Virtualization is one of the essential characteristics of a cloud computing environment because it allows for dynamic sharing of physical resources and allows several applications to run on various platforms called VMs on a server. A provider can ensure the quality of service (QoS) provided to diverse users while maximizing resource usage and lowering power consumption using the hybridization concept. A novel mathematical model based on the balancing measure and Particle Swarm Optimization is proposed in [24] to solve the TS in CC. To validate the performance of the proposed method, several evaluation measures are considered. The results showed that the proposed method preserved the Makespan and execution time. A novel Discrete Symbiotic Organism method is suggested in [6] to tackle the task scheduling problems on various cloud devices. The suggested method is validated by t-test, and the results revealed that the suggested method got improvements significantly compared to other methods.

This paper presents a task scheduler based on many discrete variations of the particle swarm (PSO) algorithm for job scheduling in cloud computing [30]. These techniques were presented in three well-known metaheuristics on task scheduling problems to assess their performance. The outcomes of the experiments demonstrate the productivity and efficacy of the offered methodologies. For the suggested PSO-based scheduler, a practical approach is to employ the exponential decreasing technique to produce an optimal scheduler. A novel Energy-aware Task Scheduling method is introduced in [36] to reduce the energy cost for smartphones while providing severe time restrictions and the possible restriction for utilizations. The results showed that the proposed method significantly decreases energy cost than the conventional scheduling system. A new task scheduling-based Whale Optimizer is introduced in [42] using a multi-objective optimization design. The suggested method efficiently schedules the given tasks to the VMs while producing the minimum Makespan and cost values. Some of the common optimization method can be found in [7, 14]. These methods can be used to solve any problem and the TS also can be solved suing these method and pother such as in [10, 33, 37–39]. So, finding a new efficient method is not easy as each problem has some specific features and characterizations [13, 28, 46, 47, 49]. Also, the variance between the problems parameters are wildly known and each problem has a set of different parameters that should be adjusted carefully [16, 22, 26, 51].

A swarm intelligence method-based job scheduler for IoT applications is suggested in this paper, called IAO. As a hybrid technique based on a new transition mechanism, the proposed method combines traditional Aquila Optimizer (AO) and Particle Swarm Optimizer (PSO). The suggested hybrid technique (IAO) combines the Aquila Optimizer and the Particle Swarm Optimizer to avoid disadvantages. These flaws are imprisoned in the local search area and low solution diversity. The suggested transition method is designed to acquire suitable transitions between search operators in order to maintain improvements. It switches between them when any algorithm becomes stopped or when the variety of solutions reduces. Several scenarios are undertaken and evaluated to confirm the suggested method's capacity to handle the task scheduling problem; these scenarios involve varying numbers of tasks (i.e., 600, 1000, and 2000). In terms of Max, Mean, Min of the Expected Complete Time (ECT), Friedman ranking test, and Wilcoxon signed-rank test, the acquired findings are compared to other well-known techniques. Compared to comparable approaches, the suggested method (IAO) produced superior results and showed promise; it may be an ideal scheduling strategy for handling any relevant scheduling problem.

2 Task Scheduling Problem and Its Notations

Task scheduling is considered as a hard problem, to locate various tasks into virtual machines (VMs) efficiently [2, 3, 19]. The efficient method can determine the minimal execution time for the given tasks. Cloud computing (CC) normally has N virtual machines (VMs) as presented in Eq. (1).

$$CC = [VM_1, VM_2, VM_3, ..., VM_i, ..., VM_N] \tag{1}$$

where VM_i is the i_{th} VM and N is the used VMs in the given scenarios.

The attributes of VM_i is presented as in Eq. (2).

$$VM_i = [SIDV_i, mips_i] \tag{2}$$

where $SIDV$ is used as an identifier of the i_{th} used VM and $mips_i$ is the millions-of-instructions-per-second. The tasks (Ts) are mathematically presented as follows.

$$Ts = [Task_1, Task_2, ..., Task_i, ..., Task_{Dim}] \tag{3}$$

where N_{Dim} is the given tasks by the end users through the experiments. The i_{th} task is presented as in Eq. (4).

$$Task_i = [SIDT_i, Task - length_i, ECT_i, LI_i] \tag{4}$$

where $SIDT_i$ is the unique number for the i_{th} task and $task - length_i$ is length of the i_{th} task. ECT_i is the expected executing time the i_{th} task; LI_i is the task priority [23].

Completion time (CT) is the running time after executing the entire task on VM calculated by Eq. (5).

$$CT_i = \sum_{j=1}^{n} \frac{T_{i.length}}{VM_{i.pesnumber} \times VM_{i.mips}}, \tag{5}$$

Makespan measure is the overall time needed to execute all the given tasks, which is calculated by Eq. (6).

$$makespan = \max_{1 \leq i \leq m} \{CT_i\} \tag{6}$$

Execution cost (EC) is the cost of the CC user calculated by Eq. (7) [31].

$$EC = \sum_{i=1}^{m} price_i * CT_i \tag{7}$$

3 The Proposed Swarm Intelligence Scheduler Method

3.1 Aquila Optimizer (AO)

In the Aquila Optimizer (AO), the optimization process is concluded on a set of solutions (X) as presented in Eq. (8). These solutions are created randomly between the maximum (UB) and minimum (LB) decision values. In each iteration, the best solution is determined.

$$X = \begin{bmatrix} x_{1,1} & \cdots & x_{1,j} & x_{1,Dim-1} & x_{1,Dim} \\ x_{2,1} & \cdots & x_{2,j} & \cdots & x_{2,Dim} \\ \cdots & \cdots & x_{i,j} & \cdots & \cdots \\ \vdots & \vdots & \vdots & \vdots & \vdots \\ x_{N-1,1} & \cdots & x_{N-1,j} & \cdots & x_{N-1,Dim} \\ x_{N,1} & \cdots & x_{N,j} & x_{N,Dim-1} & x_{N,Dim} \end{bmatrix} \tag{8}$$

where X is the used candidate solutions, N is the number of solutions in X, and these solutions are created by Eq. (9),

X_i is the decision of the ith solution, Dim is the number of used locations in the tested problem, and N is the number of used candidate solutions.

$$X_{ij} = rand \times (UB_j - LB_j) + LB_j, \tag{9}$$

$$i = 1, 2, 4, 5, \ldots\ldots, N \tag{10}$$

$$j = 1, 2, 3, 4, 5, \ldots., Dim \tag{11}$$

where $rand$ is a random value.

Mathematical model of AO The mathematical presentation of the AO is presented as follows.

- The first part (X_1) is mathematically presented as given in Eq. (12).

$$X_1(t+1) = X_{best}(t) \times \left(1 - \frac{t}{T}\right) + (X_M(t) - X_{best}(t) * rand), \tag{12}$$

where, $X_{best}(t)$ presents the best solution, $(\frac{1-t}{T})$ is a control search mechanism, and $X_M(t)$ is determined by Eq. (13) to find the mean value of the used solutions. t and T are the used iteration and the maximum iterations, respectively.

$$X_M(t) = \frac{1}{N}\sum_{i=1}^{N} X_i(t), \ \forall j = 1, 2, \ldots, Dim \tag{13}$$

- The second part (X_2) is mathematically presented as given in Eq. (14)

$$X_2(t+1) = X_{best}(t) \times Levy(Dim) + X_R(t) + (y - x) * rand, \quad (14)$$

where $Levy(D)$ is calculated using Eq. (15), and $X_R(t)$ is a selected random candidate solution.

$$Levy(D) = s \times \frac{u \times \sigma}{|\upsilon|^{\frac{1}{\beta}}} \quad (15)$$

where s is a parameter adjusted to 0.01, u and υ are random values. σ is determined by Eq. (16).

$$\sigma = \left(\frac{\Gamma(1+\beta) \times sine(\frac{\pi\beta}{2})}{\Gamma(\frac{1+\beta}{2}) \times \beta \times 2^{(\frac{\beta-1}{2})}} \right) \quad (16)$$

where β is adjusted to 1.5, in Eq. (14), y and x are utilized to show the spiral form during the search process, which is as follows.

$$y = r \times cos(\theta) \quad (17)$$

$$x = r \times sin(\theta) \quad (18)$$

where,

$$r = r_1 + U \times D_1 \quad (19)$$

$$\theta = -\omega \times D_1 + \theta_1 \quad (20)$$

$$\theta 1 = \frac{3 \times \pi}{2} \quad (21)$$

r_1 a value in range [1 20], and U is fixed to 0.00565, D_1 a value in range [1 Dim], and ω is fixed to0.005. All parameters values are taken from the original paper [18].

- The third part (X_3) is mathematically presented as given in Eq. (22)

$$X_3(t+1) = (X_{best}(t) - X_M(t)) \times \alpha - rand + ((UB - LB) \times rand + LB) \times \delta, \quad (22)$$

where α and δ are parameters fixed 0.1

- The fourth part (X_4) is mathematically presented as given in Eq. (23).

$$X_4(t+1) = QF \times X_{best}(t) - (G_1 \times X(t) \times rand) - G_2 \times Levy(D) + rand \times G_1, \quad (23)$$

where QF is the quality function utilized to adjust the search processes, which is in Eq. (24), G_1 is generated by Eq. (25), G_2 is generated by Eq. (26), and $X(t)$ is the current solution.

$$QF(t) = t^{(\frac{2\times rand()-1}{(1-T)^2})} \tag{24}$$

$$G_1 = 2 \times rand() - 1 \tag{25}$$

$$G_2 = 2 \times (1 - \frac{t}{T}) \tag{26}$$

3.2 *Particle Swarm Optimizer (PSO)*

Particle Swarm Optimizer (PSO) utilized the global-best concept to find the best solution and it is updated in each iteration [34].

Two main strategies are used in PSO to update the candidate solutions; velocity as in Eq. (28) and position update as in Eq. (27).

$$X(t+1) = X_{ij} + V_{ij} \tag{27}$$

where,

$$V_{i,j} = w * \times V_{ij} + c_1 \times rand_1 \times (LB_I - x_{i,j}) + c_2 \times rand_2 \times (GB_I - X_{i,j}), \tag{28}$$

LB_I is the best-local solution, and GB_I is the best-global solution. The inertia weight value is calculated using Eq. (29).

$$w = w_{max} - w_{min} \times (\frac{I_{max} - I}{I_{max}}) + w_{min} \tag{29}$$

where w_{max} and w_{min} are the max and min values, which are normal fixed to 0.2 and 0.9, respectively.

3.3 *The Proposed IAO Scheduler*

In this section, the proposed IAO method is presented. In the proposed IAO, two main search methods (i.e., Aquila Optimizer and Particle Swarm Optimizer) are employed based on a novel transition mechanism (TM). Equation 30 presents the proposed transition mechanism, which is used to equilibrium the search process and keep the diversity of the solutions. This cause in averting the local search trapped and improving the quality of the candidate solutions

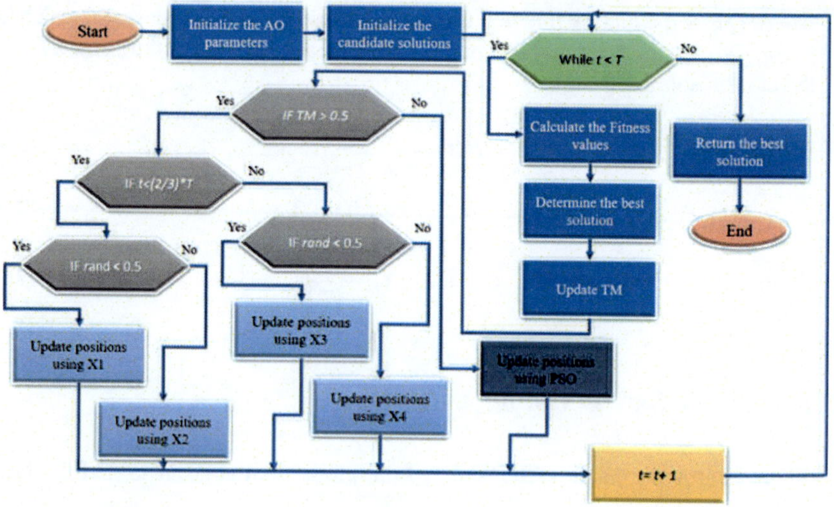

Fig. 1 The flowchart of the proposed method

$$TM = \frac{1}{2} Sin(\pi + 2\pi \times \frac{t}{T}) + rand \tag{30}$$

Figure 1 shows the general procedure of the proposed IAO method. It is clear that the proposed method starts with initialize the solutions and define the parameters, including the transition parameter (TM). Then, if the TM $> t$, the optimization process depends on the AO operators; otherwise, it depends on the PSO operators. The changes in the updating strategies improve the solution diversity and avoiding any trapped in the search areas, especially the local search area.

4 Experiments Results and Discussion

Several scenarios (i.e., 600, 1000, and 2000 tasks) are tested to evaluate the performance of the proposed method in solving the task scheduling problem in cloud computing applications. To prove the effectiveness of the proposed method, several well-known methods are used in the comparisons, including Aquila Optimizer (AO) [18], Genetic Algorithm (GA) [48], Particle Swarm Optimizer (PSO) [34], and Differential Evolution (DE) [40]. The parameter values are the same as in the original papers. In terms of Max ECT, Mean ECT, Min ECT, Friedman ranking test, and Wilcoxon signed-rank test, the suggested technique is compared to well-known methods. we choose many comparative methods and scenarios to efficiently prove the excellent ability of the given method.

Table 1 The results of the comparative methods for various tasks in terms of the fitness function values (ECT)

Tasks	Measure	AO	GA	PSO	DE	IAO
600	Max	253.1455	195.6450	165.1940	510.6400	160.3620
	Mean	128.7274	92.6637	95.7374	429.1955	90.6637
	Min	124.3250	90.1541	91.3652	325.3671	88.6178
	Rank	4	2	3	5	1
	p-value	5.15E-03	3.65E-03	6.48E-03	6.54E-04	NaN
	Sign	1	1	1	1	NaN
1000	Max	195.2650	186.6470	174.6540	432.6700	179.6580
	Mean	98.9958	97.9958	96.9958	397.5399	91.9958
	Min	91.2350	89.2540	90.2541	125.6150	88.1514
	Rank	4	3	2	5	1
	p-value	2.58E-03	3.65E-03	5.49E-04	6.55E-04	NaN
	Sign	1	1	1	1	NaN
2000	Max	253.6550	232.1450	211.6810	501.2530	203.3250
	Mean	103.1168	105.2156	101.0366	471.3632	93.0235
	Min	93.6550	95.3690	92.4850	256.3250	89.3256
	Rank	3	4	2	5	1
	p-value	1.54E-04	6.56E-04	5.12E-04	5.45E-05	NaN
	Sign	1	1	1	1	NaN
Friedman test						
Mean rank		3.6666	3.0000	2.3333	5.0000	1.0000
Final ranking		4	3	2	5	1
Wilcoxon test						
(W – F – L)		(0 – 0 – 0)	(0 – 0 – 0)	(0 – 0 – 0)	(0 – 0 – 0)	(0 – 0 – 3)

The outcomes of the comparison approach for various tasks in terms of fitness function values are shown in Table 1. Several metrics are offered to clearly display the collected findings. As demonstrated in Table 1, the suggested IAO outperformed all existing comparing approaches in all test circumstances (i.e., AO, GA, PSO, and DE). The results show that the suggested IAO's performance is clearly promising. The proposed technique yielded the highest ECT values in all of the examples studied. For 600 tasks, it received an average of 90.6637, 91.9958 for 1000 tasks, and 93.0235 for 2000 tasks. The suggested IAO received the highest rating in the Friedman ranking test, followed by PSO, GA, AO, and DE. Furthermore, according to the Wilcoxon signed-rank test, the suggested major improvements in all of the evaluated cases win in three out of three cases.

Figures 2, 3, and 4 shows the convergence curves of the comparative methods according to the ECT measure using 600, 1000, and 2000 tasks, respectively. It is clear that the proposed method converged to the best solution smoothly and got the

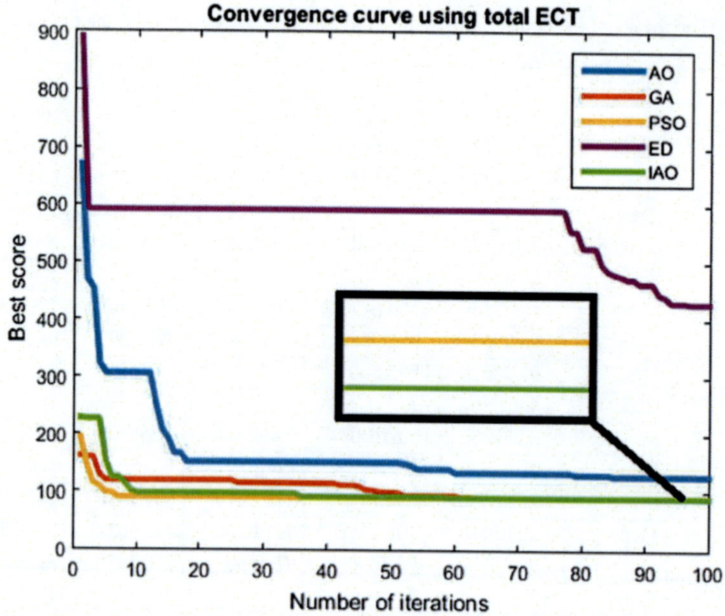

Fig. 2 The convergence curves of the comparative methods according to the ECT measure using 600 tasks

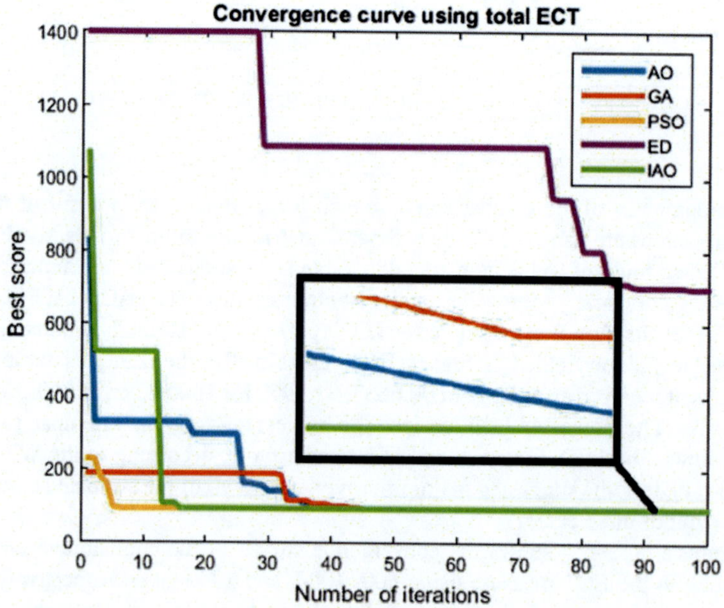

Fig. 3 The convergence curves of the comparative methods according to the ECT measure using 1000 tasks

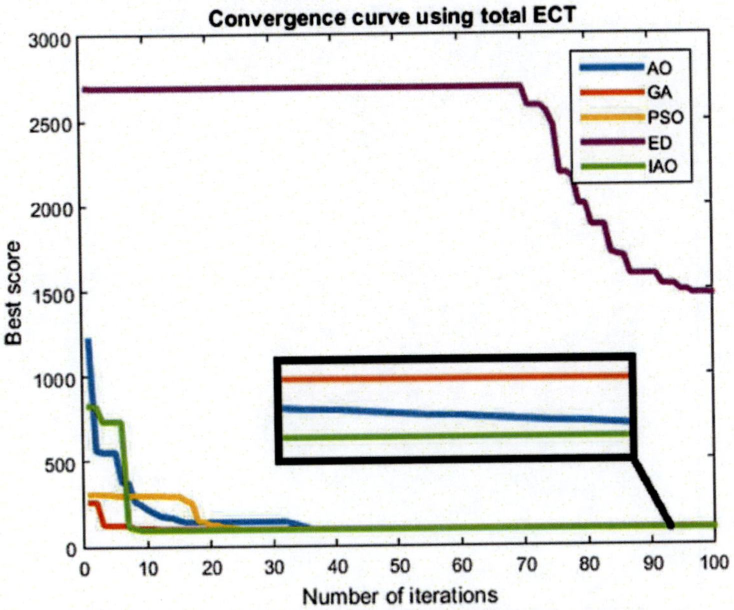

Fig. 4 The convergence curves of the comparative methods according to the ECT measure using 2000 tasks

best results in all the tested cases. From these Figures, the proposed IAO proved its ability to solve the small, middle, and big number of tasks efficiently. Figures 5, 6, 7, and 8 show the total cost obtained by the proposed method alongside with the Makespan values. From these Figures, we observed that the proposed method would probably reduce the cost values and increase the Makespan values. We concluded that the proposed IAO finds the optimal solutions for the given scenarios (i.e., 40, 160, 300, and 500 tasks) effectively by increasing the Makespan measure values and getting the minimal cost values. We found that the deal with this class of problem is not easy and need a really powerfully method to handle its details.

We concluded that the proposed IAO is an efficient method according to the obtained results. The modifications that have been performed demonstrated a clear improvement in the performance of the conventional method. The proposed IAO deals efficiently with a small and large number of tasks. The proposed method give good results compared to other methods and this encourage us also to test it on solving other problems.

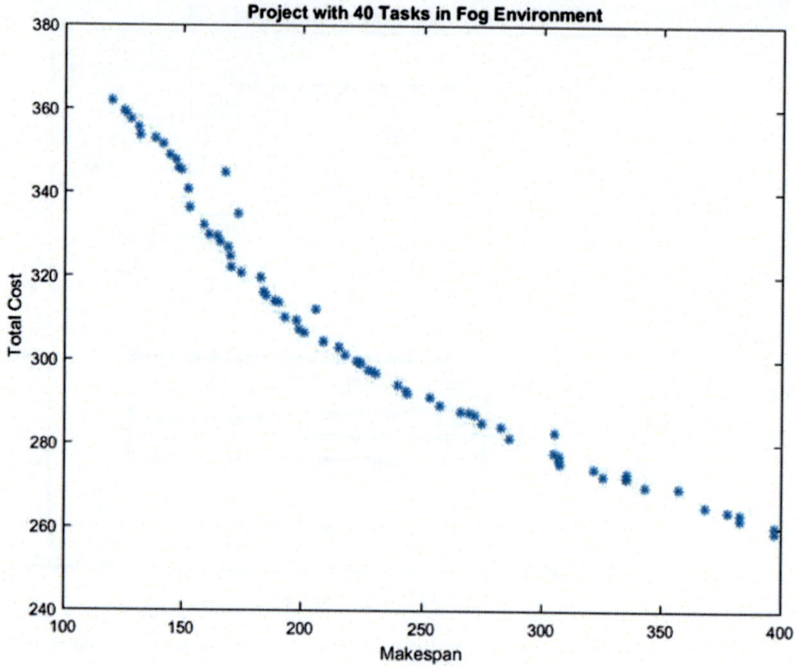

Fig. 5 The cost results of the proposed IAO with Makespan values for 40 tasks

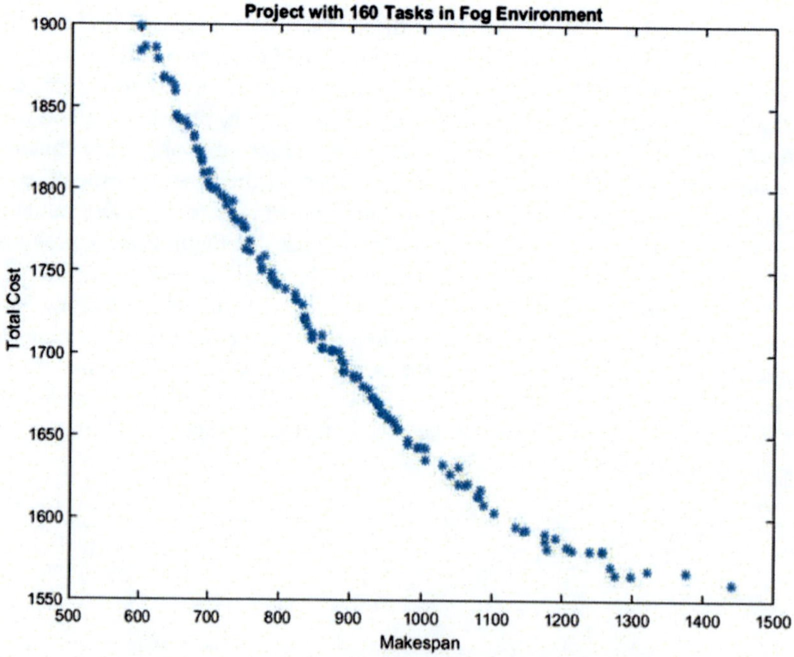

Fig. 6 The cost results of the proposed IAO with Makespan values for 160 tasks

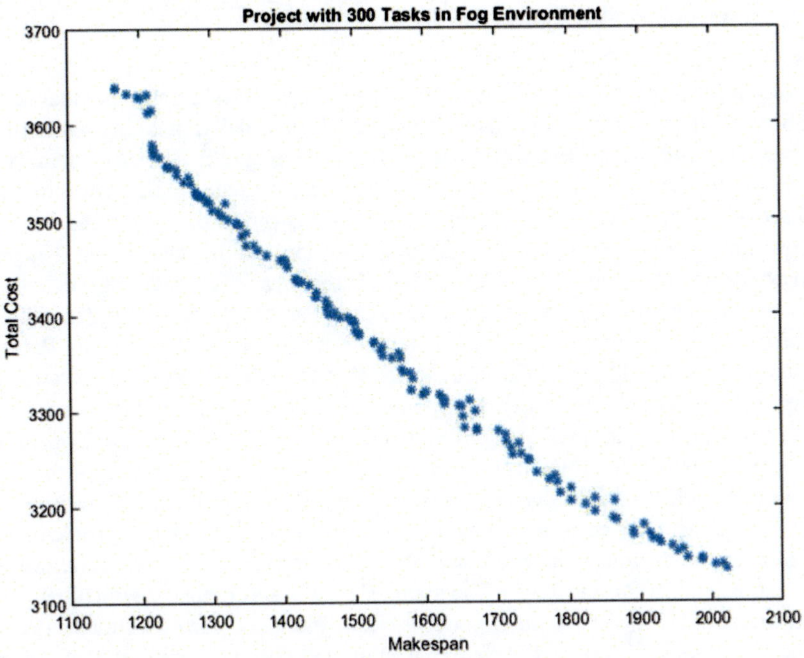

Fig. 7 The cost results of the proposed IAO with Makespan values for 300 tasks

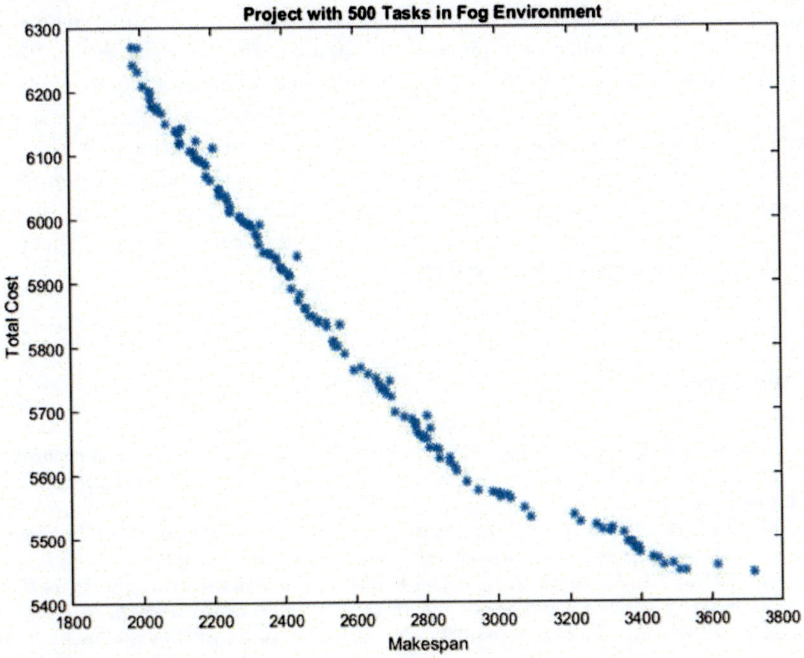

Fig. 8 The cost results of the proposed IAO with Makespan values for 500 tasks

5 Conclusion and Potential Future Works

In recent years, the Internet of Things (IoT) has become one of the most popular applications. This technology uses many intelligent devices daily, making connections, scheduling, and other activity between them. So, dealing with a massive number of devices and their tasks is a very hard problem, and it needs an efficient scheduler to arrange it. A novel swarm intelligence method is proposed for solving the task scheduling problem in IoT cloud computing (CC) applications, called IAO. The proposed method uses two main search methods based on a novel transition mechanism as a hybrid method; Aquila Optimizer (AO) and Particle Swarm Optimizer (PSO). The proposed hybrid method (IAO) combined the Aquila Optimizer and Particle Swarm Optimizer to avoid the weaknesses that face them; these weaknesses are trapped in the local search area and have low solution diversity. The proposed transition mechanism is proposed to equilibrium the changes between the search operators in order to keep the improvements; it changes between them when any algorithm gets stuck or the solutions diversity decreases. To validate the performance of the proposed method against other comparative methods to solve the task scheduling problem, several scenarios are conducted and tested; these scenarios contain a various number of tasks (i.e., 600, 1000, and 2000). The obtained results are compared with other well-known methods (i.e., Genetic Algorithm (GA), Particle Swarm Optimizer (PSO), Aquila Optimizer (AO), and Differential Evolution (DE)) in terms of Max, Mean, Min of the Expected Complete Time (ECT), Friedman ranking test, and Wilcoxon signed-rank test. The proposed IAO method got better results and promising than other comparative methods; it can be considered an efficient scheduling method for solving any related scheduling problem. In future work, two main directions are available; the first is to study the proposed algorithm behaviour and try to improve it further using other search operators, which can fix any realized problem. The section direction is to test the proposed method in solving other optimization problems such as job shop scheduling, course time table scheduling, nurse scheduling problem, travel salesman problem, parameter identification, text clustering problem, images segmenting problems, feature selection problems, advanced industrial engineering problems, benchmark functions, and others.

References

1. A. Abbassi, R.B. Mehrez, B. Touaiti, L. Abualigah, E. Touti: Parameterization of photovoltaic solar cell double-diode model based on improved arithmetic optimization algorithm. Optik, 168600 (2022)
2. M. Abd Elaziz, L. Abualigah, I. Attiya: Advanced optimization technique for scheduling iot tasks in cloud-fog computing environments. Future Gener. Comput. Syst. (2021)
3. M. Abd Elaziz, L. Abualigah, R.A. Ibrahim, I. Attiya: Iot workflow scheduling using intelligent arithmetic optimization algorithm in fog computing. Comput. Intell. Neurosci. **2021** (2021)
4. M. Abd Elaziz, A.A. Ewees, D. Yousri, L. Abualigah, M.A. Al-qaness: Modified marine predators algorithm for feature selection: case study metabolomics. Knowl. Inf. Syst. 1–27 (2022)

5. M. Abd Elaziz, S. Xiong, K. Jayasena, L. Li, Task scheduling in cloud computing based on hybrid moth search algorithm and differential evolution. Knowl.-Based Syst. **169**, 39–52 (2019)
6. M. Abdullahi, M.A. Ngadi et al., Symbiotic organism search optimization based task scheduling in cloud computing environment. Futur. Gener. Comput. Syst. **56**, 640–650 (2016)
7. L. Abualigah, M. Abd Elaziz, P. Sumari, Z.W. Geem, A.H. Gandomi, Reptile search algorithm (rsa): A nature-inspired meta-heuristic optimizer. Expert Syst. Appl. **191**, 116158 (2022)
8. L. Abualigah, M. Alkhrabsheh: Amended hybrid multi-verse optimizer with genetic algorithm for solving task scheduling problem in cloud computing. J. Supercomput. 1–26 (2021)
9. L. Abualigah, M. Alkhrabsheh, Amended hybrid multi-verse optimizer with genetic algorithm for solving task scheduling problem in cloud computing. J. Supercomput. **78**(1), 740–765 (2022)
10. L. Abualigah, A. Diabat: Chaotic binary group search optimizer for feature selection. Expert Syst. Appl. 116368 (2021)
11. L. Abualigah, A. Diabat: Improved multi-core arithmetic optimization algorithm-based ensemble mutation for multidisciplinary applications. J. Intell. Manuf. 1–42 (2022)
12. L. Abualigah, A. Diabat, M. Abd Elaziz: Intelligent workflow scheduling for big data applications in iot cloud computing environments. Cluster Comput. 1–20 (2021)
13. L. Abualigah, A. Diabat, M.A. Elaziz: Improved slime mould algorithm by opposition-based learning and levy flight distribution for global optimization and advances in real-world engineering problems. J. Ambient Intell. Humaniz. Comput. 1–40 (2021)
14. L. Abualigah, A. Diabat, S. Mirjalili, M. Abd Elaziz, A.H. Gandomi, The arithmetic optimization algorithm. Comput. Methods Appl. Mech. Eng. **376**, 113609 (2021)
15. L. Abualigah, A. Diabat, P. Sumari, A.H. Gandomi: Applications, deployments, and integration of internet of drones (iod): a review. IEEE Sensors J. (2021)
16. L. Abualigah, A. Diabat, P. Sumari, A.H. Gandomi, A novel evolutionary arithmetic optimization algorithm for multilevel thresholding segmentation of covid-19 ct images. Processes **9**(7), 1155 (2021)
17. L. Abualigah, M.A. Elaziz, A.M. Khasawneh, M. Alshinwan, R.A. Ibrahim, M.A. Al-qaness, S. Mirjalili, P. Sumari, A.H. Gandomi: Meta-heuristic optimization algorithms for solving real-world mechanical engineering design problems: a comprehensive survey, applications, comparative analysis, and results. Neural Comput. Appl. 1–30 (2022)
18. L. Abualigah, D. Yousri, M. Abd Elaziz, A.A. Ewees, M.A. Al-qaness, A.H. Gandomi, Aquila optimizer: a novel meta-heuristic optimization algorithm. Comput. Ind. Eng. **157**, 107250 (2021)
19. L.M. Abualigah, A. Diabat, A novel hybrid antlion optimization algorithm for multi-objective task scheduling problems in cloud computing environments. Clust. Comput. **24**(1), 205–223 (2021)
20. M. Al Shinwan, L. Abualigah, T.D. Huy, A. Younes Shdefat, M. Altalhi, C. Kim, S. El-Sappagh, M. Abd Elaziz, K.S. Kwak, An efficient 5g data plan approach based on partially distributed mobility architecture. Sensors **22**(1), 349 (2022)
21. B. Alsalibi, S. Mirjalili, L. Abualigah, A.H. Gandomi et al.: A comprehensive survey on the recent variants and applications of membrane-inspired evolutionary algorithms. Arch. Comput. Methods Eng. 1–17 (2022)
22. A.M. Altabeeb, A.M. Mohsen, L. Abualigah, A. Ghallab, Solving capacitated vehicle routing problem using cooperative firefly algorithm. Appl. Soft Comput. **108**, 107403 (2021)
23. D.A. Amer, G. Attiya, I. Zeidan, A.A. Nasr: Elite learning harris hawks optimizer for multi-objective task scheduling in cloud computing. J. Supercomput. 1–26 (2021)
24. A. Awad, N. El-Hefnawy, H. Abdel_kader: Enhanced particle swarm optimization for task scheduling in cloud computing environments. Proc. Comput. Sci. **65**, 920–929 (2015)
25. X. Chen, L. Cheng, C. Liu, Q. Liu, J. Liu, Y. Mao, J. Murphy, A woa-based optimization approach for task scheduling in cloud computing systems. IEEE Syst. J. **14**(3), 3117–3128 (2020)
26. F. Gul, I. Mir, L. Abualigah, P. Sumari, Multi-robot space exploration: an augmented arithmetic approach. IEEE Access **9**, 107738–107750 (2021)

27. M.H. Hassan, S. Kamel, L. Abualigah, A. Eid, Development and application of slime mould algorithm for optimal economic emission dispatch. Expert Syst. Appl. **182**, 115205 (2021)
28. E.H. Houssein, M. Dirar, L. Abualigah, W.M. Mohamed: An efficient equilibrium optimizer with support vector regression for stock market prediction. Neural Comput. Appl. 1–36 (2021)
29. X. Huang, C. Li, H. Chen, D. An: Task scheduling in cloud computing using particle swarm optimization with time varying inertia weight strategies. Cluster Comput. 1–11 (2019)
30. X. Huang, C. Li, H. Chen, D. An, Task scheduling in cloud computing using particle swarm optimization with time varying inertia weight strategies. Clust. Comput. **23**(2), 1137–1147 (2020)
31. I.M. Ibrahim et al., Task scheduling algorithms in cloud computing: a review. Turk. J. Comput. Math. Educ. (TURCOMAT) **12**(4), 1041–1053 (2021)
32. M. Jamei, M. Karbasi, M. Mosharaf-Dehkordi, I.A. Olumegbon, L. Abualigah, Z. Said, A. Asadi: Estimating the density of hybrid nanofluids for thermal energy application: application of non-parametric and evolutionary polynomial regression data-intelligent techniques. Measurement, 110524 (2021)
33. Y. Jiang, Q. Luo, Y. Wei, L. Abualigah, Y. Zhou, An efficient binary gradient-based optimizer for feature selection. Math. Biosci. Eng. **18**, 3813–3854 (2021)
34. J. Kennedy, R. Eberhart: Particle swarm optimization. In: Proceedings of ICNN'95-international conference on neural networks, vol 4, pp. 1942–1948. IEEE (1995)
35. A.S. Kumar, M. Venkatesan, Task scheduling in a cloud computing environment using hgpso algorithm. Clust. Comput. **22**(1), 2179–2185 (2019)
36. Y. Li, M. Chen, W. Dai, M. Qiu, Energy optimization with dynamic task scheduling mobile cloud computing. IEEE Syst. J. **11**(1), 96–105 (2015)
37. S. Lin, H. Jia, L. Abualigah, M. Altalhi, Enhanced slime mould algorithm for multilevel thresholding image segmentation using entropy measures. Entropy **23**(12), 1700 (2021)
38. M.H. Nadimi-Shahraki, A. Fatahi, H. Zamani, S. Mirjalili, L. Abualigah, M. Abd Elaziz, Migration-based moth-flame optimization algorithm. Processes **9**(12), 2276 (2021)
39. M.H. Nadimi-Shahraki, S. Taghian, S. Mirjalili, A.A. Ewees, L. Abualigah, M. Abd Elaziz, Mtv-mfo: Multi-trial vector-based moth-flame optimization algorithm. Symmetry **13**(12), 2388 (2021)
40. K.V. Price: Differential evolution. In: Handbook of optimization (Springer, 2013), pp. 187–214
41. H.M. Ridha,H. Hizam, S. Mirjalili, M.L. Othman, M.E. Ya'acob, L. Abualigah: A novel theoretical and practical methodology for extracting the parameters of the single and double diode photovoltaic models (December 2021). IEEE Access (2022)
42. K. Sreenu, M. Sreelatha, W-scheduler: whale optimization for task scheduling in cloud computing. Clust. Comput. **22**(1), 1087–1098 (2019)
43. S. Srichandan, T.A. Kumar, S. Bibhudatta, Task scheduling for cloud computing using multi-objective hybrid bacteria foraging algorithm. Future Comput. Inf. J. **3**(2), 210–230 (2018)
44. Z. Tong, H. Chen, X. Deng, K. Li, K. Li, A novel task scheduling scheme in a cloud computing environment using hybrid biogeography-based optimization. Soft. Comput. **23**(21), 11035–11054 (2019)
45. M.S. Turgut, O.E. Turgut, L. Abualigah: Chaotic quasi-oppositional arithmetic optimization algorithm for thermo-economic design of a shell and tube condenser running with different refrigerant mixture pairs. Neural Comput. Appl. 1–33 (2022)
46. S. Wang, H. Jia, L. Abualigah, Q. Liu, R. Zheng, An improved hybrid aquila optimizer and harris hawks algorithm for solving industrial engineering optimization problems. Processes **9**(9), 1551 (2021)
47. S. Wang, Q. Liu, Y. Liu, H. Jia, L. Abualigah, R. Zheng, D. Wu: A hybrid ssa and sma with mutation opposition-based learning for constrained engineering problems. Comput. Intell. Neurosci. **2021** (2021)
48. D. Whitley, A genetic algorithm tutorial. Stat. Comput. **4**(2), 65–85 (1994)
49. R. Zheng, H. Jia, L. Abualigah, Q. Liu, S. Wang, Deep ensemble of slime mold algorithm and arithmetic optimization algorithm for global optimization. Processes **9**(10), 1774 (2021)

50. R. Zheng, H. Jia, L. Abualigah, Q. Liu, S. Wang, An improved arithmetic optimization algorithm with forced switching mechanism for global optimization problems. Math. Biosci. Eng. **19**(1), 473–512 (2022)
51. A. Ziadeh, L. Abualigah, M.A. Elaziz, C.B. Şahin, A.A. Almazroi, M. Omari, Augmented grasshopper optimization algorithm by differential evolution: a power scheduling application in smart homes. Multimedia Tools Appl. **80**(21), 31569–31597 (2021)
52. R.A. Zitar, L. Abualigah: Application of red deer algorithm in optimizing complex functions. In: 2021 14th International congress on image and signal processing, biomedical engineering and informatics (CISP-BMEI), pp. 1–6. (IEEE, 2021)
53. R.A. Zitar, L. Abualigah, N.A. Al-Dmour: Review and analysis for the red deer algorithm. J. Ambient Intell. Humaniz. Comput. 1–11 (2021)

Correction to: Hybrid Adaptive Moth-Flame Optimizer and Opposition-Based Learning for Training Multilayer Perceptrons

Benedict Jun Ma

Correction to:
Chapter "Hybrid Adaptive Moth-Flame Optimizer and Opposition-Based Learning for Training Multilayer Perceptrons" in: E. H. Houssein et al. (eds.), *Integrating Meta-Heuristics and Machine Learning for Real-World Optimization Problems*, Studies in Computational Intelligence 1038, https://doi.org/10.1007/978-3-030-99079-4_11

The chapter was inadvertently published with incorrect figures. The author provided new figures for figs 1 to 6 that have been replaced now.

The book and chapter is updated with the change.

The updated version of this chapter can be found at
https://doi.org/10.1007/978-3-030-99079-4_11

E. H. Houssein et al. (eds.), *Integrating Meta-Heuristics and Machine Learning for Real-World Optimization Problems*, Studies in Computational Intelligence 1038,
https://doi.org/10.1007/978-3-030-99079-4_20

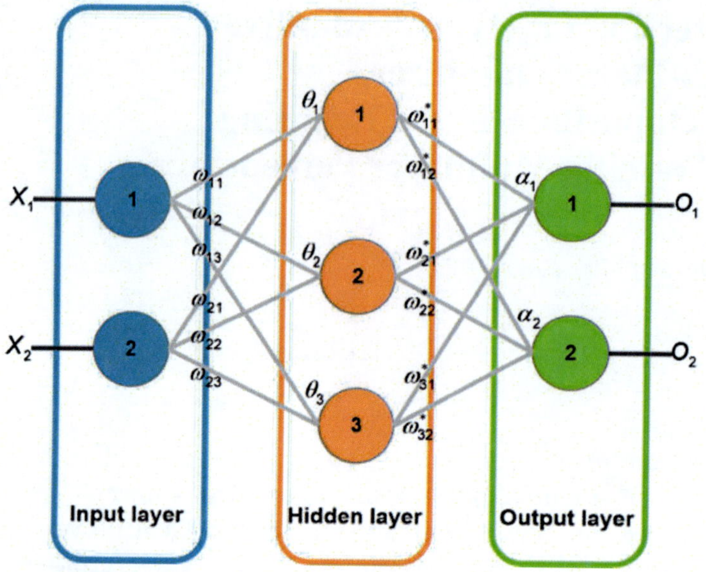

Fig. 1 MLP with the structure of 2-3-2

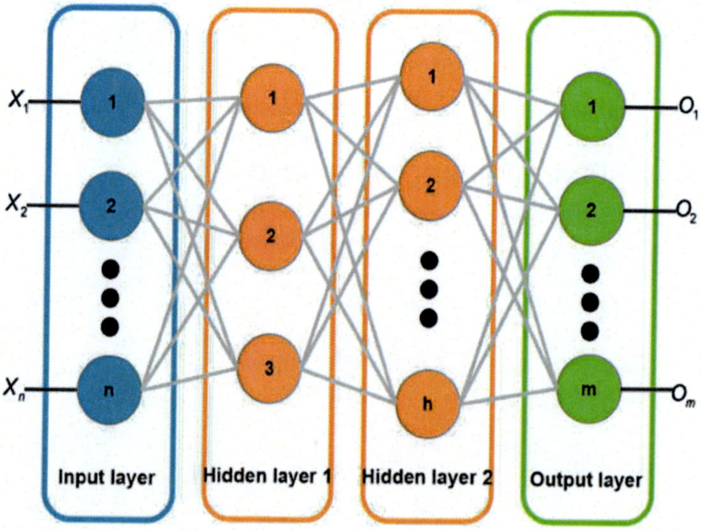

Fig. 2 MLP with the structure of n-3-h-m

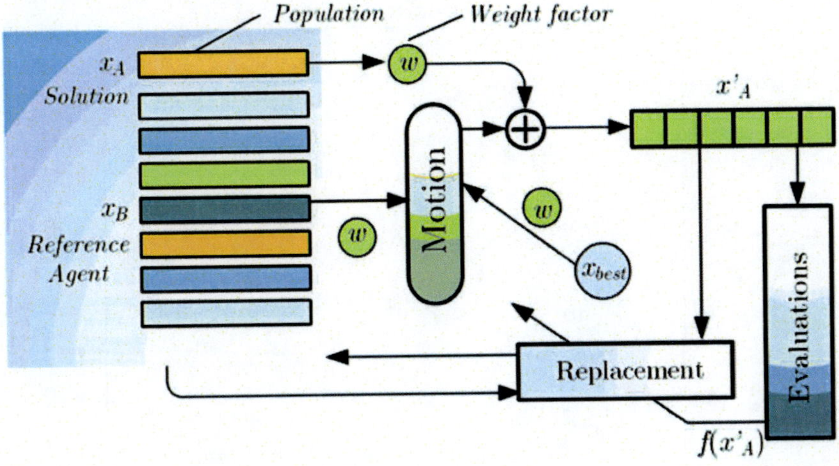

Fig. 3 The general structure of MAs

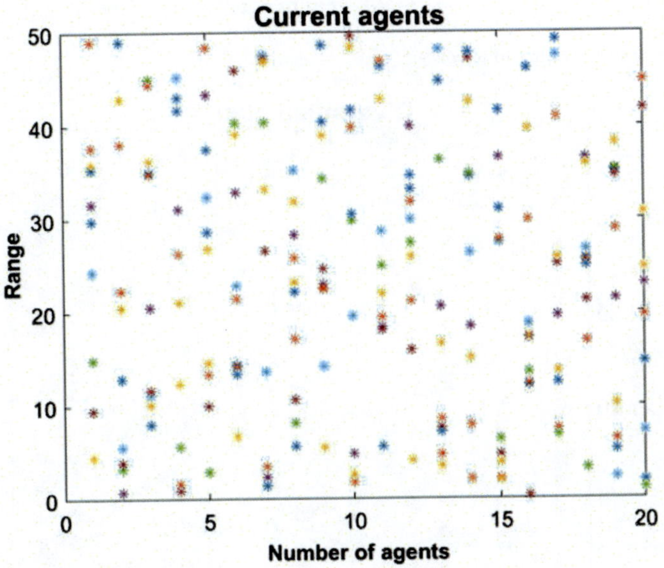

Fig. 4 Current positions of agents

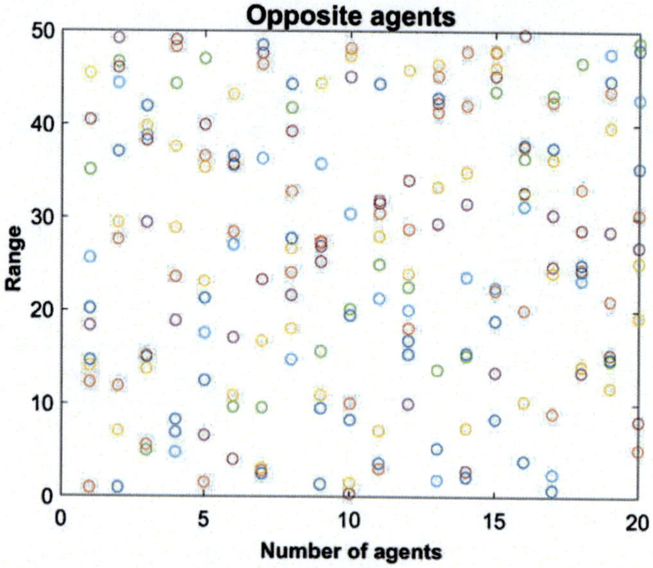

Fig. 5 Opposite positions of agents

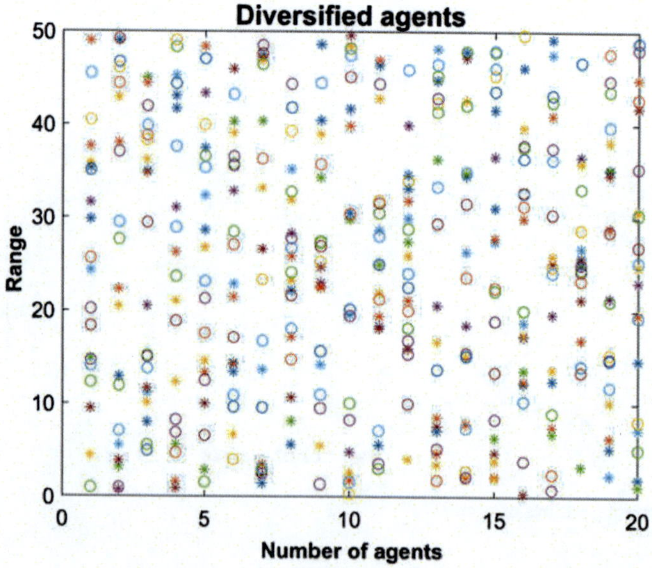

Fig. 6 Both current and opposite positions of agents